Theoretical Physics

Josef Honerkamp   Hartmann Römer

Theoretical Physics

A Classical Approach

Translated by H. Pollack
With 141 Figures and 39 Problems

Springer-Verlag
Berlin Heidelberg New York
London Paris Tokyo
Hong Kong Barcelona
Budapest

Professor Dr. Josef Honerkamp
Professor Dr. Hartmann Römer

Albert-Ludwigs-Universität, Fakultät für Physik,
Hermann-Herder-Straße 3, D-79104 Freiburg, Germany

Translator:

Howard Pollack

715 South Washington Street, Bloomington, IN 47401, USA

Title of the original German edition: *Klassische Theoretische Physik,* 3. Auflage
(Springer-Lehrbuch)
ISBN-13:978-3-642-77986-2
© Springer-Verlag Berlin Heidelberg 1986, 1989 and 1993

ISBN-13:978-3-642-77986-2   e-ISBN-13:978-3-642-77984-8
DOI: 10.1007/978-3-642-77984-8

Library of Congress Cataloging-in-Publication Data. Honerkamp, J. [Klassische Theoretische Physik. English]
Theoretical physics: a classical approach / Josef Honerkamp, Hartmann Römer; translated by H. Pollack. p. cm.
Translation of: Klassische Theoretische Physik. Includes bibliographical references and index.
ISBN-13:978-3-642-77986-2 (New York: alk. paper).   1. Mathematical physics.
I. Römer, H. (Hartmann) II. Title. QC20.H5713   1993   530'.1'51–dc20   92-42563

This work is subject to copyright. All rights are reserved, whether the whole or part of the material is concerned, specifically the rights of translation, reprinting, reuse of illustrations, recitation, broadcasting, reproduction on microfilm or in any other way, and storage in data banks. Duplication of this publication or parts thereof is permitted only under the provisions of the German Copyright Law of September 9, 1965, in its current version, and permission for use must always be obtained from Springer-Verlag. Violations are liable for prosecution under the German Copyright Law.

© Springer-Verlag Berlin Heidelberg 1993
Softcover reprint of the hardcover 1st edition 1993

The use of general descriptive names, registered names, trademarks, etc. in this publication does not imply, even in the absence of a specific statement, that such names are exempt from the relevant protective laws and regulations and therefore free for general use.

Production editor: A. Kübler
Typesetting: Macmillan India Ltd., India
56/3140 - 5 4 3 2 1 0 – Printed on acid-free paper

Preface

This introduction to classical theoretical physics emerged from a course for students in the third and fourth semester, which the authors have given several times at the University of Freiburg (Germany).

The goal of the course is to give the student a comprehensive and coherent overview of the principal areas of classical theoretical physics. In line with this goal, the content, the terminology, and the mathematical techniques of theoretical physics are all presented along with applications, to serve as a solid foundation for further courses in the basic areas of experimental and theoretical physics.

In conceiving the course, the authors had four interdependent goals in mind:

- the presentation of a consistent overview, even at this elementary level
- the establishment of a well-balanced interactive relationship between physical content and mathematical methods
- a demonstration of the important applications of physics, and
- an acquisition of the most important mathematical techniques needed to solve specific problems.

In relation to the first point, it was necessary to limit the amount of material treated. This introductory course was not intended to preempt a later, primarily theoretical, course. On the other hand, we aimed for a certain completeness in the presentation of the basic principles and concepts of classical theoretical physics, which would serve as a lasting basis for later work. Emphasis was placed on presenting the material clearly and coherently, in the form of a clearly thought out (but not formalistic) introduction to the fundamental concepts and methods. To achieve clarity, the presentations, with few exceptions, go from the general to the particular. The conceptual framework is prepared first and is not developed much further in the examples. Nevertheless, after the structural fundamentals have been clearly explained, the carefully chosen examples play an essential role in each section of this course. Using these examples, the material which we have explained earlier becomes concrete and is demonstrated in a meaningful way.

In addition, we have provided a number of summaries, reviews of earlier material, and tastes of what is to come. This places the subject matter in a larger context, and anticipates further developments, all of which helps to provide a broad perspective on the entire material.

We also demonstrate, in many cases, how particular mathematical concepts and structures appear in different physical fields and contexts with different physical interpretations, for example in our treatment of the elementary results of linear algebra. We deliberately present mathematical concepts in a familiar manner, as they might be introduced in lectures in analysis and linear algebra. In this context, they are already familiar to the student, and this should help the student to recognize them in a physical context. Thus, mathematical knowledge is utilized. We have found that knowledge and understanding of these areas in physics as well as in mathematics have profited from this method.

We cannot talk of an appropriate interaction between physics and mathematics if physics is seen only as an example of the realization of mathematical structures, or if conceptual exactness is confused with formalistic pedantry. Much is done to combat such a misunderstanding which often arises among students, particularly among talented ones. Physical and mathematical arguments are often developed in parallel, and carefully held apart from each other. Wherever possible, the physical origins of mathematical assumptions are revealed.

Thus, it is not only from lack of space that mathematical proofs are often avowedly incomplete, or even omitted; rather, this corresponds to our intention. For example, the theory of distributions is developed as far as possible within the conceptual framework of linear algebra, ignoring mathematical subtleties.

Here, again, the many examples we use are significant. We use not only dry, highly idealized systems, chosen for their easy treatment, such as the simple pendulum, but rather the manifold of physical phenomena, including examples from applied branches of physics like geophysics and physical chemistry. We discuss the examples as completely as possible, with particular emphasis on the physical interpretations of the results obtained. Thus, the connection is made between the physical situation, the mathematical formulation and discussion, and the intuitive physical results. It is here that the close relationship between mathematical deduction and intuitive interpretation, which is the essence of theoretical physics, emerges clearly.

These thoroughly discussed examples also serve the last primary goal: they demonstrate the value of mathematical-technical dexterity in the solution of problems.

This technical and methodological knowledge represents, so to speak, the tools of the trade. Familiarity with these techniques does not come from just listening to the lectures or reading, or even following the individual steps in the argument, it must also proceed from individual practice. It is essential for progress towards mastery that the student learns to use the equations, to find possible methods of solution, to go through the calculations in a problem, to interpret a result in its physical meaning, and to examine its plausibility himself or herself.

This is naturally the purpose of the exercises which always accompany an introductory theoretical course. For reasons of space, we have given only a small collection of 39 worked-through homework problems. Comprehensive collections of such exercises already exist in great numbers.

We should offer one word of explanation as to why this presentation of the fundamental principles of physics is limited to classical physics and thus leaves out modern, important, and "exciting" areas like relativity and quantum mechanics.

First, in the opinions of the authors, the addition of this material would have made it impossible to present the course in two semesters – without simultaneously losing sight of the goal of active mastery of the basics as well as an overview of the entire subject matter.

Furthermore, the classical fields of physics have the advantage that they work within the realm of phenomena more easily accessible to immediate intuitive observation. The interaction between formal deduction and intuitive interpretation, which is tremendously important in theoretical physics, is best practised within the framework of classical physics. Only with a greater sense of security can the student then progress into a realm where intuitive understanding is less forthcoming.

We attempted to avoid unnecessary one-sidedness in the selection of material in the areas of classical physics represented. Thus, for example, statistical mechanics and thermodynamics, as well as the fundamentals of fluid mechanics, have received treatment here, based on their importance particularly for applied physics. As we have already stated, students should receive a sound foundation of knowledge as an initial preparation for further research in fields like quantum mechanics, relativity theory, fluid dynamics, analytical mechanics, irreversible thermodynamics, or the theory of dynamical systems.

Finally, we wish to thank all those who have contributed to the publication of this book. In particular, we want to name Mrs. H. Kranz, Mrs. E. Rupp, Mrs. E. Ruf, and Mrs. W. Wanoth, who wrote out the long, difficult manuscript and never lost patience during the countless corrections.

We thank Mrs. I. Weber and Mrs. B. Müller for drawing the figures. We also express our gratitude to the participants in our course "Introduction to theoretical physics", in which this concept was first tested, for their many suggestions: also to those who took care of the accompanying exercises, above all Dr. H.C. Oettinger and Mr. R. Seitz, as well as Mr. P. Biller, Dr. H. Hess, Dr. M. Marcu, Mr. J. Müller, Mr. G. Mutschler, and Dr. A. Saglio de Simonis. Mr. A. Geidel, Dr. H. Simonis, Mr. F.K. Schmatzer, and Mr. M. Zähringer gave us valuable assistance in proofreading.

Freiburg, June 1993   *J. Honerkamp · H. Römer*

Contents

1. Introduction

The title *Theoretical Physics* might give the impression that besides physics there is a completely different field of theoretical physics with its own special concerns.

In reality, in this book as in every physics course, a canon of physical phenomena will be described and explained. The label "theoretical physics" indicates only a slight displacement of the point of view:

The theoretician works more with the formal construction of physics and thus focusses on the corresponding basic concepts and on the understanding and the structure of the basic equations which describe physical phenomena. Examples of such basic equations are Newton's laws, Maxwell's equations, and Schrödinger's equation. The investigation of the fundamental equations, the process of solving them, a discussion of the solutions themselves, and finally the derivation and interpretation of their physical consequences is the primary task of theoretical physics.

Fundamental equations are so important because many phenomena and experimentally observable laws can be derived from them. A whole class of phenomena can thus be explained within the framework of a theory based on fundamental equations.

In this book, we will treat such classes of phenomena. First, in Chaps. 2 and 3, the movement of material bodies will be studied, in the special case where the physical extension of the body has no influence on the motion, as in the motion of the planets around the sun or for certain motions on an inclined plane. If these bodies are then idealized as point masses, we speak of particle mechanics. Thus in Chap. 2, Newtonian mechanics will be discussed, and basic themes such as the conservation laws for the individual mechanical quantities, Kepler's Laws, as well as the general motion in a central force field will be explained in detail.

For the case in which not all of the forces are immediately known, Lagrangian mechanics will be introduced in Chap. 3. While we will present a detailed study of classical mechanics in its Newtonian and Lagrangian form, with regard to Hamiltonian mechanics, we will discuss only the Hamilton function and Hamilton's equations. All further themes, such as canonical transformations or the Hamilton–Jacobi method are left to a further course on classical mechanics, which can also treat in detail topics such as perturbation theory, the KAM-theorem, the behavior of chaotic systems, etc.

The move from particles to rigid bodies is carried out in Chap. 4. The methods of describing the position and orientation of rigid bodies are laid out in detail and selected examples are used to elucidate typical calculations of the behavior of rigid bodies under the influence of external forces.

In the short Chap. 5, the motion of bodies in non-inertial systems is studied using the methods of Chap. 4. Here, we study the Coriolis force as well as the centrifugal force as so-called fictitious forces and we investigate Foucault's pendulum.

Though analytical results can be obtained only for systems with a small number of particles by means of a thorough consideration of the interactions between particles, the N-particle problem can be solved easily if the interactions between particles can be approximated in quadratic form. This leads in Chap. 6 to the area of linear vibrations. Although this area is introduced within the framework of classical mechanics, we also do not avoid showing the universal character of this approximation and the appearance of these systems of linear differential equations in other branches of physics and engineering. The methods used to treat such systems are explained in full detail. In addition, mathematical methods such as Fourier series expansions, Fourier transforms, and concepts like Green's functions are introduced at this point.

Finally, in Chap. 7, many-particle systems are discussed. A macroscopic body is considered as a system of $\sim 10^{23}$ particles (molecules) whose interactions are understood here in the framework of classical mechanics. This leads into our treatment of classical statistical mechanics, whose basic concepts are developed. We are then able to introduce the oft feared thermodynamic formalism clearly and intuitively (even though we do not fail to show the attraction of a purely phenomenological approach, which is introduced in Sect. 7.11).

Chapter 8 then leads into the applications of thermodynamics. Important physical phenomena known to students from everyday experience such as phase transitions, changes in freezing or boiling points due to different dilutions of a solution, osmosis, etc. are treated here and the corresponding laws are derived with the help of the thermodynamic concepts and laws which we have discussed.

After the static characteristics of macroscopic thermodynamic systems have been discussed in Chaps. 7 and 8, Chap. 9 is devoted to the dynamic characteristics of such systems. After an overview of the entire field of mechanics of deformable media, including its branches in engineering, we then derive the fundamental equations of fluid mechanics, one of its most important sub-fields. Diffusion, heat conduction, and flow of fluids are such important phenomena in the work of physicists, that their theoretical basis cannot be learned early enough. These digressions into the theory of applied physics should communicate to the student that physics is the mother of many neighboring scientific and engineering disciplines, and that a broad education in physics is a great help, not least for later interdisciplinary work.

After deriving so many partial differential equations, in Chap. 10, another of the more mathematical chapters, we discuss methods by which these equations are treated. We present the procedures used to solve these linear partial differential equations as well as an introduction to special functions in physics, such as the Legendre and Bessel functions.

Thus equipped, the reader will now be able to tackle the mathematical challenges of the following introduction to electrodynamics in Chaps. 11 to 14 without problem.

In Chaps. 11 and 12, we treat electrostatics and magnetostatics, while Chap. 13 covers the full time dependent Maxwell equations. In Chap. 14, the macroscopic Maxwell equations for fields in continuous media are derived. In these chapters on electrodynamics, only the simplest applications of Maxwell's equations are covered, but they are selected in such a way that the most essential concepts and phenomena necessary for further study are stressed. In these chapters, a new conception different from other books, is least developed. In any case, they present a short introduction to electrodynamics, reduced to the essentials.

In the six appendices, important mathematical concepts and techniques of calculation are presented, in particular, an introduction to tensor calculus, to the theory of Fourier transformation and distributions, to vector analysis, and to the use of curvilinear coordinates.

2. Newtonian Mechanics

The task of mechanics is the quantitative description and calculation of the motion of material bodies. This is achieved in two steps:

First, a conceptual and formal framework is established in order to describe quantitatively the changes in location and form of bodies (*kinematics*), and then a procedure is constructed which allows, at least in principle, the motions of bodies to be calculated and predicted (*dynamics*).

We will consider first *particle mechanics*. This theory describes situations in which the spatial extent and the possible changes in form of material bodies play no essential role. These situations occur most frequently when the physical dimensions of the bodies are small in comparison to the distances between them as well as to the distance which they travel. We then represent the bodies as *point-masses* without extension (or *particles*). Whether such an idealization is possible and worthwhile depends on the physical circumstances and on the nature of the questions asked. For example, in celestial mechanics the earth can be very well approximated as a point-mass, whereas this same approximation applied to geology and geography would be meaningless.

The mechanics of extended bodies will occupy us later, when we present the theories of rigid bodies and continuous media. It will turn out that extended systems can be formally treated as systems composed of a large number of point-masses.

2.1 Space and Time in Classical Mechanics

In order to describe the motion of particles quantitatively, we need mathematical models for *space* and *time*.

Time is described as the set of all "points in time" represented by the set of real numbers $\mathbb{R}$. $\mathbb{R}$ is an ordered set; this corresponds to the ordering of points in time according to before and after or past, present, and future. In classical mechanics, time is thought of as a universal: each "point-event," that is, each event which occurs during a negligibly short time, can be uniquely identified with a point in time in $\mathbb{R}$, and the time points of different "point-events" can, without restriction, be compared with each another.

This time scale can be realized and made physically measurable, at least in principle, using a system of standard clocks, which are synchronized with one another and run at the same rate. Within the framework of classical mechanics, the synchronization of these clocks presents no problems. It can occur, for example, by means of a calibrating clock which is then moved around and compared with all other clocks. This conception of time, which seems so apparent and obvious based on our perception of the world, is called into question by the theory of relativity.

We also need a mathematical model of the space in which our point-masses move. The "points" of space are the possible locations of a point-mass. The choice – supported by a vast amount of experience and observations – of the *affine space*[1] E^3 is a good model for physical space in classical mechanics. This structure is well known in mathematics (see, for example, [Greub, Moore, Yaqub and Robinson]).

We start, in this model, with two different sets of basic objects. First, a set A is given, whose elements are called points and which represent all the possible locations of a particle. In addition to these points, we are given a three dimensional real *vector space*[2] V^3 with vectors $x, y, \ldots$.

For the set E^3 the following is assumed:

a) Associated with each ordered pair of points (P, Q), there is a vector x from V^3, which is written $\overrightarrow{PQ}$.

b) Conversely, for each point P and each vector x from V^3, there is a uniquely determined point Q, such that $\overrightarrow{PQ} = x$. "Any vector can be extended from the point P."

c) For every three points P, Q, R, we have

$$\overrightarrow{PQ} + \overrightarrow{QR} = \overrightarrow{PR} \ . \tag{2.1.1}$$

A set of points with such a structure is called a *three dimensional real affine space*.

It is easy to show that:

$$\overrightarrow{PP} = 0 \quad \text{and} \quad \overrightarrow{PQ} = -\overrightarrow{QP} \ . \tag{2.1.2}$$

The choice of E^3 indicates certain other facts:

– Space is *homogeneous*, that is, no one space-point is privileged over the others (as opposed to the vector space V^3 which possesses one singular element, namely the zero vector).

---

[1] *Affine space, affinis,* (lat.), bordering, related. Affine transformations (similarity transformations) are those transformations which can be produced from a combination of displacements, rotations, and (uniform) contractions and expansions. They form a group which is identical to the group of all invertible, single-valued, linear, inhomogeneous coordinate transformations. An affine space is characterized by the invariance of its structure under the group of affine transformations.

[2] Vector (lat.) a new construction from *vehere*: to move (something), thus something like "mover". Here we are thinking of a displacement or a velocity vector. The most important characteristic of vector is that it has direction. Contrasted to the vector is the scalar (from the Latin *scala*, meaning scale), an undirected quantity.

Similarly, we also have "tensors", (from the Latin *tendere*: to strain) see Appendix C. A tensor field, for example, can describe the stress condition of a continuous medium.

– Space is *isotropic*[3], that is, there is no privileged direction.
– Terms like "line" and "plane" have a well-defined meaning and satisfy the laws of elementary geometry.

If a given point O is chosen as the origin or reference point, every point P of the affine space is uniquely labelled by the vector $\overrightarrow{OP} = r$. The vector r is called the *position vector* of P with respect to the reference point O.

If we choose a basis (e_i) ($i = 1, 2, 3$) of the vector space V^3, we can then represent the position vector $r = \overrightarrow{OP}$ as

$$\overrightarrow{OP} = \sum_{i=1}^{3} x_i e_i \ . \tag{2.1.3}$$

Thus, the point P can also be characterized by the three-tuple of numbers (x_1, x_2, x_3). The x_i are called *coordinates* with respect to the *affine coordinate system* defined by the choice of the origin *and* the basis vectors (O, e_1, e_2, e_3). If a point of reference has been chosen, yet a particular basis in the vector space has not been fixed, we have a so-called *system of reference*.

This concept of coordinate system is consistent with and, in fact, a formalization of a method which, in principle, allows us to actually determine the position of a mass point:

At the point O, the site of the observer, a rigid system of axes is set up, with fixed unit markings on the axes, which line up with the system e_1, e_2, e_3 of basis vectors. The coordinates of a point are then determined by parallel projection onto the axes. In many cases, O, e_1, e_2, and e_3 will be time-independent, however it is often useful or necessary to introduce time dependent reference or coordinate systems.

It is important to realize that the coordinates of a point depend on the choice of basis. Let two different affine coordinate systems be given by

$$(O, e_1, e_2, e_3) \quad \text{and} \quad (O', e_1', e_2', e_3')$$

and let

$$e_i = \sum_{k=1}^{3} e_k' D_{ki} \tag{2.1.4}$$

be the expansion of the basis vectors e_i with respect to the basis (e_k') ($k = 1, 2, 3$). Let the coordinates of a point P be given by (x_1, x_2, x_3) and (x_1', x_2', x_3') respectively, so that

$$\overrightarrow{OP} = \sum_{i=1}^{3} x_i e_i \quad \text{and} \quad \overrightarrow{O'P} = \sum_{i=1}^{3} x_i' e_i' \ .$$

Then, if we write out $\overrightarrow{O'O}$ as $\sum_{k=1}^{3} c_k e_k'$, we have

$$\overrightarrow{O'P} = \overrightarrow{O'O} + \overrightarrow{OP} = \sum_{k=1}^{3} c_k e_k' + \sum_{k,i=1}^{3} e_k' D_{ki} x_i \ . \tag{2.1.5}$$

---

[3] *Isotropic* (from Greek, "equally turning"): The equivalence of all directions.

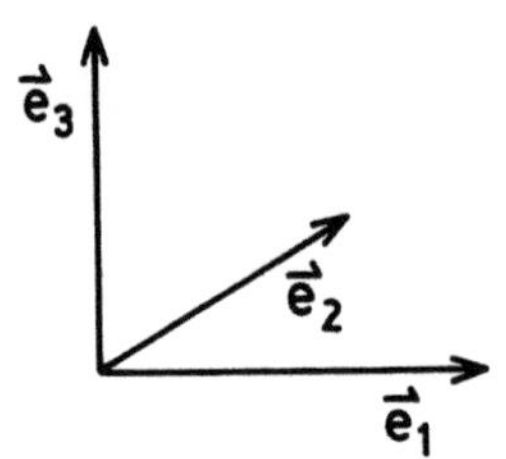

Fig. 2.1.1. An orthonormal system which we will say possesses positive or right-handed orientation. The basis vector e_2 points into the page

Thus the relation between coordinates with respect to two different coordinate systems is given by:

$$x'_k = c_k + \sum_{i=1}^{3} D_{ki} x_i \ .$$
(2.1.6)

A basis (e_1, e_2, e_3) of V^3 defines an *orientation* of the space E^3. Another basis

$$e'_i = \sum_{k=1}^{3} e_k D_{ki}$$

is then said to be of the same orientation if $\det(D_{ki}) > 0$, and is said to be oppositely oriented if $\det(D_{ki}) < 0$. We will always assume a consistent orientation for all of our bases, which we will call the positive orientation (Fig. 2.1.1).

In order to be able to formalize the possibility of measuring lengths and angles, we now introduce a *scalar product* in the vector space V^3, $(x, y) \mapsto x \cdot y \in \mathbb{R}$ such that for all $x, y, z \in V^3$, $\alpha \in \mathbb{R}$ the following properties are satisfied:

$$x \cdot y = y \cdot x \ ,$$
(2.1.7)

$$x \cdot (y + z) = x \cdot y + x \cdot z \ ,$$
(2.1.8)

$$x \cdot (\alpha y) = \alpha x \cdot y \ ,$$
(2.1.9)

$$x \cdot x \geq 0 \ ; \quad x \cdot x = 0 \Leftrightarrow x = 0 \ .$$
(2.1.10)

Then, E^3 is called a *Euclidean*[4] *affine space*.

We can then make the following definitions:

a) The *distance* between the two points P, Q is defined as:

$$\overline{PQ} := |\overrightarrow{PQ}| = \sqrt{\overrightarrow{PQ} \cdot \overrightarrow{PQ}} \ .$$

This satisfies the so-called *triangle inequality*:

$$\overline{PQ} \leq \overline{PR} + \overline{RQ} \ .$$
(2.1.11)

---

[4] *Euclid* (ca. 300 B.C.). Euclid's *Elementa* (later completed by Hilbert) contains an axiomatic development of geometry which was for two thousand years the model of exact mathematics.

b) The *angle* between two vectors is defined by:

$$\cos \sphericalangle (x, y) := \frac{x \cdot y}{|x| \cdot |y|} \, ,$$

c) A positively oriented *orthonormal basis* is a positively oriented basis e_1, e_2, e_3 of V^3 such that $e_i \cdot e_j = \delta_{ij}$.

d) A *vector product* $(x, y) \mapsto x \times y \in V^3$ is defined by the following properties:

$$x \times y = - y \times x \tag{2.1.12}$$

$$x \times (y + z) = (x \times y) + (x \times z) \tag{2.1.13}$$

$$x \times (\alpha y) = \alpha x \times y \tag{2.1.14}$$

$$e_1 \times e_2 = e_3 \tag{2.1.15}$$

$$e_2 \times e_3 = e_1 \tag{2.1.16}$$

$$e_3 \times e_1 = e_2 \tag{2.1.17}$$

for some positively oriented orthonormal basis.

It is easy to show that this definition is independent of the positively oriented orthonormal basis chosen.

i) We can also write the vector product of the basis vectors e_i in the form:

$$e_i \times e_j = \sum_k \varepsilon_{ijk} e_k \, .$$

The symbol ε_{ijk} is defined as follows:

$\varepsilon_{ijk} = 0,$ when any two indices are equal

$\varepsilon_{ijk} = 1,$ if i, j, k is an even permutation of 1, 2, 3

$\varepsilon_{ijk} = - 1,$ if i, j, k is an odd permutation of 1, 2, 3

The following properties can be shown:

$$\sum_{i=1}^{3} \varepsilon_{ijk} \varepsilon_{irs} = \delta_{jr} \delta_{ks} - \delta_{js} \delta_{kr} \, , \tag{2.1.18}$$

$$a \times (b \times c) = b(a \cdot c) - c(a \cdot b) \, , \tag{2.1.19}$$

$$a \cdot (b \times c) = \det(a, b, c) \, . \tag{2.1.20}$$

ii) We consider the affine transformations which leave the distance between each pair of points unchanged. In mathematics, these are called the *displacements*.

$$P \mapsto P', \quad Q \mapsto Q' \quad \text{such that} \quad \overline{PQ} = \overline{P'Q'} \, .$$

For the corresponding position vectors, with $De_i := \sum_k e_k D_{ki}$ this is equivalent to:

$$x \mapsto Dx + a = x' \, ,$$

$$y \mapsto Dy + a = y' , \quad \text{and}$$

$$(y - x) \cdot (y - x) = D(y - x) \cdot D(y - x) , \quad \text{or}$$

$$y \cdot y - 2x \cdot y + x \cdot x = DyDy - 2Dx \cdot Dy + Dx \cdot Dx ,$$

and thus

$$x \cdot y = Dx \cdot Dy \quad \text{for all} \quad x, y \in V^3 .$$

In particular for $x = e_i$, $y = e_j$ with $e_i \cdot e_j = \delta_{ij}$:

$$De_i \cdot De_j = \delta_{ij} = \sum_{k=1}^{3} D_{ki} D_{kj} .$$

Thus, in relation to an orthonormal basis, the matrix D_{ij} of a displacement must be an *orthogonal matrix*, that is, it must satisfy the condition

$$\sum_{k=1}^{3} D_{ki} D_{kj} = \delta_{ij} .$$

Writing this in the form of a matrix equation, we have

$$D^{\mathrm{T}} D = 1 , \quad \text{where} \quad (D^{\mathrm{T}})_{ik} = D_{ki}$$

is the transpose of D, and $(1)_{ik} = \delta_{ik}$ is the identity matrix, which represents the identity mapping.

In order to save time writing and to make the formulas easier to read, we will usually leave out the summation sign when performing sums over vector indices. Thus we write

$$x_i e_i \text{ instead of } \sum_{i=1}^{3} x_i e_i , \quad x_i y_i \text{ instead of } \sum_{i=1}^{3} x_i y_i ,$$

or

$$e_i \times e_j = \varepsilon_{ijk} e_k \quad \text{instead of} \quad e_i \times e_j = \sum_{k} \varepsilon_{ijk} e_k .$$

We thus assume the *Einstein*[5] *summation convention*: any repeated vector index indicates a sum. The position of the index (x^i or x_i) is of no significance, so long as orthonormal bases are used.

Given an affine coordinate system (O, e_1, e_2, e_3), the location of any point-mass at any time can be characterized by the position vector $r(t) = x_i(t)e_i$. In the following discussions, unless otherwise indicated, we will always choose a Cartesian coordinate system, so that $e_i \cdot e_j = \delta_{ij}$.

With the progress of time, the position $r(t)$ of the point-mass describes a *trajectory* in E^3, that is, a mapping $\mathbb{R} \to E^3$. If at every point in time the same

---

[5] *Einstein, Albert* (*1879 Ulm, d. 1955 Princeton). His achievements are well-known:
1905: Theory of special relativity, theory of Brownian motion, photoelectric effect (light quanta).
1915: Theory of general relativity.

coordinate system is used, then the development in time is completely specified by the coordinate functions $x_i(t)$.

We now require that these functions $x_i(t)$ be differentiable to at least second order with respect to time. Then we call:

$$\dot{r}(t) = \frac{d}{dt} r(t) = v(t) \quad \text{the } \textit{velocity} \ ,$$

$$\ddot{r}(t) = \frac{d}{dt} a(t) = v(t) \quad \text{the } \textit{acceleration} \ ,$$

of the point-mass. The mathematical definition of these derivatives as quotients of differentials is in exact agreement with the actual physical procedures used to measure velocity and acceleration. Note that as opposed to the position vector, the velocity and the acceleration vectors are independent of the choice of a (time independent) origin (reference point).

As a first example, consider the trajectory of a point-mass in *uniform linear motion*. It is given by the equation:

$$r(t) = r_0 + v_0 t \tag{2.1.21}$$

with constant vectors r_0 and v_0. In this case, we have

$$v(t) = v_0, \quad \text{and} \quad a(t) = 0 \ .$$

The graph of this trajectory is a straight line. The velocity of the point-mass is constant in magnitude as well as direction.

2.2 Newton's Laws

In the seventeenth century, a new perspective on the motion of material bodies was discovered. The high point of this development, which occupied the best educated minds of the time, was provided by *Isaac Newton*[6]. In his work "Principia", published in 1687, he formulated his three epoch-making laws which signalled the beginning of the scientific age. Newton's most important

---

[6] *Newton, Isaac* (*1643 Woolsthorpe, d. 1727 Kensington). Considered by many to be the greatest of all physicists. The founder of mechanics and celestial mechanics, pioneering work in optics.

   1686: *Philosophiae naturalis principia mathematica*, containing fundamental ideas he had developed in the years 1665–1667 during a stay in his hometown Woolsthorpe, where he had gone to flee the plague. Includes the derivation of Kepler's laws from the law of gravitation as well as the development of calculus (independently discovered by Leibniz, which led later to violent debates about who had discovered it first).

   1704: *Optics.* Newton became a professor at Cambridge starting in 1669. In 1696, he became the director of the government mint, and from 1703 on, he was President of the Royal Society.

discovery was that it was not uniform linear motion which required explanation, but rather divergence from such motion. He traced these deviations, caused by the outside influences of the environment, to the *forces* which material bodies exert on one another. The form of these forces, for example their dependence on the distance between bodies, was then left to be postulated.

Thus Newton's *first law* states that a stationary body, or a body in uniform linear motion, will remain in that state if it is not subject to outside influences, that is, if no forces act upon it. This law is also called the principle of inertia and has to be attributed to Galilei. Newton's first law implies the following postulate: there is a "zero element" in the set of possible forces or influences upon a material body. If this occurs in an actual physical situation, then this situation is also an example of the "zero class" of possible motions. This "zero class" includes bodies at rest, but also – and this is precisely the novel element – bodies in uniform linear motion, that is, motion for which $a = 0$.

Any non-zero force, i.e. any non-negligible influence, thus leads to a non-vanishing acceleration, and therefore to a change in the motion.

Newton's first law is well-defined only if we specify a certain system of reference in which it holds. Obviously, it cannot be satisfied in all systems of reference. If it holds in a system S, it cannot hold in a system S' which is accelerating relative to S, because then with respect to S' the point-masses at rest in S will be subject to an acceleration, even if no forces act upon them.

Coordinate systems in which Newton's first law is valid are called *inertial systems*[7]. It is in no way immediately clear that such systems exist. It is true that by means of a time dependent coordinate transformation we can make any given trajectory $r(t)$ into a uniform linear trajectory. However, the first law demands more, namely, that *all* trajectories of point-masses not subject to external forces must be uniform and linear. Despite this strong condition, it turns out that any coordinate system which is in uniform linear motion with respect to the fixed stars and which does not rotate is, to a very good approximation, an inertial system. A coordinate system with reference point on the surface of the earth is somewhat less close to an inertial system, because the earth's rotation provides an acceleration with respect to the sun. We will later study more precisely the deviation of this reference system from an inertial system. Note that if a system S is inertial, then any other system which moves in a uniform linear manner with respect to S and does not rotate with respect to S is also an inertial system. The origin O' of the system S' then moves with uniform linear motion with respect to the system S, while the directions indicated by the axes in S and those in S' coincide at all times.

If the trajectory of a particle in S is given by $r(t)$, then (because time is an absolute in all systems in classical mechanics) in the system S' the trajectory has the form

$$r'(t) = r(t) + v_0 t + r_0$$

---

[7] *Inertial system* (fr. lat. *inertia*: laziness). A system of reference in which the law of inertia holds.

with constant vectors r_0, v_0. In this equation, v_0 represents the relative velocity of S and S'. The transformation of the trajectory between S and S' is called a *Galilean transformation*[8]. In mechanics, the *principle of relativity* is valid, which states that all inertial systems are physically identical, and it is thus impossible to distinguish a particular inertial system using mechanical measurements. Formally, this means that the laws of classical mechanics must be invariant under Galilean transformations. The principle of relativity does not apply only to mechanics, but has been found to apply generally to all of physics. Of course, the assumption mentioned about the absoluteness of time does not stand up to exact measurement. The exact transformation from one inertial system to another is not given by a Galilean transformation, but rather by a *Lorentz transformation*[9]. However, for speeds which are small compared to the speed of light, the Galilean transformation provides an excellent approximation.

Newton's first law tells us what happens if no force is exerted upon a particle. The *second* law explains how the postulated forces influence the motion of material bodies. The statement is as follows: If we write the force as a function of time as $K(t)$, then

$$ma(t) = K(t) \; , \tag{2.2.1}$$

that is, the acceleration at any moment in time is proportional to the force. The proportionality constant m is a property of the material body on which the forces are acting. m is called the *inertial mass* of the body.

If we know that the forces exerted on different bodies are equal in magnitude, we can deduce their relative masses from their accelerations. (When we come to the third law, we will see that this particular situation occurs frequently.)

From $|m_1 a_1| = |m_2 a_2|$, it then follows that

$$\frac{m_2}{m_1} = \frac{|a_1|}{|a_2|} \; . \tag{2.2.2}$$

Establishing a standard mass, we can then ascertain the mass of any material body in terms of the standard. The truly remarkable feature of the second law is that such a complicated entity like the net total influence which the environment has on a particle can be expressed in terms of a single vector function $K(t)$, and that the reaction of the particle to this force is determined solely by the particle's mass.

---

[8] *Galilei, Galileo* (*1564 Pisa, d. 1642 Arcetri near Florence). His best known accomplishments are the discovery of the laws of free fall, which led, by extrapolation to smaller accelerations, to the discovery of the law of inertia, the building of a telescope, and the discoveries he made using it: moons of Jupiter, the phases of Venus, and the individual stars which make up the Milky Way. He published these results in 1610: *Siderius Nuncius*. In 1616, he was warned by the Church about his support of the Copernican system. In 1632, he published *Dialogues Concerning the two Chief World Systems*, for which he was tried and later forced to recant. 1638 saw *Discorsi*, his major work in physics.

[9] *Lorentz, Hendrik Antoon* (*1853 Arnhem, d. 1928 Haarlem). Known especially for his "electron theory," a theory of matter with applications to the electrodynamics of moving bodies. Worked also on thermodynamics and the kinetic theory of gases. Began in 1918 to plan the drainage of the Zuider-Zee.

From experiment, we find further that:

a) The mass of a body is always positive and is an *extensive quantity*[10], that is, a body composed of two parts of masses m_1 and m_2 has mass $m_1 + m_2$ (by contrast, neither speed nor temperature is an extensive quantity).

b) Forces add like vectors (in force parallelograms): if there are two independent influences on a particle, the one with force $\boldsymbol{K}_1$ the other with force $\boldsymbol{K}_2$, then the resultant force of the combined influences is given by the vector sum $\boldsymbol{K}_1 + \boldsymbol{K}_2$.

Having found from the second law the exact effect of force on the motion of a particle, we can now calculate this motion, that is, we can calculate the location of the particle $r(t)$ at any time t from the equation:

$$m\ddot{r}(t) = \boldsymbol{K}(t) \tag{2.2.3}$$

as long as we know

(a) the force $\boldsymbol{K}(t)$ as a function of time, and

(b) the initial values $r(0)$ and $\dot{r}(0) = \boldsymbol{v}(0)$ of position and velocity at a particular initial time $t = t_0$, say, when $t = 0$.

In general, though, the force $\boldsymbol{K}(t)$ is not directly known. In principle, the force $\boldsymbol{K}(t)$ which a particle experiences at a given time t can in principle depend on its entire prior behavior. In practice, though, simple force laws seem to hold. The force on a particle at time t is already determined by a few quantities, such as the velocity and location of the particle at the time t:

$$\boldsymbol{K}(t) = \boldsymbol{F}(r(t), \dot{r}(t), t) \ . \tag{2.2.4}$$

In this case, Newton's second law becomes:

$$m\ddot{r}(t) = \boldsymbol{F}(r(t), \dot{r}(t), t) \ . \tag{2.2.5}$$

This is called the *equation of motion* of the particle. Since it contains the ordinary second derivative of $r(t)$, it is an ordinary differential equation of second order. The solution of such an equation is in general unique, given the initial value of $r(t)$ and of its first derivative. All of the solutions of the equation of motion thus form a set of trajectories, each of which is determined by the values of six parameters, the components of the initial values $r(t_0)$ and $\boldsymbol{v}(t_0)$ at a particular time t_0.

The fact that not only the initial position of a particle but also the initial velocity must be specified in order to determine its later trajectory also corresponds with experience. The flight path of a ball depends both on where the ball is released as well as on the speed with which it is released.

---

[10] *Extensive* (lat.) from *extensio*: extent, size.

An especially important case, which occurs frequently, is when the force depends only on the particle's instantaneous position. Then, we have:

$$K(t) = F(r(t)) \ . \tag{2.2.6}$$

In this case, the function $F: E^3 \to V^3$ which associates to each point in space the force that a particle there would experience is called a *force field*.

The force field $F(r)$ is not to be confused with the actual force $K(t)$. The actual force $K(t)$ is obtained by substituting the trajectory of the particle $r(t)$ into the force field law $F(r)$, or, in mathematical language, by taking the composition of the mappings $F: r \mapsto F(r)$ and $r: t \mapsto r(t)$.

The discovery of the equations of motion as well as their solutions and their corresponding physical interpretations is an essential goal of classical mechanics.

Newton's second law can, of course, be used in reverse. If we measure a particle's trajectory, we can deduce the force which causes it without any prior knowledge about its source. As we will see later, Newton himself used this method to derive the law of gravitation. From this method, it can be shown that one and the same force law can be responsible for a variety of different phenomena. The law of gravitation explains planetary motion in the same way as it explains an apple falling to the earth. It is the fact that such force laws are universal which gives Newton's second law its full importance.

We now turn to Newton's *third law* which makes a statement about the mutual forces between different bodies:

If one body exerts a force $K_{21}(t)$ on a second body, then the second exerts a force $K_{12}(t)$ on the first, which is equal in magnitude, but opposite in direction.

More generally, for a system with N bodies, if the k-th body exerts the force K_{ik} on the i-th body, then we have:

$$K_{ik} = - K_{ki} \ . \tag{2.2.7}$$

The law can also be formulated by the phrase: "every action has an equal and opposite reaction."

2.3   A Few Important Force Laws

It so happens that many of the forces which occur in nature can be derived from a relatively small number of force laws. Here, again, we just consider forces $K(t)$ which depend only on $r(t)$, $\dot{r}(t)$, and t. Then,

$$K(t) = F(r(t), \dot{r}(t), t) \ .$$

In this section, we will present several such force laws, beginning with those of the form

$$K(t) = F(r(t), t) \; ,$$

i.e., force laws for which there are force fields (which we will eventually allow to be time dependent).

i) In the simplest case, consider a force $K(t)$ exerted on a particle which depends neither on its position nor on time,

$$K(t) = F_0 = \text{const} \; . \tag{2.3.1}$$

In this case, we call the force field *homogeneous* and *time independent*. In very small regions of space and time, it often happens that force fields are homogeneous to a good approximation. For example, the gravitational field of the earth on the earth's surface is virtually homogeneous for distances less than 10 miles and over large periods of time.

The most general solution of the equation of motion

$$m\ddot{r}(t) = F_0 \tag{2.3.2}$$

can be found immediately by integrating twice:

$$r(t) = \frac{1}{2m} F_0 t^2 + v_0 t + r_0 \; . \tag{2.3.3}$$

In this equation, v_0 and r_0 are the initial values of the velocity and position, and the solution $r(t)$ of the equation of motion is uniquely determined by these values. It is also easy to find the general solution of the equation of motion for homogeneous force fields which are time dependent, such as

$$m\ddot{r} = f(t) \; . \tag{2.3.4}$$

Integration yields:

$$r(t) = r_0 + v_0 t + \frac{1}{m} \int_0^t dt' \int_0^{t'} dt'' f(t'') \; . \tag{2.3.5}$$

ii) A second case, which is slightly less trivial and is of greater importance, is a force field $F(r, t)$ which is linearly dependent on r. As a model, we take the linear time-independent force field

$$F(r) = - Dr \tag{2.3.6}$$

with constant D. This is said to be a *harmonic* force law. Such a law is obeyed, for example, by the resisting force produced by a spring stretched from its equilibrium position or by a pendulum, if its displacement is not too large. The corresponding linear equation of motion,

$$m\ddot{r}(t) + Dr(t) = 0 \tag{2.3.7}$$

is a second-order linear differential equation with constant coefficients. Such differential equations always occur, as we will see in Chap. 6, in mechanical systems near an equilibrium condition. This equation can be solved in closed form using elementary methods. Therein lies the tremendous practical value of linear force laws.

If another homogeneous force field is added to this, the new equation of motion

$$m\ddot{r}(t) + Dr(t) = f(t) \; , \tag{2.3.8}$$

can also be solved in closed form. Equations of this type are found in situations involving forced harmonic motion and resonance phenomena.

iii) Gravitational forces were recognized and formally described by Newton as unified phenomena. He discovered that all bodies, merely on account of their mass, exert forces on one another. His theory, now considered elementary, describes gravitational interactions with such precision that small deviations from his predictions were not found until much later and, indeed, it was only in this century that these deviations could be explained by Einstein's theory of general relativity.

The first general characteristic of the force $K_m(t)$ exerted by other masses on a particle of (inertial) mass m is that it is proportional to this inertial mass:

$$K_m(t) = mG(t) \; . \tag{2.3.9}$$

This is an extremely remarkable characteristic of the gravitational field, because we might have anticipated that the gravitational force felt by a particle would depend on a different characteristic of the particle, which we might have called its "gravitational mass." This fundamental property of gravitation, which has been shown to be exact in experiments, is called "the principle of equivalence of inertial and gravitational mass." The general theory of relativity starts with this principle, though in the Newtonian theory of gravitation it is merely postulated without a perceived need for explanation.

From the principle of equivalence, we see that the mass immediately cancels from both sides of the equation of motion

$$m\ddot{r}(t) = mG(t) \; . \tag{2.3.10}$$

If we assume that the mass m of the particle is so small that we can ignore the reverse effect of its gravitational pull on the much heavier bodies responsible for the gravitational force $K_m(t)$ it feels, then the motion of the particle will be independent of m. In such a situation, the force exerted by a body with very large mass on a particle with relatively small mass m is referred to as an *external gravitational force*, where the word "external" signifies that the force exerted by the smaller mass on the greater can be ignored. Under the influence of an external gravitational field, the motion of a particle will be independent of its own mass.

As an example, consider the weight of an object on the earth's surface, which is the attractive force exerted by the earth on the object. We can consider earth's gravitational field to be a homogeneous external gravitational field for spatial areas which are not too large, and we thus obtain:

$$m\ddot{\boldsymbol{r}}(t) = m\boldsymbol{g} \tag{2.3.11}$$

We find the approximate value $|\boldsymbol{g}| = 9.81$ m/s^2, with a variation of about 0.5% over the surface of the Earth.

The fact that the mass of the object can be cancelled from both sides of this equation implies that bodies of all masses fall at exactly the same speed towards the earth.

After making these general remarks about gravitational forces, we will now present the exact expression for $\boldsymbol{G}(t)$.

Following Newton's theory, all gravitational forces arise from a force field and we can derive all of them from a single elementary force law which determines the gravitational attraction between two masses:

Any two point-masses, located, say, at P_1 and P_2, exert an attraction on each other. The force exercised by the mass M_2 at point P_2 on the mass M_1 at point P_1 is then given by (see Fig. 2.3.1):

$$\boldsymbol{F}_{12} = \gamma \frac{M_1 M_2}{r^3} \boldsymbol{r} \ , \quad \text{with} \quad \boldsymbol{r} = \overrightarrow{P_1 P_2} \ ,$$

$$\gamma = 6.67 \times 10^{-11}\,\mathrm{m}^3\,\mathrm{kg}^{-1}\,\mathrm{s}^{-2} \ . \tag{2.3.12}$$

Here, γ is a fundamental constant of nature, called the gravitational constant. The equivalence of gravitational and inertial mass is already present in this form of the force law. The force law implies the following:

a) $|\boldsymbol{F}_{12}| \sim 1/r^2$,

b) $\boldsymbol{F}_{12} \parallel \overrightarrow{P_1 P_2}$,

c) $\boldsymbol{F}_{12} = -\boldsymbol{F}_{21}$ (action = reaction) .

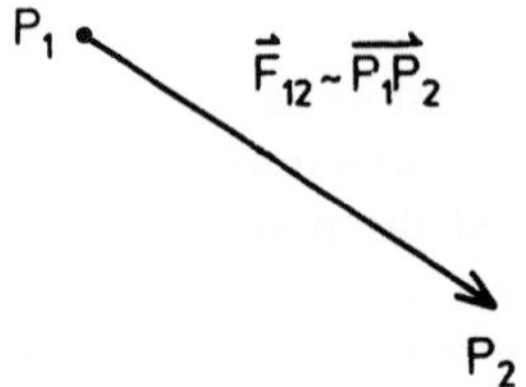

Fig. 2.3.1. The direction of the force $\boldsymbol{F}_{12}$, exerted by the point-mass at location P_2 on the point-mass at location P_1. The force is attractive

Newton arrived at this force law using the following reasoning. He started with Kepler's third law which states that the cube of the radius of a planet's orbit is proportional to the square of its period of revolution:

$$r^3/T^2 = \text{constant} .$$

In a circular orbit, the frequency is given by $\omega = 2\pi/T$, and thus

$$r^3\omega^2 = \text{const} \quad \text{or} \quad r\omega^2 = \text{const} \cdot 1/r^2 .$$

In a circular orbit, if the e_1, e_2-plane is the plane of the circle, we have:

$$r(t) = re_1\cos t + re_2\sin t , \quad \text{and thus}$$

$$\ddot{r} = -\omega^2 r; \quad \text{therefore} \quad |a| = \omega^2 r .$$

At any point on a circular orbit, then, we always have an acceleration in the direction of the center. This acceleration must be caused by the attraction of the two masses. Then, since experimentally we find the relationship $\omega^2 r = \text{constant} \cdot r^{-2}$, this attraction must be proportional to $1/r^2$.

Of course, such an argument can only present an intuitive derivation of the dependence of the force on the distance between the masses. For example, this argument relies on the assumption of perfectly circular planetary orbits. The true success of Newton's law of gravitation as an accurate description of the attraction between masses is only demonstrated by its usefulness and its agreement with experiment. For example, from this one law all three of Kepler's laws (among other things) can be derived. In addition, it accurately predicts the acceleration due to the earth's gravitational field.

In the reference system of a particle in a circular orbit, a *non-inertial system*, the attractive force is exactly compensated by a centrifugal force $mr\omega^2$, so that in this system, the particle does not experience any net force. Centrifugal forces, like the Coriolis force which we will study later, do not arise from other masses, rather they appear only in non-inertial systems. In the non-inertial system of a particle in a circular orbit, Newton's first law does not hold, because in this system the particle stays at rest despite the gravitational force exerted upon it. We can, however, interpret this by the action of the *centrifugal force*, which will balance out exactly the gravitation force. This centrifugal force, in classical mechanics, is a force of another type, called a *fictitious force* which is never present in an inertial system of reference. Fictitious forces arise from the transition to a non-inertial system, and therefore they never have the universal character of forces exerted by bodies on other bodies. These latter universal forces can be enumerated and, depending on the physical situation, are always at work in inertial systems and must therefore appear in the equation of motion. Fictitious forces are more complicated byproducts of the deviation of a certain system of reference from being inertial. A formulation of Newton's laws in an arbitrary non-inertial system would presuppose that a classification of these non-inertial forces in terms of the deviation of the given system from an inertial system can be given. But this already presupposes a knowledge of inertial systems, which immediately takes us back to the original Newtonian conception, namely the formulation of the equation of motion in an inertial system, which will then only include forces exerted by one body on another.

iv) *Velocity-dependent force laws* also play an important role in physics. The following fundamental law applies when a particle carries an electric charge e. If such a particle is located in an electrical field $E(r, t)$ and a magnetic field $B(r, t)$, the force exerted on the particle is given by:

$$F_L(r, \dot{r}, t) = e[E(r, t) + \dot{r} \times B(r, t)] \ . \tag{2.3.13}$$

F_L is called the *Lorentz force*. This force depends on the velocity of the particle, and can also be explicitly time dependent if E and B are. Note that the fields E and B here are caused by other charged particles.

v) The force which two charges at rest, q_1 and q_2 located at points P_1 and P_2, exert on each other is given by *Coulomb's Law*[11]

$$F_{12} = -\frac{1}{4\pi\varepsilon_0} \frac{q_1 q_2}{r^3} r \quad \text{with} \quad r = \overrightarrow{P_1 P_2} \ ,$$

$$4\pi\varepsilon_0 = 1.1126 \times 10^{-10} \ \mathrm{CV^{-1} m^{-1}} \ . \tag{2.3.14}$$

This law has the same form as the law of gravitation. Consider the forces exerted on each other by two protons. They attract each other gravitationally but repel each other by the Coulomb force. The Coulomb force in this case is about 10^{36} times stronger than the gravitational force.

The reason why electric forces are so much less obvious in everyday life is that there are positive as well as negative charges. Precisely because of the strength of the electromagnetic forces, the positive and negative charges tend to compensate for each other as much as possible. Masses, on the other hand, are always positive, and so the gravitational force, unlike electric forces, cannot normally be cancelled out. This is why gravitational forces – despite being so much weaker – are much more easily observed.

vi) Finally, there is one more important type of velocity-dependent force. Motion can be influenced by friction. Experimentally, we find that this *frictional force*, at least for small speeds, is proportional to the velocity and is opposite to the direction of motion:

$$F_R = -\kappa\dot{r} \ , \quad \kappa > 0 \ . \tag{2.3.15}$$

In the case of a free fall, the equation of motion taking friction into account is thus given by:

$$m\ddot{r} = mg - \kappa\dot{r} \ . \tag{2.3.16}$$

---

[11] *Coulomb, Charles Auguste de* (*1736 Angoulême, d. 1806 Paris). In the years 1784–1789, he published important papers about electricity and magnetism. The torsion balance he invented made possible his discovery of the force law between charges at rest.

2.4 The Energy of a Particle in a Force Field

2.4.1 Line Integrals

Consider a particle of mass m, which acts under the influence of a time-independent force field. We have:

$$F(t) = F(r(t)) \ .$$

We then obtain the following equation of motion:

$$m\ddot{r}(t) = F(r(t)) \ . \tag{2.4.1}$$

If we take the scalar product of both sides of this equation with $\dot{r}$, we find:

$$m\ddot{r} \cdot \dot{r} = F(r) \cdot \dot{r} \ . \tag{2.4.2}$$

Now integrate both sides with respect to t, from t_1 to t_2. The left side of the equation becomes

$$m \int_{t_1}^{t_2} dt\, \ddot{r} \cdot \dot{r} = \frac{1}{2} m \int_{t_1}^{t_2} dt\, \frac{d}{dt}(\dot{r}^2) = \frac{1}{2} m\dot{r}^2 \Big|_{t_1}^{t_2} = T(t_2) - T(t_1) \quad \text{with} \tag{2.4.3}$$

$$T = \frac{1}{2} m\dot{r}^2 \ .$$

The right side yields the integral

$$\int_{t_1}^{t_2} F(r(t)) \cdot \frac{dr(t)}{dt}\, dt \ , \tag{2.4.4}$$

where $r(t)$ in $F(r(t))$ is a solution of the equation of motion. If we write $r(t_1) = r_1, r(t_2) = r_2$ and if C is the part of the trajectory between r_1 and r_2, we write

$$\int_{t_1}^{t_2} F(r(t)) \cdot \frac{dr(t)}{dt}\, dt = \int_{r_1, C}^{r_2} F(r) \cdot dr \ . \tag{2.4.5}$$

The expression

$$\int_{r_1, C}^{r_2} F(r) \cdot dr = A_{12}(r_1, r_2; C, F) \equiv A_{12}(C) \tag{2.4.6}$$

is called, mathematically, a *line integral*. We will show explicitly that this line integral depends only on the path between r_1 and r_2 (i.e., the direction in which it runs) and not on the actual trajectory of the particle in time, $r(t)$ between t_1 and t_2. For if we replace the time t with another parameter $\tau = \tau(t)$ with $t = t(\tau)$, then

$$\int\limits_{\tau(t_1)}^{\tau(t_2)} F(r(t(\tau))) \cdot \frac{dr(t(\tau))}{d\tau}\, d\tau = \int\limits_{\tau(t_1)}^{\tau(t_2)} F(r(t(\tau))) \cdot \frac{dr(t(\tau))}{dt}\, \frac{dt}{d\tau}\, d\tau$$

$$= \int\limits_{t_1}^{t_2} F(r(t)) \cdot \frac{dr(t)}{dt}\, dt \ . \tag{2.4.7}$$

In order to specify a particular line integral, we need:

a) the endpoints of the path,
b) the path between these two points,
c) the integrand, that is, the vector field.

Before we go into the physical meaning of the line integral, let us first discuss a few of its general properties:

i) In general, a vector field $F(r, t)$ is called *conservative*, if the line integral

$$\int\limits_{r_1,C}^{r_2} F(r, t) \cdot dr = \int\limits_{0}^{\sigma} F(r(\sigma'), t) \cdot \frac{dr(\sigma')}{d\sigma'}\, d\sigma' \tag{2.4.8}$$

is independent of the particular path $C = \{r(\sigma') | 0 \le \sigma' \le \sigma\}$ between r_1 and r_2, and thus only depends on $r(0) = r_1$ and $r(\sigma) = r_2$. Note that t here plays the role of one (or several) parameters and is held constant during integration along the path.

ii) A vector field $F(r, t)$ is conservative if and only if its line integral vanishes along any *closed path*.

This is obvious, since if C_1 and C_2 are two paths from r_1 to r_2 (Fig. 2.4.1), and F is a conservative field, we have

$$\int\limits_{r_1,C_1}^{r_2} F \cdot dr = \int\limits_{r_1,C_2}^{r_2} F \cdot dr = - \int\limits_{r_2,-C_2}^{r_1} F \cdot dr \ , \tag{2.4.9}$$

so that

$$\int\limits_{C_1 \cup - C_2} F \cdot dr = 0 \ . \tag{2.4.10}$$

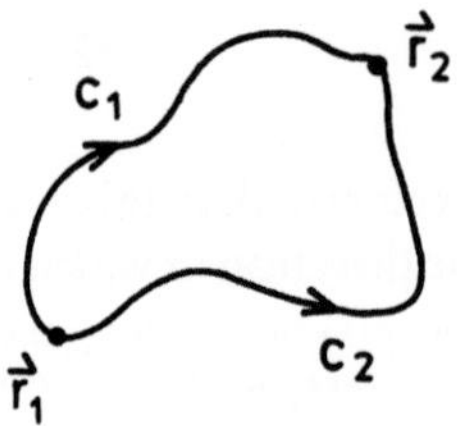

Fig. 2.4.1. Two paths from r_1 to r_2

If, conversely, the line integral of a given vector field over any closed path vanishes, we can show that any line integral of the field is path independent. Consider two paths C_1 and C_2 which join the points r_1 and r_2. These two paths together form a single closed path, and the above equations, read in reverse order, show that the integral is independent of path.

iii) A vector field F is conservative if and only if there exists a *scalar field* $U(r, t)$ with

$$F = - \nabla U(r, t) \equiv - \operatorname{grad} U(r, t) \ . \tag{2.4.11}$$

In this equation, the *gradient* is a vector field defined by:

$$\nabla U(r, t) = \left(\frac{\partial U}{\partial x_1}, \frac{\partial U}{\partial x_2}, \frac{\partial U}{\partial x_3} \right) = \sum_{i=1}^{3} \frac{\partial U}{\partial x_i} e_i \ , \tag{2.4.12}$$

where x_1, x_2, x_3 are the coordinates relative to an orthonormal basis. (The minus sign in the above formula is a convention.) This scalar field U is unique up to a constant.

Proof. a) Assume that there is such a $U(r, t)$. We want to show that the value of the integral

$$\int_{r_1}^{r_2} F \cdot dr$$

is independent of path over which the integral is calculated. Let C be any path with

$$C = \{ r(\sigma') | 0 \leq \sigma' \leq \sigma \} \ , \quad r(0) = r_1 \ , \quad r(\sigma) = r_2 \ .$$

Then,

$$\int_{r_1, C}^{r_2} F \cdot dr = - \int_{0}^{\sigma} \nabla U(r(\sigma'), t) \cdot \frac{dr(\sigma')}{d\sigma'} \, d\sigma' = - \int_{0}^{\sigma} d\sigma' \frac{d}{d\sigma'} U(r(\sigma'), t)$$

$$= - [U(r_2) - U(r_1)] \ . \tag{2.4.13}$$

This result is obviously independent of the particular path C.

b) Conversely, if the integral

$$\int_{r_1}^{r_2} F \cdot dr$$

is independent of path, we define

$$U(r) := - \int_{r_1, C}^{r} F \cdot dr \tag{2.4.14}$$

for a fixed but arbitrary C and r_1. With

$$C = \{r(\sigma') \,|\, 0 \le \sigma' \le \sigma\} \;, \qquad r(0) = r_1 \;, \qquad r(\sigma) = r \;,$$

we then have,

$$U(r(\sigma)) = -\int_0^\sigma F(r(\sigma')) \cdot \frac{dr(\sigma')}{d\sigma'}\, d\sigma' \;, \tag{2.4.15}$$

and thus

$$\frac{d}{d\sigma}\, U(r(\sigma)) = \frac{dr(\sigma)}{d\sigma} \cdot \nabla U = - F(r(\sigma)) \cdot \frac{dr(\sigma)}{d\sigma}$$

or

$$(F + \nabla U) \cdot \frac{dr(\sigma)}{d\sigma} = 0 \;.$$

Since this formula holds for any path C, we then conclude that

$$F = - \nabla U \;.$$

If we were to choose another point r_1' instead of r_1 as the initial point of our path, we have

$$\hat{U} = \int_{r_1'}^{r} F \cdot dr = \int_{r_1}^{r} F \cdot dr + \int_{r_1}^{r_1} F \cdot dr = U + c \;.$$

On the other hand,

$$\nabla \hat{U} = \nabla(U + c) = \nabla U \;.$$

Thus, we find that U is determined by F up to an arbitrary constant.
U is called the *potential field* or *potential* corresponding to F.

iv) If F is conservative, we have:

$$\nabla \times F = 0 \;, \tag{2.4.16}$$

$\nabla \times F$ (called the *"curl of F"*) here is a vector field defined by:

$$\nabla \times F = \left(\frac{\partial F_3}{\partial x_2} - \frac{\partial F_2}{\partial x_3}, \frac{\partial F_1}{\partial x_3} - \frac{\partial F_3}{\partial x_1}, \frac{\partial F_2}{\partial x_1} - \frac{\partial F_1}{\partial x_2} \right) \tag{2.4.17}$$

or by

$$(\nabla \times F)_i = \varepsilon_{ijk} \frac{\partial F_k}{\partial x_j} \tag{2.4.18}$$

in a right-handed orthonormal coordinate system. If $F = -\nabla U$, then

$$F_k = -\frac{\partial U}{\partial x_k} \quad \text{and} \quad \frac{\partial F_k}{\partial x_j} - \frac{\partial F_j}{\partial x_k} = -\frac{\partial^2 U}{\partial x_k \partial x_j} + \frac{\partial^2 U}{\partial x_j \partial x_k} = 0$$

by the symmetry of second-order partial derivatives, so we have shown that

$$\nabla \times F = 0 \ .$$

v) Conversely, it is clear that if $\nabla \times F \neq 0$, then F is not conservative and the integral $\int_{r_1}^{r_2} F \cdot dr$ is not path-independent.

vi) It is also possible to show (see Appendix F) that if

$$\nabla \times F = 0$$

in a simply connected region of E^3, then in that region there is always a U such that

$$F = -\nabla U \ .$$

2.4.2 Work and Energy

Let us return to our line integrals of the force field $F(r)$.

If the force field is a conservative vector field, we have

$$F(r) = -\nabla U(r) \quad \text{and} \quad \int_{r_1}^{r_2} F(r) \cdot dr = U(r_1) - U(r_2) \ , \tag{2.4.19}$$

and therefore

$$T(t_2) + U(r(t_2)) = T(t_1) + U(r(t_1)) \ . \tag{2.4.20}$$

This means that the quantity

$$E = \tfrac{1}{2} m \dot{r}^2(t) + U(r(t)) \tag{2.4.21}$$

is constant with respect to time, if the trajectory $r(t)$ is a solution of the equation of motion. E is called the *energy*, T is called the *kinetic energy*, and $U(r)$ is called the *potential energy* of the particle at the point r.[12]

---

[12] *Energy* (from Greek); originally a philosophical concept, used by Aristotle as a synonym for entelechy, later used as a technical term in physics meaning inherent work or the ability to do work.

  Kinetic energy: energy of motion from the Greek *kinein* meaning motion. Potential energy: something like "energy of possibility" from the Latin *potentia* meaning power or possibility. If you have two weights, the one situated at a higher elevation has the greater possibility of doing work, it has more "energy stored in it".

In Newtonian mechanics, then, the energy of a particle subject to a conservative force is conserved; during the motion of this particle energy is neither lost nor gained. The value of this fixed quantity can be determined at $t = 0$, for example, using the initial conditions: $E = \frac{1}{2} m \dot{r}^2(0) + U(r(0))$.

In general, even in the case of non-conservative forces, we refer to the integral

$$W = \int\limits_{r_1,C}^{r_2} F \cdot dr$$

as the *work done* by the force on the particle along the path C between the points r_1 and r_2. In the case of a conservative force, W is negative the change in the potential energy of the particle.

For the Lorentz force $F_L(t) = e[E(r, t) + r(t) \times B(r, t)]$, $F_L(t) \cdot \dot{r}(t) = e\dot{r}(t) \cdot E(r(t))$. The magnetic field thus performs no work. If the field $E(r)$ is conservative, the work done by the Lorentz force can also be expressed as the change in potential energy.

Examples. i) A homogeneous force field is conservative. The potential corresponding to

$$F \equiv A \quad \text{is} \quad U = - A \cdot r + \text{const} .$$

This is clear, since

$$\nabla(A \cdot r) = \left(\frac{\partial}{\partial x_1} A \cdot r, \frac{\partial}{\partial x_2} A \cdot r, \frac{\partial}{\partial x_3} A \cdot r \right) = (A_1, A_2, A_3) = A . \tag{2.4.22}$$

With $A = mg$, we have $U = - mg \cdot r$, and if we define the z-axis as the $- g$ direction, then

$$U = + mgz . \tag{2.4.23}$$

ii) If $F(r) = f(r) r/r$, $r = |r|$, then

$$U(r) = - \int_{r_0}^{r} f(r') dr' = U(r) , \quad \text{with} \quad U' = -f(r) , \tag{2.4.24}$$

since then

$$- \nabla U(r) = - \frac{dU}{dr} \cdot \nabla r = f(r) \nabla r \tag{2.4.25}$$

and $\nabla r = r/r$.

In particular, gravitational and harmonic forces are conservative, with the potentials

$$U(r) = - \frac{\gamma M_1 M_2}{r} \tag{2.4.26}$$

for the gravitational force and

$$U(r) = \tfrac{1}{2}Dr^2 \tag{2.4.27}$$

for the harmonic force $F(r) = -Dr$.

A force field of the form

$$F(r) = g(r)r/r \tag{2.4.28}$$

is called a *spherically symmetric central force*. In general, a force field which always acts along the line between a point and a fixed center O is called a *central force*. A general central force law then has the form:

$$F(r) = g(r)r/r \ . \tag{2.4.29}$$

For spherically symmetric forces, which as we have seen are always conservative, the strength of the force depends only on the distance r from the center. It is easy to see that central forces which are not spherically symmetric cannot be conservative, since in this case we can immediately find a closed path along which the work integral does not vanish.

iii) An example of a nonconservative force is given by

$$F = (y, -x, 0) \quad \text{with} \quad r = (x, y, z) \ . \tag{2.4.30}$$

The curl $(\nabla \times F) = (0, 0, -2)$, is non-zero.

In Fig. 2.4.2, we have drawn the vector F at a few points. We see immediately that

$$\int_A^B F \cdot dr$$

is dependent on the path chosen, since on the right side of the circle F is parallel to dr, and on the left side it is antiparallel to dr.

iv) For one-dimensional motion, $F(x)$ is always conservative, that is, we can always find a $U(x)$ such that

$$F(x) = (-d/dx)U(x) \ , \tag{2.4.31}$$

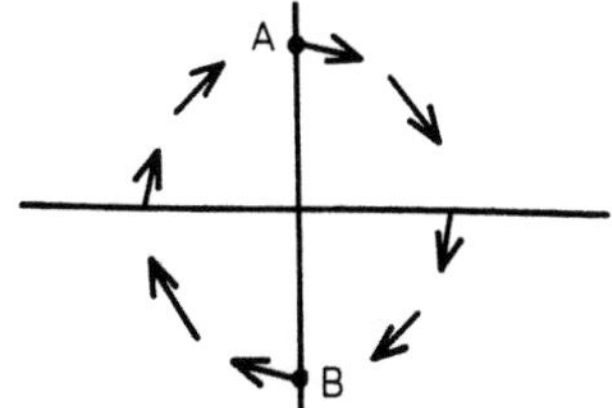

Fig. 2.4.2. A vector field with non-zero curl

namely the integral of $-F(x)$. Then, we have

$$\tfrac{1}{2}m\dot{x}^2 + U(x) = E = \text{const} , \quad \text{and thus} \tag{2.4.32}$$

$$\dot{x} = \pm \sqrt{(2/m)[E - U(x(t))]} \quad \text{or} \tag{2.4.33}$$

$$\pm \int_{t_0}^{t} \frac{\dot{x}\,dt'}{\sqrt{(2/m)[E - U(x(t'))]}} = \pm \int_{x_0}^{x} \frac{dx'}{\sqrt{(2/m)[E - U(x')]}}$$

$$= \int_{t_0}^{t} dt' = t - t_0 , \tag{2.4.34}$$

where we have written $x_0 = x(t_0)$ and $x = x(t)$.

In this way, the equation of motion can be solved immediately using the fact that E is a conserved quantity. The solution is then of the form

$$x = x(t; E, x_0) .$$

The two parameters E and x_0 which determine the solution replace the usual initial values $x(0) = x_0$, $\dot{x}(0) = v_0$. The connection between E and v_0 is given by

$$v_0 = \sqrt{(2/m)[E - U(x_0)]} . \tag{2.4.35}$$

Since the kinetic energy T can never be negative, it follows that $E = T + U \geq U$, and thus the total energy is never smaller than the potential energy, and they can only be equal if $\dot{x} = 0$. Given a graph of the function $U(x)$, we can immediately identify the possible regions in which a particle can travel for a given value of the energy E (Fig. 2.4.3).

In particular, if x_0 is a minimum of U and $E_0 = U(x_0) + |\varepsilon|$ is a small amount larger than $U(x_0)$, then the particle will always stay in the neighborhood of the point of equilibrium x_0. On the other hand, if x_1 is a maximum of U, then any small change in energy from $E = U(x_1)$ to $E_1 = U(x_1) + \varepsilon$ will cause the particle to move far away from x_1.

Thus, the minimum points of the potential energy correspond to stable points of equilibrium, the maxima to unstable points of equilibrium. These

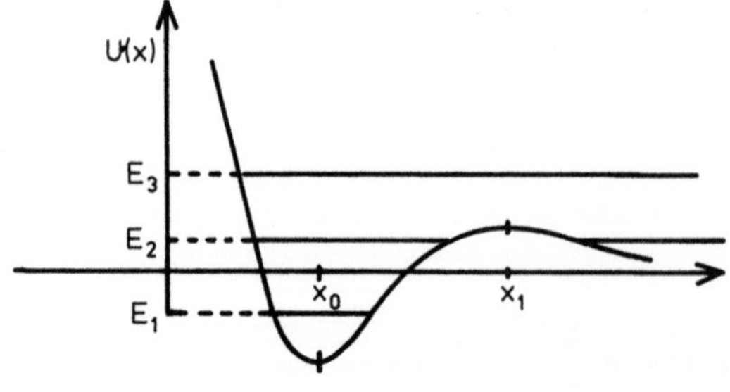

Fig. 2.4.3. An example of a potential function for one-dimensional motion with the allowed regions for different energies

considerations are also valid for motion in a higher number of dimensions. A point of equilibrium then is stable if it corresponds to a minimum of the potential energy.

2.5 Several Interacting Particles

In the previous section, we considered only the case of a single particle, studied its equation of motion, and discussed energy as a conserved quantity.

Now, let us consider N particles in similar fashion. The equation of motion for the i-th particle is as follows:

$$m_i \ddot{r}_i(t) = F_i(t) , \quad i = 1, \ldots, N ,$$
(2.5.1)

where m_i is the mass of the i-th particle and $F_i(t)$ is the force acting on this particle.

If this force on the i-th particle is given by

$$F_i(t) = F_i(r_1(t), \ldots, r_N(t))$$
(2.5.2)

then the equations of motion

$$m_i \ddot{r}_i(t) = F_i(r_1(t), \ldots, r_N(t))$$
(2.5.3)

represent a system of $3N$ differential equations; as initial conditions, we might be given $r_i(0)$ and $\dot{r}_i(0)$. There are $6N$ initial conditions, and these conditions determine a unique solution of this system of equations.

Multiplying by $\dot{r}_i$, summing over i from 1 to N, and then integrating from t_1 to t_2 yields:

$$\int_{t_1}^{t_2} dt \sum_{i=1}^{N} m_i \ddot{r}_i \cdot \dot{r}_i = \int_{t_1}^{t_2} \sum_{i=1}^{N} F_i(r_1(t), \ldots, r_N(t)) \cdot \frac{dr_i(t)}{dt} dt$$

or also

$$T(t_2) - T(t_1) = \int_{t_1}^{t_2} \sum_{i=1}^{N} F_i(r_1(t), \ldots, r_N(t)) \cdot \frac{dr_i(t)}{dt} dt$$
(2.5.4)

with

$$T(t) = \frac{1}{2} \sum_{i=1}^{N} m_i \dot{r}_i^2(t) .$$
(2.5.5)

$T(t)$ is called the *kinetic energy of the system of N particles*. It would be quite interesting if also in this general case of N particles, the forces could still be derived from a single potential.

To discuss this possibility, let us consider the $3N$ dimensional vector space of all $3N$ vector coordinates:

$$Z = V^3 \oplus \ldots \oplus V^3 \ (N \text{ times})$$

$$\tag{2.5.6}$$

$$Z = \{\underline{z} = (r_1, \ldots, r_N) \mid r_i \in V^3\} \ .$$

The positions of all N particles can then be described by points in Z, by combining the position vectors $r_1, \ldots, r_N$ into a single vector $\underline{z} = (r_1, \ldots, r_N)$ in Z.

This space $Z = V^{3N}$, containing all of the possible arrangements of the N-particle system, is called the *configuration space* of the system.

In Z, we define a scalar product as follows: If $\underline{z} = (r_1, \ldots, r_N)$ and $\underline{z}' = (r_1', \ldots, r_N')$ then we define

$$\underline{z} \cdot \underline{z}' = \sum_{i=1}^{N} r_i \cdot r_i' \ . \tag{2.5.7}$$

The N trajectories $r_i(t)$ $(i = 1, \ldots, N)$ correspond to a single trajectory $t \mapsto \underline{z}(t) = (r_1(t), \ldots, r_N(t)) \in Z$. In the same way, we combine the force fields $F_i(r_1, \ldots, r_N)$ into a single force field $\underline{F}: V^{3N} \mapsto V^{3N}$:

$$\underline{F}(\underline{z}) = (F_1(r_1, \ldots, r_N), \ldots, F_N(r_1, \ldots, r_N)) \ . \tag{2.5.8}$$

Then,

$$\int_{t_1}^{t_2} \sum_{i=1}^{N} F_i(r_1(t), \ldots, r_N(t)) \cdot \frac{dr_i(t)}{dt} \, dt = \int_{t_1}^{t_2} \underline{F}(\underline{z}(t)) \cdot \frac{d\underline{z}}{dt} \, dt$$

$$= \int_{\underline{z}_1, C}^{\underline{z}_2} \underline{F} \cdot d\underline{z} \ . \tag{2.5.9}$$

In Sect. 2.4.1, we made a few mathematical comments about line integrals of vector fields. These comments were formulated in terms of a three dimensional vector space, but they can immediately be generalized to a space of arbitrary dimension.

A force $\underline{F}(\underline{z})$ in the vector space Z is called conservative if the line integral

$$\int_{\underline{z}_1, C}^{\underline{z}_2} \underline{F} \cdot d\underline{z}$$

is independent of the path C in Z, that is, it depends only on $\underline{z}_1$ and $\underline{z}_2$.

In the same way, we find:

A force field $\underline{F}(\underline{z})$ is conservative if and only if there exists a potential $U(\underline{z})$ such that

$$\underline{F} = -\underline{\nabla} U \ , \quad \text{that is} \tag{2.5.10}$$

$$\underline{F}_i = -\nabla_i U(r_1, \ldots, r_N) \tag{2.5.11}$$

where $\mathbf{V}_i$ is simply the gradient with respect to the variables r_i; this is often written

$$\mathbf{V}_i = \frac{\partial}{\partial \mathbf{r}_i} \ . \tag{2.5.12}$$

If the force $\mathbf{F}$ is conservative and time-independent, it follows that the quantity

$$E = T(t) + U(r_1(t), \ldots, r_N(t)) \tag{2.5.13}$$

is conserved if the trajectory $\underline{z}(t) = (r_1(t), \ldots, r_N(t))$ is a solution of the equation of motion. E is called the *total energy* of the system of N particles.

Examples. i) We consider two particles such that each exerts a force on the other whose strength depends only on $|r_1 - r_2| = |r| = r$ and this force is exerted along the line connecting the two particles (see Fig. 2.5.1). Then, with $r = r_1 - r_2$,

$$m_1 \ddot{r}_1 = F_1(r_1, r_2) = -f(r)\frac{r}{r} \tag{2.5.14}$$

$$m_2 \ddot{r}_2 = F_2(r_1, r_2) = f(r)\frac{r}{r} \ . \tag{2.5.15}$$

Thus, in accordance with Newton's third law, F_1 is negative F_2.

Claim. (F_1, F_2) is a conservative force.

Let $U(r)$ be an indefinite integral of $f(r)$, then

$$F_i = -\mathbf{V}_i U(r) = -\frac{\partial}{\partial \mathbf{r}_i} U(|r_1 - r_2|) \ . \tag{2.5.16}$$

Proof. We have,

$$-\mathbf{V}_i U(r) = -U'(r)\mathbf{V}_i r = -f(r)\mathbf{V}_i r \ . \tag{2.5.17}$$

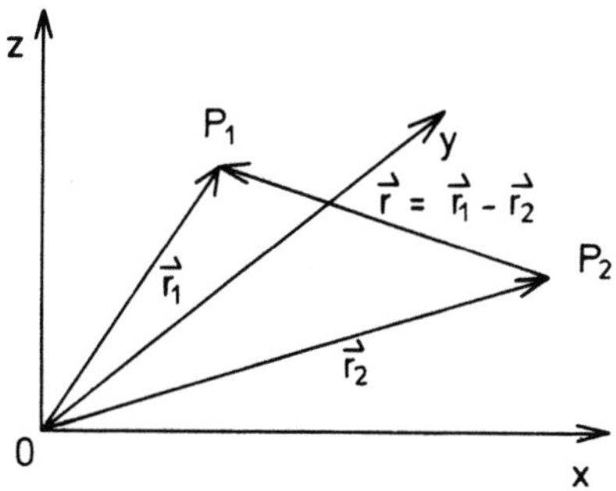

Fig. 2.5.1. The interaction of two particles

Now with

$$r_i = (x_i, y_i, z_i) \ , \quad \nabla_1 r = \left(\frac{\partial}{\partial x_1}, \frac{\partial}{\partial y_1}, \frac{\partial}{\partial z_1} \right) r \ ,$$

$$r = \sqrt{(x_1 - x_2)^2 + (y_1 - y_2)^2 + (z_1 - z_2)^2} \ ,$$

$$\nabla_1 r = \frac{(x_1 - x_2, y_1 - y_2, z_1 - z_2)}{r} :$$

$$\nabla_1 |r_1 - r_2| = \frac{(r_1 - r_2)}{r} = \frac{r}{r} \ . \tag{2.5.18}$$

Similarly,

$$\nabla_2 |r_1 - r_2| = \nabla_2 r = -\frac{(r_1 - r_2)}{r} = -\frac{r}{r} \ . \tag{2.5.19}$$

End of proof.

For

$$f(r) = \gamma \frac{M_1 M_2}{r^2}, \quad \text{we have} \quad U(r) = -\gamma \frac{M_1 M_2}{r} \ .$$

With this potential, we obtain the equations of motion for two particles which attract each other according to Newton's law of gravitation. $U(r)$ is the *gravitational potential*. Then,

$$E = \tfrac{1}{2} M_1 \dot{r}_1^2 + \tfrac{1}{2} M_2 \dot{r}_2^2 + U(|r_1 - r_2|) \tag{2.5.20}$$

is a constant in time.

ii) We will now practice the procedure of finding the equation of motion: Let three particles be given which interact with each other as in example (i).

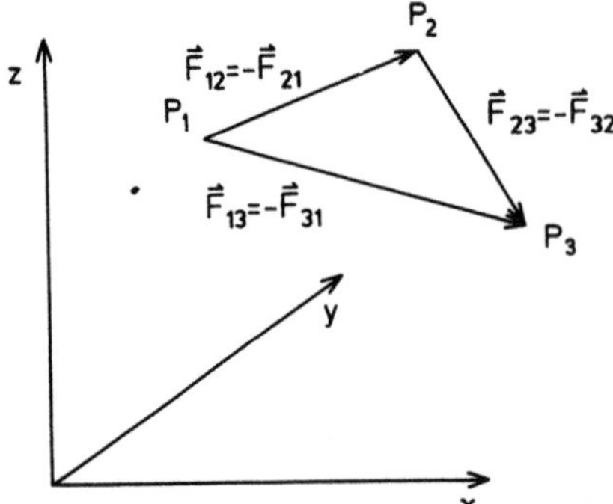

Fig. 2.5.2. The particles with the gravitational force which they exert upon each other

Then, we have (Fig. 2.5.2):

$$m_1 \ddot{r}_1 = F_{12}(r_1 - r_2) + F_{13}(r_1 - r_3)$$

$$m_2 \ddot{r}_2 = F_{21}(r_1 - r_2) + F_{23}(r_2 - r_3) \tag{2.5.21}$$

$$m_3 \ddot{r}_3 = F_{31}(r_1 - r_3) + F_{32}(r_2 - r_3) \ .$$

F_{ij} is the force that is exerted by particle j on particle i, it depends on the vector $r_i - r_j$.

According to Newton's third law, we have

$$F_{ij}(r_i - r_j) = - F_{ji}(r_i - r_j) \ . \tag{2.5.22}$$

If the 9-dimensional force field

$$(F_{12} + F_{13}, F_{21} + F_{23}, F_{31} + F_{32})$$

is conservative, there must be a corresponding potential $U(r_1, r_2, r_3)$. In the case of gravitational attraction, such a potential is given by

$$U(r_1, r_2, r_3) = - \gamma \frac{M_1 M_2}{|r_1 - r_2|} - \gamma \frac{M_2 M_3}{|r_2 - r_3|} - \gamma \frac{M_3 M_1}{|r_3 - r_1|}$$

$$= \sum_{i<j} U_{ij}(|r_i - r_j|) \ . \tag{2.5.23}$$

We can check explicitly that by taking the gradient of this potential, we obtain the right side of the equation of motion, for example:

$$- \nabla_1 U = F_{12}(r_1 - r_2) + F_{13}(r_1 - r_3)$$

$$= - \gamma \frac{M_1 M_2}{|r_1 - r_2|^3} (r_1 - r_2) - \gamma \frac{M_1 M_3}{|r_1 - r_3|^3} (r_1 - r_3) \ . \tag{2.5.24}$$

Remark. These examples also reflect physically important general cases. Consider, for example, the following:

In general, $F_i(r_1, \ldots, r_N)$, the force on the i-th particle, is of the form:

$$F_i(r_1, \ldots, r_N) = F_i^{(e)}(r_i) + \sum_{\substack{j=1 \\ j \neq i}}^{N} F_{ij}(r_i, r_j) \ , \tag{2.5.25}$$

that is, the i-th particle is subject to an *external force* $F_i^{(e)}$ which depends solely on the position of the particle r_i, as well as to the forces F_{ij} exerted on it by the other particles. The forces exerted by other particles of the system are usually called *internal forces*.

If these forces F_{ij} are of the form

$$F_{ij} = \frac{r_i - r_j}{|r_i - r_j|} f_{ij}(|r_i - r_j|) \tag{2.5.26}$$

and are conservative, along with the external forces $F_i^{(e)}$, then the force field $F = (F_1, \ldots, F_N)$ is conservative with potential

$$U(r_1, \ldots, r_N) = \sum_{j=1}^{N} V_j^{(e)}(r_j) + \sum_{\substack{l,k=1 \\ l<k}}^{N} V_{lk}(|r_l - r_k|) \tag{2.5.27}$$

where $-V_{ik}$ is the integral of $f_{ik} = f_{ki}$, since then

$$F_i = -\nabla_i U$$

$$= -\nabla_i V_i^{(e)}(r_i) - \sum_{i<k} \frac{r_i - r_k}{|r_i - r_k|} V_{ik}'(|r_i - r_k|) + \sum_{l<i} \frac{r_l - r_i}{|r_l - r_i|} V_{li}'(|r_l - r_i|)$$

$$= -\nabla_i V_i^{(e)}(r_i) - \sum_{\substack{k=1 \\ k \neq i}}^{N} \frac{r_i - r_k}{|r_i - r_k|} V_{ik}'(|r_i - r_k|) \;, \tag{2.5.28}$$

because $V_{ij} = V_{ji}$. Thus, F_i has same structure as the examples.

2.6   Momentum and Momentum Conservation

Besides the position, velocity, and energy of a particle, we can define another useful quantity to describe its motion, namely, the *momentum*,[13] given by

$$p_i(t) := m_i \dot{r}_i(t) \;. \tag{2.6.1}$$

In a system of N particles, the total momentum $P(t)$ is defined as

$$P := \sum_{i=1}^{N} p_i = \sum_{i=1}^{N} m_i \dot{r}_i = \frac{d}{dt} \sum_{i=1}^{N} m_i r_i \;. \tag{2.6.2}$$

The position vector R of the *center of mass* is defined as:

$$R := \sum_{i=1}^{N} m_i r_i \bigg/ \sum_{i=1}^{N} m_i \;. \tag{2.6.3}$$

(The denominator $\sum m_i$ is chosen so that if the position vectors are all displaced $r_i: r_i \mapsto r_i + a$, the center of mass R will also be displaced by the same vector $a: R \mapsto R + a$). Then,

$$P = M\dot{R} \tag{2.6.4}$$

with $M := \sum_{i=1}^{N} m_i$, the total mass of the system.

---

[13] Momentum (lat.) weight, importance.

Thus, we can view P also as the momentum of the system of N particles, thought of as a collective particle of mass $M = m$ located at the center of mass.

With the help of the concept of momentum, Newton's second law can also be formulated

$$\dot{p}_i(t) = F_i(t) \tag{2.6.5}$$

that is, the force exerted on the i-th particle produces a change in the momentum of this particle in accordance with the above equation. This form of the second law is in fact more general, and can be shown to hold also, when the mass of the particle changes during its motion (as is the case for a rocket, which expels its burnt fuel). If we combine all of the individual momenta into one momentum $\underline{p}(t) := (p_1(t), \ldots, p_N(t))$, the Newtonian equation of motion reads

$$\underline{\dot{p}}(t) = \underline{F}(t) \ . \tag{2.6.6}$$

For many general considerations, it is also useful to combine the vectors $\underline{z} \in V^{3N}$ and $\underline{p} \in V^{3N}$ into a single vector $(\underline{z}, \underline{p}) \in V^{6N}$. The 6$N$-dimensional space of all positions and momenta is called the *phase space* of the N-particle system. Each trajectory $t \mapsto \underline{z}(t)$ in configuration space corresponds exactly to a trajectory

$$t \mapsto (\underline{z}(t), \underline{p}(t))$$

in phase space. Each set of initial values $\underline{z}(0)$, $\underline{p}(0)$ corresponds to an initial point in phase space through which the phase space trajectory passes at $t = 0$; and, through each point in phase space given as an initial condition, there is exactly one phase space trajectory which is a solution of the equation of motion.

If we are given the equations of motion in the form

$$\dot{p}_i = F_i^{(e)} + \sum_{\substack{j=1 \\ j \neq i}}^{N} F_{ij} \quad \text{with} \quad F_{ij} = -F_{ji} \ , \quad \text{it then follows that} \tag{2.6.7}$$

$$\sum_{i=1}^{N} \dot{p}_i = \dot{P} = \sum_{i=1}^{N} F_i^{(e)} \ . \tag{2.6.8}$$

The net change in the total momentum is then equal to the sum of the external forces. If this sum vanishes, we have

$$\dot{P} = M\ddot{R} = 0 \ .$$

The total momentum is thus a conserved quantity if the sum of the external forces vanishes. In this case, the center of mass undergoes uniform linear motion.

Of course, this is guaranteed if each of the external forces vanishes, that is, if

$$F_i^{(e)} = 0$$

for $i = 1, \ldots, N$. In this case, one speaks of a *closed system*.

Ignoring the internal motion of the individual particles, the system as a whole then behaves like a free particle. If there are non-vanishing external forces, the acceleration of the center of mass is then determined by the total mass $M = \sum m_i$ and the net external force

$$F^{(e)} = \sum_{i=1}^{N} F_i^{(e)} \ . \tag{2.6.9}$$

This can be understood as an expression of the additivity of masses.

Examples. In the two-body problem

$$m_1 \ddot{r}_1 = -f(|r_1 - r_2|) \frac{r_1 - r_2}{|r_1 - r_2|} \ ,$$

$$\tag{2.6.10}$$

$$m_2 \ddot{r}_2 = f(|r_1 - r_2|) \frac{r_1 - r_2}{|r_1 - r_2|}$$

there are no external forces (i.e., the system is closed). Adding these equations yields

$$m_1 \ddot{r}_1 + m_2 \ddot{r}_2 = M\ddot{R} = \frac{d}{dt}(p_1 + p_2) = \dot{P} = 0 \ . \tag{2.6.11}$$

The equation of motion for the center of mass has the solution

$$R(t) = R_0 + V_0 t \ . \tag{2.6.12}$$

The total momentum $P = M V_0$ is constant.

This solves three differential equations out of the six contained in the equations of motion. In order to solve the other three equations, it is useful to change coordinates, in order to find a new equation for $r = r_1 - r_2$:

$$\ddot{r}_1 = -\frac{1}{m_1} f(r) \frac{r}{r} \ . \quad \ddot{r}_2 = \frac{1}{m_2} f(r) \frac{r}{r} \ , \tag{2.6.13}$$

and thus

$$\ddot{r} = \ddot{r}_1 - \ddot{r}_2 = \left(-\frac{1}{m_1} - \frac{1}{m_2} \right) f(r) \frac{r}{r} = -\frac{1}{\mu} f(r) \frac{r}{r} \tag{2.6.14}$$

where

$$\frac{1}{\mu} = \frac{1}{m_1} + \frac{1}{m_2} \quad \text{or} \quad \mu = \frac{m_1 m_2}{m_1 + m_2} \ . \tag{2.6.15}$$

μ is called the *reduced mass* of the two particle system. If $m_1 \approx m_2$, then

$\mu \approx m_1/2$. If $m_1 \ll m_2$, then

$$\mu = \frac{m_1}{(1 + m_1/m_2)} \approx m_1 \ .$$

For example, the reduced mass of the earth-sun system is approximately the mass of the earth.

The equation

$$\mu\ddot{r} = -f(r)\frac{r}{r} \tag{2.6.16}$$

is called the equation of motion for the *relative motion* of the two particles. The simple motion of the center of the mass has been separated out, and we are left to study the complicated equation describing the relative position vector r. This equation is analogous to the physical situation in which the center of force remains at rest at the origin and a single particle of mass μ moves under the influence of a force.

The energy

$$E = \tfrac{1}{2}m_1\dot{r}_1^2 + \tfrac{1}{2}m_2\dot{r}_2^2 + U(r) \tag{2.6.17}$$

can be separated into two pieces, the energy corresponding to the motion of the center of mass and the energy of the relative motion:

$$E = E_{\mathrm{cm}} + E_{\mathrm{rel}} \quad \text{with}$$

$$E_{\mathrm{em}} = \tfrac{1}{2}M\dot{R}^2 \ , \qquad E_{\mathrm{rel}} = \tfrac{1}{2}\mu\dot{r}^2 + U(r) \ . \tag{2.6.18}$$

Proof. We have:

$$\frac{1}{2}M\dot{R}^2 = \frac{1}{2}M\frac{1}{M^2}(m_1\dot{r}_1 + m_2\dot{r}_2)^2 \ , \qquad \frac{1}{2}\mu\dot{r}^2 = \frac{1}{2}\frac{m_1 m_2}{M}(\dot{r}_1 - \dot{r}_2)^2 \ .$$

Adding both equations, we get

$$\dot{r}_1^2\left(\frac{m_1^2}{2M} + \frac{m_1 m_2}{2M}\right) + \dot{r}_2^2\left(\frac{m_2^2}{2M} + \frac{m_1 m_2}{2M}\right) + \dot{r}_1\cdot\dot{r}_2\left(\frac{m_1 m_2}{M} - \frac{m_1 m_2}{M}\right) = \frac{1}{2}m_1\dot{r}_1^2 + \frac{1}{2}m_2\dot{r}_2^2 \ .$$

We see that this separation into two parts is an identity corresponding to the substitution of the two coordinates R and r for the coordinates r_1 and r_2. It is important to realize, though, that in the case when $\dot{R} = $ constant, both parts of the energy E_{cm} and E_{rel} are conserved separately.

Quite generally, for any closed system it is possible to consider the motion of the center of mass and the relative motions of the particles separately by adding all the equations of motion and introducing the total momentum P. If we introduce vectors running from the center of mass $\boldsymbol{R}$ to the particles $\boldsymbol{r}_i$ as follows:

$$r_i = R + x_i , \quad \text{it follows that}$$

$$\tag{2.6.19}$$

$$r_i - r_j = x_i - x_j \quad \text{and} \quad \sum_{i=1}^{N} m_i x_i = 0 ,$$

and thus

$$T = \frac{1}{2}\sum_i m_i \dot{r}_i^2 = \frac{1}{2}\sum_i m_i(\dot{R} + \dot{x}_i)^2 = \frac{1}{2}\sum_i m_i \dot{R}^2 + \frac{1}{2}\sum_i m_i \dot{x}_i^2 , \quad \text{i.e.} \tag{2.6.20}$$

$$E = T + \sum_{\substack{i,j=1 \\ i<j}}^{N} V_{ij}(|r_i - r_j|) = E_{\mathrm{cm}} + E_{\mathrm{rel}} \quad \text{with} \tag{2.6.21}$$

$$E_{\mathrm{cm}} = \tfrac{1}{2}M\dot{R}^2 , \tag{2.6.22}$$

$$E_{\mathrm{rel}} = \frac{1}{2}\sum_{i=1}^{N} m_i \dot{x}_i^2 + \sum_{i<j} V_{ij}(|x_i - x_j|) , \tag{2.6.23}$$

and the equations of motion then read:

$$M\ddot{R} = 0 , \tag{2.6.24}$$

$$m_i \ddot{x}_i = \sum_{\substack{j=1 \\ j \neq i}}^{N} F_{ij}(x_i - x_j) , \quad (i = 1, \ldots, N) . \tag{2.6.25}$$

Adding the N equations of motion for the relative coordinates, and using the fact that

$$\sum_i m_i x_i = 0 \quad \text{and} \quad \sum_{i,j=1}^{N} F_{ij} = 0$$

we find

$$\sum_i m_i \ddot{x}_i = \sum_{i,j=1}^{N} F_{ij} = 0 . \tag{2.6.26}$$

These equations are therefore linearly dependent, and thus one of them, say, the one for the vector x_N, follows from the others. By separating out the equation for the center of mass, we have thus effectively reduced a closed N-particle system into a system with $(N - 1)$ particles.

Remarks. i) In the two-body problem, using this method we find:

$$m_1 \ddot{x}_1 = F_1(x_1 - x_2) .$$

However, $x_1 - x_2 = r$ and $x_1 = (m_2/M)r$, and thus $\mu\ddot{r} = F_1(r)$ follows as above.

ii) In general, if there are external forces it is not possible to consider the motion of the center of mass separately. In this case

$$V_i^{(e)}(r_i) = V_i^{(e)}(R + x) \neq V_i^{(e)}(R) + W(x_1, \ldots, x_N) \tag{2.6.27}$$

i.e., it is impossible in general to separate the potential energy (in contrast to the kinetic energy) into two pieces, one describing the center of mass, using only center of mass coordinates, and the other describing the relative motion, using only relative coordinates. In some cases, though, the external forces are only weakly dependent on the positions of the particles, so that we may write:

$$F_i^{(e)}(r_i) \approx F_i^{(e)}(R) \ . \tag{2.6.28}$$

We can use this approximation if the external force changes only a small amount over distances comparable to the relative separation of the particles. In this case, the center of mass vector satisfies:

$$M\ddot{R} = \sum_{i=1}^{N} F_i^{(e)}(R) = F^{(e)}(R) \ . \tag{2.6.29}$$

We see then, that a small system of particles can be approximated ideally by a single particle, if we are not interested in the internal motion of the particles. This, in turn, provides a justification for the concept of a point-mass.

A particularly interesting case is that of external gravitational forces, because they are proportional to the mass. Then,

$$F_i^{(e)}(r_i) = m_i G(r_i) = m_i G(R + x_i) \approx m_i G(R) \tag{2.6.30}$$

for weakly varying fields or for systems of particles which do not extend over large distances. The approximate equation of motion for the center of mass then reads:

$$M\ddot{R} = \sum_{i=1}^{N} F_i^{(e)} = \sum_{i=1}^{N} m_i G(R) = MG(R) \ ,$$

and for the relative coordinates we have

$$m_i \ddot{x}_i = F_i(x_1, \ldots, x_N) \ . \tag{2.6.31}$$

Thus, if we have a system of particles in an external gravitational field, and if the system is small enough so that the force does not change appreciably over its extent, it follows that the motion of the center of mass is independent of the total mass of the system.

For inhomogeneous external fields it is not possible to separate out the motion of the center of mass in a mathematically exact way. We can replace an inhomogeneous gravitational field by a homogeneous field and then calculate

the amount of error produced using a Taylor expansion of the external force
with respect to x_i:

$$\sum_{i=1}^{N} F_i^{(e)}(R + x_i) = \sum_{i=1}^{N} m_i G(R + x_i)$$

$$= MG(R) + \sum_{i=1}^{N} m_i x_i \cdot \nabla G(R) + \frac{1}{2} \sum_{i=1}^{N} m_i (x_i \cdot \nabla)^2 G(R) + \dots \ .$$

$$(2.6.32)$$

Since

$$\sum_{i=1}^{N} m_i x_i = 0$$

the second term vanishes. Thus, deviations from homogeneity of external gravitational fields, which lead to *tidal forces* for large bodies, are present only at the
second order of the expansion. This result is only valid in general, though, for
gravitational fields. For other types of fields, first-order effects of inhomogeneity
are to be expected.

iii) In a *collision*, particles collide with each other and exchange momentum.
If the total momentum before the collision is given by

$$P = \sum_{i=1}^{N} m_i v_i \tag{2.6.33}$$

and after, by

$$P' = \sum_{i=1}^{N} m_i v_i' \tag{2.6.34}$$

then as a result of momentum conservation, we must have

$$\sum_{i=1}^{N} m_i v_i = \sum_{i=1}^{N} m_i v_i' \ . \tag{2.6.35}$$

If, during the collision process, the particles are altered, so that afterwards there
may even be a different number of particles or they may have different masses,
then in this more general case momentum conservation becomes:

$$\sum_{i=1}^{N} m_i v_i = \sum_{i=1}^{N'} m_i' v_i' \ . \tag{2.6.36}$$

This must be satisfied in every inertial system, so that we must also have:

$$\sum_{i=1}^{N} m_i (v_i + V) = \sum_{i=1}^{N'} m_i' (v_i' + V) \tag{2.6.37}$$

for an arbitrary V, and thus

$$\sum_{i=1}^{N} m_i = \sum_{i=1}^{N'} m_i' \ . \tag{2.6.38}$$

Hence the total mass of the system cannot change, even if there are changes in the number and type of particles.

2.7 Angular Momentum

Another extremely important quantity in mechanics is *angular momentum.*

Consider a particle, and let r be its position vector with respect to the origin O. Then, the angular momentum at the time t with respect to the origin O is defined by the vector product

$$L(t) := r(t) \times p(t) = mr(t) \times \dot{r}(t) \ . \tag{2.7.1}$$

The meaning of this quantity is more apparent when we calculate the time derivative of $L(t)$:

$$\dot{L}(t) = m\dot{r}(t) \times \dot{r}(t) + mr(t) \times \ddot{r}(t) = r(t) \times m\ddot{r}(t)$$

$$= r(t) \times F(t) =: N(t) \ . \tag{2.7.2}$$

The quantity $N(t) = r(t) \times F(t)$ is called the *torque*[14] of the force $F(t)$ (with respect to the origin O). A torque produces a change in angular moment, just as a force causes a change in linear momentum. If the force can be written in terms of a force field $F(r)$, we have $N(t) = r(t) \times F(r(t))$. Clearly angular momentum is a conserved quantity if and only if for all times t, $N(t) = 0$. This is the case whenever $F(r)$ is parallel to r, that is, whenever $F(r)$ is a central force with respect to origin O. Under this condition, $L(t) = mr(t) \times \dot{r}(t)$ does not vary with respect to time for any solution $r(t)$ of the equation of motion, and its value is already determined by the initial conditions.

It is sufficient, then, for $F(r)$ to be of the form

$$F(r) = f(r)r/r \ ,$$

the central·force in this case does not need to be spherically symmetric. It is also apparent that conservation of angular momentum even holds for time-dependent central fields, as long as the origin stays fixed.

---

[14] Torque from Latin "torquere" to twist.

For a system of N particles, the total angular momentum is simply defined as

$$L(t) := \sum_{i=1}^{N} L_i(t) \; ; \tag{2.7.3}$$

$$L_i(t) = m_i r_i \times \dot{r}_i = r_i \times p_i \; . \tag{2.7.4}$$

We next want to find the time dependence of L. We have,

$$\dot{L} = \sum_{i=1}^{N} \dot{L}_i = \sum_{i=1}^{N} m_i \dot{r}_i \times \dot{r}_i + \sum_{i=1}^{N} m_i r_i \times \ddot{r}_i \; . \tag{2.7.5}$$

With

$$m_i \ddot{r}_i = F_i^{(e)}(r_i) + \sum_{\substack{j=1 \\ j \neq i}}^{N} F_{ij}(r_1, \ldots, r_N)$$

it follows that

$$\dot{L} = \sum_{i=1}^{N} r_i \times F_i^{(e)} + \sum_{\substack{i,j=1 \\ j \neq i}}^{N} r_i \times F_{ij} \; . \tag{2.7.6}$$

The two sums on the right side can also be written:

$$\sum_{\substack{i,j=1 \\ j \neq i}}^{N} r_i \times F_{ij} = \frac{1}{2} \sum_{i,j=1}^{N} (r_i \times F_{ij} + r_j \times F_{ji}) = \frac{1}{2} \sum_{i,j=1}^{N} (r_i - r_j) \times F_{ij} \; , \tag{2.7.7}$$

since $F_{ij} = -F_{ji}$.

If now $F_{ij} \parallel r_i - r_j$, then $(r_i - r_j) \times F_{ij}$ vanishes, and we obtain

$$\dot{L} = \sum_{i=1}^{N} r_i \times F_i^{(e)} =: \sum_{i=1}^{N} N_i^{(e)} = N^{(e)} \; . \tag{2.7.8}$$

Hence, if the internal forces F_{ij} are exerted along the lines which connect the particles i and j, which is true in all practical cases, then the torque due to these forces is completely cancelled out, and the change in the total angular momentum is determined solely by the total torque of the external forces.

We see also that in a closed system, the total angular momentum is a conserved quantity.

Remarks. i) The angular momentum of a particle is a quantity which, like the position vector, depends on the origin of the reference system. For a closed system of N particles, however, the angular momentum is conserved regardless of which reference point is used, since only differences of position vectors, which are independent of origin, appeared in the proof of the time independence of angular momentum.

ii) It is also possible to consider the angular momentum with respect to another point P, having position vector r_0. The angular momentum and torque with respect to an origin at P are then given by

$$L^P = \sum_{i=1}^{N} (r_i - r_0) \times p_i = \sum_{i=1}^{N} (r_i \times p_i) - \sum_{i=1}^{N} (r_0 \times p_i)$$

$$= L - r_0 \times P \tag{2.7.9}$$

and

$$N^P = \sum_{i=1}^{N} (r_i - r_0) \times F_i^{(e)} = N - \sum_{i=1}^{N} r_0 \times F_i^{(e)} \ . \tag{2.7.10}$$

In addition,

$$\dot{L}^P = N^P \ . \tag{2.7.11}$$

If we have

$$\sum_{i=1}^{N} F_i^{(e)} = 0 \ ,$$

as in any closed system, the torque is independent of the choice of origin, hence $N^P = N$. This condition is also satisfied if we have two external forces which satisfy the condition $F_1^{(e)} = -F_2^{(e)}$. These two forces are then referred to as a *couple*. In addition, the total angular momentum of a closed system, calculated with respect to any fixed origin, is a conserved quantity.

iii) The total angular momentum and total torque of a system can be divided up, like the kinetic energy, into a center of mass component and a relative component.
From

$$L = \sum_{i=1}^{N} m_i r_i \times \dot{r}_i = \sum_{i=1}^{N} m_i (R + x_i) \times (\dot{R} + \dot{x}_i)$$

it follows that:

$$L = R \times P + \sum_{i=1}^{N} m_i x_i \times \dot{R} + R \times \sum_{i=1}^{N} m_i \dot{x}_i + \sum_{i=1}^{N} m_i x_i \times \dot{x}_i \ .$$

Since $\sum_{i=1}^{N} m_i x_i = 0$, we find that

$$L = L_{cm} + L_{rel} \ , \quad \text{where} \tag{2.7.12}$$

$$L_{cm} = R \times P \ , \quad L_{rel} = \sum_{i=1}^{N} m_i x_i \times \dot{x}_i \ . \tag{2.7.13}$$

The center of mass component of the angular momentum is the angular momentum of the center of mass of the system calculated with respect to the origin; the relative component is the total angular momentum of the system calculated with respect to the center of mass. For the total torque, we find

$$N = \sum_{i=1}^{N} r_i \times F_i^{(e)} = \sum_{i=1}^{N} (R + x_i) \times F_i^{(e)}$$

$$= \sum_i R \times F_i^{(e)} + \sum_i x_i \times F_i^{(e)} = N_{\text{cm}} + N_{\text{rel}} \ . \tag{2.7.14}$$

We calculate:

$$\dot{L}_{\text{cm}} = \dot{R} \times P + R \times \dot{P} = M\dot{R} \times \dot{R} + R \times \dot{P}$$

$$= R \times \dot{P} = \sum_{i=1}^{N} R \times K_i^{(e)} = N_{\text{cm}} \ . \tag{2.7.15}$$

Then, from $\dot{L} = N$, we have $\dot{L}_{\text{rel}} = N_{\text{rel}}$. From this, it follows that: For a closed system, not only is the total angular momentum $L = L_{\text{cm}} + L_{\text{rel}}$ a conserved quantity, but also the components of angular momentum corresponding to the center of mass and the relative motion of the particles are individually conserved.

In particular, for a two-body system,

$$L_{\text{rel}} = m_1 x_1 \times \dot{x}_1 + m_2 x_2 \times \dot{x}_2 \ . \tag{2.7.16}$$

Since

$$x_1 = \frac{m_2}{M} r \ , \qquad x_2 = -\frac{m_1}{M} r$$

it follows that

$$L_{\text{rel}} = \left(\frac{m_1 m_2^2}{M^2} + \frac{m_2 m_1^2}{M^2} \right) r \times \dot{r} = \mu r \times \dot{r} \ . \tag{2.7.17}$$

As expected, L_{rel} here corresponds to the angular momentum of a particle with mass μ.

2.8   The Two-Body Problem

We have seen in our general considerations that in a closed system of two particles, the following equation for the relative position vector r remains after the equation for the center of mass has been separated out:

$$\mu \ddot{\boldsymbol{r}} = -f(r)\frac{\boldsymbol{r}}{r} \; . \tag{2.8.1}$$

Here we have assumed again a spherically symmetric force field. The energy of the relative motion was given by

$$E_{\mathrm{rel}} = \tfrac{1}{2}\mu \dot{r}^2 + U(r) \tag{2.8.2}$$

and the angular momentum was

$$\boldsymbol{L}_{\mathrm{rel}} = \mu \boldsymbol{r} \times \dot{\boldsymbol{r}} \; . \tag{2.8.3}$$

It is clear in this case that

$$\dot{\boldsymbol{L}}_{\mathrm{rel}} = \mu \boldsymbol{r} \times \ddot{\boldsymbol{r}} = \boldsymbol{0} \; . \tag{2.8.4}$$

We can then deduce:

i) $\boldsymbol{L}_{\mathrm{rel}}$ is a constant in time. Since $\boldsymbol{r}(t)$ points from one particle to the other, and $\boldsymbol{L}_{\mathrm{rel}}$ is perpendicular to $\dot{\boldsymbol{r}}(t)$ (and $\boldsymbol{r}(t)$), the motion of the other particle must occur in a plane perpendicular to $\boldsymbol{L}_{\mathrm{rel}}$. If we choose to have $\boldsymbol{L}_{\mathrm{rel}}$ point in the z-direction, then $\boldsymbol{r}$ (as well as $\dot{\boldsymbol{r}}$) must always lie in the xy-plane. Thus, we have already determined two of the six initial conditions: $z(0) = \dot{z}(0) = 0$.

ii) Consider the vector $\boldsymbol{r}(t)$. In an infinitesimal time interval dt it sweeps out an area dA (Fig. 2.8.1).
For dA, we have

$$dA = \tfrac{1}{2}|\boldsymbol{r} \times \dot{\boldsymbol{r}}| \, dt \; , \tag{2.8.5}$$

since this area is a triangle with two sides formed by the vectors $\boldsymbol{a} = \boldsymbol{r}(t)$ and $\boldsymbol{b} = \dot{\boldsymbol{r}}(t)\,dt$.
The area of a triangle is given by

$$A = \tfrac{1}{2}|\boldsymbol{a}||\boldsymbol{b}|\sin \measuredangle (\boldsymbol{a}, \boldsymbol{b}) = \tfrac{1}{2}|\boldsymbol{a} \times \boldsymbol{b}| \; . \tag{2.8.6}$$

Thus, we have

$$dA = \frac{1}{2\mu}|\boldsymbol{L}_{\mathrm{rel}}| \, dt \; , \tag{2.8.7}$$

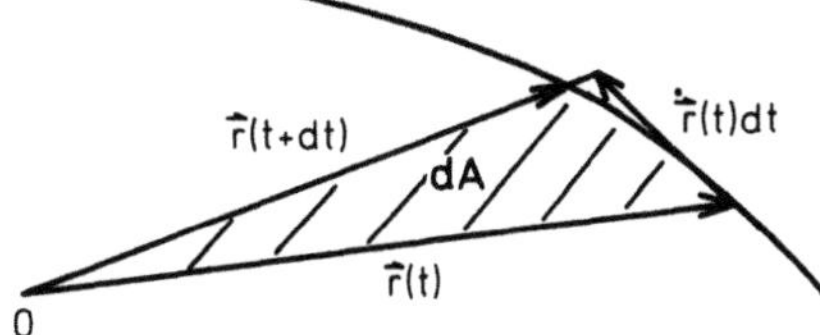

Fig. 2.8.1. The area dA swept out by the relative position vector $\boldsymbol{r}(t)$ in the time interval $[t, t + dt]$

and since $|L_{\mathrm{rel}}|$ is constant, it follows that:

The areas swept out by the radius vector during equal amounts of time are equal.

We recognize this statement as Kepler's second law of planetary motion, which we have derived as a direct consequence of the conservation of angular momentum. This law is therefore not just valid for the two-body problem with a gravitational interaction, but also for arbitrary central fields, for instance, for the harmonic field:

$$F(r) = -Dr \ . \tag{2.8.8}$$

iii) If we introduce polar coordinates in the xy-plane, so that

$$x = r\cos\varphi \ , \qquad y = r\sin\varphi \ ,$$

and

$$\dot{x} = \dot{r}\cos\varphi - r\dot{\varphi}\sin\varphi \ , \qquad \dot{y} = \dot{r}\sin\varphi + r\dot{\varphi}\cos\varphi \ ,$$

and then on the one hand

$$\dot{r}^2 = \dot{x}^2 + \dot{y}^2 = \dot{r}^2 + r^2\dot{\varphi}^2 \ , \tag{2.8.9}$$

and on the other hand,

$$L_{\mathrm{rel}} = (0,0,l),$$

$$l = \mu(x\dot{y} - y\dot{x}) \tag{2.8.10}$$

$$= \mu r\cos\varphi(\dot{r}\sin\varphi + r\dot{\varphi}\cos\varphi) - \mu r\sin\varphi(\dot{r}\cos\varphi - r\dot{\varphi}\sin\varphi)$$

$$= \mu r^2\dot{\varphi} \ , \qquad \text{so that}$$

$$\dot{\varphi} = l/\mu r^2 \ , \qquad \text{and thus} \tag{2.8.11}$$

$$\frac{1}{2}\mu\dot{r}^2 = \frac{1}{2}\mu\dot{r}^2 + \frac{l^2}{2\mu r^2} \ . \tag{2.8.12}$$

For the conserved quantity $E = \frac{1}{2}\mu\dot{r}^2 + U(r)$, we then obtain

$$E = \frac{1}{2}\mu\dot{r}^2 + \frac{l^2}{2\mu r^2} + U(r) = \frac{1}{2}\mu\dot{r}^2 + U_{\mathrm{eff}}(r) \tag{2.8.13}$$

with

$$U_{\mathrm{eff}}(r) = U(r) + \frac{l^2}{2\mu r^2} \ . \tag{2.8.14}$$

This expression for E has exactly the same form as the one for a one-dimensional system with the potential U_{eff}. We have therefore effectively reduced the problem of finding the function $r(t)$ for a given value of the angular momentum of relative motion $L_{\text{rel}} = l$ to solving a one-dimensional equation. The term $l^2/2\mu r^2$ in U_{eff} is called the *centrifugal term* or the *centrifugal barrier*. We can now calculate $r(t)$ using the methods of Sect. 2.4.2, by solving for $\dot{r}$ and then integrating:

$$\dot{r} = \pm\sqrt{(2/\mu)[E - U_{\text{eff}}(r)]} \; , \quad \text{so that} \tag{2.8.15}$$

$$\pm \int_{r_0}^{r} \frac{dr'}{\sqrt{(2/\mu)[E - U_{\text{eff}}(r')]}} = \int_{t_0}^{t} dt' = t - t_0 \; . \tag{2.8.16}$$

We then find

$$r = r(t; E, l^2, r_0) \quad \text{with} \tag{2.8.17}$$

$$r_0 = r(t_0; E, l^2, r_0) \; , \tag{2.8.18}$$

and $\varphi(t)$ can be found from

$$\dot{\varphi} = \frac{l}{\mu r^2(t)} \quad \text{viz.} \tag{2.8.19}$$

$$\varphi - \varphi_0 = \int_{t_0}^{t} dt' \, \frac{l}{\mu r^2(t')} = \Phi(t, l, E) \; . \tag{2.8.20}$$

We then find

$$\varphi(t) = \varphi_0 + \Phi(t, l, E) \quad \text{with} \tag{2.8.21}$$

$$\varphi(t_0) = \varphi_0 \; . \tag{2.8.22}$$

The equation $\dot{\varphi} = l/\mu r^2$ also tells us that as a consequence of the conservation of angular momentum, $\dot{\varphi}$ always has the same sign, so that the direction of revolution can never change. From the two functions $\varphi(t)$ and $r(t)$, the trajectory $r(t)$ of the relative motion is now determined, effectively as the solution of a one-particle problem. The values of r_0, φ_0, E, and l should be regarded as parameters for this trajectory, equivalent to the four missing initial values. E and l are related to r_0 and $\dot{\varphi}_0$ by the simple formulas

$$l = \mu r_0^2 \dot{\varphi}_0 \; , \quad E = E_{\text{rel}} = \tfrac{1}{2}\mu \dot{r}_0^2 + U_{\text{eff}}(r_0) \; . \tag{2.8.23}$$

If we are interested in the orbit itself rather than the trajectory, we should calculate $r = r(\varphi)$. If we know $r(\varphi)$ and $\varphi(t)$ then we would also be able to calculate $r(t)$ from $r(\varphi(t))$. Thus, we have

$$\dot{r} = \frac{dr}{dt} = \frac{dr}{d\varphi}\frac{d\varphi}{dt} \; , \quad \text{so that} \quad \dot{r} = \frac{dr}{d\varphi}\dot{\varphi} \; , \tag{2.8.24}$$

and then

$$\frac{dr}{d\varphi} = \frac{\dot{r}}{\dot{\varphi}} = \frac{\pm\sqrt{(2/\mu)[E - U_{\mathrm{eff}}(r)]}}{l/\mu r^2} = \pm\frac{\sqrt{2\mu}}{l} r^2 \sqrt{[E - U_{\mathrm{eff}}(r)]} \qquad (2.8.25)$$

From this we find

$$\varphi - \varphi_0 = \pm\frac{l}{\sqrt{2\mu}} \int_{r_0}^{r} dr' \frac{1}{r'^2 \sqrt{[E - U_{\mathrm{eff}}(r')]}} , \qquad (2.8.26)$$

where now φ_0 and r_0 are assumed to be given.

iv) Considering $U_{\mathrm{eff}}(r)$ as a function of r, we can qualitatively describe the progress of the trajectory.

The procedure is exactly analogous to one-dimensional motion. From the graph of the function U_{eff}, we can immediately read off the allowed values of the variable r corresponding to given values of E_{rel} and l. Minima and maxima of U_{eff} now correspond to stable and unstable circular orbits.

It is interesting to examine the following cases in some detail:

a) Assume that the graph of the effective potential looks like Fig. 2.8.2. Here we have assumed that $l \neq 0$ and that $U(r)$ is unbounded from above as $r \to \infty$ and that it is not too strongly singular as $r \to 0$ so that for $r \to 0$ the centrifugal term dominates.

Let $E_{\mathrm{rel}} = E_0$ be given: Since

$$E_0 = \tfrac{1}{2}\mu\dot{r}^2 + U_{\mathrm{eff}}(r)$$

then for every point on the trajectory we must have $U_{\mathrm{eff}}(r) \leq E_0$. The quantity

$$T = E_0 - U_{\mathrm{eff}} \geq 0 \qquad (2.8.27)$$

is the kinetic energy of the "radial" motion.

At $r = r_{\mathrm{min}}$ and $r = r_{\mathrm{max}}$, we have $U_{\mathrm{eff}}(r) = E_0$ and thus $\dot{r} = 0$. However, since $l \neq 0$, and thus $\dot{\varphi} \neq 0$ (note that $\dot{\varphi} = l/\mu r^2$), this does not mean that at these points the velocity of the particle $\dot{r} = 0$.

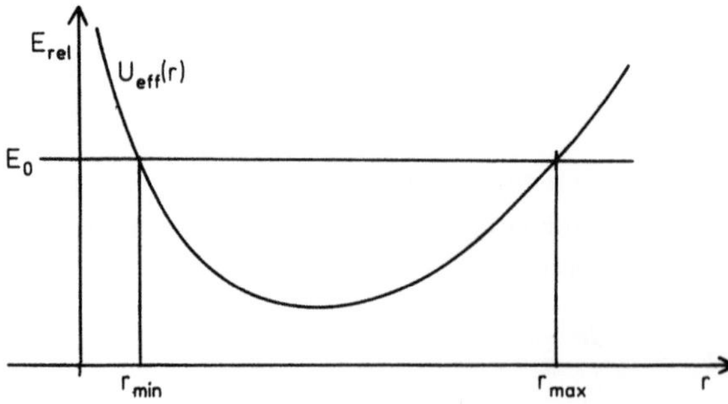

Fig. 2.8.2. A possible graph of $U_{\mathrm{eff}}(r)$ in which the distance r for a fixed value of the energy E_0 must lie between r_{min} and r_{max}

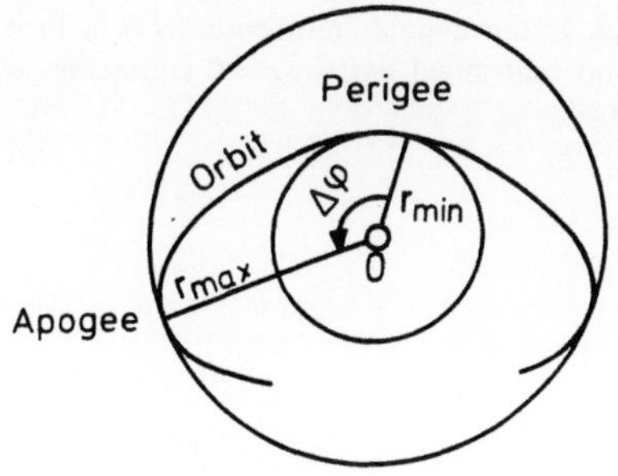

Fig. 2.8.3. A typical trajectory for the effective potential of Fig. 2.8.2

For r, we have

$$r_{\min} \leq r \leq r_{\max} \ , \tag{2.8.28}$$

i.e., the motion is bounded and takes place in between two circles in the plane of motion (Fig. 2.8.3). Since $\dot{\varphi}$ must always have the same sign, φ is a monotonically increasing function of t, while r oscillates between $r_{\min}$ and $r_{\max}$.

A point with $r = r_{\min}$ is called a *perigee* and a point with $r = r_{\max}$ is called an *apogee* (in the case of motion around the sun, a point of closest or farthest approach is referred to as a perihelion or an aphelion, respectively). The orbit need not be closed. The angle traced out between a perigee and the following apogee is easily seen to be

$$\Delta\varphi = \pm \frac{1}{\sqrt{2\mu}} \int_{r_{\min}}^{r_{\max}} \frac{dr'}{r'^2 \sqrt{E_0 - U_{\mathrm{eff}}(r')]}} \ , \tag{2.8.29}$$

and $2\Delta\varphi$ is the angle between two perigees. If $n\Delta\varphi = \pi m$, where m and n are whole numbers, then the orbit is closed.

b) There is a minimal $E_0 = E_{\min}$, such that

$$r_{\min} = r_{\max} \ , \qquad \text{i.e.,} \qquad r = \mathrm{const} = r_0 \ .$$

The orbit is a circle, and since

$$\varphi(t) = \frac{l}{\mu r_0^2} t + \varphi_0 \tag{2.8.30}$$

the particle moves uniformly in the circular orbit.

c) If $l = 0$, the centrifugal barrier vanishes. Even the value $r = 0$ is possible (Fig. 2.8.4), if $U(r)$ is bounded from above as $r \to 0$.

For $l = 0$, we find also that $r \parallel \dot{r}$ and $\dot{\varphi} = 0$. The motion is centrally directed (either towards the center, or radially outward).

d) If $U(r) \to U_0$ as $r \to \infty$, then by adding a constant we can always make $U(r) \to 0$ as $r \to \infty$.

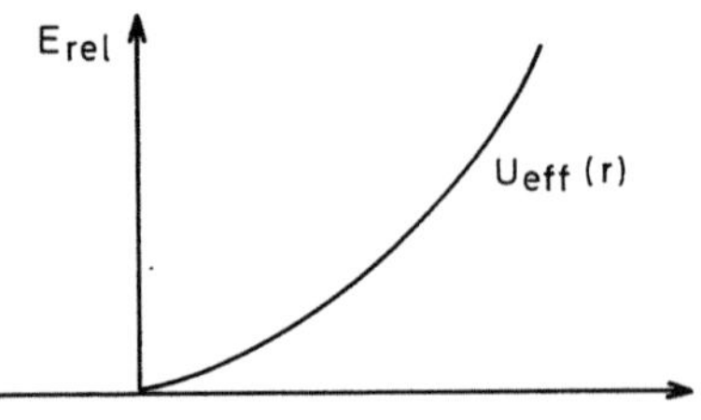

Fig. 2.8.4. If the angular momentum $l = 0$, then there is no centrifugal barrier. $r = 0$ is possible in this case

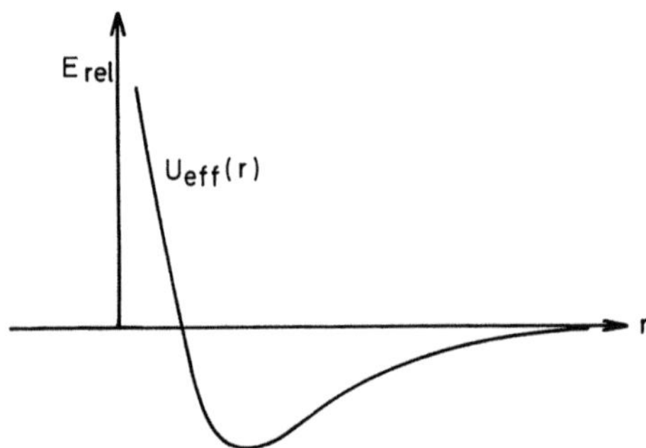

Fig. 2.8.5. The graph of $U_{\text{eff}}(r)$ (with $l \neq 0$) for the case that the effective potential remains finite as $r \to \infty$

Then $U_{\text{eff}}(r)$ has a graph as shown in Fig. 2.8.5, and we must distinguish between two following cases:

d_1) $E_0 < 0$: (only possible if U_{eff} also takes on negative values): the orbits are bounded. Everything said above in cases (a–c) is true here as well.

d_2) $E_0 > 0$: r is not bounded from above. Here, $r_{\text{max}} = \infty$, but there is still a perigee r_{min}. The orbit looks like Fig. 2.8.6 or Fig. 2.8.7, depending on whether the force is attractive or repulsive.

As $r \to \infty$, we have $\dot{\varphi} \to 0$, and the integral

$$\varphi(r) - \varphi_0 = \pm \frac{l}{\sqrt{2\mu}} \int_{r_0}^{r} \frac{dr'}{r'^2 \sqrt{[E_0 - U_{\text{eff}}(r')]}} \tag{2.8.31}$$

approaches a finite limit value as $r \to \infty$. The orbit thus asymptotically approaches a straight line. The angle between the perigee and an asymptote is given by

$$\Delta\varphi = \frac{l}{\sqrt{2\mu}} \int_{r_{\text{min}}}^{\infty} \frac{dr'}{r'^2 \sqrt{[E_0 - U_{\text{eff}}(r')]}} \, . \tag{2.8.32}$$

As $r \to \infty$, we have further that

$$|\dot{r}| \to \sqrt{2E_0/\mu} \, . \tag{2.8.33}$$

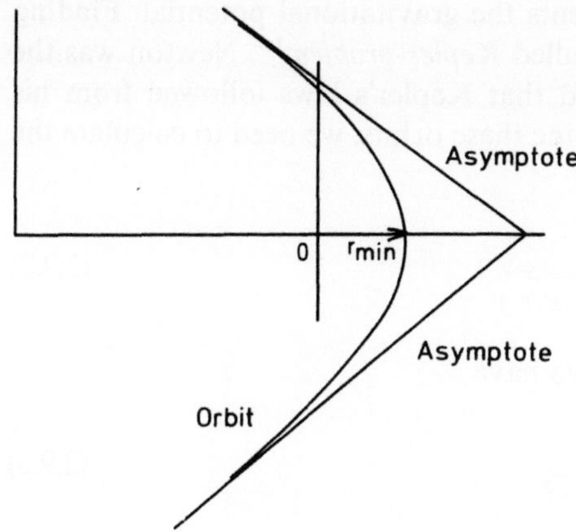

Fig. 2.8.6. Orbit for an attractive potential in the case $E_0 \geq 0$

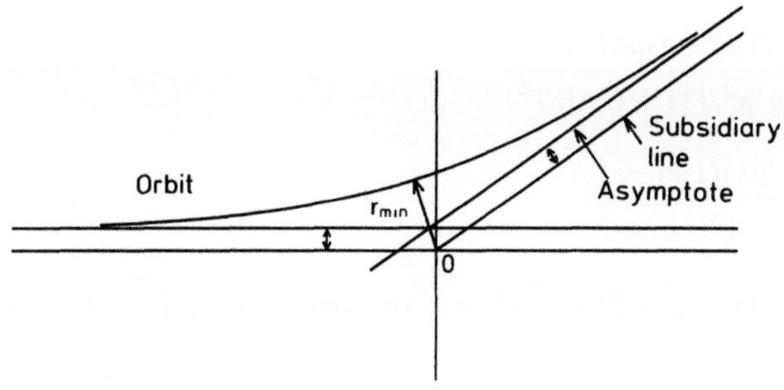

Fig. 2.8.7. Orbit for a repulsive potential in the case $E_0 \geq 0$

which can be found from

$$\dot{r} = \pm\sqrt{(2/\mu)[E_0 - U_{\text{eff}}(r)]} \tag{2.8.34}$$

since $U_{\text{eff}}(r) \to 0$ as $r \to \infty$. Since also $\dot{\varphi}(r) \to 0$ as $r \to \infty$, the motion becomes uniform and linear as $r \to \infty$ (and thus as $t \to \infty$) along the asymptotic direction of the orbit with a speed v_∞, which can be found from

$$E_{\text{rel}} = E_0 = \tfrac{1}{2}\mu v_\infty^2 \tag{2.8.35}$$

d_3) $E_0 = 0$: We will not discuss this boundary case here.

2.9 The Kepler Problem

In this section, we will calculate the trajectories and orbits of a closed two-particle system with potential

$$U(r) = -\frac{\kappa}{r}. \tag{2.9.1}$$

If we set $\kappa = \gamma m_1 m_2$, then $U(r)$ represents the gravitational potential. Finding these trajectories and orbits is the so-called *Kepler-problem*[15]. Newton was the first to solve this problem. He showed that Kepler's laws followed from his universal law of gravitation. To determine these orbits, we need to calculate the integral

$$\varphi = \frac{l}{\sqrt{2\mu}} \int^r dr' \frac{1}{r'^2 \sqrt{(E - l^2/2\mu r'^2 + \kappa/r')}} \,. \tag{2.9.2}$$

With $r' = 1/s$, so that $-dr'/r'^2 = ds$, we have

$$\varphi = - \int^{1/r} ds \frac{1}{\sqrt{2\mu E/l^2 + 2\mu\kappa s/l^2 - s^2}} \,. \tag{2.9.3}$$

From the usual integration techniques, we find

$$\varphi - \varphi_0 = \arccos \frac{(1/r) - (\mu\kappa/l^2)}{\sqrt{(\mu^2\kappa^2/l^4) + (2\mu E/l^2)}}$$

$$= \arccos \frac{(l^2/\mu\kappa)(1/r) - 1}{\sqrt{1 + (2l^2 E/\mu\kappa^2)}} \tag{2.9.4}$$

(from the integral $\int dx/\sqrt{c + 2bx - x^2} = -\arccos(x - b)/\sqrt{b^2 + c}$ for $b^2 + c^2 > 0$).

Since the kinetic energy is always positive, it is always true that $E \geq U_{\text{eff,min}}$, and for the minimal value $U_{\text{eff,min}}$ of U_{eff}, we find that

$$U_{\text{eff,min}} = -\frac{\mu\kappa^2}{2l^2} \,. \tag{2.9.5}$$

Thus, we always have

$$b^2 + c \sim 1 + \frac{2l^2 E}{\mu\kappa^2} > 0 \quad \text{for} \quad E > U_{\text{eff,min}} \,.$$

(in the special case $E = U_{\text{eff,min}}$, we have $\dot{r} = 0$, $r = \text{constant}$). If we introduce

$$p = l^2/\mu\kappa, \quad \varepsilon = \sqrt{1 + (2l^2 E/\mu\kappa^2)} \,, \quad \text{then} \tag{2.9.6}$$

$$\varphi - \varphi_0 = \arccos\left(\frac{p/r - 1}{\varepsilon}\right) \quad \text{or} \tag{2.9.7}$$

---

[15] *Kepler, Johannes* (*1571 Weil der Stadt, d. 1630 Regensburg). After 1600, he lived as a mathematician and astronomer in Prague. He discovered his laws of planetary motion primarily from his efforts studying the orbit of Mars. The first two laws were published in 1609 in "Astronomia Nova", while the third was only discovered in 1618 and published in "Harmonices mundi".

$$\varepsilon \cos (\varphi - \varphi_0) = \frac{p}{r} - 1 \quad \text{or}$$

$$r = \frac{p}{(1 + \varepsilon \cos \varphi)} \quad \text{with} \quad \varphi_0 = 0 \ . \tag{2.9.8}$$

What we have just found is the equation in polar coordinates of a conic section with a focus at the origin, and indeed (see Appendix B)

for $\varepsilon < 1$, i.e. $E < 0$ we have an ellipse;
for $\varepsilon = 1$, i.e. $E = 0$, we have a parabola;
for $\varepsilon > 1$, i.e. $E > 0$, we have a hyperbola.

If $E < 0$, then, as we know, the motion is bounded, and in this case the orbit is even closed.

The Different Cases

   i) $\varepsilon < 1$: r oscillates between

$$r_{\min} = p/(1 + \varepsilon) \ , \quad \text{when} \quad \varphi = 0 \tag{2.9.9}$$

and

$$r_{\max} = p/(1 - \varepsilon) \ , \quad \text{when} \quad \varphi = \pi \tag{2.9.10}$$

(see Fig. 2.9.1). For $\varepsilon = 0$, the orbit is a circle. The semimajor axis of the ellipse satisfies:

$$2a = r_{\min} + r_{\max} = p/(1 + \varepsilon) + p/(1 - \varepsilon) = 2p/(1 - \varepsilon^2)$$

so that

$$a = \frac{p}{(1 - \varepsilon^2)} = \frac{l^2}{\mu\kappa} \left(-\frac{\mu\kappa^2}{2l^2 E} \right) = \frac{\kappa}{2|E|} \ . \tag{2.9.11}$$

To find the semiminor axis b, we note that

$$\varepsilon^2 = \frac{a^2 - b^2}{a^2} \ , \quad \text{and so} \tag{2.9.12}$$

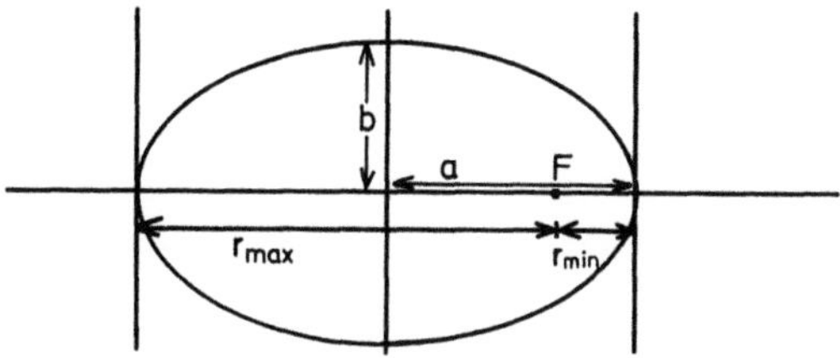

Fig. 2.9.1. Orbit of a particle in the Kepler-problem for $E < 0$. The orbit is an ellipse with semimajor and semiminor axes a and b

$$b^2 = a^2(1 - \varepsilon^2) = \frac{\kappa^2}{4E^2}\left(-\frac{2l^2 E}{\mu\kappa^2}\right) = \frac{l^2}{2|E|\mu} = p \cdot a \; . \tag{2.9.13}$$

We have thus shown Kepler's first law:

Planets describe elliptical orbits around the sun, of which the sun is located in one of the two foci.

In addition, we know how the semimajor and semiminor axes depend on the energy and momentum of the relative motion.

It should be noted that the excentricities of the planetary orbits are very small. From astronomical measurements, we find that for Mercury $\varepsilon = 0.206$, for the Earth $\varepsilon = 0.017$, and for Mars $\varepsilon = 0.093$. With the exception of Pluto (which was then unknown) and Mercury (which is difficult to observe), Mars has the largest excentricity, so that it was the most suitable planet for Kepler to use in the discovery of his first law.

We have already demonstrated Kepler's second law, which states that the radius vector of the orbit sweeps out equal areas in equal amounts of time in Sect. 2.8, as a consequence of the conservation of angular momentum.

Thus, we need only prove Kepler's third law, which states that the square of the period of revolution goes as the cube of the semimajor axis of the orbit.

To show this, we return to the relation

$$dA = \frac{1}{2\mu}|L|dt \; .$$

We know that in one period the area of an ellipse

$$A = \pi ab \tag{2.9.14}$$

is swept out. Thus, we have

$$\pi ab = \frac{1}{2\mu} lT \; , \quad \text{or}$$

$$T = \frac{2\pi\mu}{l} ab = \frac{2\pi\mu}{l} a\sqrt{pa} = \frac{2\pi\mu}{l} a^{3/2} \frac{1}{\sqrt{\mu\kappa}} \tag{2.9.15}$$

$$= 2\pi \sqrt{\frac{\mu}{\kappa}} a^{3/2} \; , \quad \text{or} \tag{2.9.16}$$

$$T^2 = \frac{4\pi^2 \mu}{\gamma m_1 m_2} a^3 = \frac{4\pi^2}{\gamma(m_1 + m_2)} a^3 \; . \tag{2.9.17}$$

Since one of these masses (say m_1) is the mass of the sun, and thus is far greater than any of the planetary masses, we have, to a good approximation:

$$T^2 = \frac{4\pi^2}{\gamma m_1} a^3 \ . \tag{2.9.18}$$

For systems composed of bodies of comparable mass, for example, for double stars, it is always possible to find the sum of the masses from the orbital period.

We have shown that all of Kepler's laws can be derived from Newton's law of gravitation.

ii) $\varepsilon = 1$, i.e. $E = 0$. This is a boundary case, in which the orbit is a parabola. We will not investigate this case further.

iii) $\varepsilon > 1$. In this case, we see that $1 + \varepsilon \cos \varphi$ can be both positive and negative (Fig. 2.9.2).

Since in the equation

$$r = p/(1 + \varepsilon \cos \varphi)$$

r is always positive, but $p = l^2/\mu\kappa$ can be both positive and negative depending on the sign of κ, we must distinguish between the two cases:

a) $\kappa > 0$, i.e. $p > 0$ as in the case $\varepsilon < 1$, i.e., the force is attractive.

At the perigee, we have $r = r_{\min} = p/(1 + \varepsilon)$ and $\varphi = 0$. The asymptotes are given by

$$1 + \varepsilon \cos \varphi_2 = 0 \tag{2.9.19}$$

and we have $|\varphi_2| > \pi/2$ (Figs. 2.9.2, 3).

b) $\kappa > 0$, then the force is repulsive, as in the case of two charges with the same sign. Then $p < 0$, and so it must always be true that

$$1 + \varepsilon \cos \varphi < 0 \ .$$

The allowed values of angle, which correspond to positive values of $1/r$ are exactly complementary to the allowed angles in case (a) (Fig. 2.9.2).

The orbits are now hyperbolas which do not include the center of force, but which approach it and then fall back (Fig. 2.9.4).

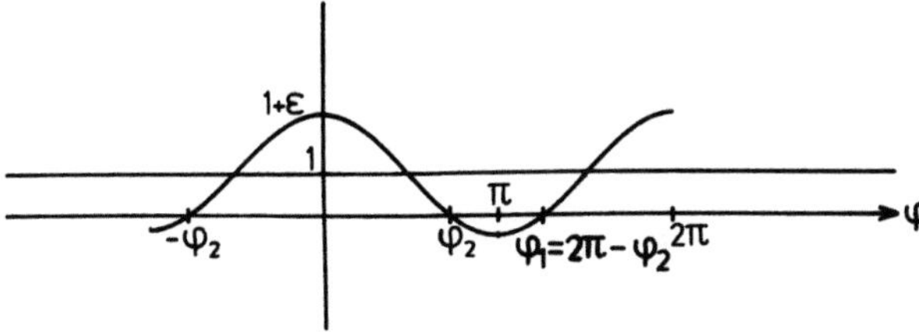

Fig. 2.9.2. The graph of the function $(1 + \varepsilon \cos \varphi)$ for $\varepsilon > 1$

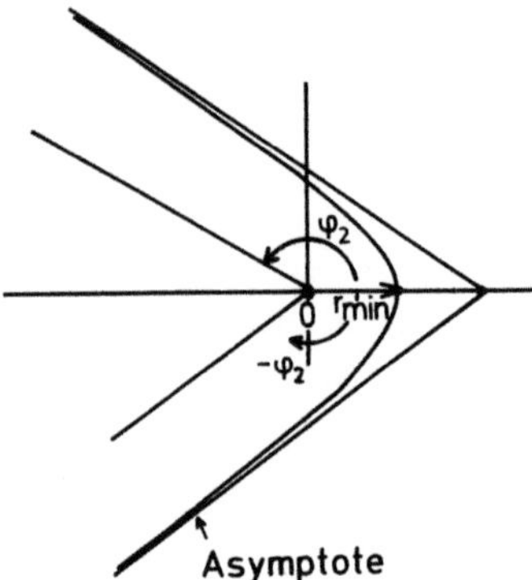

Fig. 2.9.3. Orbit of a particle in the Kepler problem (for an attractive potential) in the case $E > 0$

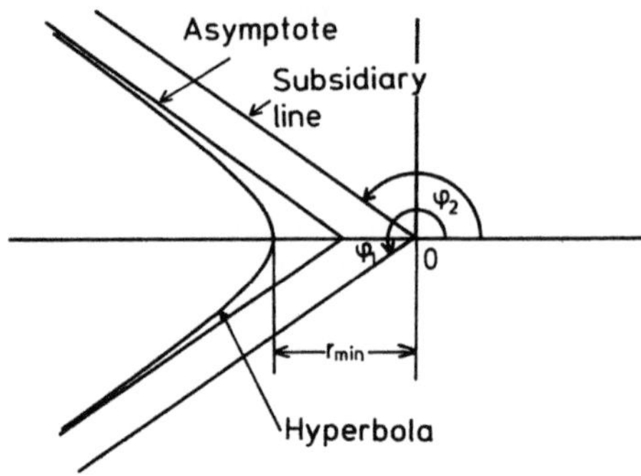

Fig. 2.9.4. Orbit of a particle in the Kepler problem (repulsive potential) for $E > 0$

We will discuss these orbits, which correspond to scattering from a repulsive potential, later in greater depth.

Remarks. i) The fact that for the potential

$$U(r) = -\kappa/r$$

bounded motions always occur in closed orbits, might lead us to suspect that there is another conserved quantity besides E and L. This could be a vector, for example, that points from the center of force to the perigee. This vector being constant would mean that the perigee does not move.

In actuality, the perihelion of a planet does move slightly. This motion is greatest in the case of Mercury.

From theoretical considerations, we obtain the values

5557.62″ ± 0.20″ per century, of which 5025″ occurs due to the precession of the equinox and 532″ arises from disturbances produced by other planets.

43.03″ per century from the effects of the theory of general relativity.

The sum of these contributions agrees well with the observed value.

To find the conserved vector, we rearrange the equation of motion, and the definitions of angular momentum and energy:

$$\mu\ddot{\mathbf{r}} + \kappa\,\frac{\mathbf{r}}{r^3} = 0 \;, \qquad \mathbf{L} = \mu\mathbf{r}\times\dot{\mathbf{r}} \;, \qquad E = \frac{1}{2}\mu\dot{r}^2 - \frac{\kappa}{r} \;, \tag{2.9.20}$$

into

$$\ddot{\mathbf{r}} + \beta\,\frac{\mathbf{r}}{r^3} = 0 \;, \qquad \mathbf{C} = \mathbf{r}\times\dot{\mathbf{r}} \;, \qquad B = \dot{r}^2 - \frac{2\beta}{r} \;, \tag{2.9.21}$$

with

$$\beta = \frac{\kappa}{\mu} = \gamma(m_1 + m_2) \ , \qquad C = \frac{L}{\mu} \ , \qquad B = \frac{2E}{\mu} \ , \tag{2.9.22}$$

and then take the vector product of the equation of motion with C:

$$0 = \ddot{\boldsymbol{r}} \times \boldsymbol{C} + \beta \frac{\boldsymbol{r}}{r^3} \times \boldsymbol{C} = \frac{d}{dt}(\dot{\boldsymbol{r}} \times \boldsymbol{C}) + \frac{\beta}{r^3} \boldsymbol{r} \times (\boldsymbol{r} \times \dot{\boldsymbol{r}})$$

$$= \frac{d}{dt}(\dot{\boldsymbol{r}} \times \boldsymbol{C}) + \frac{\beta}{r^3}\left[\boldsymbol{r}(\boldsymbol{r}\cdot\dot{\boldsymbol{r}}) - r^2\dot{\boldsymbol{r}}\right]$$

$$= \frac{d}{dt}\left(\dot{\boldsymbol{r}} \times \boldsymbol{C} - \beta \frac{\boldsymbol{r}}{r}\right) \ , \qquad \text{since} \tag{2.9.23}$$

$$\frac{d}{dt}\frac{\boldsymbol{r}}{r} = \frac{\dot{\boldsymbol{r}}}{r} - \boldsymbol{r}\frac{\boldsymbol{r}\cdot\dot{\boldsymbol{r}}}{r^3} \quad \text{is} \ ,$$

because from $r^2 = \boldsymbol{r}^2$ it follows that $r\cdot\dot{r} = \boldsymbol{r}\cdot\dot{\boldsymbol{r}}$.
Thus,

$$\boldsymbol{A} = \dot{\boldsymbol{r}} \times \boldsymbol{C} - \beta \frac{\boldsymbol{r}}{r} \tag{2.9.24}$$

is a vector constant in time which lies in the plane of motion. Since $\boldsymbol{A}$ is constant, it can be determined by its value at any time, including at the perigee. At this point in the motion, $\boldsymbol{r}/r$ and $\dot{\boldsymbol{r}} \times \boldsymbol{C}$ lie in the direction of the perigee. Thus, $\boldsymbol{A}$ always points in the direction of the perigee. The vector $\boldsymbol{A}$ is called the *Lenz–Runge Vector*[16]. The fact that $\boldsymbol{A}$ is time-independent is a special characteristic of the Kepler problem and does not hold in general for all central forces. Further, we have

$$\boldsymbol{A}\cdot\boldsymbol{C} = 0 \ ,$$

since $\dot{\boldsymbol{r}} \times \boldsymbol{C}$ and $\boldsymbol{r}/r$ are perpendicular to $\boldsymbol{C}$. In addition,

$$\boldsymbol{A}^2 = \left(\dot{\boldsymbol{r}} \times \boldsymbol{C} - \beta \frac{\boldsymbol{r}}{r}\right)^2 = (\dot{\boldsymbol{r}} \times \boldsymbol{C})^2 - 2\beta\left(\frac{\boldsymbol{r}}{r} \times \dot{\boldsymbol{r}}\right)\cdot\boldsymbol{C} + \beta^2$$

$$= \dot{r}^2 C^2 - \frac{2\beta}{r}C^2 + \beta^2 = BC^2 + \beta^2 \ , \tag{2.9.25}$$

---

[16] *Lenz–Runge Vector.* Runge, Carl David Tolmé (*1856 Bremen, d. 1927 Göttingen), mathematician and physicist, specialized in number theory, analysis, and spectroscopy.

William Lenz (*1988, d. 1957) since 1921 Professor of theoretical physics in Hamburg. Worked on Maxwell's theory, statistical mechanics, theory of atoms and molecules. E. Ising, whose name is known from the Ising model in statistical mechanics, was his student.

so that the magnitude of the vector A is determined by the energy and the angular momentum. Clearly it is true that

$$\frac{A^2}{\beta^2} = 1 + \frac{2E}{\mu}\frac{L^2}{\mu^2}\frac{1}{\beta^2} = 1 + \frac{2l^2 E}{\mu\kappa^2} \ ,$$
(2.9.26)

which means that $|A|/\beta$ is identical to ε.

If we now construct

$$A \cdot r = A \cdot r \cos\varphi \quad \text{with} \quad \varphi = \sphericalangle(A, r) \ ,$$

then we have

$$A \cdot r = r \cdot \left[(\dot{r} \times C) - \beta\frac{r}{r} \right] = (r \times \dot{r}) \cdot C - \beta r$$

or

$$rA\cos\varphi = C^2 - \beta r \ , \quad \text{which means}$$
(2.9.27)

$$r(1 + \varepsilon \cos\varphi) = C^2/\beta \quad \text{or}$$

$$r = \frac{p}{1 + \varepsilon \cos\varphi} \ , \quad \text{since}$$
(2.9.28)

$$\frac{C^2}{\beta} = \frac{L^2}{\mu^2}\cdot\frac{\mu}{\kappa} = \frac{l^2}{\kappa\mu} = p \ .$$
(2.9.29)

Thus, we find again the same conic sections which we had found above without using the conserved quantity A.

ii) Until now, we have considered only the relative motion. Since

$$r_1 = R + x_1 = R + \frac{m_2}{M}r \ , \quad r_2 = R + x_2 = R - \frac{m_1}{M}r$$
(2.9.30)

we can immediately find the complete motion in the two-body problem. For $E_{\text{rel}} < 0$, for example, the two particles follow elliptical orbits around their common center of mass, which itself moves with uniform linear motion and which always lies on the line connecting the two particles (Fig. 2.9.5).

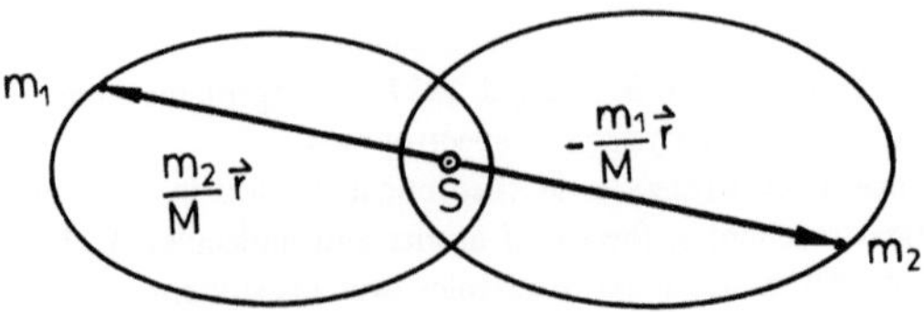

Fig. 2.9.5. The motion of two masses around a common center of mass

2.10 Scattering

As we saw in Sect. 2.8, it is easy to qualitatively find the orbit in the two-body problem. In particular, in the case

$$U(r) = -\kappa/r \quad \text{and} \quad E_{\text{rel}} > 0$$

we find hyperbolic trajectories. This case, which is called the *scattering* of two particles, can also be explored quantitatively. In general, the following occurs during a *scattering process*:

Free particles move towards each other, and as they near each other, they interact. This interaction may cause changes in the direction, momentum, and even the nature of the particles. They then move apart, so that after some period of time their interaction ceases. They are then free particles again.

A large number of the experiments performed in modern physics, especially in atomic, molecular, nuclear, and particle physics are scattering experiments, in which we attempt to learn about the interactions between the particles by examining the effects of scattering. What we know today, say, about elementary particles, is almost entirely derived from the results of scattering experiments.

Here we consider within the framework of Newtonian mechanics only the simplest scattering process: an *elastic* collision of two particles with a spherically symmetric interaction potential, in which the total mechanical energy is a conserved quantity and the particles are the same before and after the collision. If we represent the *velocities of the particles long before* and *long after* the collision by v_1, v_2 and v'_1, v'_2 respectively, then by energy and momentum conservation, we have

$$\tfrac{1}{2}m_1 v_1^2 + \tfrac{1}{2}m_2 v_2^2 = \tfrac{1}{2}m_1 v_1'^2 + \tfrac{1}{2}m_2 v_2'^2 \tag{2.10.1}$$

and

$$m_1 v_1 + m_2 v_2 = m_1 v'_1 + m_2 v'_2 \; . \tag{2.10.2}$$

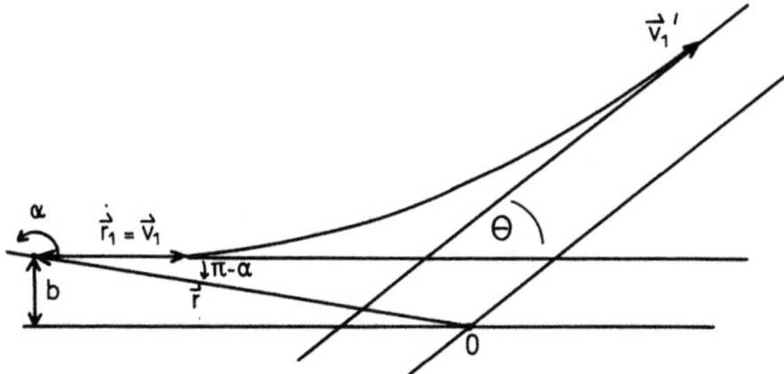

Fig. 2.10.1. Particle 1 is deflected by the repulsive force of particle 2 at O. The impact parameter b is the distance by which particle 1 would have missed particle 2 if its path had not been influenced by any forces

From the given initial values of the velocities v_1 and v_2, we are to find the six components v'_1 and v'_2 of the velocities after the collision. The conservation laws for momentum and energy give us four equations, so that there are still two independent quantities to determine, for example, the direction of v'_1.

Of greatest interest here is one particular angle, the angle between v_1 and v'_1, the angle by which the first particle is deflected (Fig. 2.10.1). For if we rotate the velocity vectors v'_1 and v'_2 of the particles after the collision about the direction of the total momentum P, then by the rotational invariance of the potential, we obtain essentially the same collision. The final configuration of the collision process is thus determined by a single scattering angle.

2.10.1   Relative Motion in the Scattering Process

In order to make calculations involving the elastic scattering of two particles, we first separate out the center of mass motion and then consider the time dependence of the relative coordinate

$$r(t) = r_1(t) - r_2(t) \ .$$

For reasons of simplicity, we first limit ourselves to a repulsive potential ($\kappa < 0$).

i) We know that the angular momentum L_{rel} and the energy E_{rel} are conserved quantities. We want to express L_{rel} and E_{rel} in terms of other, more physically accessible quantities:

Long before and after the collision, the particles move freely, so that as $t \to -\infty$

$$\frac{r(t)}{t} \to v = v_1 - v_2 \tag{2.10.3}$$

and as $t \to +\infty$

$$\frac{r(t)}{t} \to v' = v'_1 - v'_2 \ . \tag{2.10.4}$$

The value of the conserved quantity

$$E_{\mathrm{rel}} = \tfrac{1}{2}\mu v^2(t) + U(r(t))$$

remains the same at all times. It is easiest to calculate it at $t \to \pm \infty$, since at these times the particles are so far separated from each other that we can ignore the potential energy of their interaction. We have then

$$E_{\mathrm{rel}} = E_{\mathrm{rel}}(\pm \infty) = \tfrac{1}{2}\mu v^2 = \tfrac{1}{2}\mu v'^2 \tag{2.10.5}$$

which is simply determined by the relative velocity $\dot{r} = v_1 - v_2$. Hence,

$$|v| = |v'| \ : \tag{2.10.6}$$

Clearly, it is also true that $E_{\text{rel}} > 0$, as we would expect for an unbounded motion.

The angular momentum is given by

$$l = |L_{\text{rel}}| = \mu|r \times \dot{r}| = \mu r |\dot{r}| \sin\alpha \quad \text{with} \quad \alpha = \measuredangle(r, \dot{r}) \ . \tag{2.10.7}$$

If we consider the situation as $t \to -\infty$, we find (Fig. 2.10.1)

$$r \sin\alpha = r \sin(\pi - \alpha) =: b \ . \tag{2.10.8}$$

b is called the *impact parameter*. b is the distance by which particle 1 would have missed particle 2 if it had not been deflected by any forces. We then have

$$l = |L_{\text{rel}}| = \mu b |v| \quad \text{and} \tag{2.10.9}$$

$$E_{\text{rel}} = \tfrac{1}{2}\mu v^2 \ . \tag{2.10.10}$$

Thus we can express the conserved quantities l and E_{rel} in terms of the more physically accessible quantities $|v|$ and b.

ii) We now inquire about the scattering angle θ of the relative vector, that is, we want to find the angle between v and v', the angle by which the relative velocity is deflected in the collision (Fig. 2.10.2). If we indicate the angle in between perigee and asymptote by Φ, then obviously $\theta = \pi - 2\Phi$.

Then, we find for the scattering angle:

$$\theta = \pi - 2\frac{l}{\sqrt{2\mu}} \int\limits_{r_{\text{min}}}^{\infty} \frac{dr'}{r'^2 \sqrt{[E - U_{\text{eff}}(r')]}} \ . \tag{2.10.11}$$

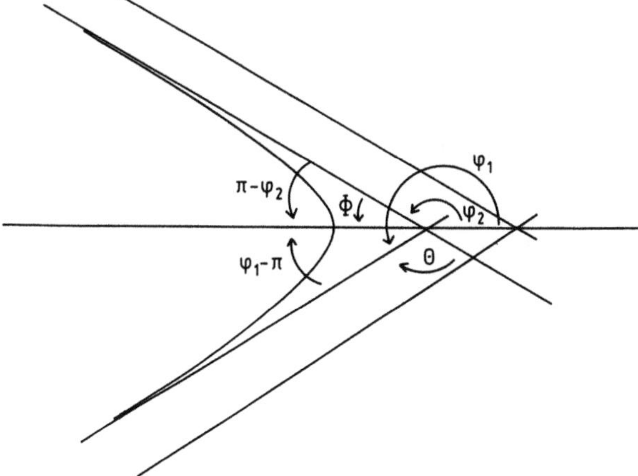

Fig. 2.10.2. The scattering angle is determined by the deflection of the relative velocity. We have $\Phi = \pi - \varphi_2 = \varphi_1 - \pi$, and $2\Phi + \theta = \pi$

iii) Let the target particle now be an N-fold charged nucleus, and let the incident particle be an He-nucleus (α-particle, *Rutherford scattering*[17]). The repulsive forces can be described by the potential:

$$U(r) = \frac{1}{4\pi\varepsilon_0} \frac{q_1 q_2}{r} \; ,$$

(q_2: the charge of the nucleus, q_1: the charge of the α-particle).

From the above equation, we can calculate the scattering angle. In this case, though, we already know the trajectory. It is a hyperbola with the polar equation (Sect. 2.9)

$$r = \frac{p}{(1 + \varepsilon \cos \varphi)} \tag{2.10.12}$$

and it then follows (Fig. 2.10.2) that

$$\varphi_2 \le \varphi \le \varphi_1 \quad \text{with}$$

$$1 + \varepsilon \cos \varphi_i = 0 \; , \quad i = 1, 2 \; .$$

It is also true that

$$\pi - \varphi_2 = \varphi_1 - \pi = \Phi \; .$$

Furthermore,

$$\varphi_1 - \varphi_2 = \pi - \theta \; , \quad \text{so that with}$$

$$\varphi_1 - \pi = \pi - \varphi_2 \quad \text{and}$$

$$\pi - \varphi_2 = \varphi_2 - \theta \quad \text{or}$$

$$\varphi_2 = \frac{\pi + \theta}{2} \quad \text{and thus} \tag{2.10.13}$$

$$-\frac{1}{\varepsilon} = \cos \varphi_2 = \cos\left(\frac{\pi + \theta}{2}\right) = -\sin\left(\frac{\theta}{2}\right) \; .$$

It follows that

$$\sin\left(\frac{\theta}{2}\right) = \frac{1}{\varepsilon} = \frac{1}{\sqrt{1 + (2El^2/\mu\kappa^2)}} \; . \tag{2.10.14}$$

---

[17] *Rutherford, Ernest* (*1871 Spring Grove, New Zealand, d. 1937 Cambridge). His most important discovery, published in 1911, was that the mass of an atom is concentrated in a small, dense nucleus. This discovery was made in experiments involving the scattering of α-particles by atoms. He won the Nobel Prize for Chemistry in 1908.

From the trigonometric identity:

$$\sin\left(\frac{\theta}{2}\right) = \frac{1}{\sqrt{1 + \cot^2(\theta/2)}}$$

it follows that

$$\cot^2(\theta/2) = 2El^2/\mu\kappa^2$$

$$= 2\tfrac{1}{2}\mu v^2 \mu^2 v^2 b^2/\mu\kappa^2 = \mu^2 v^4 b^2/\kappa^2$$

or

$$\tan(\theta/2) = \frac{|\kappa|}{\mu b v^2} . \tag{2.10.15}$$

Since θ is between 0 and π, $\tan(\theta/2)$ is always positive, so that the right side should always be chosen to be positive.

We note:

a) The smaller b is, the greater $\tan(\theta/2)$ and thus θ, as $b \to 0$ approaches π. This represents a head-on collision, in which particle 1 is aimed directly at particle 2, reaches an $r_{\min}$, and then turns around and returns in the same direction it comes from. The angular momentum vanishes. $r_{\min}$ can be calculated from

$$E = \tfrac{1}{2}\mu v^2 + U(r) .$$

We have

$$E = \tfrac{1}{2}\mu v^2 \quad \text{before the collision}$$

$$E = U(r_{\min}) = -\kappa/r_{\min} \quad \text{at the turning point} .$$

Then, we have for $b = 0$:

$$r_{\min} = \frac{2|\kappa|}{\mu v^2} . \tag{2.10.16}$$

It is easy to show by calculation, and also apparent from observation of the graphs of $U_{\text{eff}}(r)$ for different values of l (which is proportional to b), that for collisions with $b > 0$, the corresponding distance $r_{\min}(b)$ is greater than $r_{\min}(0)$.

The greater v is, the closer the two charged particles come to each other. If the initial energy is so large that $r_{\min}$ approaches the radius of the charged particles, the experimental results show a deviation from the scattering equation for $\tan(\theta/2)$, because then the charged bodies can no longer be treated as pointlike particles.

Example. If particle 1 is an α-particle with $v = 1.61 \times 10^9$ cm/s and particle 2 is a copper nucleus, then $r_{\min} = 1.55 \times 10^{-12}$ cm. The scattering equation is still found to be valid, hence atomic nuclei are smaller than 10^{-12} cm.

b) The larger $q_1 q_2$ is, the larger again is the angle θ, so that if we compare $e^- e^-$ scattering with Rutherford scattering for the charges $q_1, q_2 = Z q_1, Z \gg 1$, then the angles of deflection are much greater in the second case. Rutherford concluded from this in 1911 that the positively charged constituents of an atomic nucleus are clustered together in a very small center.

2.10.2 The Center of Mass System and the Laboratory System

Until now, we have limited ourselves again to a study of the relative motion. Now, we remind ourselves that the distances x_i of the two particles from the center of mass R are given by:

$$x_1 = \frac{m_2}{M} r , \qquad x_2 = -\frac{m_1}{M} r .$$

The velocities of the two particles are then:

$$\dot{r}_1(t) = \frac{m_2}{M} \dot{r}(t) + \dot{R}(t) , \tag{2.10.17a}$$

$$\dot{r}_2(t) = -\frac{m_1}{M} \dot{r}(t) + \dot{R}(t) . \tag{2.10.17b}$$

The velocities and deflection angles which are measured will depend on the velocity of the center of mass $\dot{R}$, and thus on the inertial frame of reference chosen. However, the following quantities are independent of the choice of inertial system (and are thus Galilei-invariant):

$$v = \lim_{t \to -\infty} \dot{r}(t) = v_1 - v_2 \quad \text{and} \tag{2.10.18a}$$

$$v' = \lim_{t \to +\infty} \dot{r}(t) = v'_1 - v'_2 \tag{2.10.18b}$$

with $v^2 = v'^2$ and the *momentum transfer*

$$q = p'_1 - p_1 = p_2 - p'_2 = m_1(v'_1 - v_1) , \tag{2.10.19}$$

The *energy transfer*, however,

$$\Delta E_1 = E'_1 - E_1 \tag{2.10.20}$$

depends on the choice of inertial system, because energy is quadratic in velocity, and hence the velocity of the center of mass does not cancel in the difference.

We will now discuss two inertial systems which are frequently used:

a) The Center-of-Mass System (or CMS): This is defined by $\dot{\boldsymbol{R}} \equiv 0$, so that it is the inertial system in which the center of mass is at rest.

Then, we have

$$\dot{\boldsymbol{r}}_1(t) = \frac{m_2}{M}\,\dot{\boldsymbol{r}}(t) \ , \quad \dot{\boldsymbol{r}}_2(t) = -\frac{m_1}{M}\,\dot{\boldsymbol{r}}(t) \ . \tag{2.10.21}$$

We see that the angle of deviation θ of the relative velocities calculated above is also the angle of deflection of particles 1 and 2 in the center-of-mass system. Typical particle trajectories in this system are illustrated in Fig. 2.10.3. For the velocities and energies before and after the collision, we have

$$v_1 = v_1' = \frac{m_2}{M}\,v \ , \quad v_2 = v_2' = \frac{m_1}{M}\,v \ , \tag{2.10.22a}$$

$$E_1 = E_1' = \frac{m_1 m_2^2}{2M^2}\,v^2 \ , \quad E_2 = E_2' = \frac{m_2 m_1^2}{2M^2}\,v^2 \ , \tag{2.10.22b}$$

$$E = E_1 + E_2 = E_1' + E_2' = \tfrac{1}{2}\mu v^2 = E_{\mathrm{rel}} \ . \tag{2.10.22c}$$

There is no energy transfer in this inertial system, and the momentum transfer is given by

$$q^2 = \mu^2 (v' - v)^2 = \mu^2 (v^2 + v'^2 - 2vv' \cos\theta) \ ,$$

and thus, since $|v'| = |v|$,

$$q^2 = 2\mu^2 v^2 (1 - \cos\theta) \ . \tag{2.10.23}$$

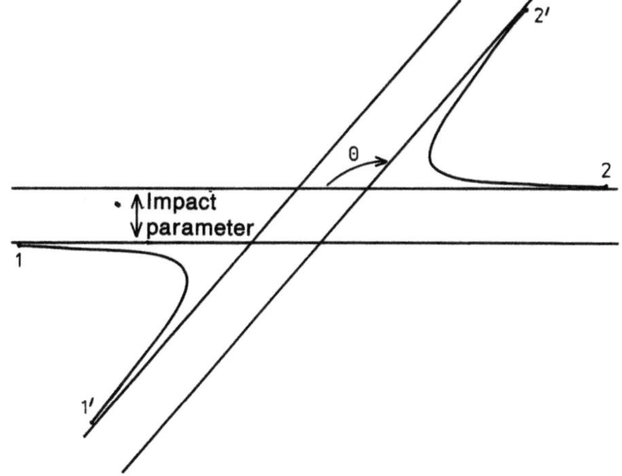

Fig. **2.10.3.** Trajectories of particles in the center-of-mass system

b) The Laboratory System: Of course, a coordinate system in which the center of mass is at rest is seldom realized in experiment. Much more often, we have the case in which one of the particles, called the *target particle*, is at rest before the collision, and the *incident particle* is shot towards it. We will assume that the target particle is particle 2. The system in which $v_2 = 0$ holds is called the *laboratory system* for obvious reasons. In the laboratory system, then (all quantities measured with respect to the laboratory system will be indicated by the subscript L):

$$\dot{R}(t) = \dot{R}(\pm\infty) = \frac{m_1}{M}\, v_{L1} \; . \tag{2.10.24}$$

We also note for later purposes that $v_{L1} = \dot{r}(-\infty) = v$.
According to these definitions, we find

$$\dot{r}_{L1}(t) = \frac{m_2}{M}\, \dot{r}(t) + \frac{m_1}{M}\, v_{L1} \; , \tag{2.10.25a}$$

$$\dot{r}_{L2}(t) = -\frac{m_1}{M}\, \dot{r}(t) + \frac{m_1}{M}\, v_{L1} \; . \tag{2.10.25b}$$

Since $\dot{r}(-\infty) = v_{L1}$, it is easy to verify that

$$\dot{r}_{L1}(-\infty) = v_{L1} \quad \text{and} \quad \dot{r}_{L2}(-\infty) = 0 \; . \tag{2.10.26}$$

The trajectories of the two particles are depicted in Fig. 2.10.4.
The scattering angle θ_2 is defined by

$$\theta_L = \sphericalangle(p_{L1}, p'_{L1}) \; . \tag{2.10.27}$$

In order to determine the relationship between this angle and the scattering angle θ in the center-of-mass system, we observe that

$$p'_{L1} = m_1 v'_{L1} = \mu \dot{r}(\infty) + \frac{m_1^2}{M}\, \dot{r}(-\infty) \; , \tag{2.10.28}$$

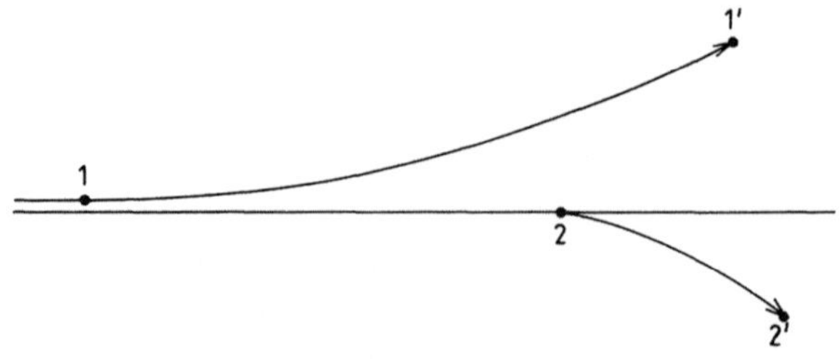

Fig. 2.10.4. Trajectories of particles in the laboratory system

so that

$$p_{L1}^{\prime 2} = \frac{m_1^2 m_2^2}{M^2}\dot{r}^2(\infty) + \frac{m_1^4}{M^2}\dot{r}^2(-\infty) + 2\frac{m_1^3 m_2}{M^2}|\dot{r}(\infty)|\cdot|\dot{r}(-\infty)|\cos\theta\ .$$

$$(2.10.29)$$

Since $|\dot{r}(\infty)| = |\dot{r}(-\infty)|$, it follows that the magnitude of the momentum of particle 1 after the collision is given by

$$p_{L1}' = \frac{m_1}{M}|\dot{r}(\infty)|\sqrt{m_2^2 + 2m_1 m_2\cos\theta + m_1^2}\ . \qquad (2.10.30)$$

With these expressions for p_{L1} and p_{L1}', it follows that

$$\cos\theta_L = \frac{p_{L1}\cdot p_{L1}'}{p_{L1} p_{L1}'}$$

$$= \frac{m_1\dot{r}(-\infty)\cdot[\mu\dot{r}(\infty) + (m_1^2/M)\dot{r}(-\infty)]}{m_1|\dot{r}(-\infty)|(m_1/M)|\dot{r}(+\infty)|\sqrt{m_2^2 + 2m_1 m_2\cos\theta + m_1^2}}$$

$$= \frac{m_1 + m_2\cos\theta}{\sqrt{m_2^2 + 2m_1 m_2\cos\theta + m_1^2}}\ . \qquad (2.10.31)$$

From the identity,

$$\tan^2\theta_L = -1 + \frac{1}{\cos^2\theta_L}$$

it follows also that

$$\tan\theta_L = \frac{m_2\sin\theta}{m_1 + m_2\cos\theta}\ . \qquad (2.10.32)$$

The momentum of the first particle after the collision is p_{L1}'. For particle 2, we use momentum conservation to find,

$$p_{L2}' = |p_{L1} - p_{L1}'| = |q|\ . \qquad (2.10.33)$$

In order to express these quantities entirely in terms of the scattering angle θ_L in the laboratory system, we use the conservation laws for energy and momentum in the form

$$m_1 v_{L1} = m_1 v_{L1}' + m_2 v_{L2}'\ , \qquad (2.10.34a)$$

$$\tfrac{1}{2}m_1 v_{L1}^2 = \tfrac{1}{2}m_1 v_{L1}'^2 + \tfrac{1}{2}m_2 v_{L2}'^2\ , \qquad (2.10.34b)$$

to derive two formulas for the momentum transfer $q = (p'_{L1} - p_{L1}) = -p'_{L2}$

$$q^2 = m_1^2(v^2 + v'^2_{L1} - 2vv'_{L1}\cos\theta_L)$$

$$= m_2^2 v'^2_{L2} = m_2(m_1 v^2 - m_1 v'^2_{L1}) \tag{2.10.35}$$

from which we can determine v'_{L1}. The list of velocities and energies before and after the collision is then given by:

$$v_{L1} = v , \quad v_{L2} = 0 , \tag{2.10.36a}$$

$$v'_{L1} = \frac{m_1}{M} v \cos\theta_L \pm \frac{m_2}{M} v \sqrt{1 - \frac{m_1^2}{m_2^2}\sin^2\theta_L} , \tag{2.10.36b}$$

where the positive sign is used for the case $m_1 < m_2$, and for the case $m_1 > m_2$ both signs can be used as we will see. Further:

$$v'_{L2} = \sqrt{\frac{m_1}{m_2}(v^2 - v'^2_{L1})} , \tag{2.10.37a}$$

$$E_{L1} = \tfrac{1}{2}m_1 v^2 , \quad E_{L2} = 0 , \tag{2.10.37b}$$

$$E'_{L1} = \tfrac{1}{2}m_1 v'^2_{L1} , \quad E'_{L2} = p'^2_{L2}/2m_2 = q^2/2m_2 . \tag{2.10.37c}$$

We will discuss specially a few boundary cases.

a) If $m_2 \gg m_1$, we can make the approximation $v'_{L1} = v_{L1} = v$, and there is no exchange of energy. This becomes clear, if we consider that for $m_1/m_2 \rightarrow 0$, the center-of-mass system and the laboratory system coincide.

b) For $m_1 = m_2$ we simply have

$$v'_{L1} = v\cos\theta_L , \quad v'_{L2} = v\sin\theta_L \quad \text{and} \tag{2.10.38}$$

$$v'_{L1} \cdot v'_{L2} = 0 , \quad \text{since}$$

$$2v'_{L1} \cdot v'_{L2} = (v'_{L1} + v'_{L2})^2 - v'^2_{L1} - v'^2_{L2} = v^2 - v'^2_{L1} - v'^2_{L2} = 0 \tag{2.10.39}$$

from energy and momentum conservation. For $\theta \rightarrow \pi$, that is, a head-on collision, $\theta_L \rightarrow \pi/2$.

c) For $m_1 > m_2$ there is a maximum scattering angle in the laboratory system, given by

$$\cdot \sin^2\theta_{L,\,\text{max}} = m_2^2/m_1^2 .$$

For $m_2/m_1 \rightarrow 0$, only forward scattering is possible.

In order to study more exactly the energy transfer from particle 1 to particle 2, we define the relative energy transfer

$$\varrho = \frac{E_{1L} - E'_{1L}}{E_{1L}} = \frac{E'_{2L}}{E_{1L}} = \frac{q^2/2m_2}{\tfrac{1}{2}m_1 v^2} = \frac{q^2}{m_1 m_2 v^2} . \tag{2.10.40}$$

If we substitute the relationship $q^2 = 2\mu^2 v^2 (1 - \cos\theta)$, we find

$$\varrho = 2\frac{m_1 m_2}{M^2}(1 - \cos\theta) \; . \tag{2.10.41}$$

For $q^2 = 0$, $\varrho = 0$ (there is no energy transfer without momentum transfer). The largest possible value of ϱ

$$\varrho_{max} = 4m_1 m_2 / M^2 \tag{2.10.42}$$

is achieved when $\theta = \pi$, that is, when there is backwards scattering in the center-of-mass system. ϱ_{max} can also be written in the form

$$\varrho_{max} = 1 - [(m_1 - m_2)^2 / M^2] \tag{2.10.43}$$

which demonstrates that we always have $0 \le \varrho_{max} \le 1$ with $\varrho_{max} = 1$ when $m_1 = m_2$. A complete transfer of energy ($E'_{L2} = E_{L1}$) is only possible for two particles of identical mass.

Remarks. i) The relationship between θ_L and θ can also be demonstrated graphically (Fig. 2.10.5), because

$$v'_{L1} = v'_1 - v_2 \; . \tag{2.10.44}$$

We see from this diagram:

$$\tan\theta_L = \frac{v'_1 \sin\theta}{v_2 + v'_1 \cos\theta} = \frac{m_2 \sin\theta}{m_1 + m_2 \cos\theta} \tag{2.10.45}$$

since

$$v'_1 = (m_2/M)v \quad \text{and} \quad v_2 = (m_1/M)v \; .$$

We see again that for $m_1 > m_2$, there is a largest possible angle of scattering $\theta_{L,max}$ in the laboratory system with

$$\theta_{L,max} \le \frac{\pi}{2} \; . \tag{2.10.46}$$

This maximum scattering angle occurs if and only if v'_1 and v'_{L1} are perpendicular to each other (Fig. 2.10.6). We find again,

$$\sin\theta_{L,max} = v'_1/v_2 = m_2/m_1 \; . \tag{2.10.47}$$

To $\theta_L < \theta_{L,max}$, there correspond two possible angles θ, as can also be seen immediately from Fig. 2.10.6. This explains the two possible signs in the equation (2.10.36b).

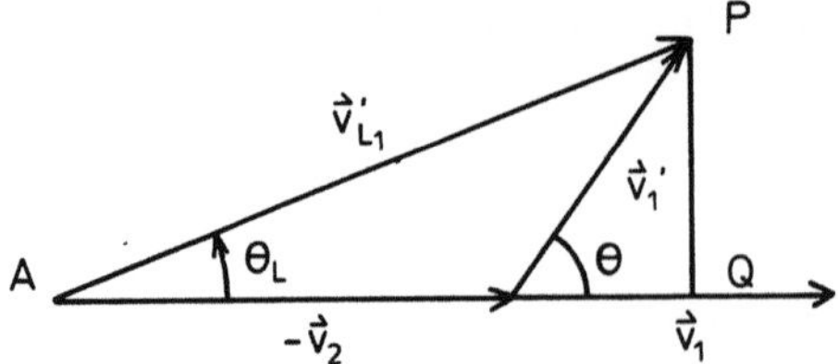

Fig. 2.10.5. Since $\boldsymbol{v}_1'$ and $-\boldsymbol{v}_2$ sum to $\boldsymbol{v}_{L1}'$, we can depict graphically the relation between θ and θ_L. The length of $\overline{PQ}$ is $v_1' \sin\theta$, while $\overline{AQ} = v_2 + v_1' \cos\theta$

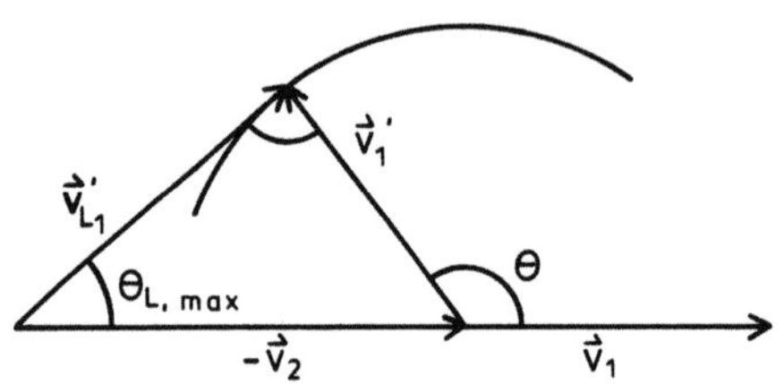

Fig. 2.10.6. If $\boldsymbol{v}_1'$ and $\boldsymbol{v}_{L1}'$ are perpendicular to each other, θ_L achieves its maximum value. Here, we are assuming that $m_1 > m_2$ so that $|\boldsymbol{v}_2| > |\boldsymbol{v}_1|$ from $m_1\boldsymbol{v}_1 + m_2\boldsymbol{v}_2 = 0$

If $m_1 = m_2$, the relationship between θ and θ_L is especially simple

$$\cos\theta_L = \frac{1 + \cos\theta}{\sqrt{2(1 + \cos\theta)}} = \cos\frac{\theta}{2}\,, \qquad \text{so that} \qquad \theta = 2\theta_L\,, \tag{2.10.48}$$

since if $m_1 = m_2$ then $v_1 = v_2$.

For $m_2/m_1 \to 0$, we have $\theta_L \to 0$: a very heavy particle will not be noticeably deflected in a collision with a light particle.

ii) Momentum and energy transfer are largest, as we have seen, when $\theta = \pi$, that is, in a head-on collision in the center-of-mass system. Then, we have

$$\cos\theta_L = \frac{m_1 - m_2}{|m_1 - m_2|} = \begin{cases} +1, & \text{if } m_1 > m_2 \\ -1, & \text{if } m_1 < m_2\,. \end{cases} \tag{2.10.49}$$

For $m_1 = m_2$, since $\theta_L = \theta/2$, for $\theta \to \pi$, i.e. $b \to 0$, we have $\theta_L \to \pi/2$, i.e., $\cos\theta_L \to 0$. In the case $b = 0$, particle 1 is at rest, anyway, after the collision, so that θ_L is no longer defined.

A head-on collision is shown in Fig. 2.10.7.

iii) In order to express θ in terms of θ_L, we simply solve the θ_L equation for θ. We find,

$$\cos\theta = \pm\cos\theta_L \sqrt{1 - \frac{m_1^2}{m_2^2}\sin^2\theta_L} - \frac{m_1}{m_2}\sin^2\theta_L\,. \tag{2.10.50}$$

The positive sign in this equation should be used for $m_1 < m_2$, so that $\theta = \pi$ for $\theta_L = \pi$. For $m_1 > m_2$, we can again use both signs.

iv) For a scattering process with

$$E_L = E_{L1} = \tfrac{1}{2}m_1 v^2$$

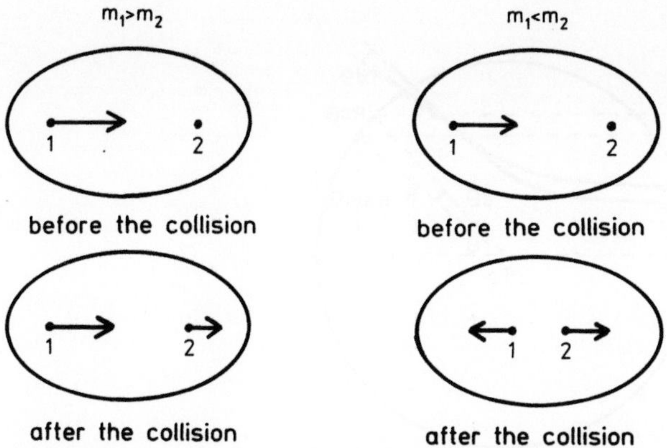

Fig. 2.10.7. In the case of a head-on collision in the center-of-mass system ($\theta = \pi$), the scattering angle in the laboratory system θ_L is either 0 or π, depending on whether $m_1 > m_2$ or $m_1 < m_2$

we find

$$E_{\mathbf{rel}} = \tfrac{1}{2}\mu v^2 \quad \text{thus}$$

$$E_{\mathbf{rel}}/E_{\mathrm{L}} = \mu/m_1 = m_2/M < 1 \ .$$

Only the internal energy $E_{\mathbf{rel}}$ can be used during the collision to transform or change the interacting particles. From this point of view, it is always counter-productive to shoot a heavy particle at a lighter target particle, since in this case m_2/M is especially small. When particles have equal mass, exactly half the laboratory energy is available as relative energy.

2.11   The Scattering Cross-Section

In many experiments, a stream of particles with fixed energy (i.e., velocity) and direction of flight, which is homogeneous over a certain cross-section, is aimed at a scattering center. Conclusions can then be drawn about the interaction potential $U(r)$ based on the deflection of the particles in the beam at the scattering center. In this process, the number of particles scattered into solid angle $d\Omega$ per unit time is measured.

   This is to say:

   We place an imaginary sphere on the scattering center (Fig. 2.11.1) and measure the number of particles which pass through the element of surface area per unit time. The direction of the scattered particles agrees with the asymptotic direction of the scattering trajectory if the sphere is big enough, and thus the scattering angle θ can be determined. Thus, we are really counting the number of

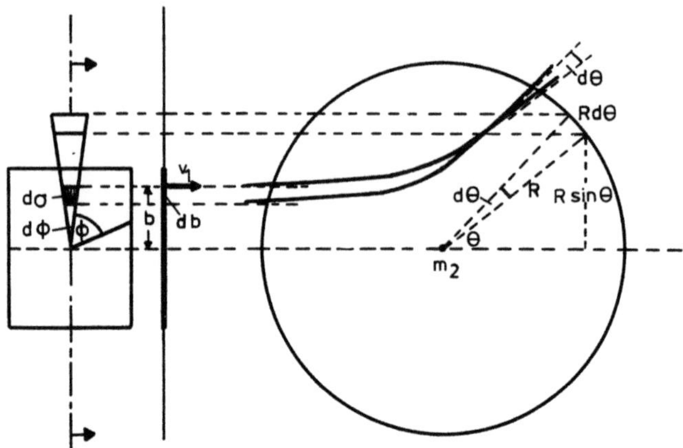

Fig. 2.11.1. The trajectories of particles with neighboring values of b and θ

particles which are scattered through an angle between

$$\theta \quad \text{and} \quad \theta + d\theta$$

and that have an azimuthal angle between

$$\varphi \quad \text{and} \quad \varphi + d\varphi \ .$$

For every θ, there is a corresponding impact parameter b, and all the particles which pass through $d\Omega$ in one second must first go through an area element

$$d\sigma = -b\,db\,d\varphi \ , \quad b = b(\theta) \ . \tag{2.11.1}$$

before being scattered. Here, the minus sign indicates that an increase of θ means a decrease of b.

If, now, the density of the incident beam of particles, that is, the number of particles which pass through a unit area per unit time, is equal to j, then

$$j\,d\sigma \text{ particles per second}$$

pass first through $d\sigma$ and then, after the collision, through solid angle $d\Omega$. Of course, this number is proportional to the number of incident particles.

The quantity characteristic of the interaction potential between the two particles is then $d\sigma$, which can, in practice, be calculated from measured data using the relationship

$$d\sigma = \frac{\text{Number of particles which are scattered through } d\Omega \text{ per second}}{\text{Number of incident particles per s per m}^2}$$

$$= \frac{d\sigma}{d\Omega}\,d\Omega \tag{2.11.2}$$

$d\sigma$ or $d\sigma/d\Omega$ is called the *differential cross section*. It has the dimension of an area.

$d\sigma$ can be calculated for a hypothetical potential, if we know $b = b(\theta)$. For the Coulomb potential, since

$$\tan\frac{\theta}{2} = \frac{|\kappa|}{\mu b v^2}$$

it follows that

$$b = \frac{|\kappa|\cot\left(\dfrac{\theta}{2}\right)}{\mu v^2}$$

and thus

$$db = -\frac{\kappa}{2\mu v^2}\,\frac{1}{\sin^2\left(\dfrac{\theta}{2}\right)}\,d\theta \;,$$

so that

$$d\sigma = -b\,db\,d\varphi$$

$$= \left(\frac{\kappa}{\mu v^2}\right)^2 \frac{1}{2\sin^2(\theta/2)}\frac{\cos(\theta/2)}{\sin(\theta/2)}\,d\theta\,d\varphi$$

$$= \left(\frac{\kappa}{\mu v^2}\right)^2 \frac{1}{4\sin^4(\theta/2)}\sin\theta\,d\theta\,d\varphi$$

$$= \left(\frac{\kappa}{2\mu v^2}\right)^2 \frac{1}{\sin^4(\theta/2)}\,d\Omega$$

or, with $E_1 = \mu v^2/2$:

$$\frac{d\sigma}{d\Omega} \equiv -\frac{b\,db\,d\varphi}{\sin\theta\,d\theta\,d\varphi} \equiv -\frac{b}{\sin\theta}\frac{db}{d\theta}$$

$$= \left(\frac{\kappa}{4E_1}\right)^2 \frac{1}{\sin^4(\theta/2)} \; . \tag{2.11.3}$$

This is the differential cross section for Coulomb scattering.

The total cross section σ_{tot} is then

$$\sigma_{\text{tot}} = \int \frac{d\sigma}{d\Omega}\,d\Omega \; . \tag{2.11.4}$$

σ_{tot} is the area through which any particle must pass to be scattered at all.

For the Coulomb potential, we find

$$\sigma_{\text{tot}} \sim 2\pi \int_0^\pi d\theta \sin\theta \, \frac{1}{\sin^4(\theta/2)} = \infty \;, \tag{2.11.5}$$

which means that if we make the cross section of the incident beam large enough so that even the scattering angle $\theta = 0$ occurs, this cross section must be infinite.

This follows clearly from the fact that we must allow $b \to \infty$ in order to achieve $\theta = 0$. The Coulomb potential is so long-range, that a deflection occurs even for very large impact parameters.

If $U(r) \equiv 0$ for $r > R$, we have $\theta = 0$ for $b > R$. The total interaction cross section is then equal to the geometric cross section πR^2.

Thus, in classical mechanics there are only two possibilities for the total cross section of a spherically symmetric potential: For all potentials, for which there is an R such that $U(r) = 0$ (or $=$ constant) for $r > R$, the total interaction cross section is equal to the geometric area; for all other potentials, the interaction cross section is infinitely large. Only in quantum mechanics is it possible to have a finite total interaction cross section for potentials which fall off quickly enough as $r \to \infty$ without vanishing for $r > R$.

Remarks. i) Using theoretical considerations, we can find the scattering angle θ which is also the scattering angle in the center-of-mass system. In the laboratory, it is usually the scattering angle in the laboratory system which is measured along with the differential cross section in the laboratory system. Suppose this is given by

$$\frac{d\sigma}{d\Omega_{\text{L}}} \quad \text{with} \quad d\Omega_{\text{L}} = d\cos\theta_{\text{L}}\, d\varphi \;,$$

then, obviously,

$$\frac{d\sigma/d\Omega}{d\sigma/d\Omega_{\text{L}}} = \frac{d\Omega_{\text{L}}}{d\Omega} = \frac{d\cos\theta_{\text{L}}}{d\cos\theta}$$

$$= \frac{m_2}{(m_1^2 + m_2^2 + 2m_1 m_2 \cos\theta)^{1/2}} - \frac{2m_1 m_2(m_1 + m_2 \cos\theta)}{2(m_1^2 + m_2^2 + 2m_1 m_2 \cos\theta)^{3/2}}$$

$$= \frac{m_2^2(m_2 + m_1 \cos\theta)}{(m_1^2 + m_2^2 + 2m_1 m_2 \cos\theta)^{3/2}} \;. \tag{2.11.6}$$

If the mass of the target is much greater than that of the incident particle, we have

$$\frac{m_1}{m_2} \ll 1 \;,$$

and this factor has value one. If both masses are equal,

$$\theta_L = \frac{\theta}{2} \, , \qquad (2.11.7)$$

and the factor is

$$\frac{1 + \cos\theta}{[2(1+\cos\theta)]^{3/2}} = \frac{1}{[8(1+\cos\theta)]^{1/2}} = \frac{1}{4\cos(\theta/2)} = \frac{1}{4\cos\theta_L} \, . \qquad (2.11.8)$$

ii) In actual experiment, the target is never composed of a single individual atom (or nucleus). If the beam of particles encounters not just one target particle, but n target particles, then we in fact measure n times as many particles scattering into the solid angle. Here, we are assuming that target is small enough with respect to the cross section of the beam and that the scattering particles are not too close to each other.

iii) So far, we have limited ourselves to a discussion of repulsive forces. For the general case of a spherically symmetric potential which vanishes quickly enough at large distances, there can be an additional complication: the relationship between the impact parameter b and the scattering angle θ might no longer be monotonic (see Fig. 2.11.2).

Then, in order to find the number of particles per second deflected between the angles $\theta \pm \Delta\theta$, we must consider the entire cross section in impact parameter space corresponding to the scattering angles in the interval $[\theta - \Delta\theta, \theta + \Delta\theta]$.

The general formula is

$$\frac{d\sigma}{d\Omega}(\theta) = \sum_a \left| b_a(\theta) \frac{db_a(\theta)}{\sin\theta \, d\theta} \right| \, , \qquad (2.11.9)$$

where the sum is to be taken over all the "branches" of the (many-valued) inverse $b(\theta)$ of the function $\theta(b)$.

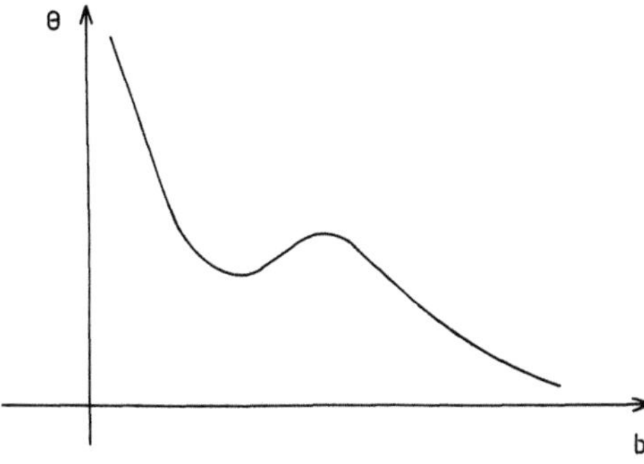

Fig. 2.11.2. If $\theta(b)$ is not monotonic and thus the function $b(\theta)$ is many-valued, we have to sum over all the appropriate cross sections in impact-parameter space

2.12 The Virial Theorem

There are few general theorems which apply to many-body problems. In this section and the next, we will present two important and useful ways of examining these problems.

The virial theorem is a statement about the time averages of the kinetic and potential energy for systems of N particles, whose interaction can be described by a potential.

We make the following definition:

The time average of a bounded function $f(t)$ is

$$\bar{f} := \lim_{A \to \infty} \frac{1}{2A} \int_{-A}^{A} f(t)\,dt \ . \tag{2.12.1}$$

We consider now

$$2T = \sum_{i=1}^{N} m_i \dot{\boldsymbol{r}}_i \cdot \dot{\boldsymbol{r}}_i = \sum_{i=1}^{N} \boldsymbol{p}_i \cdot \dot{\boldsymbol{r}}_i$$

$$= \frac{d}{dt} \sum_{i=1}^{N} \boldsymbol{p}_i \cdot \boldsymbol{r}_i - \sum_{i=1}^{N} \boldsymbol{r}_i \cdot \dot{\boldsymbol{p}}_i = \frac{d}{dt} \sum_{i=1}^{N} \boldsymbol{p}_i \cdot \boldsymbol{r}_i + \sum_{i=1}^{N} \boldsymbol{r}_i \cdot \boldsymbol{\nabla}_i U \ . \tag{2.12.2}$$

If we now construct the time average for the bounded function $T(t)$, we find

$$2\bar{T} = \lim_{A \to \infty} \frac{1}{2A} \int_{-A}^{+A} dt\, \frac{d}{dt} \sum_{i=1}^{N} \boldsymbol{p}_i \cdot \boldsymbol{r}_i + \overline{\sum_{i=1}^{N} \boldsymbol{r}_i \cdot \boldsymbol{\nabla}_i U}$$

$$= \lim_{A \to \infty} \frac{1}{2A} \sum_{i=1}^{N} \boldsymbol{p}_i \cdot \boldsymbol{r}_i \Bigg|_{-A}^{+A} + \overline{\sum_{i=1}^{N} \boldsymbol{r}_i \cdot \boldsymbol{\nabla}_i U} \ . \tag{2.12.3}$$

If $\sum_{i=1}^{N} \boldsymbol{p}_i \cdot \boldsymbol{r}_i$ is bounded in time, then we obtain

$$2\bar{T} = \overline{\sum_{i=1}^{N} \boldsymbol{r}_i \cdot \boldsymbol{\nabla}_i U} \ . \tag{2.12.4}$$

The quantity

$$\overline{\sum_{i=1}^{N} \boldsymbol{r}_i \cdot \boldsymbol{\nabla}_i U} \tag{2.12.5}$$

is called the *virial*[18] of the potential U and the above identity is called the *virial theorem*.

---

[18] *Virial* (Latin) fr. *vis*: Force

If we now additionally assume that the potential

$$U(r_1, \ldots, r_N)$$

is a *homogeneous function* of degree k, that is,

$$U(\alpha r_1, \ldots, \alpha r_N) = \alpha^k U(r_1, \ldots, r_N) , \quad (\alpha \geq 0) , \tag{2.12.6}$$

then, differentiating with respect to α, we find:

$$\frac{\partial}{\partial \alpha} U(\alpha r_1, \ldots, \alpha r_N) = \frac{\partial}{\partial \alpha} \alpha^k U(r_1, \ldots, r_N) = k\alpha^{k-1} U(r_1, \ldots, r_N)$$

$$= \sum_{i=1}^{N} \frac{\partial U}{\partial \alpha r_i} \frac{\partial \alpha r_i}{\partial \alpha} = \sum_{i=1}^{N} \frac{\partial U}{\partial \alpha r_i} \cdot r_i , \tag{2.12.7}$$

and setting $\alpha = 1$:

$$\sum_{i=1}^{N} r_i \cdot \nabla_i U = kU . \tag{2.12.8}$$

This important equation is called *Euler's*[19] *equation* for homogeneous functions. We also see immediately, that the n-th derivatives of a homogeneous function of degree k is itself a homogeneous function of degree $k - n$.

Examples

a) $U(r) = \dfrac{\kappa}{r}$ then $k = -1$ and

$$\nabla U = -\frac{\kappa}{r^2} \frac{r}{r} , \quad \text{so that}$$

$$r \cdot \nabla U = -\frac{\kappa}{r} = -U .$$

b) $U(r) = \frac{1}{2}kr^2$, so that $k = 2$

$$r \cdot \nabla U = -\frac{\kappa}{r} = -U .$$

Thus, when the potential $U(r_1, \ldots, r_N)$ is a homogeneous function of degree k, we have

$$2\bar{T} = k\bar{U} \tag{2.12.9}$$

---

[19] *Euler, Leonhard* (*1701 Basel, d. 1783 St. Petersburg. Most important work in the areas of pure and applied mathematics, especially in the calculus of variations, hydrodynamics, celestial mechanics, mechanics, acoustics, and optics. Worked 1727–1741 and 1766–1783 at the Petersburg Academy, 1741–1766 at the Prussian Academy in Berlin.

and since

$$E = \bar{E} = \bar{T} + \bar{U} = [(k/2) + 1]\bar{U} \tag{2.12.10}$$

for $k \neq -2$,

$$\bar{U} = \frac{2}{k+2} E \;, \qquad \bar{T} = \frac{k}{k+2} E \;, \tag{2.12.11}$$

which gives us the time averages of the kinetic and potential energy.

The case $k = -2$ is an instructive exception. We would seem to have $\bar{T} = -\bar{U}$, from which it would follow that $E = \bar{E} = \bar{T} + \bar{U} = 0$. From the virial theorem, it thus seems that the only possible value for the energy of the system is $E = 0$. A closer examination shows, however, that in this case the conditions of the Virial Theorem are not satisfied so that the conclusion does not follow:

In particular, for a repulsive potential which is homogeneous of degree $k = -2$, $E > 0$ and, since the orbits are unbounded,

$$\sum_{i=1}^{N} \boldsymbol{p}_i \cdot \boldsymbol{r}_i$$

is unbounded. For an attractive potential, the particles come so close together that T and U are unbounded.

Applications. i) For the harmonic oscillator, we have $k = 2$, so that

$$\bar{U} = \tfrac{1}{2}E \;, \qquad \bar{T} = \tfrac{1}{2}E \;. \tag{2.12.12}$$

Thus, the time averages of the kinetic energy and the potential energy are equal, and are in both cases equal to half the total energy. This is also true for a system of coupled harmonic oscillators.

To a good approximation, a crystal can be regarded as a system of harmonic oscillators in which the atoms vibrate around their points of equilibrium. One half of the energy of a crystal then is in the form of potential energy, while the other half is in the form of kinetic energy.

ii) For Newton's law of gravitation, we have $k = -1$, so that

$$\bar{U} = 2E \;, \qquad \bar{T} = -E \;. \tag{2.12.13}$$

Note that this applies only to the case $E < 0$, because only then is $\boldsymbol{p} \cdot \boldsymbol{r}$ bounded.

For a circular orbit with radius r_0, we have

$$U = U(r_0) = \bar{U} \quad \text{and thus}$$

$$E = \tfrac{1}{2}U(r_0) = -\frac{1}{2}\frac{\gamma m_1 m_2}{r_0} \tag{2.12.14}$$

so that $r_0 = \gamma m_1 m_2 / 2|E|$ in agreement with Sect. 2.9.

2.13   Mechanical Similarity

We will now work with the idea of mechanical similarity.

This following is meant: Let $x(t)$ be a solution of the equation of motion ·

$$m\ddot{r} + \nabla U(r) = 0 \ . \tag{2.13.1}$$

We now change m by a factor $\gamma > 0$ and U by a factor $\delta > 0$, then look for solutions of the new equation of motion

$$\gamma m\ddot{r} + \delta \nabla U(r) = 0 \ . \tag{2.13.2}$$

In particular, we seek solutions $X(t)$ which are geometrically similar to $x(t)$:

$$X(t) = \alpha x(t/\beta) \quad \text{with} \quad \alpha, \beta > 0 \ . \tag{2.13.3}$$

The orbit of X is stretched out by a factor α in comparison to that of x, but is otherwise similar, while β describes the stretching of the unit of time: an increase in β means that the corresponding points on the orbit will be reached later.

We consider first the case $\alpha = 1$, in which the paths $x(t)$ and $X(t)$ agree, and only the time dependence differs. Since $X(t)$ is a solution to the equation of motion (2.13.2), it must be true that

$$\gamma m\ddot{X}(t) + \delta \nabla U(X(t)) = 0 \quad \text{or}$$

$$(\gamma/\beta^2)m\ddot{x}(t/\beta) + \delta \nabla U(x(t/\beta)) = 0 \ .$$

Since x is a solution of the equation of motion (2.13.1), this can happen if and only if:

$$\frac{\gamma}{\beta^2} = \delta \quad \text{or} \quad \beta = \sqrt{\frac{\gamma}{\delta}} \ . \tag{2.13.4}$$

From this, it follows in particular: If the mass of a particle is increased by the factor γ with the same potential, then the motion along the orbit is slowed down by the factor $\sqrt{\gamma}$. If U is increased by the factor δ, then the orbit speeds up by the factor $\sqrt{\delta}$.

In the case $\alpha \neq 1$, we can only proceed further if we assume that U is a homogeneous function, say of degree k. Then, we have $(\nabla U)(\alpha r) = \alpha^{k-1}U(r)$, and $X(t)$ is a solution of (2.13.2) if and only if

$$\gamma/\beta^2 = \delta\alpha^{k-2} \ . \tag{2.13.5}$$

Now the cases in which $\gamma = \delta = 1$ are also interesting, because in these cases we can look for manifestly similar solutions of the same equations. The condition then becomes

$$\beta^2 = \alpha^{2-k} \, . \tag{2.13.6}$$

Examples

$k = 2$ (harmonic oscillator):

$\beta = 1$:   This shows that the period of oscillation is independent of the amplitude.

$k = -1$ (the Kepler problem):

$\beta^2 = \alpha^3$:   This demonstrates a special case of Kepler's third law: If the orbit is increased by a factor C, then the period of revolution must be increased by a factor $C^{3/2}$. These considerations do not predict the independence of the time of revolution from the length of the semimajor axis.

Finally, in the case of scattering from a potential which is a homogeneous function of degree k, these similarity considerations enable us to determine the dependence of cross section on mass, velocity of the beam, and the strength of the potential (however, we cannot find the angular dependence using similarity).

It is clear that geometrically similar orbits correspond to the same angle of deflection θ and an impact parameter increased by α. From dimensional considerations, the differential cross section $d\sigma/d\Omega$ must be proportional to α^2.

Now, $\gamma\alpha^2/\beta^2$ is the scale factor for the kinetic energy of the incident particle. From equation (2.13.5), we also find:

$$\alpha^2 = \left(\frac{\gamma}{\delta} \frac{\alpha^2}{\beta^2}\right)^{2/k} \tag{2.13.7}$$

and thus the differential cross section of scattering from a potential $U(r) = c|r|^k g(r/|r|)$ with $g(r/|r|)$ dimensionless is given by

$$\frac{d\sigma}{d\Omega} = \left(\frac{c}{E}\right)^{-2/k} f(\theta) \tag{2.13.8}$$

with some function $f(\theta)$ which cannot be determined here. For a Coulomb potential, $c = Ze^2$ and $k = -1$, so that

$$\frac{d\sigma}{d\Omega} = \left(\frac{Ze^2}{E}\right)^2 f(\theta) \, . \tag{2.13.9}$$

Finally, we note that analogous similarity considerations also apply to many-particle systems.

2.14 Some General Observations About the Many-Body Problem

The two-body problem is easy to solve in the case of a rotationally symmetric central force field, since we can fully integrate the equations analytically, by using the conserved quantities. The integration of the equations of motion of the N-body problem, however, is fundamentally more difficult. In a later section, we will go into a few of the general properties and special problems of systems with very many particles. Here, we will discuss a problem which has generated much interest, as it is of extreme importance for celestial mechanics, namely, the Kepler problem for three particles.

Since 1750, there have been more than 800 works published about this problem, whose authors include the most important mathematicians. In the year 1887, H. Bruns demonstrated in a publication of the "Royal Saxon Scientific Society" that besides the well-known conserved quantities

$$E_{\mathrm{cm}}, \ E_{\mathrm{rel}}, \ L_{\mathrm{cm}}, \ L_{\mathrm{rel}}, \ \text{and} \ P \ ,$$

no others exist which can be represented as algebraic functions of position and momentum, and which are independent of the above named quantities.

This means that in the three-body problem (and in general the N-body problem) there are not enough conserved quantities to reduce the solution of the equations of motion to a simple integral, as is the case for the two-body problem.

Problems for which there are enough conserved quantities to reduce the solution of the equations of motion to a one-dimensional integral are called *fully integrable*. Here, we cannot give this concept a more precise definition. In any case, fully integrable models are rare and often they can only be defined in a world with one spatial dimension.

The three-body problem with gravitational interactions is then not fully integrable by this definition, but it does allow certain specific solutions. Here, we will study a few of these special cases.

Let three bodies be arranged so that at all times they lie along a line at fixed distances from each other, so that the line can rotate with constant angular speed ω around an axis perpendicular to it through its center of mass S (cf. Fig. 2.14.1). Let

$$|r_2 - r_1| = |e| = \text{constant} \ . \qquad |r_3 - r_2| = \lambda |e| \qquad (2.14.1)$$

with the value of λ still to be determined.

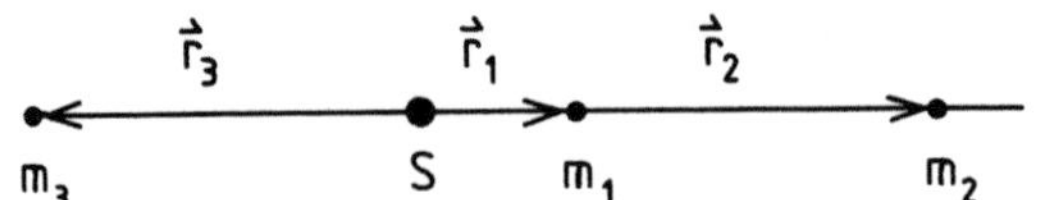

Fig. 2.14.1. The collinear arrangement of three particles

Since in this situation we have $m_1 r_1 + m_2 r_2 + m_3 r_3 = 0$, we can solve for the r_i in terms of e:

$$r_1 = -\frac{m_2 + (1 + \lambda)m_3}{M} e \;,$$

$$r_2 = \frac{m_1 - \lambda m_3}{M} e \;, \qquad\qquad\qquad (2.14.2)$$

$$r_3 = \frac{(1 + \lambda)m_1 + \lambda m_2}{M} e \;, \qquad M = m_1 + m_2 + m_3 \;.$$

The equations of motion for the vectors r_1 can now be rewritten as two independent equations for ω and λ. If we eliminate ω, we wind up with an equation of motion of fifth order in λ which has a real positive solution.

The motions of the three bodies are then collinear circular orbits. This is a special case of collinear conic section motion, which we can obtain in a similar way if we start with the condition that the distances between the masses satisfy a certain relationship but are not necessarily fixed in time. The condition is then

$$r_3 - r_2 = \lambda(r_2 - r_1) \;. \qquad\qquad\qquad (2.14.3)$$

but no longer

$$|r_2 - r_1| = |e| = \text{const} \;.$$

If one mass is small enough so that we can ignore it in comparison to the other masses, the situation is called a "restricted three-body problem." If we use the circular collinear solution in this case, the two large masses rotate around each other, and the "small" mass can then lie in three possible points L_1, L_2, or L_3 on the line through the two larger masses, depending on the order in which the masses are arranged. These three points are called *libration*[20] *points*, and they correspond to points of equilibrium in a system of reference rotating along with the masses.

There are additional solutions to the three-body problem, if we assume that the three bodies always lie in a single plane and always form an equilateral triangle. These two additional solutions L_4, L_5 are called *equilateral points* and

---

[20] *Libration* (Latin): oscillation.

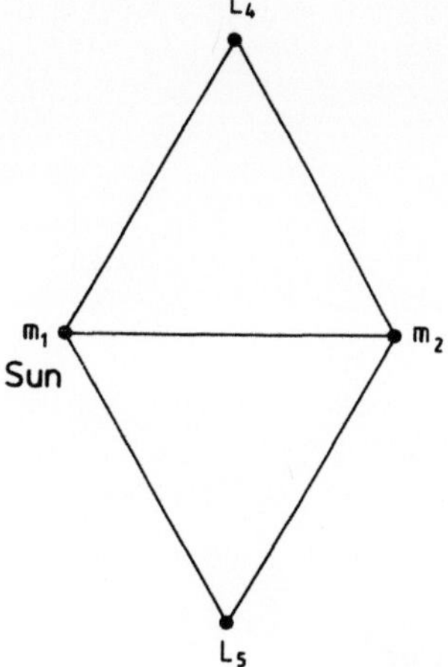

Fig. 2.14.2. The position of the equilateral points L_4, L_5

are shown in Fig. 2.14.2. In the restricted three-body problem, the small mass m can undergo stable motion around these points.

Interestingly enough, such configurations are actually present in the solar system: On February 22, 1906, M. Wolf discovered an asteroid that had a nearly circular orbit around the sun and moved in Jupiter's orbital path. This asteroid, named Achilles formed an angle of $55\tfrac{1}{2}°$ with Jupiter and the Sun. After this discovery, people examined Lagrange's theory of the libration and equilateral points, which dated from 1772. Later in 1906, another asteroid Patroclos was discovered in the area of equilateral point L_5. Today, a whole series of asteroids, named the "Trojans," is known which remain in the vicinity of the equilateral points L_4, L_5.

Problems

2.1   The Accompanying Axes. For a curve $x(\sigma)$, we define the following expressions:

– the arc length

$$s(\sigma, \sigma_0) := \int_{\sigma_0}^{\sigma} d\sigma' \sqrt{\left(\frac{dx(\sigma')}{d\sigma'}\right)^2} \; ; \quad \text{i.e.} \quad \frac{ds}{d\sigma} = \sqrt{\left(\frac{dx(\sigma)}{d\sigma}\right)^2}$$

– the unit tangent vector t

$$t(s) := \frac{dx(\sigma(s))}{ds} = \frac{d\sigma}{ds}\frac{dx}{d\sigma}$$

– the unit normal vector n and the curvature κ:

$$\frac{dt}{ds} := \kappa(s)n(s) \ , \quad \kappa(s) \geq 0$$

– the unit binormal vector b and the torsion κ:

$$b(s) = t(s) \times n(s) \ , \quad \frac{db}{ds} := -\tau(s)n(s) \ .$$

Show that:

$$\frac{dn}{ds} = -\kappa t + \tau b \ .$$

Express κ and τ in terms of $dx/d\sigma$, $d^2x/d\sigma^2$ and $d^3x/d\sigma^3$.
Calculate t, n, b, κ, and τ for the spiral curve

$$x(t) - e_1 r \cos \omega t + e_2 r \sin \omega t + e_3 kt \ .$$

Also calculate

$$\dot{x}(t) \ , \quad \ddot{x}(t) \ , \qquad \dddot{x} \ .$$

2.2 The Gravitational Field of the Earth. Over what horizontal and vertical extent near the earth's surface can we consider the earth's gravitational field to be homogeneous within 1%?

Hint: One possible way to solve this is the following:

Above its surface, the gravitational field of the Earth is given by

$$G(r) = \text{const} \frac{r}{r^3} \ .$$

Using this formula, find a linear approximation for

$$\frac{|G(r + h) - G(r)|}{|G(r)|}$$

for vectors $h \parallel r$ (vertical) and $h \perp r$ (horizontal), where $r = |r| = 6370$ km.

2.3 One-dimensional Equations of Motion. a) Consider the one-dimensional equation of motion

$$m\ddot{x} = -Ax^{k-1}$$

in the region $x > 0$, where $A > 0$ and $k \leq 0$. Calculate the collapse time $T = T(x_0)$, after which a particle with initial position $x(0) = x_0$ and initial speed

$x(0) = 0$ reaches the origin, $x(T) = 0$. We find an equation of the form

$$T(x_0) = C(k)\sqrt{\frac{m}{A}}\, x_0^{1-k/2}\ .$$

How can we derive this dependency without integrating explicitly using either the scale transformations $t \to t' = \alpha t,\ x \to x' = \beta x$ or dimensional considerations?

b) Consider the one-dimensional equation of motion

$$m\ddot{x} = -V'(x)$$

in the vicinity of a point of stable equilibrium x_0 of V ($V''(x_0) > 0$). What is the value of the period T of the motion obtained by harmonic approximation?

2.4 Motion in a Central Field. Discuss as comprehensively as possible the motion of a particle of mass m under the influence of a spherically symmetric potential of the form

$$U(r) = \frac{A}{r} + \frac{B}{r^2}$$

with arbitrary real constants A and B in analogy with the corresponding discussion of the Kepler problem ($B = 0$). As parameters for the motion in addition to A and B, you should use the conserved quantities E (energy) and l (angular momentum). To begin with, use the effective potential

$$U_{\text{eff}}(r) = \frac{\alpha}{r} + \frac{\beta}{r^2} \quad \text{with} \quad \begin{aligned} \alpha &= A \\ \beta &= B + l^2/2m \end{aligned}$$

as well as the integrals

$$\varphi = \pm\,\frac{l}{\sqrt{2m}}\int \frac{dr}{\sqrt{E - U_{\text{eff}}(r)}} \quad \text{for } l \neq 0\ .$$

$$t = \pm\,\sqrt{\frac{m}{2}}\int \frac{dr}{\sqrt{E - U_{\text{eff}}(r)}}\ ,$$

(here, the proper signs and limits of integration are to be determined case by case).

a) First, by considering U_{eff} (graphically and quantitatively) determine the allowed regions of r (in total, there are nine cases to be distinguished based on the signs $[=0,\ >0 \text{ or } <0]$ of α and β).

b) Calculate, by explicitly carrying out the integration over φ, the orbits $r = r(\varphi)$, assuming that $l \neq 0$. Discuss in particular the asymptotic behavior with respect to the variables r and φ, and determine if the asymptotic form will be attained after a finite or infinite period of time (insofar as an asymptotic form exists at all).

Hint: Possible forms of asymptotic behavior are, for example,

Limit points: $r \to \infty$ and $\varphi \to$ const.
Limit circles: $r \to$ const. and $\varphi \to \infty$
Collapse: $r \to 0$.

2.5   Precession of the Perihelion. Assume that a test particle of mass m moves in a central gravitational field with potential energy

$$V(r) = -\frac{k}{r}\left(1 + \frac{r_0}{2r}\right) ,$$

where $k = mMG$, $r_0 = \alpha GM/c^2$, $\alpha \in \mathbb{R}$, such that a closed orbit is obtained with closest approach $r_{\min} \ll r_0$. Calculate the precession of the perihelion $\Delta\varphi$ per revolution as a function of the eccentricity ε and the period T, as well as the frequency of precession of the perihelion $\Delta\varphi/T$ as a function of the eccentricity ε and of the semimajor axis a. In both cases only the linear terms in r_0/p ($p = a(1 - \varepsilon^2)$) should be considered.

3. Lagrangian Methods in Classical Mechanics

Until now, we have assumed, within the framework of Newtonian mechanics, that we know all of the forces which act on a particle when constructing its equation of motion. This knowledge was necessary in order to produce a well-defined system of differential equations.

In many cases, however, we do not know all of these forces, but we do know the effects of the unknown forces, for example, the restriction of motion to certain surfaces. In such situations, we use the Lagrangian methods of classical mechanics.

3.1 A Sketch of the Problem and Its Solution in the Case of a Pendulum

Let us consider a geometrical pendulum consisting of a point-mass hanging from a wire of length l. We assume that the wire hangs down from the origin (Fig. 3.1.1).

Clearly, there are two forces exerted on the point-mass:

a) The gravitational force $F(r)$, directed vertically downward,
b) an unknown force which forces the point-mass to stay on the spherical surface given by

$$r^2 - l^2 = 0 \; . \tag{3.1.1}$$

This force, which we will write as $f(t)$, is called a *constraining force*, and it is obviously directed along the wire towards the origin. The resultant force then produces motion on the surface of the sphere. In the equation of motion

$$m\ddot{r}(t) = F(r(t)) + f(t) \; , \tag{3.1.2}$$

$f(t)$ is unknown at first, but its effect is known: the "constraint"

$$r^2 - l^2 = 0 \; .$$

is always satisfied.

The set of possible locations of the mass-point in three-dimensional space is thus restricted to a two-dimensional surface. The known force F produces

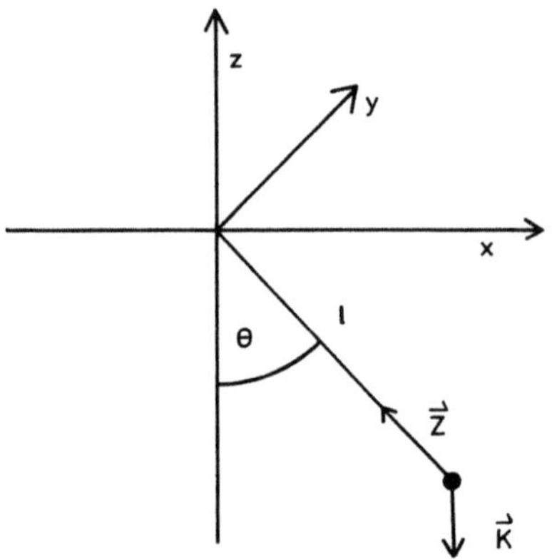

Fig. 3.1.1. A pendulum of length l

a motion on this surface, and the force of constraint $f(t)$ is always perpendicular to this motion. We can now imagine two possible ways to calculate the motion.

First, we could use the fact that the force of constraint is perpendicular to the surface, that is, it has the form

$$f(t) = \lambda(t)r(t) \tag{3.1.3}$$

and then solve the problem for $r(t)$

$$m\ddot{r}(t) = K(r(t)) + \lambda(t)r(t) , \tag{3.1.4a}$$

$$r^2(t) - l^2 = 0 . \tag{3.1.4b}$$

This gives us four equations for the four unknowns $r(t)$, $\lambda(t)$.

Second, we could "project the equation of motion onto the surface," that is, we could find vectors accompanying $r(t)$ which are, at all times, tangential to the surface at the point where the particle is located.

If we multiply both sides the equation of motion by these vectors, which are perpendicular to $f(t)$, then $f(t)$ will be eliminated. Of course, following this procedure we will not be able to find $f(t)$ itself. It is easy to find these vectors if we use coordinates which are well suited to the surface, in the sense that some of the coordinates, when varied, parametrize the surface, while the rest of the coordinates have fixed values because of the constraints.

In this case, such coordinates are easy to find; they are the polar coordinates

$$r = r(\sin \theta \cos \varphi, \sin \theta \sin \varphi, - \cos \theta) , \tag{3.1.5}$$

and the constraint is then identical to $r \equiv l$, while θ and φ can be varied freely.
From $r^2(\theta, \varphi) = l^2$, it follows that

$$r \cdot \frac{\partial r}{\partial \theta} = 0 \quad \text{and} \quad r \cdot \frac{\partial r}{\partial \varphi} = 0 . \tag{3.1.6}$$

Multiplying

$$m\ddot{r}(t) = F(r(t)) + f(t) \tag{3.1.7}$$

by $\partial r/\partial\theta$ and $\partial r/\partial\varphi$ then yields two equations which can be viewed as equations of motion for the two freely changing coordinates:

$$m\ddot{r}(\theta(t), \varphi(t)) \cdot \frac{\partial r}{\partial\theta}(\theta(t), \varphi(t)) = F[r(\theta(t), \varphi(t))] \cdot \frac{\partial r}{\partial\theta}(\theta(t), \varphi(t)) \ , \tag{3.1.8a}$$

$$m\ddot{r}(\theta(t), \varphi(t)) \cdot \frac{\partial r}{\partial\varphi}(\theta(t), \varphi(t)) = F[r(\theta(t), \varphi(t))] \cdot \frac{\partial r}{\partial\varphi}(\theta(t), \varphi(t)) \ . \tag{3.1.8b}$$

The forces of constraint, which are eliminated here, can be determined later, in backwards fashion, by substituting the solution of the equation of motion into Newton's equation in the form

$$f(t) = m\ddot{r} - F(r(t)) \ . \tag{3.1.9}$$

In the following chapter, we will formulate these two strategies in a more general way, and expand them. We will call the first strategy the *Lagrangian*[1] *method of the first type*, the second, the *Lagrangian method of the second type*.

In particular, we will show that the equations of motion for the freely varying coordinates can be derived from a function called the *Lagrangian*, which is easy to construct.

3.2   The Lagrangian Method of the First Type

We consider now a system of N particles with position vectors $r_1, \ldots, r_N$, which we again combine into one vector $\underline{z} \in \mathbb{R}^{3N}$. Let there be s independent constraints of the form

$$F_\alpha(\underline{z}, t) = 0 \ , \quad \alpha = 1, \ldots, s \ . \tag{3.2.1}$$

The independence of the s constraints simply means that none of them is a consequence of the others. In the future, we will always assume that our

---

[1] *Lagrange, Joseph-Louis* (*1736 Turin, d. 1813 Paris). French mathematician and physicist. In 1759, fundamental research in the calculus of variations. Lagrange was the successor to Euler in the Berlin Academy from 1766–1787. His most influential work was the "Méchanique analytique", a comprehensive, unified presentation of mechanics which consistently used the method of virtual displacement.

constraints are independent. For each α, the set

$$M_t^\alpha = \{\underline{z} \mid \underline{z} \in \mathbb{R}^{3N}, F_\alpha(\underline{z}, t) = 0\} \tag{3.2.2}$$

represents a $(3N - 1)$-dimensional surface M_t^α in $\mathbb{R}^{3N}$.

The manifold

$$M_t = \bigcap_{\alpha=1}^{s} M_t^\alpha \tag{3.2.3}$$

is then the set of all of the possible positions of the particles at time t. If the constraints are independent, the manifold has dimension

$$f = 3N - s \ . \tag{3.2.4}$$

f is also called the number of *degrees of freedom* of the system. (In the example of Sect. 3.1, $s = 1$, $N = 1$, and $f = 2$). Constraints of the form

$$F_\alpha(\underline{z}, t) = 0 \ , \quad \alpha = 1, \ldots, s \ ,$$

are called *holonomic*[2]. If M_t does not depend on t, then the constraints are called *scleronomic*[3], otherwise they are called *rheonomic*[4].

Non-holonomic constraints of the form

$$F_\alpha(\underline{z}, \underline{\dot{z}}, t) = 0 \tag{3.2.5}$$

will not be discussed here.

Systems with holonomic constraints should be thought of as idealized boundary cases of perfectly normal mechanical systems in which very strong elastic forces limit the possible motion of particles within the system to a region infinitesimally close to the manifold M_t for all times t. The forces of constraint are then boundary cases of normal elastic forces; in particular, in all practically important cases, they obey the law "every action has an equal and opposite reaction."

The forces of constraint $f_i(t)$, $i = 1, \ldots, N$ can all be combined into a single $3N$-dimensional constraint force $f(t)$. These forces of constraint work to restrict the motion to M_t. For holonomic and scleronomic systems, if no other forces are present each point $\underline{z} \in M_t$ is a possible point of equilibrium, and for systems which are holonomic but not necessarily scleronomic, under the same condition a displacement along M_t without resistance is always possible. This means that

---

[2] Holonomic (from the Greek), "globally lawful". All possible positions can be specified globally by given conditions, while for the non-holonomic case only the possible infinitesimal changes in state for each point of space and time can be specified.

[3] Scleronomic (from the Greek), "rigidly lawful". The constraints are independent of time.

[4] Rheonomic (from the Greek), "flowingly lawful". The constraints depend on time.

the force of constraint $f(t)$ has no components tangential to M_t and is thus perpendicular to M_t .

We call vectors tangential to the manifold M_t *virtual displacements*[5] with respect to M_t. Each vector ζ^α tangent to M_t at the point $\underline{z}_0$ can be represented as

$$\zeta^\alpha = \frac{d\underline{z}(\sigma)}{d\sigma}\bigg|_{\sigma=0}\,, \tag{3.2.6}$$

where $\underline{z}(\sigma)$ is a curve in M_t^α, which begins at the point $\underline{z}_0 \in M_t$ when $\sigma = 0$. Then, since

$$F_\alpha(\underline{z}(\sigma), t) = 0 \quad \text{we have also}$$

$$\frac{d}{d\sigma} F_\alpha(\underline{z}(\sigma), t)|_{\sigma=0} = \frac{d\underline{z}(\sigma)}{d\sigma} \cdot \underline{\nabla} F_\alpha(\underline{z}(\sigma), t)|_{\sigma=0}$$

$$= \zeta^\alpha \cdot \underline{\nabla} F_\alpha(\underline{z}_0, t) = 0 \ . \tag{3.2.7}$$

Therefore vectors normal to M_t^α are parallel to $\underline{\nabla} F_\alpha(\underline{z}, t)$. Thus, all vectors $\underline{\nabla} F_\alpha(\underline{z}, t)$ with $\underline{z} \in M_t$ are normal to $M_t = \bigcap_\alpha M_t^\alpha$, and the forces of constraint, themselves normal to M_t, can be represented as linear combinations of these vectors:

$$\underline{f}(t) = \sum_{\alpha=1}^{s} \lambda_\alpha(t) \underline{\nabla} F_\alpha(\underline{z}, t) \ . \tag{3.2.8}$$

For independent constraints, the gradients $\underline{\nabla} F_\alpha$ are linearly independent almost everywhere and the coefficients $\lambda_\alpha(t)$ are uniquely determined by $f(t)$.

In order to see this point more concretely, let us introduce f coordinates $q_1, \ldots, q_f$ which parametrize M_t. The allowed particle locations are then given by

$$\underline{z}(q_1, \ldots, q_f, t) \tag{3.2.9}$$

where $q_1, \ldots, q_f$ can be freely varied between fixed limits.
Then

$$\underline{z}(q_1, \ldots, q_f, t) \in M_t \ , \tag{3.2.10}$$

and therefore

$$\partial \underline{z}/\partial q_i \tag{3.2.11}$$

is a virtual displacement, i.e. a tangent vector to M_t.

---

[5] Virtual displacement (Latin) virtual: imagined, thought. These motions are small displacements which are imagined but not actually carried out, used to identify generalized states of equilibrium.

A general virtual displacement at the point $\underline{z}(q_1, \ldots, q_f, t)$ can then be written as

$$\delta\underline{z} = \sum_{i=1}^{f} \frac{\partial \underline{z}}{\partial q_i} \delta q_i = (\delta\boldsymbol{r}_1, \ldots, \delta\boldsymbol{r}_N) \; . \tag{3.2.12}$$

The statement that $\underline{f}(t)$ is normal to M_t, i.e.

$$\underline{f} \cdot \delta\underline{z} = \sum_{i=1}^{N} \boldsymbol{f}_i \cdot \delta\boldsymbol{r}_i = 0 \tag{3.2.13}$$

is called *d'Alembert's*[6] *principle*. Since $\delta\underline{z}$ is a vector along M_t, it is also true that the forces of constraint do no *virtual work*, that is, no work along a virtual displacement. We will see later that, in the case scleronomic constraints, actual physical motions belong to virtual displacements (as in the case of the pendulum), so that in this case the forces of constraint perform no (real) work.

If we can derive both the internal and external forces from a potential, then the equations of motion read

$$\underline{\dot{p}}(t) = -\underline{\nabla} U(\underline{z}(t), t) + \sum_{\alpha=1}^{s} \lambda_\alpha(t) \, \underline{\nabla} F_\alpha(\underline{z}(t), t) \; , \tag{3.2.14}$$

or, written out,

$$m\ddot{\boldsymbol{r}}_i(t) = -\underline{\nabla}_i U(\boldsymbol{r}_1(t), \ldots, \boldsymbol{r}_N(t), t)$$

$$+ \sum_{\alpha=1}^{s} \lambda_\alpha(t) \underline{\nabla}_i F_\alpha(\boldsymbol{r}_1(t), \ldots, \boldsymbol{r}_N(t), t) \; . \tag{3.2.15a}$$

These are $3N$ equations which, together with the s equations

$$F_\alpha(\boldsymbol{r}_1(t), \ldots, \boldsymbol{r}_N(t), t) = 0 \; , \quad \alpha = 1, \ldots, s \tag{3.2.15b}$$

determine the $3N + s$ functions $\boldsymbol{r}_1(t), \ldots, \boldsymbol{r}_N(t), \lambda_1(t), \ldots, \lambda_N(t)$. Thus the position vectors and the forces of constraint are determined at each point in time.

These $3N + s$ equations are called *Lagrange's equations of the first type.*

The calculation of the constraining forces is a very important technical problem; the entire field of statics is devoted to it. If we just want to calculate the force of constraint

$$\underline{f}_{\alpha_0} = \lambda_{\alpha_0}(t) \underline{\nabla} F_{\alpha_0}(t)$$

which belongs to the constraint $F_{\alpha_0} = 0$, we use the following procedure:

First, choose displacements $\delta_{\alpha_0}\underline{z}$ which obey all constraints $F_\alpha = 0$ with $\alpha \neq \alpha_0$, but violate $F_{\alpha_0} = 0$. Such displacements, according to our terminology, are not virtual displacements. They satisfy $\delta_{\alpha_0}\underline{z} \cdot \underline{\nabla} F_\alpha = 0$ for $\alpha \neq \alpha_0$ and $\delta_{\alpha_0}\underline{z} \cdot \underline{\nabla} F_{\alpha_0} \neq 0$.

---

[6] *D'Alembert, Jean le Rond* (*1717 Paris, d. 1783 Paris). Together with Diderot, the principle editor of the "Encyclopedia", His principle was published in 1743 in his "Traité de dynamique".

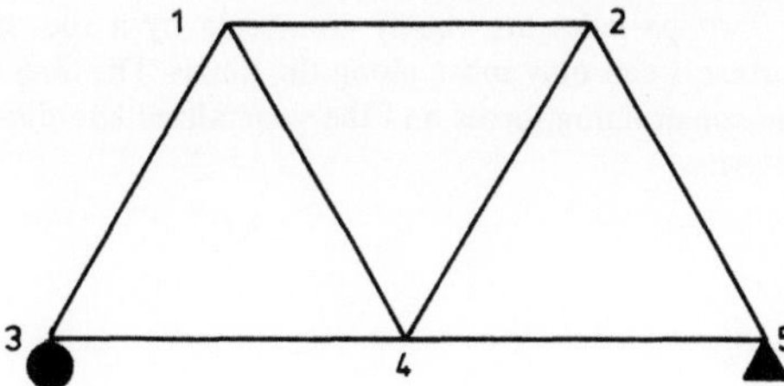

Fig. 3.2.1. Supports of a bridge with joints 1 through 5, whose positions do not change

Taking the scalar product of the equation $\underline{f} = \underline{\dot{p}} + \underline{\nabla} U$ with $\delta_{\alpha_0}\underline{z}$, we get

$$\underline{f} \cdot \delta_{\alpha_0}\underline{z} = (\underline{\dot{p}} + \underline{\nabla} U) \cdot \delta_{\alpha_0}\underline{z} \ ,$$

from which we can find $\underline{f}$.

In addition, for the case of equilibrium we also have $\underline{\dot{p}} = 0$, so that the constraining force is simply given by the variation in the potential energy under a displacement $\delta_{\alpha_0}\underline{z}$.

As a simple example of this procedure, let us consider the statics of the bridge pictured in Fig. 3.2.1.

The forces of constraint, exerted by the supporting poles and the bearings, keep the distance between the joints 1 through 5 constant. The potential energy of the entire construction is simply the potential energy of the total mass M, thought of as resting at the center of mass $\boldsymbol{R}$: $U = M\boldsymbol{g} \cdot \boldsymbol{R}$. The constraining force in the upper girder is the force keeping the distance between joints 1 and 2 constant. It obviously points in the direction of the line connecting joints 1 and 2, so we need only find its magnitude K. The work which would be done by a virtual displacement $\delta_{\alpha_0}\underline{z}$ corresponding to an increase in the length of the upper girder by a small amount δl is then given by

$$\underline{f} \cdot \delta_{\alpha_0}\underline{z} = K\delta l \ .$$

On the other hand

$$\underline{f} \cdot \delta_{\alpha_0}\underline{z} = \delta_{\alpha_0}\underline{z} \cdot \underline{\nabla} U$$

is exactly the change in the potential energy of the center of mass under the displacement $\delta_{\alpha_0}\underline{z}$, i.e.

$$\underline{f} \cdot \delta_{\alpha_0}\underline{z} = M\boldsymbol{g} \cdot \delta\boldsymbol{R}$$

Here,

$$\delta\boldsymbol{R} = \frac{d\boldsymbol{R}}{dl} \delta l$$

is the displacement of the center of mass if the length of the upper girder is changed by δl. Thus

$$K\delta l = M\boldsymbol{g} \cdot \frac{d\boldsymbol{R}}{dl} \delta l \quad \text{and} \quad K = M\boldsymbol{g} \cdot \frac{d\boldsymbol{R}}{dl} \ .$$

The constraining force can easily be calculated, then, by considering the change in height of the center of mass when the upper girder is lengthened by δl.

The derivative $d\boldsymbol{R}/dl$ can be found from the geometry of the system. It is easy to see that the center of mass sinks if the upper girder is shortened. The upper girder thus experiences a compressive stress.

By contrast, in a so-called *overdetermined system*, in which the constraints are not independent of each other, the constraining forces cannot be calculated so easily. They can be found only through

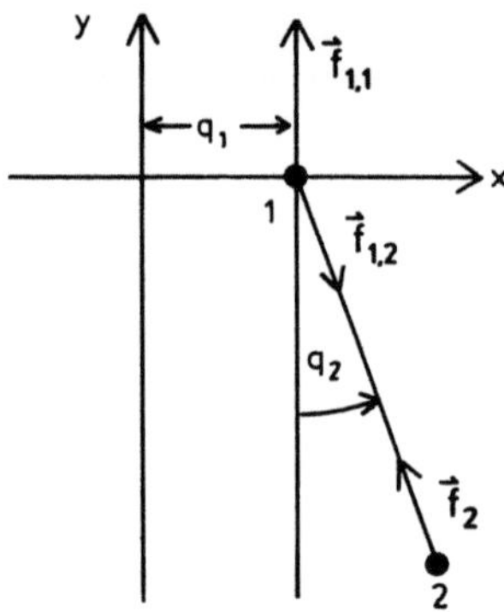

Fig. 3.2.2. Two particles are rigidly connected by a rod of lengh l. Particle 1 can only move along the x-axis. The directions of the constraining forces and the generalized coordinates are shown

a more exact analysis of the elastic properties of the system. A simple example of such a situation is a beam which has three supports, since if one of the three supports is removed, the possible motions of the beam are not changed.

Let us return to Lagrange's equations of the first type. We consider another example:

Let two particles be rigidly connected by a rod of length l, and we assume that particle 1 is constrained to move on a rail on the x-axis. We examine the motion in the xy-plane. With $r_1 = (x_1, y_1)$ and $r_2 = (x_2, y_2)$ as the coordinates of particles 1 and 2 respectively (Fig. 3.2.2), the constraints then read

$$F_1 = y_1 = 0 , \tag{3.2.16}$$

$$F_2 = (x_1 - x_2)^2 + (y_1 - y_2)^2 - l^2 = 0 . \tag{3.2.17}$$

Then the constraining forces have the form

$$f_1 = \left(\frac{\partial}{\partial x_1}, \frac{\partial}{\partial y_1} \right)(\lambda_1 F_1 + \lambda_2 F_2)$$

$$= (+2\lambda_2(x_1 - x_2), +2\lambda_2(y_1 - y_2) + \lambda_1) , \tag{3.2.18}$$

$$f_2 = \left(\frac{\partial}{\partial x_2}, \frac{\partial}{\partial y_2} \right)(\lambda_1 F_1 + \lambda_2 F_2)$$

$$= (-2\lambda_2(x_1 - x_2), -2\lambda_2(y_1 - y_2)) . \tag{3.2.19}$$

These constraining forces have a simpler form if we introduce better coordinates q_1, q_2 with

$$q_1 = x_1 , \tag{3.2.20}$$

$$q_2 = \varphi , \tag{3.2.21}$$

so that

$$x_1 = q_1 \; , \tag{3.2.22}$$

$$y_1 = 0 \; , \tag{3.2.23}$$

$$x_2 = q_1 + l \sin q_2 \equiv x_2(q_1, q_2) \; , \tag{3.2.24}$$

$$y_2 = -l \cos q_2 \equiv y_2(q_1, q_2) \; . \tag{3.2.25}$$

Then, we have

$$f_1(t) = (-2\lambda_2 l \sin q_2, \; +\lambda_1 + 2\lambda_2 l \cos q_2)$$

$$= (0, \lambda_1) - 2\lambda_2 l(\sin q_2, -\cos q_2) \; , \tag{3.2.26}$$

$$f_2(t) = 2\lambda_2 l(\sin q_2, -\cos q_2) \; . \tag{3.2.27}$$

$f_1(t)$ thus has a component $(0, \lambda_1)$ in the y-direction. This arises from the constraint $F_1 = 0$. This part of the constraining forces is responsible for keeping particle 1 on the line $y = 0$. The second contribution to $f_1(t)$ is exactly opposite to $f_2(t)$. $f_2(t)$ is responsible for keeping particle 2 at distance exactly l from particle 1, so that $f_2(t)$ is the "pull" on particle 2. This "pull" is equal in magnitude to the "pull" on particle 1. The virtual displacements are then

$$\frac{\partial r_1}{\partial q_1} = (1, 0) \; , \qquad \frac{\partial r_1}{\partial q_2} = (0, 0) \; ,$$

$$\frac{\partial r_2}{\partial q_1} = (1, 0) \; , \qquad \frac{\partial r_2}{\partial q_2} = l(\cos q_2, \sin q_2) \; ,$$

so that by d'Alembert's principle:

$$f_1 \cdot \frac{\partial r_1}{\partial q_1} + f_2 \cdot \frac{\partial r_2}{\partial q_1} = 0 \; , \tag{3.2.28}$$

$$f_1 \cdot \frac{\partial r_1}{\partial q_2} + f_2 \cdot \frac{\partial r_2}{\partial q_2} = 0 \; . \tag{3.2.29}$$

Of course, both equations are satisfied by construction.

Even if we were not aware of the form of the constraining forces, we could still read off from the first equation that

The x-components of f_1 and f_2 add up to zero

and from the second equation,

f_2 points from particle 2 in the direction of particle 1.

From this, it also follows that

$$f_2(t) = \hat{\lambda}_2(\sin q_2, \ -\cos q_2) \quad \text{and} \quad f_1(t) = (-\hat{\lambda}_2 \sin q_2, \hat{\lambda}_1) \tag{3.2.30}$$

in agreement with our first calculation of the $f_i(t)$.

The Lagrangian equations of the first type

$$m_1 \ddot{r}_1(t) = m_1 g + f_1(t) \ , \tag{3.2.31}$$

$$m_2 \ddot{r}_2(t) = m_2 g + f_2(t) \ , \tag{3.2.32}$$

with the conditions of constraint

$$y_1 = 0 \ , \tag{3.2.33}$$

$$(x_1 - x_2) + (y_1 - y_2)^2 - l^2 = 0 \tag{3.2.34}$$

are then to be solved.

These equations, though, can be transformed by multiplying them by $\partial r_1 / \partial q_i$ and $\partial r_2 / \partial q_i$ respectively and then adding, so that the forces of constraint will cancel from the statement of D'Alembert's principle above. This procedure corresponds to Lagrange's method of second type which will be discussed in the following section. It corresponds to the second strategy in Sect. 3.1.

We find, then, that

$$m_1 \ddot{x}_1 + m_2 \ddot{x}_2 = 0 \quad \text{and} \tag{3.2.35}$$

$$m_2 \ddot{x}_2 \cos q_2 + (m_2 \ddot{y}_2 + m_2 g) \sin q_2 = 0 \ . \tag{3.2.36}$$

If we also use the fact that $x_2 = x_2(q_1, q_2)$, $y_2 = y_2(q_1, q_2)$, then we find two equations for $q_1(t)$ and $q_2(t)$, namely

$$(m_1 + m_2)\ddot{q}_1 + m_2 l \ddot{q}_2 \cos q_2 - m_2 l \dot{q}_2^2 \sin q_2 = 0 \ ,$$

$$m_2 \cos q_2 (\ddot{q}_1 + l \ddot{q}_2 \cos q_2 - l \dot{q}_2^2 \sin q_2) \tag{3.2.37}$$

$$+ \sin q_2 (m_2 g + m_2 l \ddot{q}_2 \sin q_2 + m_2 l \dot{q}_2^2 \cos q_2) = 0$$

or

$$(m_1 + m_2)\ddot{q}_1 = m_2 l(\dot{q}_2^2 \sin q_2 - \ddot{q}_2 \cos q_2) \ ,$$

$$\tag{3.2.38}$$

$$\ddot{q}_1 \cos q_2 + l \ddot{q}_2 + g \sin q_2 = 0 \ .$$

The solution of this equation can be reduced to a simple integration. We will limit ourselves to small values of q_1 and q_2. Then we have, up to terms of higher order in q_1 and q_2,

$$(m_1 + m_2)\ddot{q}_1 = -m_2 l \ddot{q}_2 \ , \tag{3.2.39}$$

$$\ddot{q}_1 + l \ddot{q}_2 + g q_2 = 0 \ , \quad \text{or} \tag{3.2.40}$$

$$\ddot{q}_2 \left[1 - \frac{m_2}{m_1 + m_2} \right] + \frac{g}{l} q_2 = 0 \ , \quad \text{or} \tag{3.2.41}$$

$$\ddot{q}_2 + \frac{m_1 + m_2}{m_1} \frac{g}{l} q_2 = 0 \ , \quad \text{i.e.} \tag{3.2.42}$$

$$q_2(t) = q_2^0 \cos\left[\omega(t - t_0)\right] \quad \text{with} \quad \omega^2 = \frac{m_1 + m_2}{m_1} \frac{g}{l} \ , \tag{3.2.43}$$

and then

$$q_1(t) = -\frac{m_2 l}{(m_1 + m_2)} q_2^0 \cos\left[\omega(t - t_0)\right] + \alpha_0 + \alpha_1 t \ . \tag{3.2.44}$$

The pendulum thus swings with another frequency, modified by $(m_1 + m_2)/m_1$, as does its supporting point. For $m_1 \to \infty$, $q_1(t) = \alpha_0 + \alpha_1 t$ and $\omega^2 = g/l$, as we would expect.

3.3 The Lagrangian Method of the Second Type

Next, we formulate in a general manner the procedure by which we can eliminate the constraining forces by projecting onto M_t:

After introducing suitable coordinates

$$q = (q_1, \ldots, q_f)$$

on M_t, the constraints $F_\alpha = 0$ are satisfied identically in the coordinates $q_1, \ldots, q_f$ and t:

$$F_\alpha(\underline{z}(q, t), t) \equiv 0 \ , \quad \alpha = 1, \ldots, s \ . \tag{3.3.1}$$

Multiplying the equation of motion

$$\underline{\dot{p}}(t) = -\underline{\nabla} U(\underline{z}(t), t) + \underline{f}(t)$$

by the tangent vectors $\partial \underline{z}/\partial q$ to M_t, and using the fact that $(\partial \underline{z}/\partial q_j) \cdot \underline{f} = 0$, we find

$$\frac{\partial \underline{z}}{\partial q_j} \cdot \underline{\dot{p}} = -\frac{\partial \underline{z}}{\partial q_j} \cdot \underline{\nabla} U + \frac{\partial \underline{z}}{\partial q_j} \cdot \underline{f}$$

$$= -\frac{\partial U}{\partial q_j} \left[\underline{z}(q_1, \ldots, q_f, t)\right] \ . \tag{3.3.2}$$

The right side of the equation is already quite simple. We would like now to simplify the left side:

We will give the proof of the following claim soon: With

$$T = \frac{1}{2} \sum_{i=1}^{N} m_i \dot{r}_i^2 = T(q_1, \ldots, q_f, \dot{q}_1, \ldots, \dot{q}_f, t)$$

we have

$$\frac{\partial \underline{z}}{\partial q_j} \cdot \underline{\dot{p}} \equiv \sum_{i=1}^{N} m_i \ddot{r}_i \cdot \frac{\partial r_i}{\partial q_j} = \frac{d}{dt} \frac{\partial T}{\partial \dot{q}_j} - \frac{\partial T}{\partial q_j} \, . \tag{3.3.3}$$

Then, the f differential equations for the $q_i(t)$, $i = 1, \ldots, f$ can also be written in the form

$$\frac{d}{dt} \frac{\partial L}{\partial \dot{q}_j} - \frac{\partial L}{\partial q_j} = 0 \, , \quad j = 1, \ldots, f \quad \text{with} \tag{3.3.4}$$

$$L(q_1, \ldots, q_f, \dot{q}_1, \ldots, \dot{q}_f, t)$$

$$= T(q_1, \ldots, q_f, \dot{q}_1, \ldots, \dot{q}_f, t) - U(z(q_1, \ldots, q_f, t), t) \tag{3.3.5}$$

considered as a function with independent arguments $q_1, \ldots, q_f, \dot{q}_1, \ldots, \dot{q}_f, t$.

The function $L = T - U$ is called the *Lagrangian* in the coordinates $q_1, \ldots, q_f$.

If the Lagrangian is already known (and it is here, as soon as we know the kinetic and potential energy), it is easy to derive the equations projected onto M_t. The equations which can thus be derived from the Lagrangian $L(q_1, \ldots, q_f, \dot{q}_1, \ldots, \dot{q}_f, t)$ are called *Lagrange's equations (of the second type)*. The constraints are fully eliminated by the projection onto M_t.

Proof of the Claim. We have

$$\sum_{i=1}^{N} m_i \ddot{r}_i \cdot \frac{\partial r_i}{\partial q_j} = \frac{d}{dt} \left(\sum_{i=1}^{N} m_i \dot{r}_i \cdot \frac{\partial r_i}{\partial q_j} \right) - \sum_{i=1}^{N} m_i \dot{r}_i \cdot \frac{d}{dt} \frac{\partial r_i}{\partial q_j} \, . \tag{3.3.6}$$

But then since

$$\dot{r}_i = \sum_{k=1}^{f} \frac{\partial r_i}{\partial q_k} \dot{q}_k + \frac{\partial r_i}{\partial t} \tag{3.3.7}$$

it is also true that

$$\frac{\partial \dot{r}_i}{\partial \dot{q}_k} = \frac{\partial r_i}{\partial q_k} \, , \tag{3.3.8}$$

and further

$$\frac{d}{dt}\frac{\partial r_i}{\partial q_j} = \sum_{k=1}^{f} \frac{\partial^2 r_i}{\partial q_j \partial q_k}\,\dot{q}_k + \frac{\partial^2 r_i}{\partial q_j \partial t}$$

$$= \frac{\partial}{\partial q_j}\left(\sum_k \frac{\partial r_i}{\partial q_k}\,\dot{q}_k + \frac{\partial r_i}{\partial t}\right) = \frac{\partial \dot{r}_i}{\partial q_j}\ . \tag{3.3.9}$$

Substituting (3.3.8 and 9), we find

$$\sum_{i=1}^{N} m_i \ddot{r}_i \cdot \frac{\partial r_i}{\partial q_j} = \frac{d}{dt}\left(\sum m_i \dot{r}_i \cdot \frac{\partial \dot{r}_i}{\partial \dot{q}_j}\right) - \sum m_i \dot{r}_i \cdot \frac{\partial \dot{r}_i}{\partial q_j}$$

$$= \frac{d}{dt}\frac{\partial T}{\partial \dot{q}_j} - \frac{\partial T}{\partial q_j}\ , \tag{3.3.10}$$

which was our claim.

Lagrange's equations are a system of f coupled ordinary differential equations for the desired functions $q_1(t), \ldots, q_f(t)$.

Remarks. i) If there are no constraining forces, then the q_i, $i = 1, \ldots, 3N$ can be the Cartesian coordinates or some other coordinate system derived from a transformation of the Cartesian coordinates. The Newtonian equations of motion in Cartesian coordinates can be found immediately from the Lagrangian

$$L(r_1, \ldots, r_N, \dot{r}_1, \ldots, \dot{r}_N, t) = \frac{1}{2}\sum_{i=1}^{N} m_i \dot{r}_i^2 - U(r_1, \ldots, r_N)\ , \tag{3.3.11}$$

because then

$$\frac{d}{dt}\frac{\partial L}{\partial \dot{r}_i} = \frac{d}{dt}\, m_i \dot{r}_i = m_i \ddot{r}_i \quad \text{and} \tag{3.3.12}$$

$$\frac{\partial L}{\partial r_i} = -\frac{\partial U}{\partial r_i}\ . \tag{3.3.13}$$

Thus, Lagrange's equations

$$\frac{d}{dt}\frac{\partial L}{\partial \dot{r}_i} - \frac{\partial L}{\partial r_i} = 0$$

are identical to the equation of motion

$$m\ddot{r}_i + \nabla_i U = 0\ .$$

The derivation of Lagrange's equations from the Lagrangian is thus a *general method* to produce the equations of motion in terms of arbitrary coordinates $q_1, \ldots, q_f$ even in the presence of holonomic constraints.

ii) The coordinates $q_1, \ldots, q_f$ of M_t can be chosen arbitrarily, as long as they parametrize M_t.

If we pass from one set of coordinates $q_1, \ldots, q_f$ to another set $\bar{q}_1, \ldots, \bar{q}_f$ through an invertible single-valued transformation

$$q_i = q_i(\bar{q}_1, \ldots, \bar{q}_f, t) , \quad \text{with} \tag{3.3.14}$$

$$\dot{q}_i = \sum_{j=1}^{f} \frac{\partial q_i}{\partial \bar{q}_j} \dot{\bar{q}}_j + \frac{\partial q_i}{\partial t}$$

$$\equiv \dot{q}_i(\bar{q}_1, \ldots, \bar{q}_f, \dot{\bar{q}}_1, \ldots, \dot{\bar{q}}_j, t) , \tag{3.3.15}$$

then we consider the new Lagrangian

$$\bar{L}(\bar{q}_1, \ldots, \bar{q}_f, \dot{\bar{q}}_1, \ldots, \dot{\bar{q}}_f, t) = L(q(\bar{q}, t), \dot{q}(\bar{q}, \dot{\bar{q}}, t), t) , \tag{3.3.16}$$

formed by substituting $q(\bar{q}, t)$ and $\dot{q}(\bar{q}, \dot{\bar{q}}, t)$ into $L(q, \dot{q}, t)$. The equations of motion for $\bar{q}(t)$ are then

$$\frac{d}{dt} \frac{\partial \bar{L}}{\partial \dot{\bar{q}}_j} - \frac{\partial \bar{L}}{\partial \bar{q}_j} = 0 , \quad j = 1, \ldots, f . \tag{3.3.17}$$

This procedure is the easiest way to transform the equations of motion from one coordinate system to another.

iii) We speak of *generalized forces* Q_i, if these can be derived from a *generalized potential* $U(q_1, \ldots, q_f, \dot{q}_1, \ldots, \dot{q}_f, t)$ using the equation

$$Q_i = -\frac{\partial U}{\partial q_i} + \frac{d}{dt} \frac{\partial U}{\partial \dot{q}_i} . \tag{3.3.18}$$

Let us now consider again the Cartesian coordinates r_i. Then repeating the derivation of Lagrange's equations from the equation of motion, we obtain

$$m_i \ddot{r}_i = -\frac{\partial U}{\partial r_i} + \frac{d}{dt} \frac{\partial U}{\partial \dot{r}_i} + f_i . \tag{3.3.19}$$

Now we eliminate the constraining forces from the right side of (3.3.19):

$$\sum_{i=1}^{N} \frac{\partial r_i}{\partial q_j} \cdot \left(-\frac{\partial U}{\partial r_i} + \frac{d}{dt} \frac{\partial U}{\partial \dot{r}_i} \right)$$

$$= \sum_{i=1}^{N} \left[-\frac{\partial U}{\partial r_i} \cdot \frac{\partial r_i}{\partial q_j} + \frac{d}{dt} \left(\frac{\partial U}{\partial \dot{r}_i} \cdot \frac{\partial r_i}{\partial q_j} \right) - \frac{\partial U}{\partial \dot{r}_i} \cdot \frac{d}{dt} \frac{\partial r_i}{\partial q_j} \right]$$

$$= \sum_{i=1}^{N} \left[-\frac{\partial U}{\partial r_i} \cdot \frac{\partial r_i}{\partial q_j} + \frac{d}{dt} \left(\frac{\partial U}{\partial \dot{r}_i} \cdot \frac{\partial \dot{r}_i}{\partial \dot{q}_j} \right) - \frac{\partial U}{\partial \dot{r}_i} \cdot \frac{\partial \dot{r}_i}{\partial q_j} \right]$$

$$= \frac{d}{dt} \frac{\partial U}{\partial \dot{q}_j} - \frac{\partial U}{\partial q_j} , \tag{3.3.20}$$

and then we find again, with the left side of (3.3.19):

$$\frac{d}{dt}\frac{\partial L}{\partial \dot{q}_j} - \frac{\partial L}{\partial q_j} = 0, \quad j = 1,\ldots,f \tag{3.3.21}$$

with $L = T - U$.

Example. Let a particle with charge e be located in an electromagnetic field with potentials (ϕ, A). Then the Lagrangian is given by

$$L(r, \dot{r}, t) = \tfrac{1}{2}m\dot{r}^2 - e\phi(r, t) + eA(r, t)\cdot\dot{r} \ . \tag{3.3.22}$$

We need to show that Lagrange's equations are identical to the equations of motion

$$m\ddot{r}(t) = e(E + \dot{r}\times B) \quad \text{with} \tag{3.3.23}$$

$$B = \nabla\times A \ , \quad E = -\nabla\phi - \frac{\partial}{\partial t}A \ ,$$

which can be checked through direct calculation. The potential

$$U(r, \dot{r}, t) = e\phi(r, t) - eA(r, t)\cdot\dot{r} \tag{3.3.24}$$

is the most important generalized potential.

iv) By introducing the coordinates $q_1,\ldots,q_f$, the constraints are directly satisfied and the constraining forces are completely eliminated from Lagrange's equations. If we have found the solutions $q_1(t),\ldots,q_f(t)$ to these equations, then, as a second step, we can easily determine the constraining forces.

The trajectories are given by $\underline{z}(t) = \underline{z}(q(t), t)$, so that using $\underline{f}(t) = \dot{\underline{p}}(t) + \underline{\nabla}U(\underline{z}(q(t), t), t)$, we can now calculate the constraining forces $\underline{f}(t)$.

Example. *The Spherical Pendulum*

Let us consider a mathematical pendulum, with a string of length l, whose swinging *is not confined to a single plane.* With

$$r = (x, y, z)$$

we can introduce polar coordinates

$$x = r\sin\theta\cos\varphi \ ,$$

$$y = r\sin\theta\sin\varphi \ ,$$

$$z = -r\cos\theta \ ,$$

and the constraint $r^2 - l^2 = 0$ can be written $r = l$.

Thus, θ and φ are freely varying coordinates, and $r = r(\theta, \varphi)$. Then

$$L(\theta, \dot{\theta}, \varphi, \dot{\varphi}) = \tfrac{1}{2} m \dot{r}^2 - U(r) \qquad \text{with} \tag{3.3.25}$$

$$U(r) = mg(l + z) \, ,$$

$$\dot{r} = l(\dot{\theta} \cos \theta \cos\varphi - \dot{\varphi} \sin \theta \sin \varphi, \, \dot{\theta} \cos \theta \sin \varphi + \dot{\varphi} \sin \theta \cos \varphi, \, \dot{\theta} \sin \theta) \, .$$

Thus, we have

$$\dot{r}^2 = l^2(\dot{\theta}^2 + \dot{\varphi}^2 \sin^2 \theta) \, , \qquad \text{and thus}$$

$$L = \tfrac{1}{2} m l^2(\dot{\theta}^2 + \dot{\varphi}^2 \sin^2 \theta) - mgl(1 - \cos \theta) \, . \tag{3.3.26}$$

Then, the equation of motion for $\theta(t)$ reads

$$\frac{d}{dt} \frac{\partial L}{\partial \dot{\theta}} - \frac{\partial L}{\partial \theta} = 0 \, , \qquad \text{i.e.} \tag{3.3.27}$$

$$ml^2(\ddot{\theta} - \dot{\varphi}^2 \sin \theta \cos \theta) + mgl \sin \theta = 0 \, , \tag{3.3.28}$$

and for $\varphi(t)$

$$\frac{d}{dt} \frac{\partial L}{\partial \dot{\varphi}} = \frac{d}{dt} (ml^2 \dot{\varphi} \sin^2 \theta) = 0 \, , \tag{3.3.29}$$

since L is independent of φ.

We see immediately that

$$\frac{\partial L}{\partial \dot{\varphi}} = ml^2 \dot{\varphi} \sin^2 \theta \tag{3.3.30}$$

is a conserved quantity.

In general, a generalized coordinate on which L is independent is called *cyclic*[7]. The quantities

$$p_i = \frac{\partial L}{\partial \dot{q}_i} \tag{3.3.31}$$

are called the *generalized momenta* corresponding to q_i. In order to understand this name, consider that if $L = \tfrac{1}{2} m \dot{r}^2 - U(r)$, then

$$p = \frac{\partial L}{\partial \dot{r}} = m \dot{r} \, .$$

---

[7] Cyclic coordinate: cyclic = circular type.
The angular coordinate in a symmetrical cylindrical system is a typical example of cyclic coordinate.

The generalized momentum corresponding to a cyclic coordinate is, by this definition, a conserved quantity.

In our example

$$p_\varphi = ml^2 \sin^2 \theta \, \dot{\varphi} \tag{3.3.32}$$

is time-independent. On the other hand,

$$L_z = m(x\dot{y} - y\dot{x}) = ml^2 \sin^2 \theta \, \dot{\varphi} \ . \tag{3.3.33}$$

Thus, p_φ is just the z-component of the angular momentum.

Substituting these constants into the equation (3.3.28), we find, for the equation of motion for $\theta(t)$:

$$ml^2\ddot{\theta} - \frac{L_z^2}{ml^2 \sin^3 \theta} \cos \theta + mgl \sin \theta = 0 \tag{3.3.34}$$

or, after multiplying by $\dot{\theta}$:

$$\frac{d}{dt}\left(\frac{1}{2} ml^2\dot{\theta}^2 + \frac{L_z^2}{2ml^2 \sin^2 \theta} - mgl \cos \theta \right) = 0 \ . \tag{3.3.35}$$

Since, on the other hand, the energy is given by

$$E = T + U = \frac{1}{2} ml^2\dot{\theta}^2 + \frac{L_z^2}{2ml^2 \sin^2 \theta} + mgl(1 - \cos \theta) \tag{3.3.36}$$

(3.3.35) merely states that energy is conserved.

Now we can calculate the function $\theta(E, L_z; t)$ again by means of a simple integral, and then from the formula $L_z = ml^2 \sin^2 \theta \dot{\varphi}$ we can determine $\varphi(E, L_z; t)$. This solves the equations of motion.

We can also discuss the motion qualitatively, by studying the quantity

$$E = T + U_{\text{eff}} \qquad \text{with} \tag{3.3.37}$$

$$U_{\text{eff}}(\theta) = \frac{L_z^2}{2ml^2 \sin^2 \theta} + mgl(1 - \cos \theta) \ . \tag{3.3.38}$$

It is then possible to examine the dependence of $U_{\text{eff}}(\theta)$ on θ, while varying the parameter L_z (Fig. 3.3.1).

We distinguish the following cases:

a) $L_z = 0$, then $\dot{\varphi} = 0$, and we have motion in a single plane.

b) $L_z \neq 0$: $U_{\text{eff}}(\theta)$ is singular at $\theta = 0$ and π. The minimum of U_{eff} lies at $\theta = \theta_0 < \pi/2$. If E and L_z are given, then θ can oscillate between θ_1 and θ_2.

If the smallest value of E corresponding to a given L_z is chosen, then the pendulum moves in a circle whose radius is given by $l \sin \theta_0$. In that case, $\theta = \theta_0 = $ constant.

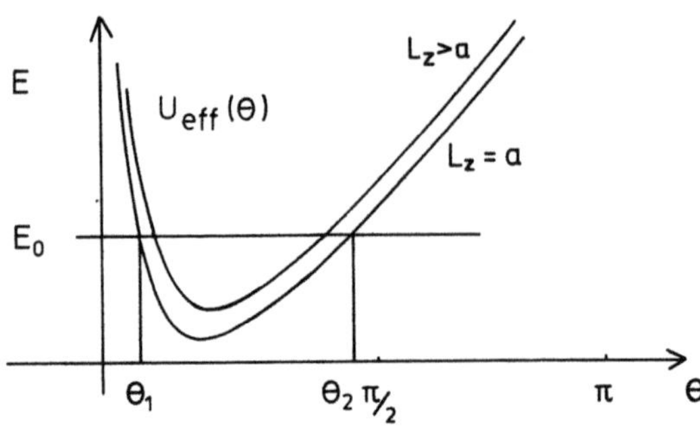

Fig. 3.3.1. Depiction of $U_{\text{eff}}(\theta)$ for various values of L_z. The minimum is located at values of θ smaller than $\pi/2$. For a given E_0 and $L_z \neq 0$, θ can oscillate between θ_1 and θ_2

The circular motion is stable under small perturbations and its angular velocity is

$$\dot{\varphi} = \frac{L_z}{ml^2 \sin^2 \theta_0} = \text{const} \ .$$

3.4 The Conservation of Energy in Motions Which are Limited by Constraints

D'Alembert's principle states that constraining forces perform no work under a virtual displacement. Virtual displacements are tangential to M_t and if the constraints are scleronomic, that is, they do not depend on time, then M_t is time-independent.

Then, changes in the position vectors of the particles during the motion

$$d\underline{z}(t) = \underline{z}(t + dt) - \underline{z}(t) = \underline{\dot{z}}(t)\,dt \tag{3.4.1}$$

are also virtual displacements, and the work that the constraining forces perform during an actual motion must vanish. The situation is different, though, if the constraints are explicitly time-dependent, i.e. rheonomic. Then

$$M_{t+dt} \neq M_t \quad \text{and} \quad d\underline{z} = \underline{\dot{z}}(t)\,dt$$

is a vector from M_t to M_{t+dt} and thus, in general, is not tangent M_t (Fig. 3.4.1). This means that actual physical motions are not virtual displacements.

We will now calculate, in general, the work performed by the constraining forces. This work includes the energy brought into the system from its environment, which exerts the constraining forces.

We start with Lagrange's equations of the first type with a time-independent potential $U(r_1, \ldots, r_N)$:

$$m_i \ddot{r}_i + \nabla_i U = \sum_{\alpha=1}^{s} \lambda_\alpha(t) \, \nabla_i F_\alpha(r_1(t), \ldots, r_N(t), t) \ . \tag{3.4.2}$$

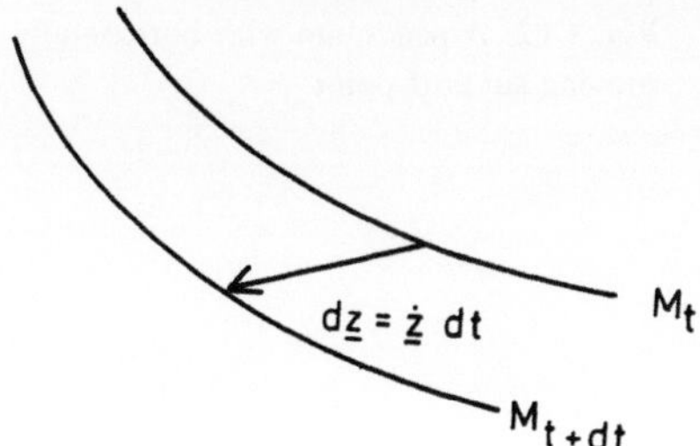

Fig. 3.4.1. If M_t is dependent on time then the change in position vector over time $d\underline{z}$ is not tangent to M_t

Multiplying by $\dot{r}_i$ and summing over i results in:

$$\frac{d}{dt}\left[\frac{1}{2}\sum_{i=1}^{N} m_i \dot{r}_i^2 + U(r_1(t), \ldots, r_N(t))\right]$$

$$= \sum_{i=1}^{N}\sum_{\alpha=1}^{s} \lambda_\alpha(t)\dot{r}_i \cdot \nabla_i F_\alpha(r_1(t), \ldots, r_N(t), t) \ . \tag{3.4.3}$$

On the other hand, from

$$F_\alpha(r_1(t), \ldots, r_N(t), t) = 0 \ , \quad \alpha = 1, \ldots, s$$

it follows that

$$\frac{d}{dt} F_\alpha = \sum_{i=1}^{N} \dot{r}_i \cdot \nabla_i F_\alpha(r_1(t), \ldots, r_N(t), t) + \frac{\partial F_\alpha}{\partial t} = 0 \ . \tag{3.4.4}$$

Thus, the energy of the system, given by

$$E(t) = \frac{1}{2}\sum_{i=1}^{N} m_i \dot{r}_i^2 + U(r_1(t), \ldots, r_N(t))$$

satisfies

$$\frac{dE(t)}{dt} = -\sum_{\alpha=1}^{s} \lambda_\alpha(t)\frac{\partial F_\alpha(r_1, \ldots, r_N, t)}{\partial t} \ . \tag{3.4.5}$$

We observe the following:

a) The expression for the energy is the same as in the case of motion without constraining forces. The only difference is that the only values which r_i and $\dot{r}_i$ take on are those consistent with the constraints.
b) Under the influence of holonomic scleronomic constraints, the energy of the system is conserved.
c) In the case of holonomic rheonomic constraints, the energy of the system is no longer conserved and there is energy exchanged with the environment.

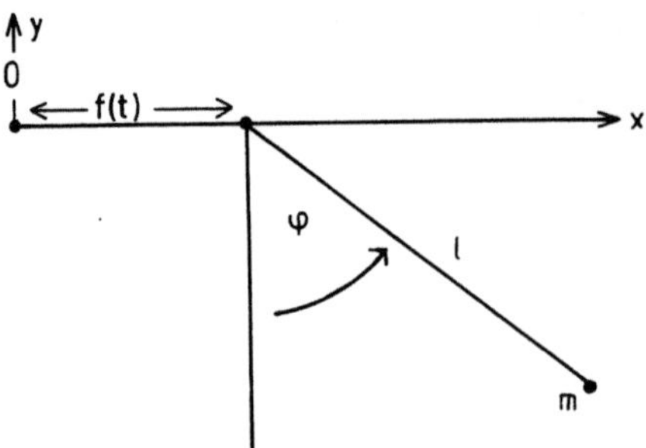

Fig. 3.4.2. A pendulum with horizontally moving support point

Examples. a) First, we consider a planar pendulum, whose supporting point can be moved horizontally (Fig. 3.4.2).

The weight, with mass m, has coordinates

$$\boldsymbol{r} = (x, y) \quad \text{with}$$

$$x = f(t) + l \sin \varphi , \quad y = -l \cos \varphi .$$

The constraint reads

$$F(x, y, t) \equiv [x - f(t)]^2 + y^2 - l^2 = 0 . \tag{3.4.7}$$

We have

$$\dot{x} = \dot{f} + l\dot{\varphi} \cos \varphi , \quad \dot{y} = l\dot{\varphi} \sin \varphi ,$$

and for the Lagrangian, we obtain

$$L = \tfrac{1}{2}m(\dot{x}^2 + \dot{y}^2) - U(x, y)$$

$$= \tfrac{1}{2}m[\dot{f}^2 + 2l\dot{f}\dot{\varphi} \cos \varphi + \dot{\varphi}^2(l^2 \cos^2 \varphi + l^2 \sin^2 \varphi)] - mgl(1 - \cos \varphi) .$$

Thus

$$L(\varphi, \dot{\varphi}, t) = \tfrac{1}{2}m[l^2\dot{\varphi}^2 + 2l\dot{f}\dot{\varphi} \cos \varphi + \dot{f}^2] + mgl \cos \varphi - mgl . \tag{3.4.8}$$

Lagrange's equation then reads

$$\frac{d}{dt} \frac{\partial L}{\partial \dot{\varphi}} - \frac{\partial L}{\partial \varphi} = 0 \quad \text{or}$$

$$\frac{d}{dt}(ml^2\dot{\varphi} + ml\dot{f}\cos \varphi) - (-ml\dot{f}\dot{\varphi} \sin \varphi - mgl \sin \varphi) = 0$$

or

$$\ddot{\varphi} + \frac{g}{l} \sin \varphi = -\frac{1}{l}\ddot{f} \cos \varphi . \tag{3.4.9}$$

It is understandable that here only $\ddot{f}(t)$ appears in the equation of motion rather than $f(t)$ or $\dot{f}(t)$, since motion in the coordinate φ cannot be influenced by uniform linear motion in the x-direction.

Since the constraint is

$$F = [x - f(t)]^2 + y^2 - l^2 = 0$$

Lagrange's equations of the first type read

$$m\ddot{x} = \lambda \frac{\partial F}{\partial x} = 2\lambda[x - f(t)] \; ,$$

$$m\ddot{y} + mg = \lambda \frac{\partial F}{\partial y} = 2\lambda y$$

and, since the constraining force always points in the direction of the supporting point, $\lambda < 0$.

On the other hand,

$$\frac{\partial F}{\partial t} = -2\lambda[f(t) - x(t)] \tag{3.4.10}$$

hence

$$\frac{dE}{dt} = -\lambda \frac{\partial F}{\partial t} = 2\lambda[f(t) - x(t)]\dot{f}(t) \; ,$$

so that

$$\frac{dE}{dt} = m\ddot{x}\dot{f}(t) \; . \tag{3.4.11}$$

We will investigate in greater detail the case in which the supporting point undergoes uniform linear motion:

Let $f(t) = vt$, then $\dot{f} = v$ and we find (Fig. 3.4.3):

$$\text{for} \quad -\frac{\pi}{2} < \varphi \leq 0 : f(t) - x(t) \geq 0 \; , \quad \text{so that} \quad dE/dt \geq 0 \; ,$$

$$\text{for} \quad 0 \leq \varphi < \frac{\pi}{2} : f(t) - x(t) \leq 0 \; , \quad \text{so that} \quad dE/dt \leq 0 \; .$$

To see that this result is plausible, consider that we can start or stop the pendulum from swinging by moving the supporting point in the proper direction.

If the system is thus displaced in a uniform linear manner with speed v in the positive x-direction, the energy of the system is lowered as it swings in the

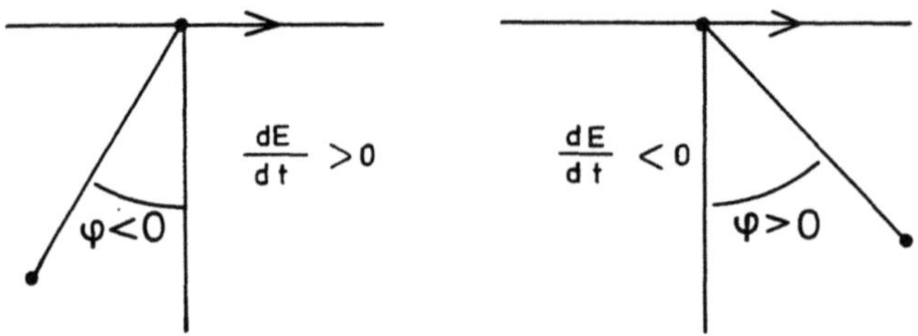

Fig. 3.4.3. For $\varphi < 0$, energy is introduced to a pendulum whose support point undergoes uniform linear motion. For $\varphi > 0$, energy is taken away

direction of motion, and it is increased when the pendulum moves in the opposite direction. The net change in energy during a full period T obviously vanishes, as we can see from

$$\int_0^T dt\, \frac{dE}{dt} = mv \int_0^T dt\, \ddot{x}(t)$$

$$= mv[\dot{x}(T) - \dot{x}(0)] = 0 \; . \tag{3.4.12}$$

If the supporting point is moved in a uniform linear manner, then the time average of the energy introduced by the constraining force vanishes. The function $f(t)$ can also be chosen in such a way that the energy of the system always increases. We can do this, for example, by ensuring that the support point always moves to the right when the pendulum is swinging to the left. This is possible with a suitable function $f(t)$. In this way, we can produce typical resonance phenomena. The calculation of such a "non-linear oscillation" under a periodic stimulation

$$\ddot{\varphi} + \frac{g}{l} \sin \varphi = -\frac{1}{l} \ddot{f}(t) \cos \varphi$$

is complicated and cannot be carried out here.

b) The pendulum with variable string length: Consider a boy on a swing. He changes his center of mass as he swings back and forth. As he swings all the way back, he increases the "string length". Passing back through the equilibrium point, he sits up and thus shortens the effective pendulum length at the very moment that the constraining force is greatest. Thus, every time he swings, energy can be introduced.

The boy who is swinging here is using a form of energy which has not yet been considered: he is using his muscles, that is, he is using chemical energy. But that is not enough to change his center of mass. It is essential that the supporting point should be fixed, so that in the total system swing + boy, the boy can change his position relative to the swing.

We can easily show quantitatively the mechanism by which energy is introduced:

The constraint for a pendulum of variable length is given by

$$F \equiv l^2(t) - r^2 = 0 \ . \tag{3.4.13}$$

Then,

$$f = \lambda(\partial F/\partial r) = -2\lambda r \ , \tag{3.4.14}$$

so that $\lambda > 0$, since f always points towards the supporting point. But also

$$\frac{\partial F}{\partial t} = 2l\dot{l} \quad \text{so that}$$

$$\frac{dE}{dt} = -2\lambda(t)l(t)\dot{l}(t) \ . \tag{3.4.15}$$

Thus, when the length is decreased, i.e. when $\dot{l}(t) < 0$, $dE/dt > 0$.
On the other hand, if the length of the pendulum is increased, the energy decreases. If, then, we shorten the length at those times when $\lambda(t)$ is largest, which occurs as we swing through the equilibrium point, then the pendulum will gain more energy than it will lose later when we again lengthen it (when λ is smaller).

Thus, it is possible, through the proper external stimulation of a periodic system, to influence dE/dt as a function of time during a period T in such a way that $E(T) - E(0) > 0$.

The motion of the incense burner of Santiago de Compostella (Fig. 3.4.4) works in a similar way to that of the swing, in that the acolytes alter its length according to the correct rhythm.

c) As a final example, let us consider the yo-yo. In this case, too, we can calculate the mechanism by which energy is introduced to the system.

We consider a rigid body, namely, a small round rod of radius r, with somewhat larger disks attached to its ends. Between the disks, on the rod,

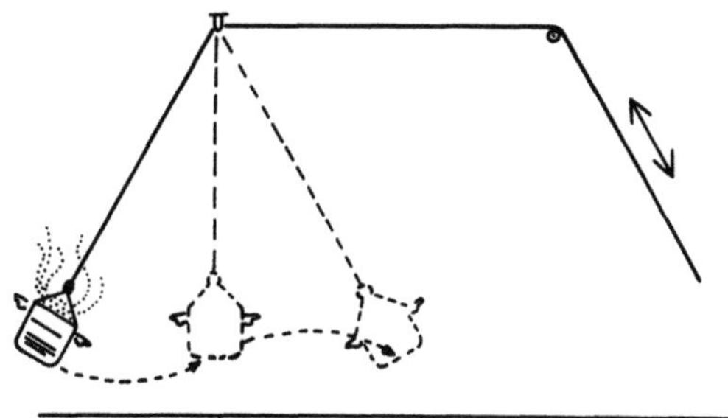

Fig. 3.4.4. The incense burner of Santiago de Compostella

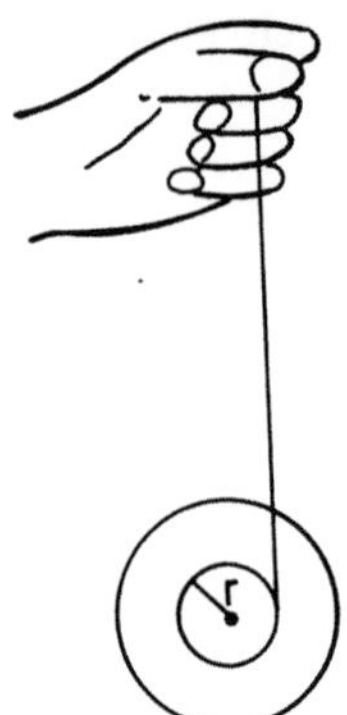

Fig. 3.4.5. The yo-yo

a string or thread is fastened and wound around. If the string is held tight at its free end, the yo-yo falls, turns itself, and unwinds the string (Fig. 3.4.5).

We have not yet considered rigid bodies, but we can already say:

The coordinates of the yo-yo are given by

$z(t)$, its height above the origin and
$\theta(t)$, the angle between a radial mark on the disk and the z-axis.

Let $\theta = 0$ be defined as the angle when the yo-yo is completely unwound, so that a rotation of the yo-yo in any direction will wind it up. If s is the length of the unrolled string, we have

$$s = rf(\theta) \ , \tag{3.4.16}$$

where the angle θ is counted out above 2π, and, if θ is not too close to $\theta = 0$,

$$f(\theta) = \theta + \text{constant} \qquad \text{when unwinding}$$

$$= -\theta + \text{constant} \qquad \text{when winding so that } f'(\theta) = \ \pm 1 \ .$$

If we write the height of the end of the string at time t as $z_0(t)$ and the height of the yo-yo as z, then the constraint reads

$$z = z_0 - rf(\theta) \quad \text{or} \tag{3.4.17}$$

$$F(z, \theta, t) = z + rf(\theta) - z_0(t) = 0 \ . \tag{3.4.18}$$

This is a holonomic rheonomic constraint.

The Lagrangian equation of the first type for the coordinate z is given by

$$M\ddot{z}(t) + Mg = \lambda(t)(\partial F/\partial z) = \lambda(t) \ , \tag{3.4.19}$$

where M is the mass of the yo-yo. For the change in energy, we find

$$\frac{dE}{dt} = -\lambda(t)\frac{\partial F}{\partial t} = \lambda(t)\dot{z}_0(t) \ . \tag{3.4.20}$$

The constraining force must always point upwards if the string remains tight. Thus, $\lambda(t) > 0$, hence $dE/dt > 0$, when the end of the string is moved upwards. $\lambda = 0$ occurs in the case of free fall.

To calculate $\lambda(t)$ more exactly, let us construct Lagrange's equation using the angle θ as a coordinate.

The kinetic energy of the yo-yo is given by (using a result from the next chapter):

$$T = \frac{M}{2}\dot{z}^2 + \frac{I}{2}\dot{\theta}^2 \; . \tag{3.4.21}$$

Here, I is the moment of inertia of the yo-yo about the axis of rotation. If we substitute $z = z_0(t) - rf(\theta)$ and $U = Mgz = Mg[z_0 - rf(\theta)]$, we obtain

$$L(\theta, \dot{\theta}, t) = \frac{M}{2}[\dot{z}_0(t) - r\dot{\theta}f'(\theta)]^2 + \frac{I}{2}\dot{\theta}^2 - Mg[z_0(t) - rf(\theta)] \; , \tag{3.4.22}$$

from which we get the equation of motion

$$\frac{d}{dt}[Mrf'(rf'\dot{\theta} - \dot{z}_0)] + I\ddot{\theta} - Mgrf' + M(\dot{z}_0 - r\dot{\theta}f')f''(\theta)r\dot{\theta} = 0 \; .$$

As long as θ is not in the neighborhood of 0, so that $f' = \pm 1$, this equation becomes

$$(Mr^2 + I)\ddot{\theta}Mgrf' + Mrf'\ddot{z}_0 = Mrf'(g + \ddot{z}_0) \; ; \quad (f' = \pm 1) \; . \tag{3.4.23}$$

This equation for $\theta(t)$ can be solved immediately for a given $z_0(t)$. However, in order to calculate $\lambda(t) = M(\ddot{z} + g)$ we use the fact that since $z = z_0 - rf(\theta)$ for all values of θ outside of a small interval around 0, then

$$\ddot{z} = \ddot{z}_0 - rf'\ddot{\theta} = \ddot{z}_0 - \frac{Mr^2}{Mr^2 + I}(g + \ddot{z}_0) \; . \tag{3.4.24}$$

Substitution, into Lagrange's equation of the first type yields

$$M\ddot{z} + Mg = M\frac{I}{I + Mr^2}(\ddot{z}_0 + g) = \lambda(t) \; . \tag{3.4.25}$$

For $\ddot{z}_0 + g > 0$, it follows again that $\lambda > 0$, as expected.

Further,

$$\frac{dE}{dt} = \lambda\dot{z}_0 = \frac{M}{1 + Mr^2/I}(\ddot{z}_0 + g)\dot{z}_0$$

$$= \frac{M}{1 + Mr^2/I}\frac{d}{dt}(\tfrac{1}{2}\dot{z}_0^2 + gz_0) \; . \tag{3.4.26}$$

The change in energy E is thus equal to the change in energy of a fictional particle of mass $M(1 + Mr^2/I)$ with trajectory $z_0(t)$. Even in the case of a bounded trajectory $z_0(t)$, arbitrarily high energies can be achieved.

What happens at $\theta(t) \approx 0$, when the string is almost all unwound? A more exact analysis shows that there the force of constraint is particularly large. If we then make sure that $\dot{z}_0 = 0$, as long as $\theta \approx 0$, no energy is introduced in this state and the expression for the increase in energy derived above remains valid. It is also possible to use the particularly large constraining force to deliver additional energy to the yo-yo by pulling the string powerfully upwards at the moment when the yo-yo goes through the bottom dead-point.

3.5   Non-holonomic Constraints

Until now, we have always assumed that the constraints are of the form

$$F_\alpha(\underline{z}, t) = 0 \ , \qquad \alpha = 1, \dots, s \ ,$$

where F_α can be a function of the positions of the particles $\underline{z} = (r_1, \dots, r_N)$ and the time t. We have called such constraints holonomic.

Non-holonomic constraints have the general form

$$F_\alpha(\underline{z}, \underline{\dot{z}}, t) = 0 \ . \tag{3.5.1}$$

We will not examine constraints in this general form, but only non-holonomic constraints which are linear in $\dot{q}_j$, and which can thus be written:

$$\sum_{j=1}^{f} a_{kj}(q, t)\dot{q}_j + b_k(q, t) = 0 \ , \qquad k = 1, \dots, s' \ , \tag{3.5.2}$$

where the $q_j, j = 1, \dots, f$ are generalized coordinates. Of course, this equation can also represent in concealed form the holonomic constraints

$$F_k(q_1, \dots, q_f, t) = 0 \ , \qquad k = 1, \dots, s'$$

as we can see if we differentiate:

$$\frac{dF_k}{dt} = \sum_{j=1}^{f} \frac{\partial F_k}{\partial q_j} \dot{q}_j + \frac{\partial F_k}{\partial t} = 0 \ . \tag{3.5.3}$$

If a_{kj} and b_k can be represented as

$$a_{kj} = \frac{\partial F_k}{\partial q_j} \ , \qquad b_k = \frac{\partial F_k}{\partial t} \tag{3.5.4}$$

then the constraints can again be formulated in an explicitly holonomic manner.

Necessary (and essentially sufficient) conditions for this are

$$\frac{\partial a_{ki}}{\partial q_j} = \frac{\partial a_{kj}}{\partial q_i} \ , \qquad \frac{\partial b_k}{\partial q_i} = \frac{\partial a_{ki}}{\partial t} \ . \tag{3.5.5}$$

If true non-holonomic constraints are present, it is of course impossible to find such functions F_k for a given a_{kj} and b_k. In this case, the set of all possible states of the system can no longer be described geometrically as a restriction to a manifold. Non-holonomic constraints then produce further limitations on the virtual displacements, which would otherwise be linearly independent, namely,

$$\sum_{j=1}^{f} a_{kj}\delta q_j = 0 \ , \qquad k = 1,\ldots,s' \ . \tag{3.5.6}$$

This linear dependence between the δq_j at a fixed time can be derived from the somewhat more general conditions on the differentials:

$$\sum_{j=1}^{f} a_{kj}dq_j + b_k dt = 0 \ , \qquad k = 1,\ldots,s' \ , \tag{3.5.7}$$

which, for physically realizable motions ($dq_j = \dot{q}_j dt$), yields equation (3.5.2), but for a fixed time leads to (3.5.6).

In the case of seemingly non-holonomic constraints which are holonomic, we would have

$$a_{kj} = \partial F_k/\partial q_j \quad \text{and thus} \quad \sum_{j=1}^{f} a_{kj}\delta q_j = \sum_{j=1}^{f} \frac{\partial F_k}{\partial q_j}\delta q_j = 0 \ ,$$

which would mean merely that the virtual displacements would be tangent vectors to the surface determined by $F_k(q_1,\ldots,q_f)$.

If now

$$\sum_{j=1}^{f} a_{kj}\delta q_j = 0 \quad \text{and} \tag{3.5.8}$$

$$\underline{f}\cdot\delta\underline{z} = 0 \ , \quad \text{i.e.}$$

$$\sum_{j=1}^{f} \delta q_j \frac{\partial \underline{z}}{\partial q_j}\cdot(\underline{\dot{p}} + \underline{\nabla}U) = 0 \ , \tag{3.5.9}$$

then

$$\sum_{j=1}^{f} \left\{ \delta q_j \frac{\partial \underline{z}}{\partial q_j}\cdot[\underline{\dot{p}} + \underline{\nabla}U(\underline{z},t)] - \sum_{k=1}^{s'} \lambda_k(t)a_{kj}\delta q_j \right\} = 0 \tag{3.5.10}$$

or also

$$\sum_{j=1}^{f} \delta q_j \left[\sum_{i=1}^{N} m_i \ddot{r}_i \cdot \frac{\partial r_i}{\partial q_j} + \nabla_i U(r_1, \ldots, r_N, t) \cdot \frac{\partial r_i}{\partial q_j} - \sum_{k=1}^{s'} \lambda_k a_{kj} \right] = 0 \ . \quad (3.5.11)$$

If we choose, say, the variations $\delta q_1, \ldots, \delta q_{f-s'}$ of the coordinates $q_1, \ldots, q_{f-s'}$ independently, then the other variations, $\delta q_{f-s'+1}, \ldots, \delta q_f$ are determined by the first $f-s'$ variations together with the constraints.

We then choose the $\lambda_k(t)$ such that the coefficients of δq_j for $j = f-s'+1, \ldots, f$ vanish identically. The coefficients of the other δq_j for $j = 1, \ldots, f-s'$ must also vanish in this case, since they are linearly independent. Then we must solve the equations

$$\frac{d}{dt} \frac{\partial L}{\partial \dot{q}_j} - \frac{\partial L}{\partial q_j} - \sum_{k=1}^{s'} \lambda_k(t) a_{kj} = 0 \ , \quad j = 1, \ldots, f \quad\quad (3.5.12a)$$

and

$$\sum_{j=1}^{f} a_{kj} \dot{q}_j + b_k = 0 \ , \quad k = 1, \ldots, s' \ . \quad\quad (3.5.12b)$$

These are $f + s'$ equations for $f + s'$ unknowns.

Thus the constraining forces which guarantee adherence to holonomic constraints have been eliminated, and instead a Lagrangian $L(q_1, \ldots, q_f, \dot{q}_1, \ldots, \dot{q}_f, t)$ has been introduced. But because of the additional non-holonomic constraints, the q_j cannot be varied independently of each other. This restriction can only be observed in the form of Lagrange's equations of the first kind.

We see again, that if the additional conditions are not truly non-holonomic, so that

$$a_{kj} = \frac{\partial F_k}{\partial q_j} \ , \quad b_k = \frac{\partial F_k}{\partial t}$$

then the equations have the form

$$\frac{d}{dt} \frac{\partial L}{\partial \dot{q}_j} - \frac{\partial L}{\partial q_j} - \sum_{k=1}^{s'} \lambda_k(t) \frac{\partial F_k}{\partial q_j} = 0 \ , \quad j = 1, \ldots, f \ ,$$

$$F_k(q_1, \ldots, q_f, t) = 0 \ , \quad k = 1, \ldots, s'$$

corresponding to Lagrange's equation of the first kind for

$$(q_1, \ldots, q_f) = (r_1, \ldots, r_N) \ , \quad s = s' \ , \quad f = 3N \ .$$

An example of a system with a non-holonomic constraint is a sphere which rolls on a plane without gliding. The set of possible positions of the sphere, i.e. the configuration space of the system, is described by the position of the center of the sphere and the spatial orientation of a tripod fixed on the sphere. The constraint is visible if we consider that when the sphere undergoes a small rotation, the corresponding displacement of its center is already determined. On the other hand, by rolling the sphere along suitable curves, it can be brought to any place on the plane with the tripod in any desired orientation. The totality of possible spatial positions of the sphere is thus not restricted by the non-holonomic constraint, rather restrictions apply only to the possible virtual displacements. Carrying out calculations for truly non-holonomic systems is often somewhat complicated.

Here, we will consider two simple examples in which the conditions are actually holonomic, but we will treat them as if they were non-holonomic.

a) Consider an inclined plane in the xz-plane, which slopes downwards and forms an angle φ with respect to the x-axis. Let this axis be moved vertically, and let a body of mass m slide frictionlessly on this plane (Fig. 3.5.1). For the coordinates $(x, z) = r$ of the body, the following relation always holds between the displacements δx and δz:

$$\delta z = -\tan \varphi \, \delta x \, , \tag{3.5.13}$$

since if the inclined plane is fixed, an increase of z corresponds to a decrease in x.

If $h(t)$ is the coordinate of the plane along the z-axis, then we have

$$\dot{z} + \dot{x} \tan \varphi - \dot{h}(t) = 0 \, . \tag{3.5.14}$$

Thus, the constraint is linear in the speed. Of course, this constraint is integrable, i.e. derivable from an equation

$$F(x, z, t) = 0 \quad \text{namely} \, ,$$

$$z + x \tan \varphi - h(t) = 0 \tag{3.5.15}$$

so that in the general formulation

$$\sum_{j=1}^{f} a_{kj} \dot{q}_j + b_k = 0 \, , \quad k = 1, \dots, s' \tag{3.5.16}$$

here,

$$s' = 1, \quad a_{1z} = 1, \quad a_{1x} = \tan \varphi \, , \quad b_1 = -\dot{h}(t) \, , \tag{3.5.17}$$

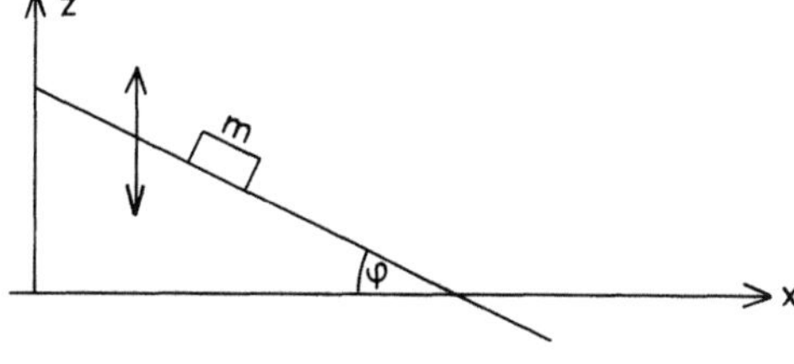

Fig. 3.5.1. A body of mass m on an inclined plane which undergoes vertical motion

and also

$$a_{1z} = \frac{\partial F}{\partial z} \; , \qquad a_{1x} = \frac{\partial F}{\partial x} \; , \qquad b_1 = \frac{\partial F}{\partial t} \tag{3.5.18}$$

with

$$F(x, z, t) = z + x \tan \varphi - h(t) = 0 \; .$$

Then, with x and z as generalized coordinates and the Lagrangian given by $L(x, z, \dot{x}, \dot{z}) = \frac{1}{2}m(\dot{x}^2 + \dot{z}^2) - mgz$, the equations read

$$m\ddot{x} - \lambda \tan \varphi = 0 \; , \tag{3.5.19a}$$

$$m\ddot{z} + mg - \lambda = 0 \; , \quad \text{and} \tag{3.5.19b}$$

$$\dot{z} + \dot{x} \tan \varphi - \dot{h} = 0 \; . \tag{3.5.19c}$$

Here, λ can be eliminated or directly determined by taking suitable linear combinations of the three equations. We want to find λ:

We obtain

$$\lambda(1 + \tan^2 \varphi) = m(\ddot{x} \tan \varphi + \ddot{z}) + mg$$

$$= m(g + \ddot{a}) \; , \tag{3.5.20}$$

so that

$$\lambda(t) = m \cos^2 \varphi(g + \ddot{a}) \; . \tag{3.5.21}$$

The constraining force

$$f = \lambda \frac{\partial F}{\partial r} = \lambda(\tan \varphi, 1) \tag{3.5.22}$$

is perpendicular to the plane and, in particular, vanishes if $\ddot{h} = -g$, that is, when the plane is in "free fall."

b) A further example is a rolling cyclinder on as inclined plane (Fig. 3.5.2).

Here, we introduce as general coordinates for the cylinder s and θ, where s is the distance traveled and θ is the angle through which the cylinder has rotated. As we will find in Chap. 4, the kinetic energy of the cylinder is given by

$$T = \tfrac{1}{2}M\dot{s}^2 + \tfrac{1}{2}I\dot{\theta}^2 \; , \tag{3.5.23}$$

where M is the mass of the cylinder and I is its moment of inertia about its lengthwise axis. The potential energy is given by

$$U = U(s) = -mgs \sin \varphi + \text{constant} \tag{3.5.24}$$

so that the greater s is, the smaller U is.

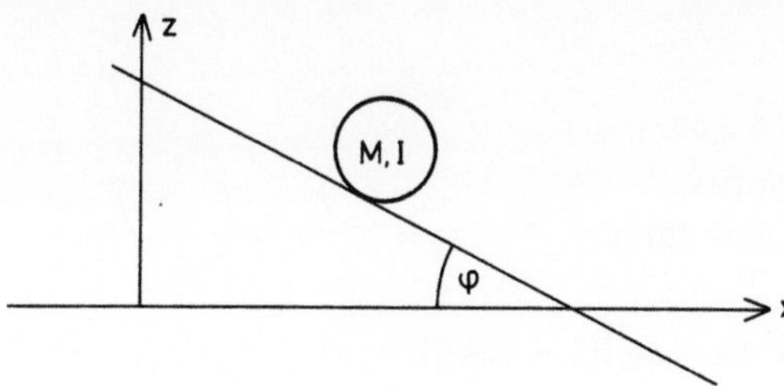

Fig. 3.5.2. A cylinder rolls on an inclined plane

Now we could produce the Lagrangian, but first we have to consider that the generalized coordinates s and θ are not independent. When the cylinder rolls, we have, if r is its radius,

$$s = r\theta \ , \quad \text{hence} \tag{3.5.25}$$

$$\delta s - r\,\delta\theta = 0 \ , \tag{3.5.26}$$

that is, $a_s = 1$, $a_\theta = -r$, and thus we obtain the following equations for the Lagrangian and the motion

$$L(s, \theta, \dot{s}, \dot{\theta}) = \tfrac{1}{2}M\dot{s}^2 + \tfrac{1}{2}I\dot{\theta}^2 + mgs\sin\varphi \ : \tag{3.5.27}$$

$$M\ddot{s} - mg\sin\varphi - \lambda = 0 \ , \tag{3.5.28a}$$

$$I\ddot{\theta} + \lambda r = 0 \ , \tag{3.5.28b}$$

$$\dot{s} = r\dot{\theta} \ . \tag{3.5.28c}$$

If we eliminate λ, then we find the following equation for $s(t)$, since $\ddot{\theta} = \ddot{s}/r$,

$$M\ddot{s} - mg\sin\varphi + I\frac{\ddot{s}}{r^2} = 0 \ . \tag{3.5.29}$$

If we had substituted s/r for θ at the beginning, we would have found the Lagrangian

$$L = \frac{1}{2}M\dot{s}^2 + \frac{1}{2}I\frac{\dot{s}^2}{r^2} + mgs\sin\varphi \tag{3.5.30}$$

and obtained the same equation of motion for $s(t)$.

3.6 Invariants and Conservation Laws

We saw in the example of the spherical pendulum in Sect. 3.3, that a coordinate q_i being cyclic leads immediately to a conservation law. If the Lagrangian is independent of q_i, we have

$$\frac{d}{dt}\frac{\partial L}{\partial \dot{q}_i} - \frac{\partial L}{\partial q_i} = \frac{d}{dt}\frac{\partial L}{\partial \dot{q}_i} = 0 \ , \tag{3.6.1}$$

and $p_i = \partial L/\partial \dot{q}_i$ is a conserved quantity.

In addition, the fact that the Lagrangian

$$L(\theta, \varphi, \dot{\theta}, \dot{\varphi}) = \frac{m}{2}\, l^2(\dot{\theta}^2 + \dot{\varphi}^2 \sin^2\theta) - mgl(1 - \cos\theta)$$

is independent of the variable φ means that it is invariant under rotations about the z-axis, because these rotations specifically change φ and leave θ fixed.

In the following, we want to examine the consequences of such an invariance of the Lagrangian, in relation to the existence of conserved quantities.

As we have done before on occasion, we will write $(q_1(t), \dots, q_f(t))$ as $q(t)$. Consider a collection of trajectories

$$q = q(t, \alpha) \ , \quad q(t, 0) = q(t) \ , \quad \alpha \in \mathbb{R} \ ,$$

be given, such that

$$L(q(t, \alpha), \dot{q}(t, \alpha), t) = L(q(t), \dot{q}(t), t) \ . \tag{3.6.2}$$

Since the left side is independent of α, it follows that

$$\frac{\partial}{\partial \alpha} L(q(t, \alpha), \dot{q}(t, \alpha), t)|_{\alpha=0} = 0 \ , \tag{3.6.3}$$

or, more explicitly,

$$\begin{aligned}
0 &= \sum_{i=1}^{f} \left(\frac{\partial L}{\partial q_i}\frac{\partial q_i}{\partial \alpha} + \frac{\partial L}{\partial \dot{q}_i}\frac{\partial \dot{q}_i}{\partial \alpha} \right)\Bigg|_{\alpha=0} \\
&= \sum_{i=1}^{f} \left(\frac{\partial L}{\partial q_i} - \frac{d}{dt}\frac{\partial L}{\partial \dot{q}_i} \right)\frac{\partial q_i}{\partial \alpha}\Bigg|_{\alpha=0} + \frac{d}{dt}\left(\sum_{i=1}^{f} \frac{\partial L}{\partial \dot{q}_i}\frac{\partial q_i}{\partial \alpha} \right)\Bigg|_{\alpha=0} \\
&= \frac{d}{dt} \sum_{i=1}^{f} \frac{\partial L}{\partial \dot{q}_i}\frac{\partial q_i}{\partial \alpha}\Bigg|_{\alpha=0} \ ,
\end{aligned} \tag{3.6.4}$$

if $q(t)$ is a solution of Lagrange's equation.

Thus, we obtain *Noether's*[8] *theorem:*

If the Lagrangian is invariant under the transformations

$$q_i(t) \mapsto q_i(t, \alpha) \ , \quad i = 1, \dots, f \ ,$$

---

[8] *Noether, Emmi* (*1882, d. 1935). German mathematician, up to now, the greatest female mathematician, important contributions to algebra.

i.e., if

$$L(q(t, \alpha), \dot{q}(t, \alpha), t) = L(q(t), \dot{q}(t), t) \ ,$$

then the quantity

$$\sum_{i=1}^{f} \frac{\partial L}{\partial \dot{q}_i} \tau_i \quad \text{with} \quad \tau_i = \frac{\partial q_i}{\partial \alpha} \bigg|_{\alpha=0}$$

is a constant with respect to time, thus a conserved quantity if $q(t)$ is a solution of Lagrange's equation.

Applications. a) Let L be invariant under *translations*

$$r_i(t) \mapsto r_i(t, \alpha) = r_i(t) + \alpha e \ , \tag{3.6.5}$$

e is a fixed, but arbitrary unit vector.

This is the case, for example, if the potential $U(r_1, \ldots, r_N)$ depends only on position differences $r_i - r_j$, so that

$$L = \frac{1}{2} \sum_{i=1}^{N} m_i \dot{r}_i^2 - U(r_1 - r_2, \ldots, r_i - r_j, \ldots, t) \ .$$

Then, if $(q_1, \ldots, q_f) = (r_1, \ldots, r_N)$,

$$\tau_i = \frac{\partial r_i(t, \alpha)}{\partial \alpha} \bigg|_{\alpha=0} = e \quad \text{and}$$

$$\sum_{j=1}^{N} \frac{\partial L}{\partial \dot{r}_j} \cdot e = \sum_{j=1}^{N} (m_j r_j) \cdot e = P \cdot e \quad \text{with} \tag{3.6.6}$$

$$P = \sum_{j=1}^{N} p_j = \sum_{j=1}^{N} m_j \dot{r}_j \ .$$

Since e was arbitrary, the vector P must be a conserved quantity.

We see the following: The total momentum is a conserved quantity if the "system is translation-invariant", that is, the corresponding Lagrangian is invariant under the translations

$$r_i(t) \mapsto r_i(t, \alpha) = r_i(t) + \alpha e \ .$$

If the invariance holds only for a specific vector e, then only $P \cdot e$, i.e. the e-component of P, is a conserved quantity.

b) Consider rotations of angle α about the n-axis, which is assumed to go through the origin.

If, say, $n = e_3$, then these rotations can be represented as

$$r(\alpha) = \begin{pmatrix} \cos\alpha & -\sin\alpha & 0 \\ \sin\alpha & \cos\alpha & 0 \\ 0 & 0 & 1 \end{pmatrix} \begin{pmatrix} x \\ y \\ z \end{pmatrix} , \qquad (3.6.7)$$

and thus

$$\left.\frac{\partial r}{\partial \alpha}\right|_{\alpha=0} = \begin{pmatrix} 0 & -1 & 0 \\ 1 & 0 & 0 \\ 0 & 0 & 0 \end{pmatrix} \begin{pmatrix} x \\ y \\ z \end{pmatrix} = e_3 \times r . \qquad (3.6.8)$$

In general, for a rotation about the n-axis we have:

$$\left.\frac{\partial r_i(t, \alpha)}{\partial \alpha}\right|_{\alpha=0} = n \times r_i(t) , \qquad (3.6.9)$$

as we will show in Chap. 4. Therefore, we obtain

$$\sum_{i=1}^{N} \frac{\partial L}{\partial \dot{r}_i} \cdot (n \times r_i) = \sum_{i=1}^{N} m_i \dot{r}_i \cdot (n \times r_i)$$

$$= \sum_{i=1}^{N} n \cdot (r_i \times p_i) = n \cdot L . \qquad (3.6.10)$$

The n-component of the total angular momentum

$$L = \sum_{i=1}^{N} L_i = \sum_{i=1}^{N} r_i \times p_i \qquad (3.6.11)$$

is thus a conserved quantity.

In our example from Sect. 3.3, the Lagrangian was invariant under rotations about the 3-axis, so that in this case $e_3 \cdot L = L_3$ is a conserved quantity.

From this argument, we see that the conservation of angular momentum is a result of the rotational invariance of a system.

If, in general, the potential energy $U(r_1, \ldots, r_N, t)$ depends only on $|r_i - r_j|$, then the Lagrangian is invariant under any rotation, because distances remain fixed under a rotation, along with the scalar products $\dot{r}_i^2$ in the kinetic energy. Thus, the angular momentum L will be a conserved quantity.

c) In several cases, the Lagrangian L is not invariant, but rather

$$\left.\frac{\partial}{\partial \alpha} L(q(t, \alpha), \dot{q}(t, \alpha), t)\right|_{\alpha=0} = \frac{d}{dt} f(q(t), \dot{q}(t), t) . \qquad (3.6.12)$$

Then, obviously

$$\sum_{i=1}^{N} \frac{\partial L}{\partial \dot{q}_i} \tau_i - f(q, \dot{q}, t) \tag{3.6.13}$$

is a conserved quantity.

Application. c1) We consider time translations

$$t \mapsto t + \alpha \ , \quad \text{i.e.} \quad q(t, \alpha) = q(t + \alpha) \ .$$

Then,

$$\frac{d}{d\alpha} L(q(t + \alpha), \dot{q}(t + \alpha), t)\bigg|_{\alpha=0} = \sum_{i=1}^{f} \left(\frac{\partial L}{\partial q_i} \dot{q}_i + \frac{\partial L}{\partial \dot{q}_i} \ddot{q}_i \right)\bigg|_{\alpha=0} \ . \tag{3.6.14}$$

On the other hand, we also have

$$\frac{d}{dt} L(q(t), \dot{q}(t), t) = \frac{\partial L}{\partial t} + \sum_{i=1}^{f} \left(\frac{\partial L}{\partial q_i} \dot{q}_i + \frac{\partial L}{\partial \dot{q}_i} \ddot{q}_i \right) \ . \tag{3.6.15}$$

If the Lagrangian is not explicitly dependent on t, we have

$$\frac{d}{d\alpha} L(q(t + \alpha), \dot{q}(t + \alpha))\bigg|_{\alpha=0} = \frac{d}{dt} L(q(t), \dot{q}(t)) \ ,$$

so that

$$\sum_{i=1}^{f} \frac{\partial L}{\partial \dot{q}_i} \dot{q}_i - L(q, \dot{q}) \tag{3.6.16}$$

is a conserved quantity.

The meaning of this conserved quantity is apparent in Cartesian coordinates. Here,

$$\sum_{i=1}^{N} \frac{\partial L}{\partial \dot{r}_i} \dot{r}_i - L = \sum_{i=1}^{N} m_i \dot{r}_i^2 - \frac{1}{2} \sum_{i=1}^{N} m_i \dot{r}_i^2 + U(r_1, \ldots, r_N)$$

$$= T + U = E \ . \tag{3.6.17}$$

Thus the conservation of energy is a consequence of invariance under time translations.

In general, with $r_i = r_i(q_1, \ldots, q_f)$, i.e. for holonomic, scleronomic constraints:

$$T = \frac{1}{2} \sum_{i=1}^{N} m_i \dot{r}_i^2 = \frac{1}{2} \sum_{i=1}^{N} m_i \sum_{k, j=1}^{f} \frac{\partial r_i}{\partial q_k} \cdot \frac{\partial r_i}{\partial q_j} \dot{q}_k \dot{q}_j = \frac{1}{2} \sum_{k, j=1}^{f} g_{kj} \dot{q}_k \dot{q}_j \tag{3.6.18}$$

with

$$g_{kj} = \sum_{i=1}^{N} m_i \frac{\partial \mathbf{r}_i}{\partial q_k} \cdot \frac{\partial \mathbf{r}_i}{\partial q_j} \; . \tag{3.6.19}$$

Then, if U does not depend on $\dot{q}$

$$\sum_{i=1}^{f} \frac{\partial L}{\partial \dot{q}_i} \dot{q}_i = \sum_{i=1}^{f} \frac{\partial T}{\partial \dot{q}_i} \dot{q}_i = \sum_{k,j=1}^{f} g_{kj} \dot{q}_k \dot{q}_j = 2T \; , \tag{3.6.20}$$

and again we find that the conserved quantity

$$\sum_{i=1}^{f} \frac{\partial L}{\partial \dot{q}_i} \dot{q}_i - L = 2T - T + U = T + U = E \tag{3.6.21}$$

is the energy, expressed in terms of q and $\dot{q}$.

c2) We consider the transformation to a system of reference in uniform linear motion

$$\mathbf{r}_i(t, \alpha) = \mathbf{r}_i(t) + \alpha \mathbf{v}_0 t \; . \tag{3.6.22}$$

Then

$$\left. \frac{\partial \mathbf{r}_i}{\partial \alpha} \right|_{\alpha=0} = \boldsymbol{\tau}_i = \mathbf{v}_0 t \; .$$

This transformation is called *the special Galilean transformation.*

If the Lagrangian is invariant under spatial translations, then

$$L(\mathbf{r}_1(t, \alpha), \ldots, \mathbf{r}_N(t, \alpha), t) = \sum_{i=1}^{N} \frac{1}{2} m_i (\dot{\mathbf{r}}_i + \alpha \mathbf{v}_0)^2 - U(\mathbf{r}_1, \ldots, \mathbf{r}_N, t) \; , \tag{3.6.23}$$

i.e., the α-dependence enters only into the kinetic energy term. Then

$$\left. \frac{dL}{d\alpha} \right|_{\alpha=0} = \sum_{i=1}^{N} m_i \dot{\mathbf{r}}_i \cdot \mathbf{v}_0 = \frac{d}{dt} \left(\mathbf{v}_0 \cdot \sum_{i=1}^{N} m_i \mathbf{r}_i \right) \; .$$

Thus: The quantity

$$\sum_{i=1}^{N} m_i \dot{\mathbf{r}}_i \cdot \mathbf{v}_0 t - \mathbf{v}_0 \cdot \sum_{i=1}^{N} m_i \mathbf{r}_i = \mathbf{v}_0 \cdot (\mathbf{P}t - M\mathbf{R}) \quad \text{with}$$

$$\mathbf{R} = \frac{1}{M} \sum_{i=1}^{N} m_i \mathbf{r}_i \tag{3.6.24}$$

is a constant in time.

Since $\boldsymbol{v}_0$ is arbitrary, it follows that the center of mass vector $\boldsymbol{R}$ satisfies

$$\boldsymbol{R}(t) = \frac{\boldsymbol{P}}{M} t + \boldsymbol{R}_0 \ . \tag{3.6.25}$$

With the given form of the kinetic energy, the invariance of the potential energy under spatial translations leads to the quasi-invariance under special Galilean transformations and thus to the uniform linear motion of the center of mass, which also follows from the conservation of total momentum.

The connection between symmetry and conservation laws is of fundamental importance in all of physics. Especially in particle physics, conservation laws provide a framework for the construction of theories. In this field, experimental observations of the scattering of (elementary) particles lead to conservation laws for electric charge, baryon number, isospin, etc., and the theory must be built in such a way that it has these laws as a consequence. The "Lagrangian" which defines the theory,which is then a "function" of fields, must be invariant under the corresponding transformations of the fields.

3.7 The Hamiltonian

In this section, we will describe the Hamiltonian formulation of mechanics. This is the point of departure for most advanced applications of theoretical mechanics and for the transition to quantum mechanics.

3.7.1 Lagrange's Equations and Hamilton's Equations

The Lagrangian $L(q, \dot{q}, t)$ is a function of generalized coordinates and their time derivatives. In these variables, the equations of motions read

$$\frac{d}{dt} \frac{\partial L}{\partial \dot{q}_i} - \frac{\partial L}{\partial q_i} = 0 \ , \quad i = 1, \ldots, f \ .$$

The generalized momenta

$$p_i = \frac{\partial L}{\partial \dot{q}_i} = p_i(q, \dot{q}, t) \tag{3.7.1}$$

are also functions of q, $\dot{q}$, and t.

We now assume that these equations for p_i can be solved for $\dot{q}_i$ for arbitrary fixed values of q_i and t, so that the $\dot{q}_i$ can be expressed as functions of the generalized coordinates q and momenta p:

$$\dot{q} = \dot{q}(q, p, t) \ .$$

In Cartesian coordinates, this is certainly true, as we have

$$p_i = \frac{\partial L}{\partial \dot{r}_i} = m_i \dot{r}_i \quad \text{and thus}$$

$$\dot{r}_i = \frac{p_i}{m_i} .$$

We now construct the *Hamiltonian*[9]

$$H(q, p, t) = \sum_{i=1}^{f} p_i \dot{q}_i(q, p, t) - L(q, \dot{q}(q, p, t), t) . \tag{3.7.2}$$

In Cartesian coordinates with

$$L = \frac{1}{2} \sum_{i=1}^{N} m_i \dot{r}_i^2 - U(r_1, \ldots, r_N)$$

then

$$H(r_1, \ldots, r_N, p_1, \ldots, p_N) = \sum_{i=1}^{N} p_i \cdot \dot{r}_i - L$$

$$= \sum_{i=1}^{N} \frac{p_i^2}{2m_i} + U(r_1, \ldots, r_N) \tag{3.7.3}$$

is the energy, expressed in terms of p_i and r_i.

We will now prove:

Lagrange's equations are equivalent to Hamilton's equations:

$$\frac{\partial H}{\partial p_i}(q, p, t) = \dot{q}_i(t) , \tag{3.7.4}$$

$$\frac{\partial H}{\partial q_i}(q, p, t) = -\dot{p}_i(t) . \tag{3.7.5}$$

The Hamiltonian thus determines the time dependence of the coordinates and the momenta.

---

[9] *Hamilton, Sir William Rowan* (*1805 Dublin, d. 1865 Dunsink). Irish physicist and mathematician. Hamilton's formulation of mechanics was introduced in 1835 as an extension of his fundamental theory of geometrical optics (1827). In geometrical optics, Fermat's principle appears in place of the principle of least action. In his lifetime, he was known especially as the discoverer of conic refraction.

Proof. From the definition of H,

$$\frac{\partial H}{\partial p_j} = \dot{q}_j(q, p, t) + \sum_{i=1}^{f} \left(p_i \frac{\partial \dot{q}_i}{\partial p_j} - \frac{\partial L}{\partial \dot{q}_i} \frac{\partial \dot{q}_i}{\partial p_j} \right) = \dot{q}_j \, , \tag{3.7.6}$$

since $p_i = \partial L / \partial \dot{q}_i$. Further,

$$\frac{\partial H}{\partial q_j} = \sum_{i=1}^{f} p_i \frac{\partial \dot{q}_i}{\partial q_j} - \frac{\partial L}{\partial q_j} - \sum_{i=1}^{f} \frac{\partial L}{\partial \dot{q}_i} \frac{\partial \dot{q}_i}{\partial q_j} = - \frac{\partial L}{\partial q_j} \, . \tag{3.7.7}$$

From Lagrange's equations, it then follows that

$$\frac{\partial H}{\partial q_j} = - \frac{d}{dt} \frac{\partial L}{\partial \dot{q}_j} = - \dot{p}_j \, . \tag{3.7.8}$$

Thus we have derived Hamilton's equations from Lagrange's equations.
Conversely, from Hamilton's equations and

$$L(q, \dot{q}, t) = \sum_{i=1}^{f} p_i(q, \dot{q}, t) \dot{q}_i - H(q, p(q, \dot{q}, t), t) \tag{3.7.9}$$

it follows analogously that

$$\frac{\partial L}{\partial q_j} = - \frac{\partial H}{\partial q_j} \quad \text{and} \quad \frac{\partial L}{\partial \dot{q}_j} = p_j \, ,$$

and therefore

$$\dot{p}_j \equiv \frac{d}{dt} \frac{\partial L}{\partial \dot{q}_j} = - \frac{\partial H}{\partial q_j} = \frac{\partial L}{\partial q_j} \, , \tag{3.7.10}$$

and thus we arrive at Lagrange's equations.

We have just proved the equivalence of Hamilton's and Lagrange's equations.

A system with f degrees of freedom can therefore be described by the $2f$ coordinates and momenta

$$q_1, \ldots, q_f \, , \quad p_1, \ldots, p_f \, .$$

The space of these $2f$ coordinates and momenta is called *phase space*. We already introduced this designation for N-particle systems described by Cartesian coordinates. The time evolution of the system is given by the phase space trajectories

$$t \mapsto (q_1(t), \ldots, q_f(t), p_1(t), \ldots, p_f(t))$$

and is determined by the solution of Hamilton's equations, which constitute a system of $2f$ ordinary differential equations of first order. As initial conditions, we assume that the $2f$ values

$$q_1, \ldots, q_f , \quad p_1, \ldots, p_f \quad \text{for} \quad t = t_0$$

are given, so that the number of initial values is exactly the same as for Lagrange's equations which use the values

$$q_1, \ldots, q_f , \quad \dot{q}_1, \ldots, \dot{q}_f \quad \text{for} \quad t = t_0$$

Through every point of phase space, there is exactly one trajectory which is a solution of Hamilton's equations. Thus the state of a system of N particles with f degrees of freedom can be uniquely characterized by a single corresponding point in phase space. In configuration space, the position of all particles at any time can be represented, but not the velocities of the individual particles.

Example. For a particle with one degree of freedom, phase space is two-dimensional. We have

$$L = \frac{m}{2} \dot{q}^2 - U(q) \quad \text{and} \tag{3.7.11}$$

$$\dot{q} = p/m \quad \text{thus} \tag{3.7.12}$$

$$H = \frac{p^2}{2m} + U(q) . \tag{3.7.13}$$

If the potential is given by, say, $U(q) = \frac{1}{2} m\omega^2 q^2$, then

$$H = \frac{p^2}{2m} + \frac{1}{2} m\omega^2 q^2 , \tag{3.7.14}$$

and the equations of motion for $q(t)$ and $p(t)$ are

$$\dot{q} = \frac{\partial H}{\partial p} = \frac{p}{m} , \quad \dot{p} = -\frac{\partial H}{\partial q} = -m\omega^2 q \equiv -\frac{\partial U}{\partial q}$$

or again

$$m\ddot{q} + \frac{\partial U}{\partial q} = 0 , \tag{3.7.15}$$

so that here

$$\ddot{q} + \omega^2 q = 0 . \tag{3.7.16}$$

Here, the Hamiltonian is identical to the energy expressed in terms of q and p. All of the states of the system in phase space for a fixed value of E lie on an ellipse given by the equation

$$E = \frac{p^2}{2m} + \frac{1}{2}m\omega^2 q^2$$

with the semiaxes

$$a = \sqrt{\frac{2E}{m\omega^2}} \,, \qquad b = \sqrt{2mE} \,.$$

Since energy is a conserved quantity, the trajectories in phase space cannot leave the ellipse.

The time dependence of any function of the generalized coordinates and momenta is in general given by

$$\frac{d}{dt} A(q(t), p(t), t) = \frac{\partial A}{\partial t} + \sum_{i=1}^{f} \left(\frac{\partial A}{\partial q_i} \dot{q}_i + \frac{\partial A}{\partial p_i} \dot{p}_i \right)$$

$$= \frac{\partial A}{\partial t} + \sum_{i=1}^{f} \left(\frac{\partial A}{\partial q_i} \frac{\partial H}{\partial p_i} - \frac{\partial A}{\partial p_i} \frac{\partial H}{\partial q_i} \right) . \tag{3.7.17}$$

The sum on the right side of the equation is abbreviated

$$\{A, H\} = \sum_{i=1}^{f} \left(\frac{\partial A}{\partial q_i} \frac{\partial H}{\partial p_i} - \frac{\partial A}{\partial p_i} \frac{\partial H}{\partial q_i} \right) \tag{3.7.18}$$

and $\{,\}$ is called the *Poisson*[10] *bracket*.

For a quantity $A(q, p)$, it is therefore true that:

A is a conserved quantity if and only if its Poisson bracket $\{A, H\}$ with the Hamiltonian vanishes. If the potential is, for example, rotationally symmetric, then $\{L, H\} = 0$, and thus L is a conserved quantity.

It is easy to show the following identities for the Poisson bracket:

a) $\{A, B\} = -\{B, A\}$, $\qquad$ (3.7.19)

b) $\{A, B + C\} = \{A, B\} + \{A, C\}$, $\qquad$ (3.7.20)

c) $\{A, BC\} = \{A, B\}C + \{A, C\}B$, $\qquad$ (3.7.21)

d) $\{A, \{B, C\}\} + \{B, \{C, A\}\} + \{C, \{A, B\}\} = 0$. $\qquad$ (3.7.22)

---

[10] *Poisson, Siméon-Denis* (*1781 Pithiviers/Loiret, d. 1840 Sceaux). French physicist and mathematician, professor at the Ecole Polytechnique from 1806. Contributions to celestial mechanics, theory of electricity, thermodynamics, probability theory, and differential geometry.

Relation (d) is called the *Jacobi*[11] *identity*. If A and B are conserved quantities, so that $\{H, A\} = \{H, B\} = 0$, it follows from the Jacobi identity that $\{H, \{A, B\}\} = -\{A, \{B, H\}\} - \{B, \{H, A\}\} = 0$. Thus, if A and B are conserved, then so is $\{A, B\}$.

3.7.2 Aside on the Further Development of Theoretical Mechanics and the Theory of Dynamical Systems

In this section, we will briefly state some of the problems which arise when theoretical mechanics is taken further or generalized.

i) In classical mechanics, besides the Newtonian and Lagrangian formulation there is also the Hamiltonian formulation. In this formulation, the first step is to find the proper Hamiltonian which corresponds to the physical situation. In order to solve Hamilton's equations, so-called *canonical*[12] *transformations* $(q, p, H) \mapsto (Q, P, K)$ are often used. These transformations leave the Poisson brackets and Hamilton's equations invariant. Methods exist to classify these transformations and, under certain circumstances, to calculate one in which the new Hamiltonian $K(Q, P)$ vanishes. The Hamiltonian formulation is also the starting point for *perturbation theory* in classical mechanics.

For our purposes, acquaintance with the Hamiltonian and Hamilton's equations is sufficient.

ii) In statistical mechanics, in which we attempt to calculate the characteristics of macroscopic bodies as a consequence of their microscopic constituent particles, the microscopic interactions are defined through a Hamiltonian. This of course assumes that such microscopic interactions can be described in the framework of classical mechanics (this is only true for limiting cases). This procedure leads to classical statistical mechanics, which we will discuss later (Chap. 7).

iii) The construction of the fundamental equation of quantum mechanics, namely, the Schrödinger equation, is also done using the Hamiltonian. In this construction, the quantities $q, p, H(q, p)$ are made into operators which act in a so-called Hilbert space. The Poisson bracket in classical mechanics corresponds in quantum mechanics to the commutator $[A, B] := AB - BA$. In quantum mechanics, if two quantities A and B commute as operators, i.e.,

$$[A, B] = 0 \ ,$$

then they can both be measured sharply at the same time.

---

[11] *Jacobi, Carl Gustav Jakob* (*1804 Potsdam, d. 1851 Berlin). One of the greatest mathematicians of the nineteenth century. Important contributions to algebra, elliptic function theory, partial differential equations, mechanics, and celestial mechanics.

[12] canonical (Greek, Latin) canon: rule, principle; exemplary, model, standard.

iv) Systems of differential equations of the form

$$\dot{x}_i = F_i(x_1, \ldots, x_n) \;, \qquad i = 1, \ldots, n \;,$$

are called *dynamical systems*. They are basic for the description of processes in fields as diverse as mechanics, irreversible thermodynamics, the theory of chemical reactions, population theory, or sociology [Haken, Lichtenberg et al., Schuster].

Hamilton's equations, with a time-independent Hamiltonian, form a special type of dynamical system, in which the following properties hold

$$n = 2f \;, \qquad x_i = q_i$$

and $\quad x_{i+f} = p_i \quad$ for $\quad i = 1, \ldots, f \;,$

as well as

$$F_i = \frac{\partial H}{\partial p_i} \quad \text{and} \quad F_{i+f} = -\frac{\partial H}{\partial q_i}, \qquad i = 1, \ldots, f \;.$$

Recently, much attention has been given to the investigation of properties of dynamical systems. We will outline a few of the important questions:

– *Critical points* are points x^0, for which

$$F_i(x^0) = 0 \;, \qquad i = 1, \ldots, n \;.$$

The critical points of a Hamiltonian system with Hamiltonian

$$H(p, q) = \sum_{i=1}^{N} \frac{p_i^2}{2m} + U(q)$$

are clearly the equilibrium points of the system.

It is possible to describe and classify the different forms of critical points as well as the behavior of trajectories in their vicinity.

– If there are s independently conserved quantities given by the continuously differentiable functions $G_1(x), \ldots, G_s(x)$, then the trajectories always remain on $(n - s)$-dimensional submanifolds of phase space, which are determined by the initial values. In order to discuss the dynamical system, the conserved quantities need to be found. It can be shown, for example, that a *fully integrable Hamiltonian system* is characterized by the existence of f independent conserved quantities, which satisfy $\{G_i, G_j\} = 0 \; (i, j = 1, \ldots, f)$.

– It is interesting to analyze the behavior of a system in phase space (or on the submanifolds mentioned above) for large values of t. Among other phenomena, there are the following (and these do not exclude one another):

a) Ergodic[13] **Behavior:** The trajectory comes arbitrarily close to each point of the submanifold. The behavior of a Hamiltonian system on the energy plane $H = E = $ constant is certainly non-ergodic, for example, if other independent constants of motion exist. Hamiltonian systems with many degrees in freedom, as they appear in statistical mechanics, are assumed in general to be ergodic.

b) Periodic Behavior

c) Quasi-periodic Behavior: Superposition of periodic motions with incommensurable periods.

d) Chaotic Behavior: Complicated, seemingly irregular motion which is neither periodic nor quasi-periodic, with a very sensitive dependence on the initial conditions. The investigation of this form of motion is currently an active area of research and is important, for example, in the theory of turbulence in hydrodynamics.

e) Attractors: These are sets of points, which the trajectories approach more and more closely after long periods of time. There can be several attractors, and which one the trajectory approaches will depend on the initial conditions. The so called *domains of attraction* of the attractors can penetrate each other in a complicated manner. The system with $\dot{q} = p, \dot{p} + \alpha p = 0$ has the attractor $\{(q, p)|p = 0\}$ and a damped pendulum has the single-point attractor $\{(q, p)|(q, p) = (0, 0)\}$. For two-dimensional systems $(n = 2)$ with compact phase space in $\mathbb{R}^2$, all possible attractors are known. There can be (Fig. 3.7.1):

α) *Asymptotically Stable Limit Points:* The trajectories for large values of t move towards a point.

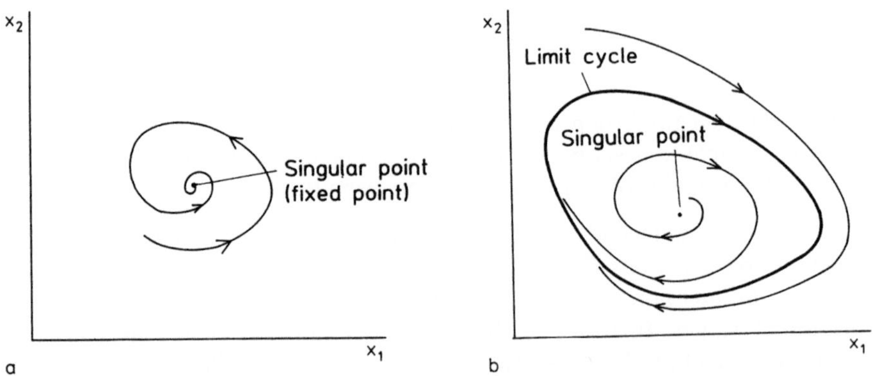

Fig. 3.7.1a, b. Attractors for systems with two degrees of freedom. (**a**) Fixed point, (**b**) limit cycle

---

[13] ergodic (Greek) from *ergon*: work. A mechanical system is called ergodic if, for states with a given energy, the time average is equal to the microcanonical average. See Chap. 7 for more details.

β) *Limit Cycles*: The trajectories for large values of t move towards a compact one-dimensional subset. The behavior approaches periodic behavior with given period.

For higher dimensions, there are entirely new phenomena.

The behavior of the attractors can nonetheless be examined.

f) Volume Distortion: An interesting characteristic of dynamical systems, which can be investigated with simple methods, is their distortion of volumes.

By this is meant the following:

We consider at time t a region W_t in the phase space of the dynamical system. At the time $t + \tau$, each point $x(t) \in W_t$ will have been transformed to point $x(t + \tau) = g(\tau, x(t))$ and thus from the region W_t, a new region $W_{t+\tau}$ will have been formed (Fig. 3.7.2).

We now want to compare the volumes V_t and $V_{t+\tau}$ of the regions W_t and $W_{t+\tau}$. Clearly, we have

$$V_t = \int_{W_t} d^n x(t) \quad \text{and} \tag{3.7.23}$$

$$V_{t+\tau} = \int_{W_{t+\tau}} d^n x(t + \tau) \ . \tag{3.7.24}$$

By introducing the quantities $x(t)$ as coordinates for $W_{t+\tau}$, we obtain the well-known formula for coordinate transformations in volume integrals

$$V_{t+\tau} = \int_{W_t} d^n x(t) \det\left[\frac{\partial g_i(\tau, x(t))}{\partial x_j(t)} \right] \ . \tag{3.7.25}$$

Thus, the determinant

$$D(t, \tau) = \det\left(\frac{\partial g_i}{\partial x_j} \right) \ . \tag{3.7.26}$$

is of paramount importance for the comparison of volumes.

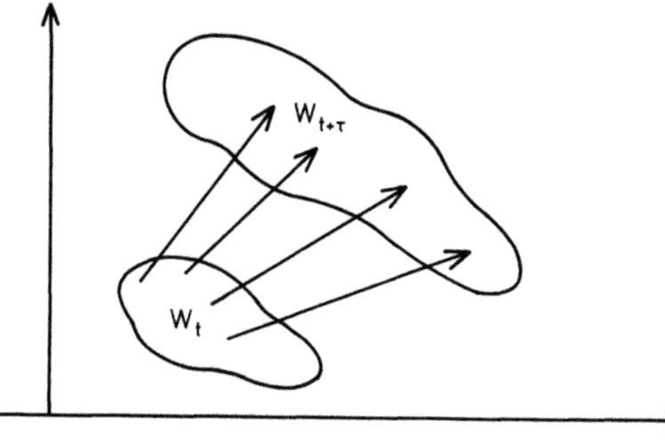

Fig. 3.7.2. A region W_t of points in phase space changes with time

Now, a Hamiltonian system must satisfy *Liouville's*[14] *Theorem*:

The volume of an arbitrary region in phase space stays constant as it evolves in time.

Liouville's Theorem is fundamental for the theory of statistical mechanics, and it reveals an interesting but hidden state of affairs which is made visible only in the Hamiltonian formulation of mechanics.

Although the volume of a region in phase space does not change, its form can become so complex (frayed, so to speak) over time that for large values of τ, every point in phase space will be in the neighborhood of some point of $W_{t+\tau}$.

To prove Liouville's Theorem, we calculate

$$\frac{d}{dt} V_t = \frac{d}{d\tau} V_{t+\tau}\bigg|_{\tau=0} = \int_{W_t} d^n x(t) \frac{\partial D(t, \tau)}{\partial \tau}\bigg|_{\tau=0} . \tag{3.7.27}$$

To evaluate the integrand, we must expand $D(t, \tau)$ (for an arbitrary t) to first order in τ for small values of τ:

From the equation of motion

$$\dot{x}_i(t) = F_i(t, x(t))$$

we have for an arbitrary dynamical system

$$x_i(t + \tau) = g_i(\tau, x(t))$$

$$= x_i(t) + \tau F_i(t, x(t)) + O(\tau^2) \tag{3.7.28}$$

and

$$\frac{\partial g_i}{\partial x_j} = \delta_{ij} + \tau \frac{\partial F_i(t, x(t))}{\partial x_j(t)} + O(\tau^2) . \tag{3.7.29}$$

If we use the fact that for a non-singular matrix A

$$\ln\left[\det(A)\right] = \text{tr}\left[\ln(A)\right] , \tag{3.7.30}$$

we find

$$D = \det\left(\frac{\partial g_i}{\partial x_j}\right) = 1 + \tau \sum_{i=1}^{n} \frac{\partial F_i(t, x(t))}{\partial x_i} + O(\tau^2) \tag{3.7.31}$$

and

$$\dot{D} = \frac{\partial}{\partial \tau} D\bigg|_{\tau=0} = \sum_{i=1}^{n} \frac{\partial F_i}{\partial x_i} . \tag{3.7.32}$$

---

[14] *Liouville, Joseph* (*1809 St.-Omer, d. 1882 Paris). French mathematician. Important works on function theory, analysis, differential geometry, number theory, and statistical mechanics.

The sign of $\dot{D}$ determines whether V_t increases or decreases at time t. If the dynamical system is a Hamiltonian system, then F_i has the special form

$$F = \left(\frac{\partial H}{\partial p_1}, \ldots, \frac{\partial H}{\partial p_f}, -\frac{\partial H}{\partial q_1}, \ldots, -\frac{\partial H}{\partial q_f} \right) . \tag{3.7.33}$$

For a Hamiltonian system, we then have

$$\sum_{i=1}^{n=2f} \frac{\partial F_i}{\partial x_i} = \sum_{i=1}^{f} \left(\frac{\partial^2 H}{\partial q_i \partial p_i} - \frac{\partial^2 H}{\partial p_i \partial q_i} \right) = 0 \tag{3.7.34}$$

for all times t, thus: $V_t = $ constant for all t, and thus we have proved Liouville's theorem.

One direct consequence of Liouville's theorem is that two-dimensional Hamiltonian systems ($n = 2f = 2$) cannot have asymptotically stable boundary points or boundary cycles. In particular, the systems with friction discussed above are clearly not Hamiltonian.

3.8   The Hamiltonian Principle of Stationary Action

In this section we will briefly present another very important interpretation of Lagrange's equations.

To every piece γ of a trajectory $\underline{z}(t)$, $t_1 \leq t \leq t_2$, for a given Lagrangian we will assign a value for an *action* $S[\gamma] \in \mathbb{R}$. It will turn out that the trajectories which satisfy Lagrange's equations can be characterized as having stationary values of action.

3.8.1   Functionals and Functional Derivatives

First we will explain the concept of a *functional* and its derivatives:

Functionals are mappings which are defined on sets of functions. For our considerations, sets of trajectories on an interval $[t_1, t_2]$ are especially important. Each such trajectory assigns points $x(t) \in \mathbb{R}^n$ in an n-dimensional vector space to times t with $t_1 \leq t \leq t_2$. In particular, we consider the following sets of trajectories:

B: The set of all smooth trajectories $\gamma \colon [t_1, t_2] \to \mathbb{R}^n, t \mapsto x(t)$
$B_{x_1 x_2}$: Set of all trajectories from B with fixed endpoints $x(t_1) = x_1, x(t_2) = x_2$.

We now examine functionals

$$F \colon B \to \mathbb{R} \quad \text{or} \quad F \colon B_{x_1 x_2} \to \mathbb{R}, \quad \gamma \mapsto F[\gamma].$$

Examples of functionals:

i) $F[\gamma] = ax(t_0)$, (for $n = 1$)   (3.8.1)

ii) $F[\gamma] = \dot{x}^2(t_0)$,   (3.8.2)

iii) $F[\gamma] = \int\limits_{t_1}^{t_2} dt f(t) x(t)$,   (3.8.3)

iv) $\mathscr{L}[\gamma] = \int\limits_{t_1}^{t_2} dt \sqrt{\dot{x}^2(t)}$ (arc length)   (3.8.4)

v) $S[\gamma] = \int\limits_{t_1}^{t_2} dt \left[\sum\limits_{i=1}^{N} \frac{1}{2} m_i \dot{x}_i^2(t) \right.$

$\left. - V(x_1(t), \ldots, x_N(t)) \right]$ (also called the *action*)   (3.8.5)

vi) $A[\gamma] = \int\limits_{t_1}^{t_2} dt L(x(t), \dot{x}(t), t)$.   (3.8.6)

(vi) is a generalization of (iii–v). In this case, the (continuously differentiable) function $L: (x, \dot{x}, t) \mapsto L(x, \dot{x}, t)$ is called a *Lagrangian*).

The functionals (i) and (ii) are local, that is, $F[\gamma]$ depends only on the behavior of γ in an arbitrarily small neighborhood of a point in time t_0, whereas the other functions are non-local, that is, $F[\gamma]$ can depend on the entire behavior of γ. The functionals (i) and (iii) are linear, that is $F[c_1\gamma_1 + c_2\gamma_2] = c_1 F[\gamma_1] + c_2 F[\gamma_2], c_1, c_2 \in \mathbb{R}$.

Exactly as in the case of normal functions, we can define continuity and differentiability for functionals on arbitrary normed spaces.

F is called continuous at the "point" γ_0 if the following condition is satisfied: For every $\varepsilon > 0$ there is a $\delta > 0$ such that

$$|F[\gamma_0 + h] - F[\gamma_0]| < \varepsilon \quad \text{for} \quad \|h\| < \delta \ . \tag{3.8.7}$$

A continuous functional F is called differentiable at the "point" γ_0 if there is a linear functional $F'[\gamma_0]$ such that

$$F[\gamma_0 + h] - F[\gamma_0] = F'[\gamma_0]h + O(\|h\|^2) \ . \tag{3.8.8}$$

(The linearity of $F'[\gamma_0]$ means, of course, that

$$F'[\gamma_0](c_1 h_1 + c_2 h_2) = c_1 F'[\gamma_0]h_1 + c_2 F'[\gamma_0]h_2 \ . \tag{3.8.9}$$

This is often written

$$\delta F[\gamma_0] = F'[\gamma_0]\delta x + O(\|\delta x\|^2) \ . \tag{3.8.10}$$

We now must define the norm $\|h\|$. There are many possibilities to consider:

$$\|h\|_1 = \max_{t_1 \le t \le t_2} |h(t)| \ , \tag{3.8.11}$$

$$\|h\|_2 = \max_{t_1 \le t \le t_2} |h(t)| + \max_{t_1 \le t \le t_2} |\dot{h}(t)| \ , \tag{3.8.12}$$

$$\|h\|_3 = \int_{t_1}^{t_2} dt |h(t)| \ , \tag{3.8.13}$$

In general, we will use the norm $\|h\|_2$.

We then obtain the following important theorem:

The functional

$$A[\gamma] := \int_{t_1}^{t_2} dt L(x(t), \dot{x}(t), t) \tag{3.8.14}$$

is continuous and differentiable for all $\gamma \in B$. The derivative at the point γ is given by the linear functional

$$A'[\gamma]h = \left[\frac{\partial L}{\partial \dot{x}}(x(t), \dot{x}(t), t) \right] \cdot h(t) \Big|_{t_1}^{t_2}$$

$$+ \int_{t_1}^{t_2} dt \left[\frac{\partial L}{\partial x}(x, \dot{x}, t) - \frac{d}{dt}\frac{\partial L}{\partial \dot{x}}(x, \dot{x}, t) \right] \cdot h(t) \ . \tag{3.8.15}$$

The proof is simple:

$$A[\gamma + h] = \int_{t_1}^{t_2} dt L(x(t) + h(t), \dot{x}(t) + \dot{h}(t), t)$$

$$= A[\gamma] + \int_{t_1}^{t_2} dt \left(\frac{\partial L}{\partial x} \cdot h + \frac{\partial L}{\partial \dot{x}} \cdot \dot{h} \right) + O(\|h\|^2)$$

$$= A[\gamma] + \int_{t_1}^{t_2} dt \left[\left(\frac{\partial L}{\partial x} - \frac{d}{dt}\frac{\partial L}{\partial \dot{x}} \right) \cdot h + \frac{d}{dt}\left(\frac{\partial L}{\partial \dot{x}} \cdot h \right) \right] + O(\|h\|^2)$$

$$= A[\gamma] + \frac{\partial L}{\partial \dot{x}} \cdot h \Big|_{t_1}^{t_2} + \int_{t_1}^{t_2} dt \left(\frac{\partial L}{\partial x} - \frac{d}{dt}\frac{\partial L}{\partial \dot{x}} \right) \cdot h + O(\|h\|^2)$$

$\left(\text{Here, we have made the following abbreviation: For curves in } \mathbb{R}^n, \left(\dfrac{\partial L}{\partial x} - \dfrac{d}{dt}\dfrac{\partial L}{\partial \dot{x}} \right) \cdot h \text{ means, more}\right.$

$\left.\text{exactly, } \displaystyle\sum_{i=1}^{n} \left(\dfrac{\partial L}{\partial x_i} - \dfrac{d}{dt}\dfrac{\partial L}{\partial \dot{x}_i} \right) \cdot h_i. \right)$

If now the functional $A[\gamma]$ is restricted to curves $\gamma \in B_{x_1, x_2}$ with fixed initial and endpoint, $h(t_1) = h(t_2) = 0$, and the derivative is simply given by

$$A'[\gamma]h = \int_{t_1}^{t_2} dt \left(\frac{\partial L}{\partial x} - \frac{d}{dt} \frac{\partial L}{\partial \dot{x}} \right) \cdot h \ . \tag{3.8.16}$$

As in normal analysis, we have the theorem:

If a functional F has a local minimum (maximum) at the "point" γ_0, then $F'[\gamma_0] = 0$ (that is, $F'[\gamma_0]h = 0$ for all h or all h with $h(t_1) = h(t_2)$ if F should be restricted to B_{x_1, x_2}). γ_0 is called a stationary point of F, if $F'[\gamma_0] = 0$.
$A'[\gamma_0] = 0$ now means

$$\frac{\partial L}{\partial x_0}(x_0(t), \dot{x}_0(t), t) - \frac{d}{dt} \frac{\partial L}{\partial \dot{x}_0}(x_0(t), \dot{x}_0(t), t) = 0 \tag{3.8.17}$$

and

$$\frac{\partial L}{\partial \dot{x}_0}(x_0(t_1), \dot{x}_0(t_1), t_1) = \frac{\partial L}{\partial \dot{x}_0}(x_0(t_2), \dot{x}_0(t_2), t_2) = 0 \ . \tag{3.8.18}$$

Proof: If

$$\frac{\partial L}{\partial x} - \frac{d}{dt} \frac{\partial L}{\partial \dot{x}} \neq 0$$

at some point $t_0 \neq t_1, t_2$, then there is an interval around t_0 in which this quantity does not vanish. Then we can find a function h with $h(t_1) = h(t_2) = 0$, such that

$$\int_{t_1}^{t_2} dt \left(\frac{\partial L}{\partial x} - \frac{d}{dt} \frac{\partial L}{\partial \dot{x}} \right) \cdot h(t) \neq 0 \ .$$

Thus

$$\frac{\partial L}{\partial x} - \frac{d}{dt} \frac{\partial L}{\partial \dot{x}} = 0 \quad \text{and further} \quad \left. \frac{\partial L}{\partial \dot{x}_0} \right|_{t_1} = \left. \frac{\partial L}{\partial \dot{x}_0} \right|_{t_2} = 0 \ .$$

If A is again restricted to B_{x_1, x_2}, then the condition for A to be stationary, $A'[\gamma] = 0$, reduces to the *Euler–Lagrange equation*:

$$\frac{d}{dt} \frac{\partial L}{\partial \dot{x}_0}(x_0(t), \dot{x}_0(t), t) - \frac{\partial L}{\partial x_0}(x_0(t), \dot{x}_0(t), t) = 0 \ . \tag{3.8.19}$$

This is a differential equation of second order for the trajectory $\gamma_0: t \mapsto x_0(t)$ which makes A stationary. γ_0 must in addition satisfy the boundary conditions $x_0(t_1) = x_1$, $x_0(t_2) = x_2$.

Example. Let us determine the stationary "point" of the curve length functional

$$S: B_{x_1 x_2} \to \mathbb{R} \ , \quad S[\gamma] = \int_{t_1}^{t_2} dt \, \sqrt{\dot{x}^2(t)} \ ,$$

i.e., the shortest curve between two points. In this case, $L(x, \dot{x}, t) = \sqrt{\dot{x}^2}$. The Euler–Lagrange equation reads:

$$\frac{d}{dt} \frac{\partial L}{\partial \dot{x}} - \frac{\partial L}{\partial x} = \frac{d}{dt} \frac{\dot{x}}{|\dot{x}|} = 0 \ .$$

$\dot{x}/|\dot{x}|$ is the unit tangent vector which must therefore remain constant along the shortest curve connecting x_1 and x_2:

The shortest distance between two points is a straight line.

3.8.2 Hamilton's Principle

Theorem: A trajectory $\gamma \in B_{x_1 x_2}$ is a solution of Newton's equations of motion

$$m_i \ddot{x}_i(t) + \nabla_i U(x_1(t), \dots, x_N(t), t) = 0 \ , \tag{3.8.20}$$

if and only if it is a stationary point of the *action functional*

$$S[\gamma] = \int_{t_1}^{t_2} dt \left[\sum_{i=1}^{N} \frac{1}{2} m_i \dot{x}_i^2(t) - U(x_1(t), \dots, x_N(t), t) \right] \ . \tag{3.8.21}$$

To prove this, it suffices to show that the Euler–Lagrange equation for the functional S agrees with the Newtonian equations of motion. Indeed, with

$$L = \sum_{i=1}^{N} \frac{1}{2} m_i \dot{x}_i^2 - U(x_1, \dots, x_N) \ ,$$

$$\frac{\partial L}{\partial x_i} = \nabla_{x_i} L = -\nabla_{x_i} U \ , \quad \frac{\partial L}{\partial \dot{x}_i} = \nabla_{\dot{x}_i} L = m_i \dot{x}_i \ .$$

The equation

$$\frac{d}{dt} \frac{\partial L}{\partial \dot{x}} - \frac{\partial L}{\partial x} = 0 \quad \text{then immediately becomes } m_i \ddot{x}_i + \nabla_i U = 0.$$

A few important remarks:

i) The Lagrangian L is not uniquely determined. If we replace $L(x, \dot{x}, t)$ with

$$\tilde{L}(x, \dot{x}, t) = L(x, \dot{x}, t) + \frac{d}{dt} f(x, t)$$

$$= L(x, \dot{x}, t) + \dot{x} \cdot \frac{\partial f}{\partial x}(x, t) + \frac{\partial f}{\partial t}(x, t) \ , \tag{3.8.22}$$

then the action functional is given by

$$\tilde{S}[\gamma] = \int\limits_{t_1}^{t_2} dt \tilde{L}(x, \dot{x}, t) = \int\limits_{t_1}^{t_2} dt \left[L(x, \dot{x}, t) + \frac{d}{dt} f(x, t) \right]$$

$$= S[\gamma] + f(x(t_2), t_2) - f(x(t_1), t_1) \ . \tag{3.8.23}$$

On B_{x_1, x_2}, S and $\tilde{S}$ differ only by a constant, so that they have the same stationary points.

ii) We can investigate the nature of the stationary point of S using the second derivative of S. To simplify writing here, let us consider the one-dimensional case

$$S[\gamma_0 + h] = S[\gamma_0] + S'[\gamma_0]h$$

$$+ \int\limits_{t_1}^{t_2} dt \tfrac{1}{2} [m\dot{h}^2(t) - U''(x_0(t))h^2(t)] + O(\|h\|^3) \ . \tag{3.8.24}$$

If γ_0 is a stationary point, we have

$$S[\gamma_0 + h] = S[\gamma_0] + \int\limits_{t_1}^{t_2} dt \tfrac{1}{2} [m\dot{h}^2(t) - U''(x_0(t))h^2(t)]$$

$$+ O(\|h\|^3) \ . \tag{3.8.25}$$

If $t_2 - t_1$ is small enough, then the integral is positive because $h(t_1) = 0$. For small enough time intervals $[t_1, t_2]$, the stationary points of S are thus *minima* (the principle of least action).

iii) Hamilton's Principle is independent of coordinate system in the following sense:

If we perform a transformation of coordinates (even time-dependent) from the Cartesian coordinates x to other coordinates $q = (q_1, \ldots, q_f)$: $x = x(q, t)$, $q = q(x, t)$, then a trajectory $t \mapsto x(t)$ becomes a mapping $t \mapsto q(t) = q(x(t), t)$. The velocity curve is then given by

$$\dot{x}(t) = \frac{d}{dt} x(t) = \frac{d}{dt} x(q(t), t)$$

$$= \frac{\partial x}{\partial q}(q(t), t)\dot{q}(t) + \frac{\partial x}{\partial t}(q(t), t) \ . \tag{3.8.26}$$

Here,

$$\frac{\partial x}{\partial q}\dot{q} \text{ is again to be understood as } \sum_{i=1}^{f} \frac{\partial x}{\partial q_i}\dot{q}_i.$$

For the action, we find

$$S[\gamma] = \int_{t_1}^{t_2} dt\, L(x(t), \dot{x}(t), t)$$

$$= \int_{t_1}^{t_2} dt\, L\left(x(q(t)), \frac{\partial x}{\partial q}(q(t), t)\dot{q}(t) + \frac{\partial x}{\partial t}(q(t), t), t \right)$$

$$= \int_{t_1}^{t_2} dt\, \tilde{L}(q(t), \dot{q}(t), t) \quad \text{with}$$

$$\tilde{L}(q, \dot{q}, t) = L\left(x(q, t), \frac{\partial x}{\partial q}(q, t)\dot{q} + \frac{\partial x}{\partial t}(q, t), t \right). \tag{3.8.27}$$

The value of $S[\gamma]$ is independent of whether the trajectories are written using the coordinates q or x. Similarly, whether or not γ_0 is a stationary point is independent of the choice of coordinates.

At the same time, we have just discovered a very convenient procedure for rewriting the equations of motion in other coordinate systems.

We have thus derived Lagrange's equations for systems without constraints in an entirely new way and found another way of viewing the independence of these equations of the choice of coordinates.

3.8.3 Hamilton's Principle for Systems with Holonomic Constraints

Lagrange's equations can also be derived from the principle of stationary action for systems with holonomic constraints.

Let us designate as $B^M_{\underline{x}_1 \underline{x}_2}$ the set of trajectories $\gamma: t \mapsto \underline{x}(t)$ with endpoints $\underline{x}_1$ and $\underline{x}_2$, which in addition satisfy the constraints, so that $\underline{x}(t) \in M_t$ for all $t \in [t_1, t_2]$. Our symbols are as in Sect. 3.2.

Then we have the *theorem*:

A trajectory $\gamma_0 \in B^M_{\underline{x}_1, \underline{x}_2}$ is a solution of Lagrange's equations of the first kind if and only if it is a stationary point of the action functional

$$S: B^M_{\underline{x}_1 \underline{x}_2} \to \mathbb{R} \;,$$

$$\gamma \to S[\gamma] = \int_{t_1}^{t_2} dt \left[\sum_{i=1}^{N} \frac{1}{2} m_i \dot{x}_i^2 - U(\underline{x}(t), t) \right]. \tag{3.8.28}$$

In other words: γ_0 must be a stationary point of the action under the additional condition $\underline{x}(t) \in M_t$ for all $t \in [t_1, t_2]$.

The proof is analogous to that given above. The only difference is that this time only virtual displacements tangential to M_t are allowed as variations:

$$S[\gamma_0 + h_M] = S[\gamma_0] - \int\limits_{t_1}^{t_2} dt \sum_{i=1}^{N} [m_i \ddot{x}_i(t) + \nabla_i U(\underline{x}(t), t)] \cdot \delta_M x_i(t)$$

$$+ O(\|h_M\|^2)$$

$$= S[\gamma_0] - \int\limits_{t_1}^{t_2} dt [\dot{\underline{p}}(t) + \underline{\nabla} U(\underline{x}, t), t)] \delta_M \underline{x}(t)$$

$$+ O(\|h_M\|^2) \ . \tag{3.8.29}$$

Then, since $\delta_M \underline{x}(t)$ is always tangential to M_t, saying that S is stationary at a point γ_0 does not mean simply that $\dot{\underline{p}}(t) + \underline{\nabla} U(\underline{x}(t), t) = 0$, but rather that $\dot{\underline{p}}(t) + \underline{\nabla} U(\underline{x}(t), t)$ is perpendicular to $\bar{M}_t$.

Hence,

$$\dot{\underline{p}}(t) + \underline{\nabla} U(\underline{x}(t), t) = \sum_{\alpha=1}^{s} \lambda_\alpha(t) \underline{\nabla} F_\alpha(x(t), t) \ . \tag{3.8.30}$$

The stationary points of S under the additional condition $\underline{x}(t) \in M_t$ can also be calculated another way:

First, we parametrize M_t by introducing coordinates $q = (q_1, \ldots, q_f)$, whose number corresponds to the number $f = \dim M_t$ of degrees of freedom of the system. The allowed positions of the particles in the system are then functions $\underline{x}(q, t)$ of the parameters: for all q and t, $F_\alpha(\underline{x}(q, t), t) = 0$ $(\alpha = 1, \ldots, s)$ and as q runs through all values, then $\underline{x}(q, t)$ runs through all possible positions (at least locally). The allowable trajectories are then given by the functions $q(t)$, and the action can be calculated as follows in these coordinates:

$$S[\gamma] = \int\limits_{t_1}^{t_2} dt \hat{L}(q(t), \dot{q}(t), t) \quad \text{with}$$

$$\hat{L}(q, \dot{q}, t) = L\left(\underline{x}(q, t), \frac{\partial \underline{x}}{\partial q}(q, t) \dot{q} + \frac{\partial \underline{x}}{\partial t}(q, t), t \right) \ .$$

The additional conditions are then already satisfied by our choice of the coordinates q, and the variations occur without further constraints on q.

Thus, the trajectory $t \mapsto \underline{x}(q(t), t)$ is a solution of Lagrange's equations of the first type if and only if

$$\frac{d}{dt} \frac{\partial \hat{L}}{\partial \dot{q}}(q(t), \dot{q}(t), t) - \frac{\partial \hat{L}}{\partial q}(q(t), \dot{q}(t), t) = 0 \ . \tag{3.8.31}$$

Thus, Lagrange's equations of the first and second types are consequences of Hamilton's Principle.

Problems

3.1   The Total Time Derivative of the Lagrangian. Suppose that a Lagrangian has the form of a total time derivative:

$$L(q, \dot{q}, t) = \frac{df(q, t)}{dt} = \dot{q}\nabla f(q, t) + \frac{\partial f(q, t)}{\partial t} \ .$$

Show explicitly that the corresponding Euler–Lagrange equations are identically satisfied, i.e., are satisfied for any arbitrary function $q(t)$. Interpret this fact.

3.2   Geometrical Optics as a Variational Problem. Fermat's principle states that a ray of light takes that particular path between two points for which the optical length is the shortest. In a two-dimensional medium with index of refraction $n(x, y)$, the optical length is given by:

$$L = \int\limits_{x_1}^{x_2} dx\, n(x, y)\sqrt{1 + y'^2} \ .$$

a) Produce the Euler–Lagrange equations, where $y(x_1) = y_1$ and $y(x_2) = y_2$ are to be held fixed.
b) Solve the Euler–Lagrange equations for the cases:
   i) n is constant over all of space,
   ii) $n = y^{-1}$ $(y > 0)$.

3.3   Conservation Laws for the Three-dimensional Harmonic Oscillator. Consider the equation of motion of the three-dimensional harmonic oscillator:

$$m\ddot{x}(t) + kx(t) = 0 \ ; \qquad \omega_0^2 = \frac{k}{m} \ .$$

a) Show that the following quantities are conserved:

$$L_{ij} = m(x_i \dot{x}_j - x_j \dot{x}_i)$$

$$A_{ij} = \frac{m}{2}\dot{x}_i \dot{x}_j + \frac{k}{2}x_i x_j \quad (i, j = 1, 2, 3) \ .$$

b) Prove the following relationship between the conserved quantities:

$$(A_{ij})^2 = A_{ii}A_{jj} - \frac{\omega_0^2}{4}(L_{ij})^2$$

(no summation convention here!)

c) A symmetric matrix has in general six independent elements. Show that the components of A satisfy at least one other condition.
(*Hint*: Calculate the determinant of A).

3.4 Relativistic Particles. Let a particle have Lagrangian

$$L = -mc^2 \sqrt{1 - \frac{|\dot{\boldsymbol{x}}|^2}{c^2}} \ ,$$

where c is a constant with the dimensions of speed. Determine:

a) the canonically conjugate momentum, the Hamiltonian, the angular momentum,
b) the equations of motion, and
c) the general solution of the equations of motion.

3.5 A Periodically Excited Pendulum. Determine the Lagrangian and the equation of motion for the following (planar) systems in the homogeneous gravitational field g:

A pendulum of length l, whose point of attachment moves with circular frequency (or angular velocity) ω

a) harmonically in the vertical direction
b) harmonically in the horizontal direction
c) uniformly in a circle.

3.6 "Spontaneous Symmetry-Breaking". Discuss the following system:

A bead slides frictionlessly under the influence of gravity on a loop of wire, which in turn rotates uniformly with angular velocity ω around a vertical axis passing through its center.

Determine qualitatively how the motion depends on ω.

Discuss in particular possible positions of equilibrium, their stability, and the harmonic frequency of small oscillations around stable points of equilibrium as they depend on ω.

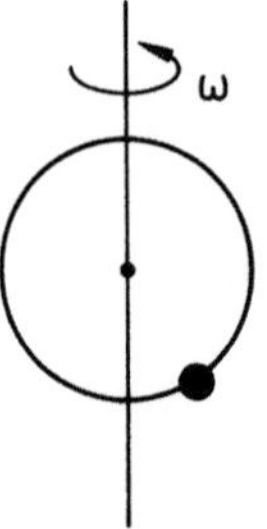

3.7 Forces of Constraint. Calculate the components of the forces of constraint indicated by the diagram with the help of d'Alembert's principle. The systems are in a homogeneous gravitational field g. The rigid beams shown each have length l and total mass M, which we assume is concentrated in the endpoints in equal amounts. The distance a is fixed in each case.

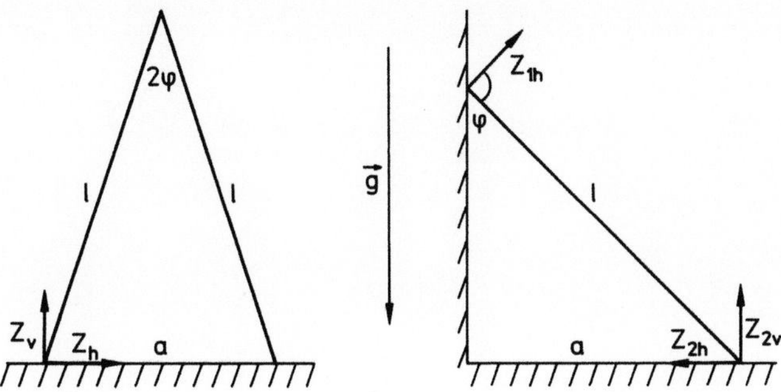

4. Rigid Bodies

So far, we have worked with the mechanics of systems of point particles. We recall that a particle is an idealized way of describing the motion of a body whose form and physical extension play no role in the physics of its motion. For, say, a rolling wheel such an idealization is not appropriate. Here we must use another idealized mechanical system, namely, the so-called *rigid body*.

A body is called rigid, if it cannot be deformed, that is, if, to a good approximation, the distances between all of its parts remain constant. We will think of a rigid body as being formed out of a large number of discrete point-masses. Rigidness means then that for the position vectors $r^{(\alpha)}$ of the individual mass-points the holonomic scleronomic constraints

$$|r^{(\alpha)} - r^{(\beta)}| - C_{\alpha\beta} = 0$$

must be satisfied for all $r^{(\alpha)}$ and $r^{(\beta)}$. The $C_{\alpha\beta}$ are constant in time.

We thus reduce the description of a rigid body to that of a special system of particles.

Of course, no body is perfectly rigid in reality. In the mechanics of continuous media, whose basic ideas will be introduced in Chap. 9, we will also go beyond this idealization.

4.1 The Kinematics of the Rigid Body

We will determine the configuration space M, i.e. the totality of possible positions of the rigid body taking into account its constraints, and find suitable coordinates in M. We now introduce a so-called *body-fixed* coordinate system:

We mark a point O_B on the body and fix a right-handed orthonormal system e_1, e_2, e_3 connected securely to the body. Each point X of the body can then be described in this system by a vector

$$\overrightarrow{O_B X} = b = b_i e_i \ . \tag{4.1.1}$$

(From now on, we will use the Einstein summation convention.) We have $\dot{b}_i = 0$, since the body is rigid.

We now consider the body in an inertial system with origin at O and coordinate axes given by the right-handed orthonormal system n_1, n_2, n_3.

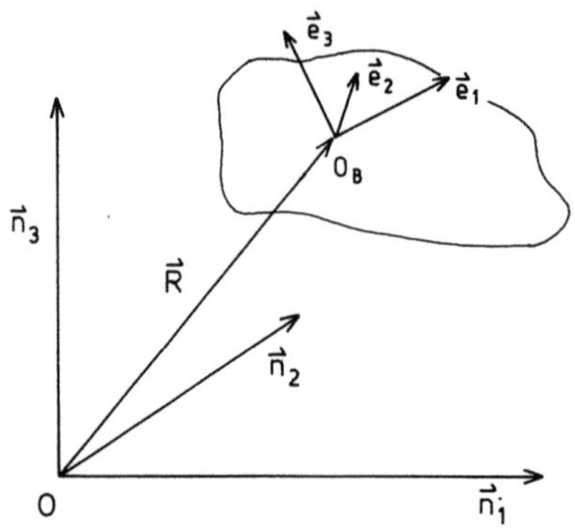

Fig. 4.1.1. Coordinate systems fixed in space and fixed on the body

If we set $\overrightarrow{OO_B} = R$, then at time t

$$\overrightarrow{OX} =: r(t) = R(t) + b_i e_i(t) \ . \tag{4.1.2}$$

The system fixed on the body follows every movement of the body. In order to know the location of the body-fixed system, and thus the location of the body, at any time t, we must therefore be able to express the vectors $e_i(t)$ in terms of the vectors n_i which do not vary with time. We write

$$e_i(t) = D(t)n_i \equiv n_j D_{ji} \tag{4.1.3}$$

and thus

$$b(t) = b_i e_i(t) = D(t)(b_i n_i) = n_j D_{ji} b_i \ . \tag{4.1.4}$$

Here, $D(t)$ is a (proper) rotation, namely the same rotation which transforms the system n_1, n_2, n_3 into $e_1(t), e_2(t), e_3(t)$.

The position of a rigid body at time t is thus uniquely described by

a) the position vector $R(t)$ of O_B and, in addition,
b) the rotation $D(t)$.

We have already seen in Sect. 2.1 that with respect to the orthonormal basis n_1, n_2, n_3, we can find an orthogonal 3×3 matrix $(D_{ij})(t)$ which corresponds to the rotation $D(t)$, so that $DD^{\mathrm{T}} = D^{\mathrm{T}}D = 1$ or

$$D_{ik}(t)D_{jk}(t) = D_{ki}(t)D_{kj}(t) = \delta_{ij} \ .$$

Further, we have det $D = 1$.

In order now to describe the rotation $D(t)$ explicitly, we need to parametrize all possible rotations. The rotation which transforms n_1, n_2, n_3 into e_1, e_2, e_3 can be divided into three rotations (Fig. 4.1.2):

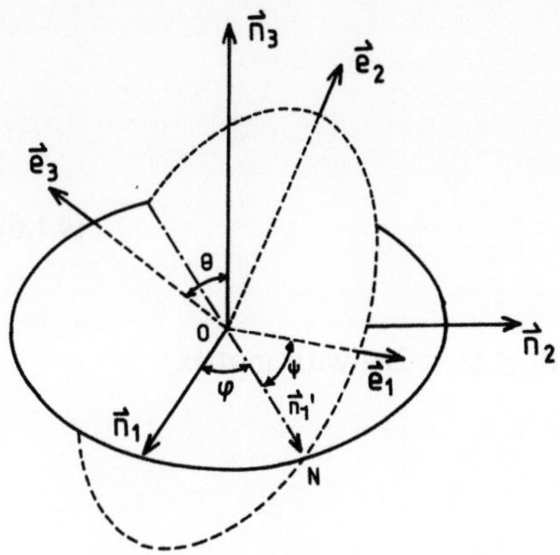

Fig. 4.1.2. Euler's angles

a) Rotation around n_3 through the angle φ $(0 \le \varphi < 2\pi)$.
n_3 remains fixed, and we call the corresponding matrix $R^3_{ij}(\varphi)$, thus

$$n'_i = n_j R^3_{ji}(\varphi) \ . \tag{4.1.5}$$

We have

$$n'_1 = \cos\varphi\, n_1 + \sin\varphi\, n_2 \ ,$$

$$n'_2 = -\sin\varphi\, n_1 + \cos\varphi\, n_2 \ ,$$

$$n'_3 = n_3 \ ,$$

thus

$$R^3(\varphi) = \begin{pmatrix} \cos\varphi & -\sin\varphi & 0 \\ \sin\varphi & \cos\varphi & 0 \\ 0 & 0 & 1 \end{pmatrix} \ .$$

It is easy to verify that $R^3 R^{3\mathrm{T}} = 1$, $\det R = 1$.

b) Rotation around n'_1 through the angle θ $(0 \le \theta < \pi)$, then we have

$$n''_1 = n'_1 \ ,$$

$$n''_2 = \cos\theta\, n'_2 + \sin\theta\, n'_3 \ ,$$

$$n''_3 = -\sin\theta\, n'_2 + \cos\theta\, n'_3 \ .$$

Thus

$$n_k'' = n_r' R_{rk}^1(\theta) \quad \text{with}$$

$$R^1(\theta) = \begin{pmatrix} 1 & 0 & 0 \\ 0 & \cos\theta & -\sin\theta \\ 0 & \sin\theta & \cos\theta \end{pmatrix} . \tag{4.1.6}$$

c) Rotation about n_3'' by the angle ψ $(0 \le \psi < 2\pi)$ with matrix

$$R^3(\psi) = \begin{pmatrix} \cos\psi & -\sin\psi & 0 \\ \sin\psi & \cos\psi & 0 \\ 0 & 0 & 1 \end{pmatrix} .$$

Then

$$e_1 = \cos\psi\, n_1'' + \sin\psi\, n_2'' \ ,$$

$$e_2 = -\sin\psi\, n_1'' + \cos\psi\, n_2'' \ ,$$

$$e_3 = n_3''' = n_3'' \ .$$

So that, all together,

$$e_i = n_j'' R_{ji}^3(\psi) = n_k' R_{kj}^1(\theta) R_{ji}^3(\psi)$$

$$= n_r R_{rk}^3(\varphi) R_{kj}^1(\theta) R_{ji}^3(\psi) = n_r D_{ri}(\varphi, \theta, \psi)$$

with

$$D_{ri} = R_{rk}^3(\varphi) R_{kj}^1(\theta) R_{ji}^3(\psi) \tag{4.1.7}$$

and explicitly

$$D =$$

$$\begin{pmatrix} \cos\varphi\cos\psi - \sin\varphi\cos\theta\sin\psi & -\cos\varphi\sin\psi - \sin\varphi\cos\theta\cos\psi & \sin\varphi\sin\theta \\ \sin\varphi\cos\psi + \cos\varphi\cos\theta\sin\psi & -\sin\varphi\sin\psi + \cos\varphi\cos\theta\cos\psi & -\cos\varphi\sin\theta \\ \sin\theta\sin\psi & \sin\theta\cos\psi & \cos\theta \end{pmatrix} . \tag{4.1.8}$$

The angles φ, θ, and ψ which parametrize each rotation, are called *Euler's angles*.

Checking the expression for the rotation matrix $D(\varphi, \theta, \psi)$:

i) $\psi = \theta = 0$: $\quad D = \begin{pmatrix} \cos\varphi & -\sin\varphi & 0 \\ \sin\varphi & \cos\varphi & 0 \\ 0 & 0 & 1 \end{pmatrix} .$

ii) $\varphi = \psi = 0$: $D = \begin{pmatrix} 1 & 0 & 0 \\ 0 & \cos\theta & -\sin\theta \\ 0 & \sin\theta & \cos\theta \end{pmatrix}$.

iii) $e_3 = n_k D_{k3} = \sin\theta \sin\varphi\, n_1 - \sin\theta \cos\varphi\, n_2 + \cos\theta\, n_3$,

so that e_3, i.e. the new 3-direction, has the coordinates

$$\begin{pmatrix} \sin\theta \sin\varphi \\ -\sin\theta \cos\varphi \\ \cos\theta \end{pmatrix}$$

in the inertial system. This is a unit vector with polar coordinates $(\theta,\ -\frac{\pi}{2} + \varphi)$, since

$$\sin\left(\varphi - \frac{\pi}{2}\right) = -\cos\varphi \ , \quad \text{and} \quad \cos\left(\varphi - \frac{\pi}{2}\right) = \sin\varphi \ .$$

For $\varphi = \frac{\pi}{2}$, for example, we have $e_3 = (\sin\theta, 0, \cos\theta)$.

The angles φ, θ, ψ now depend on t, and thus change with the motion of the body.

Now we know the relation between the systems (e_1, e_2, e_3) and (n_1, n_2, n_3):

$$e_i(t) = n_k D_{ki}(\varphi(t), \theta(t), \psi(t)) \ ,$$

and we could find the variation in time of $e_i(t)$ in terms of the time variation of φ, θ, ψ. The angles φ, θ, ψ are thus generalized coordinates, which, together with $R(t)$, completely describe the position of the rigid body.

For the velocities of the particles composing the rigid body, we find

$$\dot{r}(t) = \dot{R}(t) + b_i \dot{e}_i(t) = \dot{R}(t) + n_j \dot{D}_{ji}(t) b_i \ , \tag{4.1.9}$$

since $\dot{b}_i = 0$.

We will now prove the important theorem:

There exists a vector $\Omega(t)$, the instantaneous angular velocity, which satisfies:

$$\dot{e}_i(t) = \Omega(t) \times e_i(t) \ . \tag{4.1.10}$$

Proof. From

$$\dot{e}_i(t) = n_j \dot{D}_{ji}(t) \quad \text{it follows since}$$

$$n_j = e_k D_{jk}(t) \quad \text{that}$$

$$\dot{e}_i(t) = e_k(t) D_{jk}(t) \dot{D}_{ji}(t) = e_k \omega_{ki}$$

with $\omega_{ki} = D_{jk} \dot{D}_{ji}$.

The quantities ω_{ki} are the components of an anti-symmetric matrix, since from

$$D_{jk}D_{ji} = \delta_{ki}$$

it follows by differentiation that

$$0 = \dot{D}_{jk}D_{ji} + D_{jk}\dot{D}_{ji} = \omega_{ik} + \omega_{ki} \; .$$

Then we can introduce the three quantities Ω_r using

$$\omega_{ki} = \varepsilon_{ikr}\Omega_r \; , \quad \text{i.e.}$$

$$\omega_{12} = -\Omega_3 \; , \quad \omega_{13} = +\Omega_2 \; , \quad \omega_{23} = -\Omega_1 \; , \tag{4.1.11}$$

so that we find:

$$\dot{e}_i(t) = \varepsilon_{ikr}\Omega_r e_k \; .$$

If the angular velocity vector $\boldsymbol{\Omega}$ is defined as

$$\boldsymbol{\Omega} = \Omega_r e_r \; ,$$

so that Ω_r are the components of this vector with respect to the body coordinate system, we have

$$\dot{e}_i = \varepsilon_{ikr}\Omega_r e_k = \varepsilon_{rik}\Omega_r e_k$$

$$= \Omega_r e_r \times e_i$$

$$= \boldsymbol{\Omega} \times e_i \; , \quad \text{also}$$

$$\dot{\boldsymbol{b}} = b_i \dot{e}_i = b_i(\boldsymbol{\Omega} \times e_i) = \boldsymbol{\Omega} \times \boldsymbol{b} \; . \tag{4.1.12}$$

The direction of $\boldsymbol{\Omega}(t)$ is the direction of the momentary axis of rotation, while $|\boldsymbol{\Omega}(t)|$ is the angular speed of the rotation about this axis (Fig. 4.1.3).

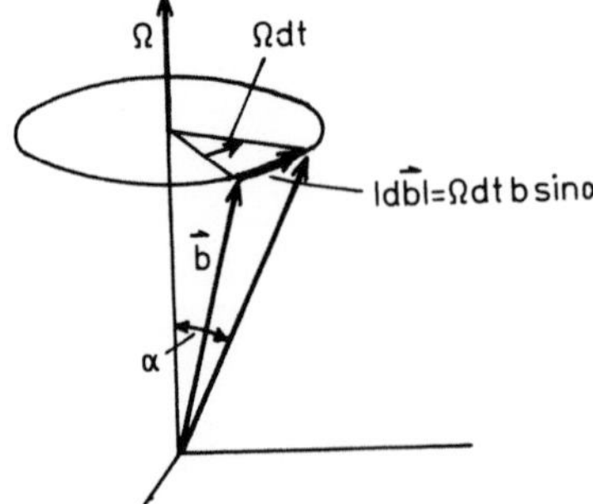

Fig. 4.1.3. The vector $\boldsymbol{\Omega}$ indicates the direction of the axis around which the point b instantaneously rotates and $|\boldsymbol{\Omega}|$ indicates the angular speed of rotation

By introducing Euler's parametrization of D_{ij}, we find the components Ω_i expressed in terms of Euler's angles and their time derivatives:

$$\Omega_1 = \dot{\theta}\cos\psi + \dot{\varphi}\sin\theta\sin\psi \ ,$$

$$\Omega_2 = -\dot{\theta}\sin\psi + \dot{\varphi}\sin\theta\cos\psi \ ,$$

$$\Omega_3 = \dot{\psi} + \dot{\varphi}\cos\theta \ . \tag{4.1.13}$$

These expressions can be read off also directly from Fig. 4.1.2.

Analogously, the components ω_i of Ω with respect to the basis n_i can also be calculated.

We have,

$$\Omega = \omega_i n_i \quad \text{with}$$

$$\omega_1 = \dot{\theta}\cos\varphi + \dot{\psi}\sin\theta\sin\varphi \ ,$$

$$\omega_2 = \dot{\theta}\sin\varphi + \dot{\psi}\sin\theta\cos\varphi \ , \tag{4.1.14}$$

$$\omega_3 = \dot{\varphi} - \dot{\psi}\cos\theta \ .$$

Remarks. i) If we think of the rotation of the body system relative to the inertial system as not dependent on time, but rather as given in terms of a parameter α, then, as we have shown, there is a vector n such that for each point with position vector $b(\alpha)$

$$\frac{db}{d\alpha} = n \times b(\alpha) \tag{4.1.15}$$

holds and n indicates the direction of the axis about which the system is rotated. With $d\alpha = \Omega\, dt$, $\Omega = \Omega n$, this formula becomes identical with (4.1.12).

Thus, the assertion made in formula (3.6.9) is proved. Of course, a direct proof would also not be difficult.

ii) The rotation $D(t)$ is, of course, independent of the choice of origin O_B, since it only indicates the orientation of the body system with respect to a system fixed in space.

Thus, $\Omega(t)$ is also independent of the choice of reference point O_B, and $\Omega(t)$ can be seen in general as the instantaneous angular velocity vector of the whole body.

iii) The set of all proper rotations forms a group, which is usually written SO(3). The configuration space of a rigid body is thus

$$M = \mathrm{SO}(3) \times \mathbb{R}^3$$

and this is exactly the group of proper motions of the Euclidean space E^3.

4.2   The Inertia Tensor and the Kinetic Energy of a Rigid Body

4.2.1   Definition and Elementary Properties of the Inertia Tensor

Let each point α on the body have components $b_j^{(\alpha)}$ in the body coordinate system. Then, the following equation holds for the position vector $r^{(\alpha)}$, $\alpha = 1, 2, \ldots, N$:

$$\dot{r}^{(\alpha)} = \dot{R} + \Omega \times b^{(\alpha)} \quad \text{with} \quad b^{(\alpha)}(t) = b_j^{(\alpha)} e_j(t) \; , \tag{4.2.1}$$

and thus the total kinetic energy is given by

$$\frac{1}{2} \sum_\alpha m_\alpha \dot{r}^{(\alpha)2} = \frac{1}{2} \sum_\alpha m_\alpha (\dot{R} + \Omega \times b^{(\alpha)})^2$$

$$= \frac{1}{2} \left(\sum_\alpha m_\alpha \right) \dot{R}^2 + \sum_\alpha m_\alpha \dot{R} \cdot (\Omega \times b^{(\alpha)}) + \frac{1}{2} \sum_\alpha m_\alpha (\Omega \times b^{(\alpha)})^2$$

$$= \frac{1}{2} M \dot{R}^2 + (\dot{R} \times \Omega) \cdot \sum_\alpha m_\alpha b^{(\alpha)}$$

$$+ \frac{1}{2} \sum_\alpha m_\alpha [\Omega^2 b^{(\alpha)2} - (\Omega \cdot b^{(\alpha)})^2] \; . \tag{4.2.2}$$

If we choose O_B to be the center of mass, then $\sum m_\alpha b^{(\alpha)} = 0$ and thus the second sum vanishes. This also vanishes if $\dot{R} = 0$, for example if the body is being held fixed at the point O_B. For the third sum, we can write

$$\tfrac{1}{2} I_{mn} \Omega_m \Omega_n \quad \text{with} \quad I_{mn} = \sum m_\alpha (\delta_{mn} b^{(\alpha)2} - b_m^{(\alpha)} b_n^{(\alpha)}) \; . \tag{4.2.3}$$

I_{mn} represents a symmetric 3×3 matrix, and we call I_{mn} the components of the *inertia tensor* (Appendix C gives a short introduction to tensor calculus).

The kinetic energy of the rigid body can then be written as

$$T = T_{\text{cm}} + T_{\text{rot}} \quad \text{where} \tag{4.2.4}$$

$$T_{\text{cm}} = \tfrac{1}{2} M \dot{R}^2 \; , \tag{4.2.5}$$

is the kinetic energy of the center of mass motion and

$$T_{\text{rot}} = \tfrac{1}{2} I_{mn} \Omega_m \Omega_n \; , \tag{4.2.6}$$

is the kinetic energy of rotational motion.

For a body with continuously distributed mass, we have

$$I_{mn} = \int d^3 b \, \varrho(b) (\delta_{mn} b^2 - b_m b_n) \; . \tag{4.2.7}$$

Note that the components of $\boldsymbol{\Omega}$ and $\boldsymbol{b}$ are always constructed with respect to a system fixed on the body.

The kinetic energy of the rotational motion depends quadratically on the components Ω_i of the angular velocity with respect to the orthonormal basis e_1, e_2, e_3. Mathematically, we can formulate this more exactly as follows (see Appendix C):

By

$$I(x, y) := I_{ij} x_i y_j$$

we can define a *bilinear form* on the Euclidean vector space V^3, which we will call the *inertia form* of the rigid body. The rotational energy is thus $E_{\mathrm{rot}} = \frac{1}{2} I(\boldsymbol{\Omega}, \boldsymbol{\Omega})$.

Obviously, $I(x, y) = I(y, x)$, so that this bilinear form is symmetric. Further, we have $E_{\mathrm{rot}} \geq 0$ for $\boldsymbol{\Omega} \neq 0$, so that it is positive semidefinite. $E_{\mathrm{rot}} = 0$ for $\boldsymbol{\Omega} \neq 0$ is only possible if $\boldsymbol{\Omega} \times \boldsymbol{b}^{(\alpha)} = 0$ for all α, which occurs only when the entire mass lies on a straight line in the $\boldsymbol{\Omega}$-direction. Aside from this degenerate case, I is positive definite.

The components I_{ij} are simply given by

$$I_{ij} = I(e_i, e_j) \ . \tag{4.2.8}$$

Still another interpretation of the inertia tensor is possible (see Appendix C):

The tensor I can also be seen as a linear operator in the Euclidean vector space V^3. For simplicity, we will also refer to the operator as I, and it is defined by

$$I \cdot x = I(x) = I(x_i e_i) = x_i I \cdot e_i = e_j I_{ji} x_i \ .$$

I_{ij} is simply the matrix of the operator I with respect to the basis e_1, e_2, e_3.

The connection between the operator and the bilinear form is easy to see:

$$I(x, y) = x \cdot (I \cdot y) \ .$$

We will see in the next section that $I \cdot \boldsymbol{\Omega}$ is the angular momentum of the rigid body. The linear mapping I is called the *inertia operator*.

Remarks. i) If the rotation occurs around a fixed axis in the direction of the unit vector $\boldsymbol{n}$, then

$$\boldsymbol{\Omega} = \Omega \boldsymbol{n} \quad \text{and}$$

$$T_{\mathrm{rot}} = \tfrac{1}{2} \Omega^2 I_{ij} n_i n_j = \tfrac{1}{2} I_n \Omega^2 \ . \tag{4.2.9}$$

Here, I_n is called the *moment of inertia* of the rigid body with respect to the axis $\boldsymbol{n}$. The inertia tensor thus determines the moments of inertia for all the axes through the point O_B.

ii) The components of I are explicitly given by

$$I_{11} = \sum_\alpha m_\alpha (b_2^{(\alpha)^2} + b_3^{(\alpha)^2}) \ ,$$

$$I_{22} = \sum_\alpha m_\alpha (b_1^{(\alpha)^2} + b_3^{(\alpha)^2}) \ ,$$

$$I_{33} = \sum_\alpha m_\alpha (b_1^{(\alpha)^2} + b_2^{(\alpha)^2}) \ ,$$

$$I_{jk} = - \sum_\alpha m_\alpha b_j^{(\alpha)} b_k^{(\alpha)} \quad \text{for} \quad j \neq k \ .$$

(4.2.10)

Obviously, it is always true that

$$I_{11} + I_{22} \geq I_{33} \ , \quad I_{22} + I_{33} \geq I_{11} \ , \quad I_{33} + I_{11} \geq I_{22} \ .$$

Here, the equal sign can only arise in the degenerate case when:

$$I_{11} + I_{22} = I_{33}$$

which means

$$\sum_\alpha m_\alpha b_3^{(\alpha)^2} = 0 \ ,$$

which is satisfied if and only if $b_3^{(\alpha)} = 0$ for all α, so that all the mass lies in the 1–2 plane. Then also

$$I_{13} = I_{23} = 0 \ .$$

iii) In general, the inertia tensor I has six independent components. It is a well-known mathematical fact that every symmetric bilinear form on a Euclidean vector space can be diagonalized by some orthogonal transformation. For our purposes, a more exact version of this assertion is:

There exists an orthonormal system of vectors e_1', e_2', e_3' such that

$$I(e_i', e_k') = 0 \quad \text{for} \quad i \neq k \ .$$

(4.2.11)

The vectors e_i' are called the principle axes of the rigid body and the quantities

$$I_i = I(e_i', e_i') \geq 0$$

(4.2.12)

are called the principle moments of inertia.

It is now clear that we want to choose our body system such that its orthogonal basis vectors e_1, e_2, e_3 lie on the principle axes.

Then the matrix I_{ij} has a particularly simple form

$$I_{ij} = I_i \delta_{ij} \quad \text{(no summation here)}$$

or

$$I_{ij} = \begin{pmatrix} I_1 & 0 & 0 \\ 0 & I_2 & 0 \\ 0 & 0 & I_3 \end{pmatrix}.$$

(4.2.13)

We will not prove here the existence of such an orthogonal system of principle axes; the proof will be given in conjunction with a procedure to find these principle axes and moments of inertia in Sect. 6.4.1.

In many cases, the location of the principle axes can be found immediately using the symmetries of the rigid body. It is easy to see the following:

If the rigid body is symmetric under reflection by a surface with normal vector n, then n is a principle axis. (The two other principle axes must then lie in the plane.)

It is not much more difficult to prove the following assertion:

If the rigid body is symmetric under a rotation about the n-axis through an angle $0 \le \alpha < 2\pi$, then n is a principle axis. The two other axes are perpendicular to n. If the rigid body is invariant under a rotation of $\alpha \ne \pi$ about the n-axis, then the principle moments of inertia with respect to the two axes perpendicular to n are equal.

iv) The set

$$\mathcal{E} = \{\zeta | I(\zeta, \zeta) = 1\}$$

is the set of angular velocities corresponding to a fixed rotational energy. If we choose the principle axes as basis vectors, then $\mathcal{E}$ has the form

$$\mathcal{E} = \{\zeta_1, \zeta_2, \zeta_3 | I_1 \zeta_1^2 + I_2 \zeta_2^2 + I_3 \zeta_3^2 = 1\} \ .$$

(4.2.14)

This is an ellipsoid whose axes point in the directions e_1, e_2, e_3, the directions of the principle axes, and have lengths

$$\frac{1}{\sqrt{I_1}}, \frac{1}{\sqrt{I_2}}, \frac{1}{\sqrt{I_3}}$$

$\mathcal{E}$ is called the *ellipsoid of inertia.*

The ellipsoid of inertia is fixed with respect to the rigid body and moves with it through space. If any two principle moments of inertia are equal, say $I_1 = I_2$, then the ellipsoid is an ellipsoid of revolution. Each axis perpendicular to e_3 is then a principle axis. If $I_1 = I_2 = I_3$, the inertia ellipsoid is a sphere and every direction is a principle axis.

v) It is often more convenient to calculate the components of the inertia tensor with respect to a coordinate system whose origin is not at the center of mass. If this origin is at point O_B, and if

$$\overrightarrow{O_B O_{B'}} = a \quad \text{then we have}$$

$$b^{(\alpha)} = b^{'(\alpha)} + a \quad \text{and}$$

$$I'_{mn} = \sum_\alpha m_\alpha (\delta_{mn} b^{'(\alpha)2} - b_m^{'(\alpha)} b_n^{'(\alpha)})$$

$$= \sum_\alpha m_\alpha [\delta_{mn} (b^{(\alpha)} - a)^2 - (b_m^{(\alpha)} - a_m)(b_n^{(\alpha)} - a_n)]$$

$$= I_{mn} + M(\delta_{mn} a^2 - a_m a_n) - 2\delta_{mn} \sum_\alpha m_\alpha b^{(\alpha)} \cdot a$$

$$+ a_m \sum_\alpha m_\alpha b_n^{(\alpha)} + a_n \sum_\alpha m_\alpha b_m^{(\alpha)} \ . \tag{4.2.15}$$

The mixed terms vanish, though, if

$$\sum_\alpha m_\alpha b^{(\alpha)} = 0 \ , \tag{4.2.16}$$

that is, if O_B is the center of mass.

An immediate consequence of this is *Steiner's*[1] *Theorem:*

If I_{mn} are the components of the inertia tensor in a body-fixed coordinate system with respect to the origin O_B at the center of mass, then the components of the tensor calculated with respect to a body system with origin at $O_{B'}$ with $\overrightarrow{O_B O_{B'}} = a$ is given by the formula

$$I'_{mn} = I_{mn} + M(\delta_{mn} a^2 - a_m a_n) \ . \tag{4.2.17}$$

Here, the additional term $M(\delta_{mn} a^2 - a_m a_n)$ is exactly the inertia tensor of a particle at point a.

The inertia tensor with respect to an axis with direction n through $O_{B'}$ is given by

$$I'_n = I'_{ij} n_i n_j = I_{ij} n_i n_j + M(\delta_{ij} a^2 - a_i a_j) n_i n_j$$

$$= I_n + M(a^2 - (n \cdot a)^2) = I_n + M(n \times a)^2$$

$$\geq I_n \ . \tag{4.2.18}$$

We note the following:

If we are given the direction n of the axis, the moment of inertia is minimal if the axis goes through the center of mass.

---

[1] *Steiner, Jakob* (*1796 Utzendorf/Canton Bern, d. 1863 Bern). Swiss mathematician, professor in Berlin after 1834. His primary area of research was synthetic geometry.

For the practical calculation of inertia tensors, we often use Steiner's theorem. Additionally, we can separate the rigid body into suitable parts and add the inertia tensors. The symmetries of the body are also used, whenever possible.

4.2.2   Calculation of Inertia Tensors

We will now demonstrate the calculation of inertia tensors.

a) Consider a molecule with two atoms lying on the z-axis (Fig. 4.2.1). Let the center of mass be

$$R = \frac{(m_1 r^{(1)} + m_2 r^{(2)})}{(m_1 + m_2)} = 0 \quad \text{with}$$

$$r^{(i)} = (0, 0, z^{(i)}) \ .$$

Then, with $z^{(1)} - z^{(2)} = l$ and $M = m_1 + m_2$, we have

$$z^{(1)} = \frac{l m_2}{M} \ , \qquad z^{(2)} = -\frac{l m_1}{M} \ .$$

Therefore

$$I_{11} = \sum_\alpha m_\alpha z^{(\alpha)^2} = m_1 \frac{l^2 m_2^2}{M^2} + m_2 \frac{l^2 m_1^2}{M^2} = \frac{l^2 m_1 m_2}{M} = l^2 \mu \ , \tag{4.2.19}$$

$$I_{22} = I_{11} \ , \tag{4.2.20}$$

$$I_{33} = \sum_\alpha m_\alpha (z^{(\alpha)^2} \delta_{33} - z^{(\alpha)} z^{(\alpha)}) = 0 \ , \tag{4.2.21}$$

$$I_{mn} = 0 \quad \text{for} \ \ n \neq m \ . \tag{4.2.22}$$

The principle axes are the x-, y-, and z-axes. The mass distribution is entirely on the z-axis, so that $I_{33} = 0$.

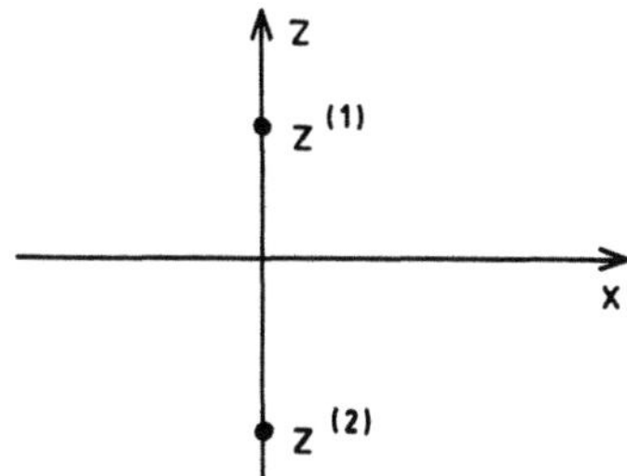

Fig. 4.2.1. The atoms of a bi-atomic molecule in a coordinate system fixed to the body

b) Homogeneous ball with density ϱ and radius R. Clearly every axis is a principle axis and $I_{11} = I_{22} = I_{33}$. It is convenient to calculate

$$I_{11} + I_{22} + I_{33} = 3I_1$$

$$= 4\pi\varrho \int_0^R dr\, r^2 (2x_1^2 + 2x_2^2 + 2x_3^2)$$

$$= 8\pi\varrho \int_0^R dr\, r^4 = \frac{8\pi}{5} \varrho R^5 \ . \tag{4.2.23}$$

Hence

$$I_1 = I_2 = I_3 = \frac{8\pi}{15} \varrho R^5 = \frac{2}{5} MR^2 \ , \tag{4.2.24}$$

where $M = 4\pi\varrho R^3/3$ is the mass of the ball.

c) Homogeneous rectangular with edge lengths a, b, c. The principle axes are parallel to the edges. One finds

$$I_{11} = I_1 = \int_{-a/2}^{+a/2} dx \int_{-b/2}^{+b/2} dy \int_{-c/2}^{+c/2} dz\, \varrho(y^2 + z^2)$$

$$= a\varrho \int_{-b/2}^{+b/2} dy \left(cy^2 + \frac{c^3}{12} \right) = a\varrho \left(c\frac{b^3}{12} + b\frac{c^3}{12} \right)$$

$$= \varrho \frac{abc}{12} (b^2 + c^2) = \frac{M}{12} (b^2 + c^2) \tag{4.2.25}$$

and analogously

$$I_2 = \frac{M}{12} (a^2 + c^2) \ , \qquad I_3 = \frac{M}{12} (a^2 + b^2) \ . \tag{4.2.26}$$

d) Homogeneous cylinder with radius R and height H. The axis of the cylinder is a principle axes. The two remaining principle moments of inertia are equal and are associated with axis perpendicular to the previous one. We take the 3-direction parallel to the axes of the cylinder and introduce cylindrical coordinates.

Then $x_1 = r\cos\varphi$, $x_2 = r\sin\varphi$, $x_3 = z$, and the volume element is given by $d^3x = r\,dr\,d\varphi\,dz$ (compare Appendix F). So,

$$I_3 = \int d^3x\, \varrho(x_1^2 + x_2^2) = \varrho \int_{-H/2}^{+H/2} dz \int_0^{2\pi} d\varphi \int_0^R dr\, r^3$$

$$= 2\pi\varrho H \frac{R^4}{4} = M \frac{R^2}{2} \quad \text{with} \quad M = \pi\varrho R^2 H \ , \tag{4.2.27}$$

$$I_2 = \int d^3x \varrho (x_1^2 + x_3^2)$$

$$= \varrho \int_{-H/2}^{+H/2} dz \int_0^{2\pi} d\varphi \int_0^R dr\, r(r^2 \cos^2 \varphi + z^2)$$

$$= \varrho \int_{-H/2}^{+H/2} dz \int_0^{2\pi} d\varphi \left(\frac{R^4 \cos^2 \varphi}{4} + R^2 \frac{z^2}{2} \right)$$

$$= \varrho \pi R^4 \frac{H}{4} + \varrho \pi R^2 \frac{H^3}{12} = \frac{M}{4}\left(R^2 + \frac{H^2}{3} \right) . \tag{4.2.28}$$

Further, $I_1 = I_2$. For $H = R\sqrt{3}$, $I_1 = I_2 = I_3$, and all axes are principle axes.

e) A homogeneous circular cone with radius R and height H. The axis of the cone is a principle axis, and the two other principle moments of inertia are equal.

We calculate the moment of inertia tensor, fixing the origin at the point of the cone:

$$I_3' = \varrho \int_0^H dz \int_0^{(R/H)z} dr\, r \int_0^{2\pi} d\varphi\, r^2 = 2\pi\varrho \int_0^H dz \frac{R^4 z^4}{4H^4}$$

$$= \pi\varrho R^4 \frac{H}{10} = 3M \frac{R^2}{10} \quad \text{with} \quad M = \pi\varrho R^2 \frac{H}{3} , \tag{4.2.29}$$

$$I_1' = \varrho \int_0^H dz \int_0^{(R/H)z} dr\, r \int_0^{2\pi} d\varphi (r^2 \sin^2 \varphi + z^2)$$

$$= \frac{3}{20} MR^2 + 2\varrho\pi \int_0^H dz \frac{R^2 z^4}{2H^2}$$

$$= \frac{3}{20} MR^2 + \frac{3}{5} MH^2 . \tag{4.2.30}$$

The center of mass lies on the axis of the cone, at a distance of $3H/4$ from the tip. Thus

$$I_3 = I_3' = \frac{3}{10} MR^2 ,$$

$$I_1 = I_2 = I_1' - \frac{9}{16} MH^2 = \frac{3}{20} M\left(R^2 + \frac{H^2}{4} \right) . \tag{4.2.31}$$

We have $I_1 = I_3$ for $H = 2R$.

4.3 The Angular Momentum of a Rigid Body, Euler's Equations

First, we will consider the angular momentum of the rigid body with reference to a body coordinate system with origin O_B.

We have,

$$L = \sum_\alpha b^{(\alpha)} \times m_\alpha \dot{b}^{(\alpha)}$$

$$= \sum_\alpha [b^{(\alpha)} \times m_\alpha (\Omega \times b^{(\alpha)})]$$

$$= \sum_\alpha m_\alpha [\Omega b^{(\alpha)2} - b^{(\alpha)}(\Omega \cdot b^{(\alpha)})]$$

$$= e_j(t) I_{jk} \Omega_k(t)$$

$$= L_j(t) e_j(t) \tag{4.3.1}$$

with

$$L_j = I_{jk}\Omega_k \ , \quad \text{i.e.} \quad L = I \cdot \Omega \ . \tag{4.3.2}$$

The angular momentum is thus a linear function of the angular velocity. It is obtained by applying the inertia tensor I to the angular velocity Ω.

The relationship between the components L_i of the angular momentum and the components Ω_i of the angular velocity with respect to the basis $e_1(t)$, $e_2(t)$, $e_3(t)$ fixed on the body is particularly simple if we choose the basis vectors to lie along the principle axes.

Then

$$I_{ik} = \delta_{ik} I_i \quad \text{and} \quad L_i = I_i \Omega_i \quad \text{(no summation here)}.$$

It is immediately obvious from this result that L and Ω are parallel if and only if Ω points in the direction of a principle axis.

The equations of motion for a rigid body are given by

$$\frac{dP}{dt} = F^{(e)} \ , \qquad \frac{dL}{dt} = N^{(e)} \ , \tag{4.3.3}$$

where P is the total momentum, $F^{(e)}$, the sum of all external forces, L, the angular momentum with respect to the center of mass, and $N^{(e)}$, the sum of all external torques.

These equations of motion are valid if the internal forces cancel each other out in the equations for momentum and angular momentum, which is the case if the internal forces $F_{\alpha\beta}$ satisfy the law of action and reaction and lie along the line connecting the points $r^{(\alpha)}$ and $r^{(\beta)}$. For a rigid body, the internal forces are constraining forces and satisfy these conditions.

We will derive the equations of motion:

For the individual particles, the equations of motion

$$m_\alpha \ddot{r}^{(\alpha)} = F_\alpha^{(e)} + f_\alpha \; , \tag{4.3.4}$$

hold, where f_α is the constraining force and $F_\alpha^{(e)}$ is the external force on the α-th particle. From d'Alembert's principle, we have

$$\sum_\alpha f_\alpha \cdot \delta r^{(\alpha)} = 0 \; ,$$

With

$$\delta r^{(\alpha)} = \delta(R + b^{(\alpha)}) = \delta R + \delta \zeta \, n \times b^{(\alpha)}$$

it follows that

$$\sum_\alpha [m_\alpha(\ddot{R} + \ddot{b}^{(\alpha)}) - F_\alpha^{(e)}] \cdot (\delta R + \delta \zeta \, n \times b^{(\alpha)}) = 0 \; . \tag{4.3.5}$$

Since δR and $\delta \zeta n$ are independent of each other, we find, if O_B is the center of mass:

$$\delta R \cdot \left(M\ddot{R} - \sum_\alpha F_\alpha^{(e)} \right) = 0 \; , \qquad \text{since}$$

$$\sum_\alpha m_\alpha b^{(\alpha)} = 0 \; , \qquad \text{so that} \tag{4.3.6}$$

$$M\ddot{R} = \sum_\alpha F_\alpha^{(e)} = F^{(e)} \; ,$$

as expected, and

$$\delta \zeta \, n \cdot \sum_\alpha b^{(\alpha)} \times [m_\alpha(\ddot{R} + \ddot{b}^{(\alpha)}) - F_\alpha^{(e)}] + \delta \zeta \, n \cdot \sum_\alpha (b^{(\alpha)} \times m_\alpha \ddot{b}^{(\alpha)} - b^{(\alpha)} \times F_\alpha^{(e)})$$

$$= \delta \zeta \, n \cdot \left[\frac{d}{dt} \sum_\alpha (b^{(\alpha)} \times m_\alpha \dot{b}^{(\alpha)} - N^{(e)} \right] = 0 \; ,$$

so that also

$$\frac{d}{dt} L = N^{(e)} \tag{4.3.7}$$

$$N^{(e)} = \sum_\alpha b^{(\alpha)} \times F_\alpha^{(e)} \; . \tag{4.3.8}$$

With

$$L = I_{ik} \Omega_k e_i(t) \; , \qquad N^{(e)} = N_i^{(e)} e_i(t)$$

the equations of motion can then be written out

$$\frac{d}{dt}L \equiv I_{ik}\dot{\Omega}_k e_i(t) + I_{ik}\Omega_k \boldsymbol{\Omega} \times e_i(t)$$

$$= N_i^{(e)} e_i(t) \tag{4.3.9}$$

or, after multiplication by $e_j(t)$:

$$N_j^{(e)} = I_{jk}\dot{\Omega}_k + I_{ik}\Omega_k \Omega_m \varepsilon_{mij} \ . \tag{4.3.10}$$

If I_{ik} is a diagonal matrix, then

$$N_1^{(e)} = I_1\dot{\Omega}_1 + (I_3 - I_2)\Omega_3\Omega_2 \ ,$$

$$N_2^{(e)} = I_2\dot{\Omega}_2 + (I_1 - I_3)\Omega_1\Omega_3 \ , \tag{4.3.11}$$

$$N_3^{(e)} = I_3\dot{\Omega}_3 + (I_2 - I_1)\Omega_2\Omega_1 \ .$$

These are *Euler's equations* for the quantities $\Omega_i(t)$.

Remarks. i) For many external forces, it is difficult to find the motion of a rigid body by solving Euler's equations. First, it is still necessary to determine the rotation $D(t)$ from the angular velocity $\boldsymbol{\Omega}(t)$, which can be done in principle by solving the differential equation

$$\dot{D}_{ij} = D_{ik}\omega_{kj} = \varepsilon_{jkr}D_{ik}\Omega_r \ ,$$

which follows directly from the definition of $\boldsymbol{\Omega}$. Second, though, the components of $F^{(e)}$ and $N^{(e)}$ with respect to the body coordinate system can only be found, in general, once $D(t)$ has been worked out.

Nevertheless, our equations of motion are fundamental and they can be solved in certain special cases.

ii) For example, Euler's equations can be solved for the case of the rigid body with no external forces:

$$\dot{P} = 0 \ , \quad \dot{L} = 0 \tag{4.3.12}$$

or, explicitly,

$$\dot{\Omega}_1 + \frac{I_3 - I_2}{I_1}\Omega_2\Omega_3 = 0 \ ,$$

$$\dot{\Omega}_2 + \frac{I_1 - I_3}{I_2}\Omega_3\Omega_1 = 0 \ , \tag{4.3.13}$$

$$\dot{\Omega}_3 + \frac{I_2 - I_1}{I_3}\Omega_1\Omega_2 = 0 \ .$$

The center of mass undergoes uniform linear motion, and for simplicity we assume it to be at rest. The system of equations (4.3.13) can be solved with elliptic functions [Landau-Lifschitz, Vol. 1]. We will not demonstrate this here.

iii) We will now describe another way to visualize the free motion of a rigid body.

Starting from Euler's equations with $N^{(e)} = 0$, we multiply by Ω_i and sum over i:

$$\sum_i I_i \dot{\Omega}_i \Omega_i = 0 = \frac{d}{dt}\left(\frac{1}{2}\sum_i I_i \Omega_i^2\right) = \dot{T}_{\text{rot}} \ . \tag{4.3.14}$$

As expected, the energy of rotation is a conserved quantity. Further, since $\dot{L} = 0$, we also have

$$\frac{d}{dt} L^2 = 0 \ . \tag{4.3.15}$$

Expressed in terms of the components L_i of L in the body coordinate system, the two conserved quantities are given by

$$L^2 = L_1^2 + L_2^2 + L_3^2 \ , \tag{4.3.16a}$$

$$T_{\text{rot}} = \frac{L_1^2}{2I_1} + \frac{L_2^2}{2I_2} + \frac{L_3^2}{2I_3} \ . \tag{4.3.16b}$$

These represent a sphere of radius L and an ellipsoid, whose axes point in the same direction as the principle axes and are of length $a_i = (2I_i T_{\text{rot}})^{1/2}$.

The time dependence of $L_i(t)$ must be such that the vector (L_1, L_2, L_3) moves along the lines formed by the intersection of the sphere and ellipsoid (Fig. 4.3.1).

From the drawing, we see that rotations about the axes corresponding to the largest and smallest moments of inertia are stable, whereas those about the axis with middle moment of inertia are unstable.

In the inertial coordinate system *fixed in space*, the motion of the rigid body can be visualized by following the motion of the ellipsoid of inertia. This description of the free motion of a rigid body goes back to *Poinsot*[2] (1834).

Here, L is time-independent and always points in the direction of the vector normal to the inertia ellipsoid at the point $\zeta = \Omega/(2T_{\text{rot}})^{1/2}$, since the vector normal to the surface described by the equation

$$F(\zeta_1, \zeta_2, \zeta_3) = I_1\zeta_1^2 + I_2\zeta_2^2 + I_3\zeta_3^2 - 1 = 0$$

---

[2] *Poinsot, Louis* (*1777 Paris, d. 1859 Paris). French mathematician. His principle area of research was mechanics, in particular, the motion of a top. He invented the term "couple" for a pair of forces F_1, F_2 with $F_1 + F_2 = 0$.

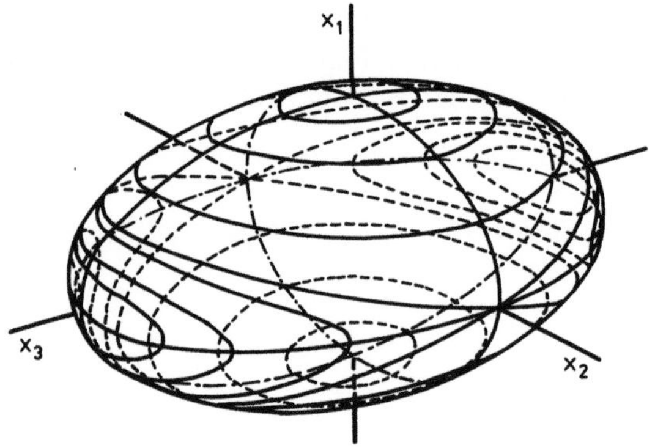

Fig. 4.3.1. The lines of intersection of the sphere and ellipsoid

is given by

$$\left(\frac{\partial F}{\partial \zeta_1}, \frac{\partial F}{\partial \zeta_2}, \frac{\partial F}{\partial \zeta_3}\right) = 2(I_1\zeta_1, I_2\zeta_2, I_3\zeta_3) \sim \boldsymbol{L} \ .$$

The inertia ellipsoid moves, then, in such a way that its normal vector at the point ζ is always perpendicular to a fixed plane (Fig. 4.3.2). The path traced out on the invariable plane is called the *herpolhode*[3] and the path traced on the inertia ellipsoid itself is called the *polhode*.

The height of the center of the inertia ellipsoid over the plane is

$$h = \zeta \cdot \frac{\boldsymbol{L}}{L} = \frac{\boldsymbol{\Omega} \cdot \boldsymbol{L}}{L\sqrt{2T_{\text{rot}}}} = \frac{\sqrt{2T_{\text{rot}}}}{L} \ , \tag{4.3.17}$$

since $\boldsymbol{\Omega} \cdot \boldsymbol{L} = 2T_{\text{rot}}$, hence h is time-independent.

Finally, ζ points in the direction of the instantaneous axis of rotation, so that the point of the ellipsoid lying on the plane is instantaneously at rest.

The inertia ellipsoid therefore rolls without gliding, with center fixed at a given distance above the fixed plane normal to the vector $\boldsymbol{L}$. In particular, the principle axes and the vector $\boldsymbol{\Omega}$ move around an axis parallel to $\boldsymbol{L}$.

An important special case is the *symmetric top*, for which $I_1 = I_2$. In this case, the inertia ellipsoid is an ellipsoid of rotation. Its 3-axis is called the *figure axis* of the top.

The vector $\boldsymbol{\Omega}$, as is evident in Fig. 4.3.3, runs along a circular cone fixed in space with axis $\boldsymbol{L}$, called the *space cone*, which has as its tip the center of the inertia ellipsoid. Another cone with the same tip and with the figure axis e_3 as its axis rolls (without slipping) on the space cone and is called the *body cone*. The line which forms the instantaneous intersection of these two cones points in the

---

[3] Polhode, herpolhode (Greek) from *hodós*: path and *hérpein*: crawl, slowly precede.

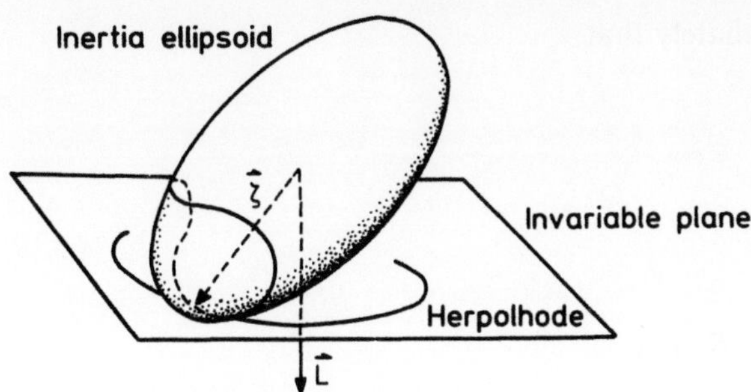

Fig. 4.3.2. Poinsot's description of the motion of a free rigid body

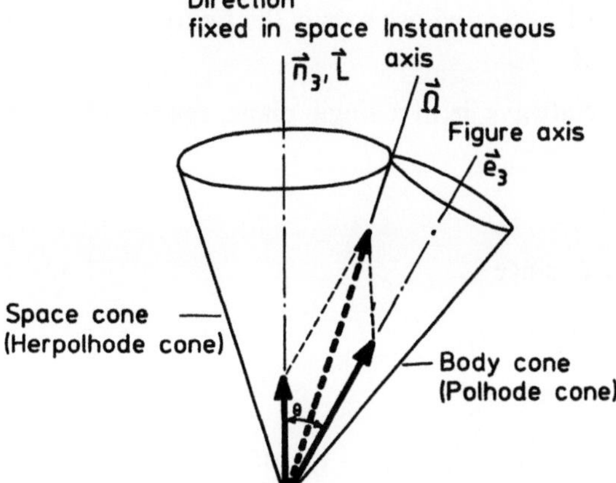

Fig. 4.3.3. Nutation in a free symmetrical top

direction of Ω. The figure axis also runs along the surface of a cone, which, in the literature, is often called the cone of precession. This circular motion of the figure axis around the fixed axis given by the vector L is often called (regular) precession. We, however, prefer to call this motion *nutation*[4], in order to save the word precession for a circular motion of L along an axis fixed in space (e.g. a vertical axis, compare Sect. 4.4 on the heavy top).

iv) Euler's equations for the free symmetrical top are:

$$\dot{\Omega}_3 = 0 \ , \quad \dot{\Omega}_1 + A\Omega_2 = 0 \ , \quad \dot{\Omega}_2 - A\Omega_1 = 0 \tag{4.3.18}$$

with

$$A = \frac{I_3 - I_1}{I_1} \, \Omega_3 \ .$$

---

[4] Nutation (Latin) from *nutare* = nod, thus nodding motion.

From this, it follows immediately that

$$\frac{d}{dt}(\Omega_1^2 + \Omega_2^2) = 0 \quad \text{and}$$

$$\frac{d}{dt}\Omega^2 = 0 \quad \text{as well as} \tag{4.3.19}$$

$$\Omega_1 = B\cos At , \quad \Omega_2 = B\sin At .$$

It follows further (which has in part already been stated in (iii)) that:

a) The instantaneous angular velocity vector $\boldsymbol{\Omega}$ moves uniformly, in the body system, around the surface of a cone whose axis is the figure axis e_3. The angular speed of this rotation is A.

b) The vectors $\boldsymbol{L}$, $e_3(t)$, and $\boldsymbol{\Omega}$ always lie in a single plane, spanned by $e_3(t)$ and

$$\boldsymbol{\Omega}_\perp = \Omega_1 e_1(t) + \Omega_2 e_2(t)$$

and contains the fixed direction $\boldsymbol{L}$, since

$$\boldsymbol{L} = I_1[\Omega_1 e_1(t) + \Omega_2 e_2(t)] + I_3\Omega_3 e_3(t)$$

$$= I_1\boldsymbol{\Omega}_\perp + I_3\Omega_3 e_3(t) . \tag{4.3.20}$$

c) The nutation of $\boldsymbol{\Omega}$ and e_3 occurs uniformly on the surface of a cone around $\boldsymbol{L}$.

The angular speed Ω_N of this nutation can be found by dividing $\boldsymbol{\Omega}$ into components in the direction of e_3 and $\boldsymbol{L}$. From

$$\boldsymbol{L} = I_1\boldsymbol{\Omega}_\perp + I_3\Omega_3 e_3(t) , \quad \text{it follows that}$$

$$\boldsymbol{\Omega} = \Omega_3 e_3(t) + \boldsymbol{\Omega}_\perp = \Omega_3 e_3(t) + \frac{[\boldsymbol{L} - I_3\Omega_3 e_3(t)]}{I_1}$$

$$= e_3(t)\Omega_3\frac{I_1 - I_3}{I_1} + \frac{\boldsymbol{L}}{I_1} . \tag{4.3.21}$$

The components in the e_3-direction again have magnitude $|A|$ and we see that

$$\Omega_N = \frac{L}{I_1} \tag{4.3.22}$$

v) For a system of N rigid bodies, we have the equations of motion

$$\dot{\boldsymbol{L}}_\alpha = \boldsymbol{N}_\alpha^{(e)} + \boldsymbol{N}_\alpha^f ,$$

$$\dot{\boldsymbol{P}}_\alpha = \boldsymbol{F}_\alpha^{(e)} + \boldsymbol{f}_\alpha , \quad \alpha = 1, \ldots, N \tag{4.3.23}$$

where the f_α are the constraining forces and the $F_\alpha^{(e)}$ are the external forces on each rigid body, and N_α^f and $N_\alpha^{(e)}$ are the corresponding torques.

Such constraining forces and corresponding torques must exist, for example, if a certain angular velocity specified in magnitude and direction is to be maintained.

Let us consider a rigid body, which rotates around a given fixed axis with direction n with fixed angular velocity Ω.

If, for the purpose of simplicity, we ignore the other external forces, we have for the force of constraint:

$$f = \dot{P} \; . \tag{4.3.24}$$

The constraining force is thus given by the acceleration of the center of mass, which is easy to calculate since it moves uniformly in a circular orbit around the axis. Hence the total force of constraint vanishes if and only if the axis of rotation goes through the center of mass. If the center of mass is not on the rotation axis, we speak of a *static imbalance* and, in this case, the bearing of the rotating axis must exert a constraining force so that the position of the axis remains steady. The "reaction" to this force then acts on the bearing itself.

Since the axis of rotation is fixed with respect to the body and $\Omega = $ constant, we also have

$$\dot{L} = I_{ik}\Omega_k\dot{e}_i = I_{ik}\Omega_k\Omega \times e_i = \Omega \times L \; , \quad \text{thus} \tag{4.3.25}$$

$$N^f = \dot{L} = \Omega \times L \; . \tag{4.3.26}$$

Then $N^f = 0$ if and only if Ω is parallel to L, that is, if the rotation is around a principle axis. In every other case there is an additional load on the bearing of the rotating axis, which is called *dynamic imbalance*.

The most important cases of damage due to the load on the bearing caused by imbalance occur when a rigid body turns around a fixed axis over a long period of time (as for automobile wheels and machine tools). Thus, it is necessary in these cases to have the system *balanced* to ensure that the center of mass lies on the axis of rotation and that this axis of rotation is a principle axis. In this process, it is necessary to change the distribution of mass, for example, by boring out material.

4.4   The Equations of Motion for the Eulerian Angles

Euler's equations of motion demand a knowledge of the components (in the body-fixed system of coordinates) of the torque of the external forces on the elements of mass. These components are not always easy to calculate. Besides, Ω is not really the quantity we want to find, rather we would like to work out $\varphi(t)$, $\theta(t)$, and $\psi(t)$, i.e., the instantaneous position of the rigid body. From the kinetic energy formula for a rigid body

$$T = \tfrac{1}{2}M\dot{R}^2 + \tfrac{1}{2}I_{ij}\Omega_i\Omega_j$$

and our knowledge of the dependence of Ω_k on the Eulerian angles and their time derivatives, we can immediately formulate the Lagrangian

$$\begin{aligned}
L &= \tfrac{1}{2}L(R^2, \dot{R}, \varphi, \theta, \psi, \dot{\varphi}, \dot{\theta}, \dot{\psi}) \\
&= \tfrac{1}{2}M\dot{R}^2 + \tfrac{1}{2}I_1(\dot{\theta}\cos\psi + \dot{\varphi}\sin\theta\sin\psi)^2 \\
&\quad + \tfrac{1}{2}I_2(-\dot{\theta}\sin\psi + \dot{\varphi}\sin\theta\cos\psi)^2 \\
&\quad + \tfrac{1}{2}I_3(\dot{\psi} + \dot{\varphi}\cos\theta)^2 - U(R, \varphi, \theta, \psi)
\end{aligned} \tag{4.4.1}$$

and then write out Lagrange's equations.

Here, we will demonstrate an interesting and important example, a symmetric top pivoted at a fixed point. The top moves in the gravitational field of the earth. Let the origin of the body coordinate system O'_B be at the end of the top which is fixed in space at the origin of the space-fixed coordinate system (Fig. 4.4.1). Then, if $I'_1 = I'_2$ and I'_3 are the principle moments of inertia with respect to O'_B, the kinetic energy is given by:

$$T = \tfrac{1}{2}I'_1(\dot{\theta}^2 + \dot{\varphi}^2\sin^2\theta) + \tfrac{1}{2}I'_3(\dot{\psi} + \dot{\varphi}\cos\theta)^2 \ , \tag{4.4.2}$$

where we have chosen $e_3(t)$ as the figure axis.

In order to calculate the potential energy, let us consider the external force which acts on an element of mass. It is given by

$$F_\alpha^{(e)} = m_\alpha g$$

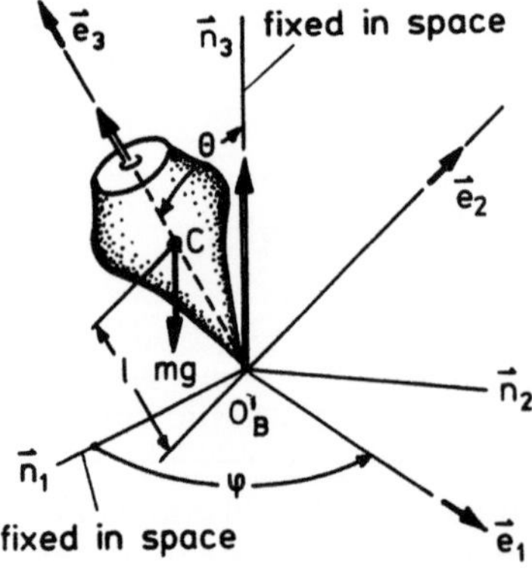

Fig. 4.4.1. The heavy symmetrical top

and therefore, following (4.3.6),

$$F^{(e)} = Mg \quad \text{or}$$

$$U = -\sum_\alpha m_\alpha g \cdot b^{(\alpha)} = -g \cdot \sum_\alpha m_\alpha b^{(\alpha)} = -Mg \cdot R \; , \tag{4.4.3}$$

where now R, the position vector of the center of mass, lies on the e_3-axis, whereas $-g$ points in the n_3-direction. Here, M again represents the total mass of the top.

Since $\angle\,(n_3, e_3) = \theta$, the potential energy can be written

$$U = Mgl\cos\theta \; , \tag{4.4.4}$$

and $l = |R|$ is a fixed distance from O'_B. We then find that the Lagrangian is given by:

$$L = L(\theta, \dot\varphi, \dot\theta, \dot\psi) = \tfrac{1}{2}I'_1(\dot\theta^2 + \dot\varphi^2\sin^2\theta)$$

$$+ \tfrac{1}{2}I'_3(\dot\psi + \dot\varphi\cos\theta)^2 - Mgl\cos\theta \; . \tag{4.4.5}$$

We see immediately:

The Lagrangian is independent of φ and ψ. We conclude that

$$p_\psi = \frac{\partial L}{\partial\dot\psi} = I'_3(\dot\psi + \dot\varphi\cos\theta) \tag{4.4.6}$$

is a conserved quantity, as is

$$p_\varphi = \frac{\partial L}{\partial\dot\varphi} = I'_1\sin^2\theta\,\dot\varphi + I'_3\cos\theta(\dot\psi + \dot\varphi\cos\theta) \; . \tag{4.4.7}$$

The fact that the generalized momenta p_ψ and p_φ are conserved quantities follows immediately from the invariance of the Lagrangian with respect to rotations around the space-fixed as well as the body-fixed 3-axis. It is easy to show that p_ψ and p_φ are the components of the angular momentum L in the e_3- and n_3-directions, respectively.

If we substitute

$$p_\psi = I'_3(\dot\psi + \dot\varphi\cos\theta)$$

into p_φ, we find

$$p_\varphi = I'_1\dot\varphi\sin^2\theta + p_\psi\cos\theta \; ,$$

so that

$$\dot{\varphi} = \frac{p_\varphi - p_\psi \cos\theta}{I_1' \sin^2\theta} \quad \text{and} \tag{4.4.8}$$

$$\dot{\psi} = \frac{p_\psi}{I_3'} - \dot{\varphi}\cos\theta \; . \tag{4.4.9}$$

Finally, since the Lagrangian does not depend explicitly on time, the energy $E = T + U$ is also a conserved quantity.

$$E = T + U = \frac{I_1'}{2}(\dot{\theta}^2 + \dot{\varphi}^2 \sin^2\theta) + \frac{I_3'}{2}(\dot{\varphi}\cos\theta + \dot{\psi})^2 + Mgl\cos\theta \; .$$

We now substitute $\dot{\varphi}$ and $\dot{\psi}$ from (4.4.8, 9) into this expression. We then find

$$E = \frac{I_1'}{2}\dot{\theta}^2 + \frac{(p_\varphi - p_\psi\cos\theta)^2}{2I_1'\sin^2\theta} + \frac{p_\psi^2}{2I_3'} + Mgl\cos\theta \tag{4.4.10}$$

or

$$E = \frac{I_1'}{2}\dot{\theta}^2 + U_{\text{eff}}(\theta) \quad \text{with} \tag{4.4.11}$$

$$U_{\text{eff}}(\theta) = \frac{(p_\varphi - p_\psi\cos\theta)^2}{2I_1'\sin^2\theta} + \frac{p_\psi^2}{2I_3'} + Mgl\cos\theta \; . \tag{4.4.12}$$

We have again effectively reduced our problem to one dimension, and thus, in principle, we have solved the motion of a symmetrical top in a gravitational field. First, we calculate $\theta(t)$ from equation (4.4.11) with the usual methods (in these cases the solution involves elliptical functions) and then we find $\varphi(t)$ and $\psi(t)$ from (4.4.8, 9) by integration. In other words: our problem is completely integrable.

It is also possible to come to a good understanding of the motion qualitatively without performing calculations.

The effective potential $U_{\text{eff}}(\theta)$ has a form like the one depicted in Fig. 4.4.2.

The poles at $\theta = 0$ and $\theta = \pi$ arise from the denominator $\sin^2\theta$. There is only one pole if $p_\varphi = \pm p_\psi$.

The angle of inclination θ of the figure axis with respect to the vertical oscillates between the values θ_1 and θ_2, which are determined by E, p_φ, and p_ψ. Since the torque

$$N^{(e)} = \sum_\alpha b^{(\alpha)} \times m_\alpha g = -Mgl e_3 \times n_3$$

is exerted on the top, and it is normal to the plane spanned by the figure axis e_3 and the vertical axis n_3, the change of L is also in this direction. From these

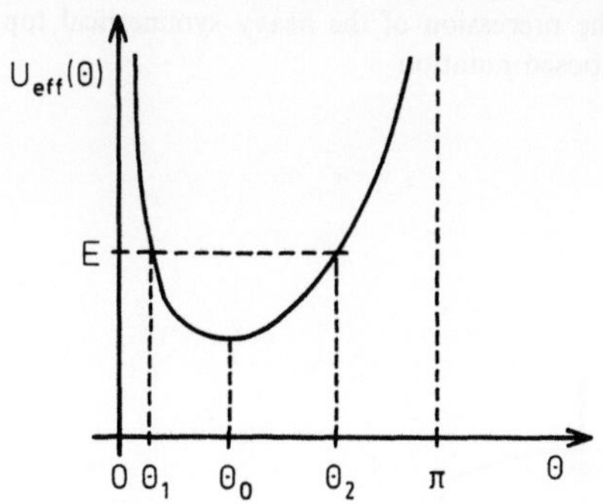

Fig. 4.4.2. The effective potential as a function of the angle θ

considerations, we see that the motion of the figure axis around the vertical axis n_3 looks like one the possibilities represented in Fig. 4.4.3.

The motion of the top is composed of three parts:

i) A motion of the angular momentum L around the vertical axis, which is called *precession*[5].
ii) A nutation of the figure axis e_3 around L. This is expressed, for example, in the time dependence of $\theta(t)$.
iii) A rotational motion $\psi(t)$ of the top around its figure axis.

Whether the wave-like motion of Fig. 4.4.3a, the winding motion of Fig. 4.4.3b, or the garland-shaped motion of Fig. 4.4.3c occurs depends on whether the sign of

$$\dot{\varphi} = \frac{p_\varphi - p_\psi \cos\theta}{I_1' \sin^2\theta} \tag{4.4.13}$$

remains fixed or changes during the motion. This is decided by the values of p_φ, p_ψ, and E. Case (c) here is the borderline case between (a) and (b).

If E corresponds exactly to the minimum θ_0 of U_{eff}, then $\theta(t) = \theta_0 = \text{con-}$stant, the inclination θ does not change, and $\dot{\varphi}$ and $\dot{\psi}$ are constant. In this special case, called *regular precession*, the figure axis along with L and Ω moves uniformly in a circular cone around n_3. The angular momentum vector L has constant magnitude, which we see from the equation

$$E = \frac{L_3^2}{2I_3} + \frac{(L_1^2 + L_2^2)}{2I_1} + U(\theta_0) \tag{4.4.14}$$

and L in this case always lies in the plane spanned by e_3 and n_3.

---

[5] Precession (Latin) from *praecedere* = to move forward, advance: the motion of the axis of a top under the influence of an external torque.

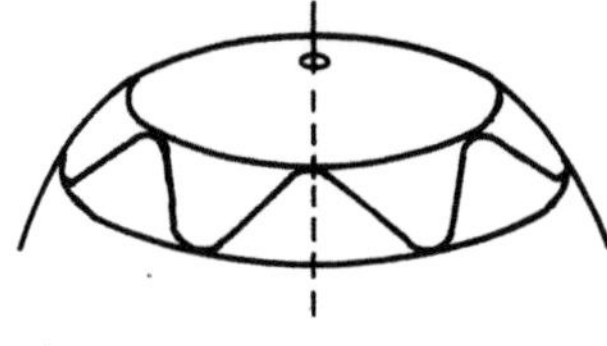

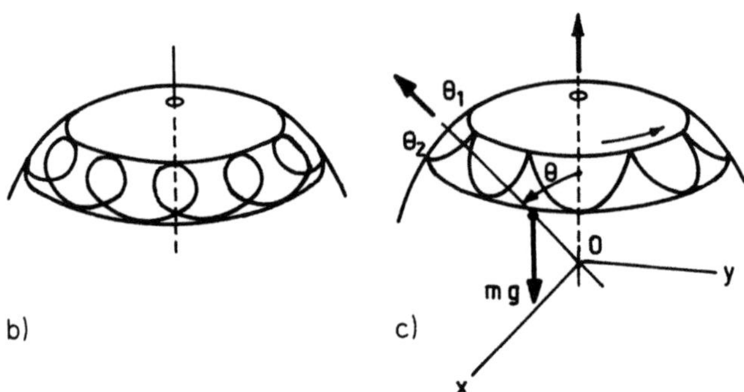

Fig. 4.4.3. The precession of the heavy symmetrical top with superimposed nutation

Remarks. i) For the free symmetrical top we have $l = 0$ (the center of mass is fixed in space) and $I'_1 = I_1$, $I'_3 = I_3$. The gravitational force term is then absent from the effective potential U_{eff}.

The angular momentum L is conserved and without loss of generality we let the n_3-direction coincide with the L-direction. That means $p_\varphi = L$ and $p_\psi = L\cos\theta$, since p_ψ is the e_3-component of L. Since p_φ and p_ψ are conserved, $\theta(t) = \theta_0$ independent of time and is given by

$$p_\psi - p_\varphi \cos\theta_0 = 0 \ .$$

Then, we have

$$\dot\varphi = \frac{p_\varphi - p_\psi \cos\theta_0}{I_1 \sin^2\theta_0} = \frac{L}{I_1}\frac{1 - \cos^2\theta_0}{\sin^2\theta_0} = \frac{L}{I_1}$$

and

$$\dot\psi = \frac{p_\psi}{I_3} - \dot\varphi\cos\theta_0 = L_3\left(\frac{1}{I_3} - \frac{1}{I_1}\right) = \frac{I_1 - I_3}{I_1}\Omega_3 \ .$$

$\dot\varphi$ agrees with the nutation frequency of the free symmetrical top $\Omega_{\text{N}} = L/I_1$ derived above, and $\dot\psi$ is identical, up to a sign (which deserves some thought), with the frequency of revolution A of Ω around e_3 in the body-fixed system compare (4.3.11).

ii) We now again consider the symmetrical top in a gravitational field. Our goal is to find the precession frequency Ω_{P} for an almost regular precession.

We limit ourselves to the so-called *fast top*, for which the rotational energy is far greater than the potential energy in the gravitational field.

In the case of approximately regular precession, L points almost exactly in the direction of the figure axis, so that $L \approx p_\psi$ and $p_\varphi \approx L \cos \theta$. If we leave out the gravitational term $Mgl \cos \theta$ in U_{eff}, then θ_0 with $p_\varphi - p_\psi \cos \theta_0 = 0$ becomes a minimum of the potential. The gravitational term in U_{eff} is now small in comparison to the two other contributions, and this term will displace the minimum θ_0 of U_{eff} only slightly. We then set $\theta = \theta_0 + x$ with $x \ll 1$ and look for the minimum $\theta_1 = \theta_0 + x_1$.

If we expand $U_{\text{eff}}(\theta)$ around θ_0 in powers of x, we find

$$U_{\text{eff}}(\theta_0 + x) = \frac{(p_\varphi - p_\psi \cos \theta_0 + p_\psi \sin \theta_0\, x + \ldots)^2}{2I'_1 \sin^2(\theta_0 + x)}$$

$$+ Mgl(\cos \theta_0 - x \sin \theta_0 - \tfrac{1}{2}x^2 \cos \theta_0 + \ldots) + \text{const}$$

$$= \left(\frac{p_\psi^2}{2I'_1} - \frac{Mgl}{2} \cos \theta_0 \right) x^2$$

$$- Mglx \sin \theta_0 + O(x^3) + \text{const} . \tag{4.4.15}$$

For the fast top,

$$\frac{p_\psi^2}{2I'_1} \approx T_{\text{rot}} \gg \frac{Mgl}{2} \cos \theta_0 \ ,$$

so that we can also ignore the term $(Mgl \cos \theta_0)/2$ in the coefficient of x^2. To lowest order, we have

$$U_{\text{eff}}(\theta_0 + x) = \frac{p_\psi^2 x^2}{2I'_1} - x Mgl \sin \theta_0 \ .$$

The minimum occurs at

$$x_1 = \frac{I'_1 Mgl}{p_\psi^2} \sin \theta_0 \ll 1 \ , \tag{4.4.16}$$

hence at

$$\theta_1 = \theta_0 + \frac{Mgl I'_1}{p_\psi^2} \sin \theta_0 \ . \tag{4.4.17}$$

The frequency of small oscillations of θ around the minimum θ_1 is thus immediately given by (here we use a few facts from Chap. 6):

$$\omega^2 = \frac{U''_{\text{eff}}(\theta_1)}{I'_1} = \frac{p_\psi^2}{I'^2_1} \approx \frac{L^2}{I'^2_1} = \Omega_N^2 \ . \tag{4.4.18}$$

As required, we have found again the nutation frequency Ω_N with I_1' replacing I_1. In order to obtain the precession frequency Ω_P, we need only substitute θ_1 into

$$\dot{\varphi} = \frac{p_\varphi - p_\psi \cos\theta}{I_1' \sin^2\theta}$$

to get:

$$\Omega_P = \frac{p_\varphi - p_\psi \cos\theta_0 + p_\psi \sin\theta_0 \, x_1 + \ldots}{I_1' \sin^2(\theta_0 + x_1)}$$

$$= \frac{p_\psi}{I_1'} \frac{x_1}{\sin\theta_0} + O(x_1^2) \ . \tag{4.4.19}$$

With the value we have found for x_1, we find, to lowest approximation

$$\Omega_P = \frac{Mgl}{L} \ . \tag{4.4.20}$$

For $l = 0$, that is, for a top with fixed center of mass, we have $\Omega_P = 0$, since a free symmetric top rotating around its figure axis has a fixed axis of rotation, from the conservation of angular momentum.

The value $\Omega_P = Mgl/L$ can also be found from the following plausibility consideration:

The precession of L around the vertical axis is an effect of the torque of the gravitational force. Then, if the motion occurs in such a way that L^2 is approximately constant and L has a constant angle θ_1 with respect to the vertical axis, we expect that the frequency Ω_P of precession will satisfy

$$\dot{L} = \Omega_P \times L = N^{(e)} = Mle_3 \times g \approx \frac{L}{L} \times Mlg \ , \tag{4.4.21}$$

thus

$$\Omega_P L = Mgl \quad \text{and} \quad \Omega_P = \frac{Mgl}{L} \ .$$

iii) We will discuss further the stability of the rotation of a symmetric top in a gravitational field about the vertical axis n_3.

The case $\theta = 0$ is only possible for $p_\varphi = p_\psi \, (= L)$, and then for $\theta \ll 1$:

$$U_{\text{eff}}(\theta) = \frac{L^2(1 - \cos\theta)^2}{2I_1' \sin^2\theta} + Mgl\cos\theta + \text{const}$$

$$= \frac{I_3'\Omega^2\theta^2}{2I_1' \cdot 4} - Mgl\frac{\theta^2}{2} + O(\theta^3) + \text{const} \ . \tag{4.4.22}$$

For the case

$$\frac{I_3'^2 \Omega^2}{4 I_1'} > Mgl \quad \text{or}$$

$$\Omega^2 > \frac{4 Mgl I_1'}{I_3'^2} \equiv \Omega_0^2 \tag{4.4.23}$$

U_{eff} now has a minimum at $\theta = 0$.

Rotation around the vertical axis is thus stable for $\Omega^2 > \Omega_0^2$ and unstable for $\Omega^2 < \Omega_0^2$.

Indeed, we observe in reality that a rapidly spinning top keeps its axis constant (a *sleeping top*) and only begins to tumble when a large enough part of its rotational energy is used up by friction, so that its angular speed goes below the critical value Ω_0.

iv) The exact analysis of the rotational motion of the earth is an especially fascinating and important application of the theory of tops. The actual situation is quite complicated and we can only indicate a few of the most important results.

The earth, to a good approximation, can be viewed as a symmetric top which rotates in such a way that the directions of e_3, L, and $\mathbf{\Omega}$ almost, but not quite exactly, coincide. Nutations should appear as variations of the height of the instantaneous pole of rotation above the horizon.

The earth, as a top, is not free, because torques from the tidal forces exerted by the sun and moon act on it (see Sect. 2.6, Remark (ii)). These torques produce a precessional motion, namely, the well-known precession of the equinoxes, which was known even to the ancient Greek astronomer Hipparchos of Nicea.[6] The period of rotation is approximately 26,000 years; it can be calculated rather reliably from the theory of symmetric tops.

Nutation leads to an observable rotation of the vectors $\mathbf{\Omega}$ and e_3 in the body-fixed system, i.e. in reference to the earth. The period of revolution of this motion is, as we have seen, $A = \Omega_3 (I_3 - I_1)/I_1$. Here, $\Omega_3 = 2\pi/\text{Day}$ and $(I_3 - I_1)/I_1 \approx 1/300$, if we view the earth as an ellipsoid of rotation with oblateness $1/300$. We then find a period of approximately 300 days.

This prediction was first made by Euler in the year 1765. The actual nutation period was not found until the year 1888 (by F. Küstner). The first exact measurement was made by S.C. Chandler in the year 1891. In the extremely complicated motion of the celestial pole, he was able to find a component with

---

[6] *Hipparchos of Nicea* (today İznik, Turkey) (*circa 190, d. circa 125 B.C.) Greek astronomer and geographer. Among other works, he produced a star catalogue, measured the distance to the sun and moon and discovered the retreat of the Spring equinox along the ecliptic.

a period of approximately 418 days. The half-angle of aperture of the corresponding cone measured only 0.3″ (seconds of arc) – this corresponds to a distance of approximately 9 meters over the surface of the earth. By comparison: the visible diameter of the full moon measures around 1800″. The discrepancy between Euler's prediction and the actual observation arises from the fact, known today, that the Earth, because of its huge size, cannot be viewed as a perfectly rigid body.

Other motions of the earth are superimposed on this nutational motion:

Of the same order of magnitude, there is a variation in the height of the pole with a period of 365 days which arises from the yearly melting of the polar ice caps.

There is a substantially larger variation in the pole height due to tidal forces from the sun and the moon. It is called (unfortunately) lunisolar nutation. Its most important component has an amplitude of 9″ and a period of about 18.6 years. Because of this large period, it can be clearly separated from the actual effects of nutation.

Problems

4.1   The Moments of Inertia of a Homogeneous Ellipoid. Calculate the principle moments of inertia of a homogeneous ellipsoid of mass m and semi-axes a_1, a_2, a_3.

4.2   Inertia Tensors. Calculate the inertia tensors of the following bodies in each case with respect to the center of gravity.

a) Ball: The mass M is distributed homogeneously in a spherical shell with inner radius R_1 and outer radius R_2.
b) Homogeneous pipe (cylindrical) with inner radius R_1, outer radius R_2, and height H.
c) A four-atom molecule whose atoms lie on the vertices of a regular tetrahedron.

4.3   Symmetries and the Inertia Tensor. a) Show that the tensor of inertia I defined by

$$I_{ij} = \int dV\, \varrho(\boldsymbol{x})(\delta_{ij}x^2 - x_i x_j)$$

transforms like a tensor, i.e. that

$$I \to I' = AIA^{\mathrm{T}} , \quad \text{i.e.} \quad I_{jk} \to I'_{jk} = A_{jl}A_{km}I_{lm} ,$$

under the rotation $A \in O(3)$

$$x \to x' = Ax , \quad \text{i.e.} \quad x_j \to x'_j = A_{jk}x_k .$$

b) Let a rigid body have one of the following symmetry properties:

i) Mirror symmetry with respect to a plane perpendicular to the 3-axis.
ii) Symmetry under rotation through the angle π about the 3-axis.
iii) Symmetry under rotation through an angle α about the 3-axis, where $\alpha \neq 0$ and $\alpha \neq \pi$.

Show (with help from a)) that the inertia tensor I then has the following properties:

i) $I_{13} = I_{31} = I_{23} = I_{32} = 0$

ii) $I_{13} = I_{31} = I_{23} = I_{32} = 0$

iii) $I_{jk} = 0 \quad \text{for} \quad j = k, I_{11} = I_{22}$.

(in cases a) and b) the 3-axis is therefore a principle axis, in case c) the 1,2,3-axes are all principle axes, and the ellipsoid of inertia is rotationally symmetric about the 3-axis).

4.4   The Top. Euler's equations for the free spinning top read:

$$I_1\dot{\Omega}_1 + (I_3 - I_2)\Omega_2\Omega_3 = 0$$

$$I_2\dot{\Omega}_2 + (I_1 - I_3)\Omega_3\Omega_1 = 0$$

$$I_3\dot{\Omega}_3 + (I_2 - I_1)\Omega_1\Omega_2 = 0$$

a) Show that a rotation about one of the principle axes is stable if and only if the corresponding principle moment of inertia is either greater than or lesser than both of the others. (Hint: Assume, say, that $|\Omega_1|$, $|\Omega_2| \ll |\Omega_3|$. Then, neglecting the term $\Omega_1\Omega_2$, solve Euler's equations in this approximation. Finally, investigate under what conditions the assumption $|\Omega_1|$, $|\Omega_2| \ll |\Omega_3|$ is consistent for all times with the solution found in this way.)

b) Using the solution found in a) for the stable case, show that the nutation period of the earth's axis is about 300 days, if we consider the earth as a rigid body in the shape of a flattened rotational ellipsoid (see problem 13) with flattening factor $\delta \approx 1/300$ (thus $a_1 = a = a_2$, $a_3 < a$, $\delta = 1 - a_3/a$).

5. Motion in a Noninertial System of Reference

So far in our study of the motion of material bodies, we have assumed that we have been working in an inertial coordinate system, that is, a system in which Newton's equation of motion holds in the form

$$m\ddot{r} = F$$

where F is the force which acts on a particle of mass m. This force can arise from other particles, or can be conveyed by an external field.

If we know the equations of motion in an inertial frame of reference, we can derive the equations in a noninertial system. In this chapter, we will discuss the so-called fictitious or inertial forces, which appear in this derivation.

5.1 Fictitious Forces in Noninertial Systems

Let the coordinates in the inertial frame of reference be given by (O, n_1, n_2, n_3) and those in the noninertial system by

$$(O_B, e_1(t), e_2(t), e_3(t)) \ .$$

Then a point P with $\overrightarrow{OP} = r$ can also be described in the non-inertial coordinate system, as we saw for the rigid body, by

$$\overrightarrow{O_B P} = b = b_i e_i(t) \ . \tag{5.1.1}$$

Then

$$r = R + b \quad \text{with} \quad R = \overrightarrow{OO_B} \ .$$

The sole difference between this description and that for the rigid body is that in this case we do not always have $\dot{b}_i = 0$.

In order to find the equations of motion in the noninertial system, let us calculate the quantities $\dot{r}$ and $\ddot{r}$. We have

$$\dot{r} = \dot{R} + \dot{b} = \dot{R} + \dot{b}_i e_i(t) + b_i \boldsymbol{\Omega} \times e_i(t)$$

$$= \dot{R} + v + \boldsymbol{\Omega} \times b \quad \text{with} \tag{5.1.2}$$

$$v = \dot{b}_i e_i(t) \; , \quad \text{and} \tag{5.1.3}$$

$$\ddot{r} = \ddot{R} + \ddot{b}_i e_i(t) + 2\dot{b}_i(\mathbf{\Omega} \times e_i) + b_i[\mathbf{\Omega} \times (\mathbf{\Omega} \times e_i)] + \dot{\mathbf{\Omega}} \times b$$

$$= \ddot{R} + a + 2\mathbf{\Omega} \times v + \mathbf{\Omega} \times (\mathbf{\Omega} \times b) + \dot{\mathbf{\Omega}} \times b \tag{5.1.4}$$

with

$$a = \ddot{b}_i e_i(t) \; . \tag{5.1.5}$$

The quantities v and a represent the velocity and acceleration as they are measured in the noninertial system. Here, $\ddot{R}$ is the acceleration that O_B experiences with respect to O. $\ddot{R}$ and $\mathbf{\Omega}$ are given in the inertial system.

The equation of motion in the inertial system

$$m\ddot{r} = -\nabla U(r) = -\frac{\partial U}{\partial r}$$

can be rewritten as

$$ma = -\frac{\partial \hat{U}}{\partial b} - m\ddot{R} - 2m(\mathbf{\Omega} \times v)$$

$$- m\mathbf{\Omega} \times (\mathbf{\Omega} \times b) - m\dot{\mathbf{\Omega}} \times b \quad \text{with} \tag{5.1.6}$$

$$\hat{U}(b) = U(R + b) \; .$$

In the noninertial system, therefore, in addition to the Newtonian forces there are additional forces, the so-called *fictitious forces*, which obviously depend on $\mathbf{\Omega}(t)$ and $R(t)$, i.e., the amount by which the system deviates from being inertial.

Before we investigate these fictitious forces, we will first show that this equation of motion can also be derived from a Lagrangian. This again has the advantage that approximations, coordinate transformations, etc. can be taken into account in the Lagrangian.

In the inertial system, we have

$$L = \tfrac{1}{2}m\dot{r}^2 - U(r) \; .$$

With

$$\dot{r} = \dot{R} + v + \mathbf{\Omega} \times b$$

it follows that

$$L = \tfrac{1}{2}m\dot{R}^2 + \tfrac{1}{2}mv^2 + \tfrac{1}{2}m(\mathbf{\Omega} \times b)^2 + m\dot{R} \cdot (v + \mathbf{\Omega} \times b)$$

$$+ mv \cdot (\mathbf{\Omega} \times b) - U(R + b) \; . \tag{5.1.7}$$

But

$$\dot{R}\cdot(v + \Omega \times b) = \dot{R}\cdot\frac{db}{dt} = \frac{d}{dt}(\dot{R}\cdot b) - b\cdot\ddot{R} \ ,$$

so that the Lagrangian is given by

$$L = \tfrac{1}{2}mv^2 + mv\cdot(\Omega \times b) + \tfrac{1}{2}m(\Omega \times b)^2 - \hat{U}(b)$$

$$+ \tfrac{1}{2}m\dot{R}^2 - mb\cdot\ddot{R} + m\frac{d}{dt}(\dot{R}\cdot b) \ . \tag{5.1.8}$$

As we have already noted, $R(t)$, $\dot{R}(t)$, and $\ddot{R}(t)$ as well as $\Omega(t)$ are given in the inertial system. Since the Lagrangian does not depend on the choice of coordinates, it is possible to think of the vectors in it as expanded either in terms of a basis of the inertial system or a basis of the noninertial system, without changing its position dependence. Of course, if we consider the Lagrangian in this way as a function of the b_i and $\dot{b}_i = v_i$, we need to determine

$$\Omega_i = \Omega\cdot e_i, \ \ddot{R}\cdot e_i, \ \dots \ .$$

Then, from Lagrange's equations

$$\frac{d}{dt}\frac{\partial L}{\partial \dot{b}_i} - \frac{\partial L}{\partial b_i} = 0 \ , \quad i = 1, 2, 3$$

we find the equations of motion for the three components b_i:

$$\frac{d}{dt}(m\dot{b}_i + m\varepsilon_{ijk}\Omega_j b_k) = [m(v \times \Omega) + m(\Omega \times b) \times \Omega - m\ddot{R}]\cdot e_i - \frac{\partial\hat{U}}{\partial b_i} \ , \quad (5.1.9)$$

where here, as always, the total derivative with respect to time

$$\frac{d}{dt}[(\dot{R}\cdot e_i)b_i]$$

does not contribute to the equation of motion (Sect. 3.8.2). We then find again

$$ma = -\frac{\partial\hat{U}}{\partial b} - m\ddot{R} - 2m(\Omega \times v) - m\Omega \times(\Omega \times b) - m(\dot{\Omega} \times b) \tag{5.1.10}$$

in agreement with the result (5.1.6) above.

The following fictitious forces have appeared:

i) The inertial force of rotation, $-m(\dot{\Omega} \times b)$, due to a time-dependent angular velocity. Note that in (5.1.10) the symbol $\dot{\Omega}$ indicates the vector

$$(\dot{\Omega}_1, \dot{\Omega}_2, \dot{\Omega}_3) \tag{5.1.11}$$

$(\dot{\Omega}_i = \dot{\boldsymbol{\Omega}} \cdot \boldsymbol{e}_i)$. In (5.1.6), we meant,

$$\dot{\boldsymbol{\Omega}} = \frac{d\boldsymbol{\Omega}}{dt} = \dot{\Omega}_i \boldsymbol{e}_i + \Omega_i(\boldsymbol{\Omega} \times \boldsymbol{e}_i)$$

which, however, is identical with (5.1.11) since $\boldsymbol{\Omega} \times \boldsymbol{\Omega}$ vanishes identically. For a rotating noninertial system fixed on the earth, $\dot{\boldsymbol{\Omega}} = \boldsymbol{0}$ to a very good approximation.

ii) The inertial force of translation is $-m\ddot{\boldsymbol{R}}$ with $\ddot{\boldsymbol{R}} = (\ddot{\boldsymbol{R}} \cdot \boldsymbol{e}_1, \ddot{\boldsymbol{R}} \cdot \boldsymbol{e}_2, \ddot{\boldsymbol{R}} \cdot \boldsymbol{e}_3)$.

If we consider the center of the earth as the origin of an inertial coordinate system (thus ignoring the motion of the earth around the sun, so that for a small period of time we consider the motion of the earth to be uniform and linear), then for the origin O_B of a noninertial system on the earth's surface, we have $\boldsymbol{R} = R\boldsymbol{e}_3$, $R = $ constant and thus

$$\dot{\boldsymbol{R}} = R\boldsymbol{\Omega} \times \boldsymbol{e}_3 \; ,$$

$$\ddot{\boldsymbol{R}} = R\boldsymbol{\Omega} \times (\boldsymbol{\Omega} \times \boldsymbol{e}_3) \; , \qquad \text{hence}$$

$$\ddot{\boldsymbol{R}} \cdot \boldsymbol{e}_i = R(\Omega_i \Omega_3 - \Omega^2 \delta_{i3}) = \mathrm{O}(R\Omega^2) \; .$$

Now, $|\boldsymbol{\Omega}| = 2\pi/\text{day} = 7.2 \times 10^{-5}\,\text{s}^{-1}$ and $R \approx 6 \times 10^6\,\text{m}$, so that $R\Omega^2 \approx 3 \times 10^{-2}\,\text{ms}^{-2}$. Compared to the gravitational acceleration on the earth $g = 9.81\,\text{ms}^{-2}$, this term is small. It represents a small correction to g which we can ignore most of the time.

iii) The *centrifugal force*[1] is given by

$$-m[\boldsymbol{\Omega} \times (\boldsymbol{\Omega} \times \boldsymbol{b})] = -m[\boldsymbol{\Omega}(\boldsymbol{\Omega} \cdot \boldsymbol{b}) - \boldsymbol{b}\Omega^2] \; . \tag{5.1.12}$$

The centrifugal force lies in the plane spanned by $\boldsymbol{b}$ and $\boldsymbol{\Omega}$ and is perpendicular to $\boldsymbol{\Omega}$, as we might expect intuitively. If $\boldsymbol{b}$ is perpendicular to $\boldsymbol{\Omega}$, we obtain the well-known term $m\Omega^2 \boldsymbol{b}$ for the centrifugal force.

On the earth's surface, the centrifugal force can either be neglected or regarded as a another slight correction to g.

iv) Finally, the *Coriolis*[2] *force*

$$-2m(\boldsymbol{\Omega} \times \boldsymbol{v}) = 2m\boldsymbol{v} \times \boldsymbol{\Omega} \tag{5.1.13}$$

depends on the velocity $\boldsymbol{v}$ in the noninertial system. For the coordinate system $(O_B, \boldsymbol{e}_1, \boldsymbol{e}_2, \boldsymbol{e}_3)$ in the *northern hemisphere*, let $\boldsymbol{e}_3$ point vertically upward and $\boldsymbol{e}_1$

---

[1] Centrifugal force (Latin) from *centrum* and *fugare*: "fleeing the center force".

[2] *Coriolis, Gustave-Gaspard* (*1792 Paris, d. 1843 Paris). Engineer and mathematician at the Ecole Polytechnique. The inertial force named after him comes from a publication of 1835.

to the east tangential to the earth's surface. Then e_2 must point northward tangential to the earth, and Ω has positive 2- and 3-components.

Each body which moves horizontally with velocity $v = v_1 e_1 + v_2 e_2$ then experiences a force

$$2m(v_1 e_1 + v_2 e_2) \times (\Omega_2 e_2 + \Omega_3 e_3)$$

$$= 2m[v_1 \Omega_2 e_3 + \Omega_3(-v_1 e_2 + v_2 e_1)] \, , \tag{5.1.14}$$

that is, the Ω_2-component produces a deviation upwards or downwards, the Ω_3-component, a deviation to the right, as seen when looking in the direction of the velocity vector.

The equation of motion in a noninertial coordinate system, fixed on the earth's surface, thus reads

$$ma = mg + 2m(v \times \Omega) \quad \text{or}$$

$$\ddot{b} = g + 2(\dot{b} \times \Omega) \quad \text{with} \quad b = (b_1, b_2, b_3) \, . \tag{5.1.15}$$

Since the angular speed of the earth's rotation $|\Omega| = 7.2 \times 10^{-5}\,\text{s}^{-1}$, the Coriolis acceleration $2(v \times \Omega)$ on the rotating earth for speeds $v \approx 7\,\text{m s}^{-1}$ is of the order of magnitude $10^{-3}\,\text{m s}^{-1}$, around 10^4 times smaller than the gravitational acceleration. If the Coriolis force can act long enough on a motion, though, there are very noticeable effects on the earth and in the atmosphere. Besides the phenomena discussed below, the larger motions of air and water masses, above all, clearly show the influence of the Coriolis force. In Chap. 9, when we discuss the equation of motion for fluids (the Navier–Stokes equation), we will return to this point.

Application. We consider the free fall of a body in this noninertial system. Ignoring the Coriolis force,

$$b(t) = b_1(t) = b_0 - \tfrac{1}{2}gt^2 e_3 \tag{5.1.16}$$

is a solution to the equation (5.1.15) with the initial conditions $b(0) = b_0$, $\dot{b}(0) = 0$.

If, taking into account the Coriolis force, we set

$$b(t) = b_1(t) + b_2(t) \quad \text{with} \quad |b_2| \ll |b_1| \, ,$$

we find

$$\ddot{b}_2 = 2(gt \times \Omega) + O(\dot{b}_2 \Omega) \, , \tag{5.1.17}$$

and thus

$$b_2(t) = \frac{t^3}{3}(g \times \Omega) \, , \tag{5.1.18}$$

where we have ignored the term of order $b_2 \Omega$ in (5.1.17).

We find for Ω at latitude φ in the northern hemisphere

$$\Omega = \Omega(0, \cos \varphi, \sin \varphi) \, ,$$

whereas we always have $g = -g(0, 0, 1)$.

We then find

$$g \times \Omega = g\Omega(\cos \varphi, 0, 0) \, . \tag{5.1.19}$$

Since e_1 points east, the deviation is eastward. The factor $g\Omega$ has the value

$$g\Omega = 9.81 \, \mathrm{m\,s^{-2}} \times 10^{-5}\mathrm{s^{-1}} \approx 7 \times 10^{-2} \, \mathrm{cm\,s^{-3}} \, .$$

A rock, which falls from height $H = 250 \, \mathrm{m}$, which occurs during time $T = \sqrt{2H/g} \sim 7.1 \, \mathrm{s}$, will thus deviate $\frac{1}{3} \times 7 \times 10^{-2} \times 7^3 \, \mathrm{cm} = 8 \, \mathrm{cm}$ towards the east at the equator, about $6 \, \mathrm{cm}$ at a somewhat higher latitude ($\varphi = 40°$).

5.2 Foucault's[3] Pendulum

We consider a pendulum in a rotating noninertial coordinate system fixed on the earth. Let the origin O_B of the noninertial system be identical with the equilibrium point of the pendulum (Fig. 5.2.1).

The Lagrangian reads

$$L = \tfrac{1}{2}mv^2 = mv \cdot (\Omega \times b) - U(b) \tag{5.2.1}$$

with

$$U(b) = -mg \cdot b = mgb_3 \, ,$$

thus

$$L = \tfrac{1}{2}m(\dot{b}_1^2 + \dot{b}_2^2 + \dot{b}_3^2) + m\Omega_1(b_2\dot{b}_3 - b_3\dot{b}_2)$$

$$+ m\Omega_2(b_3\dot{b}_1 - b_1\dot{b}_3) + m\Omega_3(b_1\dot{b}_2 - b_2\dot{b}_1) - mgb_3 \, . \tag{5.2.2}$$

---

[3] *Foucault, Jean Bernard Léon* (*1819 Paris, d. 1868 Paris). He measured the speed of light using a rotating mirror. In 1851, demonstrated his pendulum at the Pantheon in Paris.

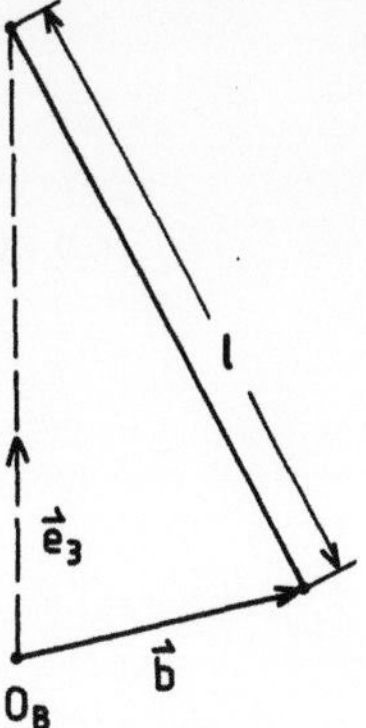

Fig. 5.2.1. Foucault's pendulum. The reference point O_B is at the resting point of the pendulum

Now since

$$(\boldsymbol{b} - l\boldsymbol{e}_3)^2 = l^2 \quad \text{so that}$$

$$b_1^2 + b_2^2 + (b_3 - l)^2 \quad \text{we have}$$

$$b_3 = l \pm \sqrt{l^2 - b_1^2 - b_2^2}$$

$$= l - l\left(1 - \frac{b_1^2 + b_2^2}{2l^2} + \ldots\right)$$

$$= \frac{b_1^2 + b_2^2}{2l} + \mathrm{O}\left(\frac{b_1^4}{l^3}\right). \tag{5.2.3}$$

For a very long pendulum, we can ignore the higher order terms b_1^4/l^3, $\dot{b}_3^2$, $\Omega_i \dot{b}_3 b_k$, and $\Omega_i b_3 \dot{b}_k$. We will then calculate $b_1(t)$ and $b_2(t)$ to the lowest order of magnitude.

We are left with the Lagrangian

$$L = \tfrac{1}{2}m(\dot{b}_1^2 + \dot{b}_2^2) + m\Omega_3(b_1\dot{b}_2 - b_2\dot{b}_1) - \frac{mg}{2l}(b_1^2 + b_2^2) \tag{5.2.4}$$

and thus the following equations of motion for $b_1(t)$ and $b_2(t)$:

$$m\ddot{b}_1 - m\Omega_3\dot{b}_2 = -m\frac{g}{l}b_1 + m\Omega_3\dot{b}_2 \,,$$

$$\tag{5.2.5}$$

$$m\ddot{b}_2 + m\Omega_3\dot{b}_1 = -m\frac{g}{l}b_2 - m\Omega_3\dot{b}_1 \,,$$

or

$$\ddot{b}_1 + \frac{g}{l} b_1 - 2\Omega_3 \dot{b}_2 = 0 \ ,$$

$$\ddot{b}_2 + \frac{g}{l} b_2 + 2\Omega_3 \dot{b}_1 = 0 \ .$$

(5.2.6)

We set $z = b_1 + ib_2$, so that

$$\ddot{z} + \frac{g}{l} z + 2i\Omega_3 \dot{z} = 0 \ .$$

This is a linear differential equation with constant coefficients. The solution of such equations will be taken up in the following chapter. Here, we will anticipate a little.

The standard procedure for solving such an equation consists in making the ansatz $z = \exp(i\omega t)$, so that we obtain an equation for ω:

$$-\omega^2 + 2i\Omega_3 i\omega + \frac{g}{l} = 0 \quad \text{or}$$

(5.2.7)

$$\omega^2 + 2\Omega_3 \omega = \frac{g}{l} \quad \text{thus}$$

(5.2.8)

$$\omega = -\Omega_3 \pm \sqrt{\frac{g}{l} + \Omega_3^2} = -\Omega_3 \pm \hat{\omega}$$

with $\hat{\omega} \approx \sqrt{g/l}$, since $\Omega_3^2 \ll g/l$. We then obtain

$$z = e^{-i\Omega_3 t}(c_1 e^{i\hat{\omega}t} + c_2 e^{-i\hat{\omega}t})$$

(5.2.9)

The two constants c_1, c_2 can be determined from the initial conditions. If, for example

$$z(0) = x_0 \ , \quad \dot{z}(0) = 0 \ , \quad \text{thus}$$

$$b_1(0) = x_0 \ , \quad b_2(0) = 0 \ , \quad \dot{b}_1(0) = \dot{b}_2(0) = 0 \ .$$

These conditions correspond to extending the pendulum, and then releasing it. Then for c_1 and c_2, we find the equations

$$c_1 + c_2 = x_0 \ , \quad (-\Omega_3 + \hat{\omega})c_1 + (-\Omega_3 - \hat{\omega})c_2 = 0$$

so that

$$c_1 = \frac{1}{2} x_0 \left(1 + \frac{\Omega_3}{\hat{\omega}}\right) \ , \quad c_2 = \frac{1}{2} x_0 \left(1 - \frac{\Omega_3}{\hat{\omega}}\right) \ ,$$

and thus, if we in addition ignore the term $\Omega_3/\hat{\omega}$

$$z(t) = x_0 e^{-i\Omega_3 t} \cos \hat{\omega} t \ . \tag{5.2.10}$$

If $\Omega_3 = 0$, we would have

$$z(t) = x_0 \cos \hat{\omega} t \ ,$$

that is, the pendulum would swing in the $b_1 b_3$-plane. For $\Omega_3 \neq 0$ or long periods of time, though, the Coriolis force term $\exp(-i\Omega_3 t)$ becomes noticeable. The pendulum experiences a small deviation to the right every time it swings, as is shown in exaggerated fashion in Fig. 5.2.2a. In the complex z-plane,

$$x_0 e^{-i\Omega_3 t} = x_0(\cos \Omega_3 t - i \sin \Omega_3 t)$$

represents a point which moves clockwise around a circle of radius x_0. For a few small swings, $\Omega_3 t$ can be treated as a constant and the term $\cos(\hat{\omega} t)$ represents an oscillation of period $2\pi/\hat{\omega} = T_1$. The plane in which the swing occurs itself rotates with period $2\pi/\Omega_3 = T_2$.

Since $\Omega_3 = \boldsymbol{\Omega} \cdot e_3 = \Omega \sin \varphi$ and $\Omega = 2\pi/\text{day}$, it follows that

$$T_2 = \frac{1 \, \text{day}}{\sin \varphi}$$

(φ is the latitude).

At the north pole, $T_2 = 1$ day, and the rotational period of the plane of the pendulum's swing is exactly one day long. At the equator, there is no Foucault effect. At a latitude in between ($40°$), we have $T_2 \approx 1.6$ days

If the pendulum is given a push at its equilibrium point, its path looks like Fig. 5.2.2b.

In the Southern hemisphere, $\boldsymbol{\Omega} \cdot e_3 = -\Omega \sin \varphi$, that is, Ω_3 is negative, and thus the plane in which the pendulum swings rotates counterclockwise.

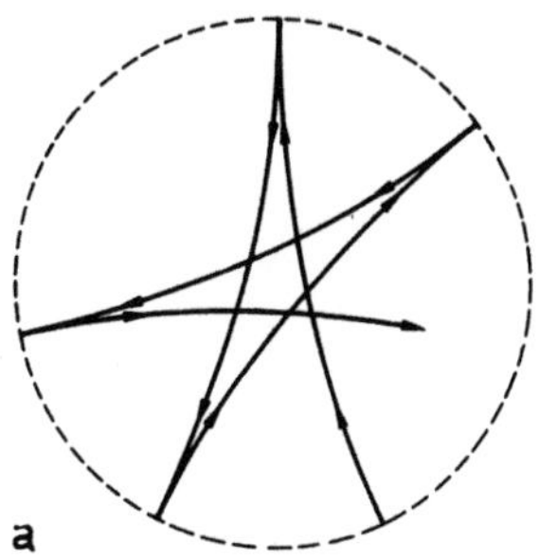
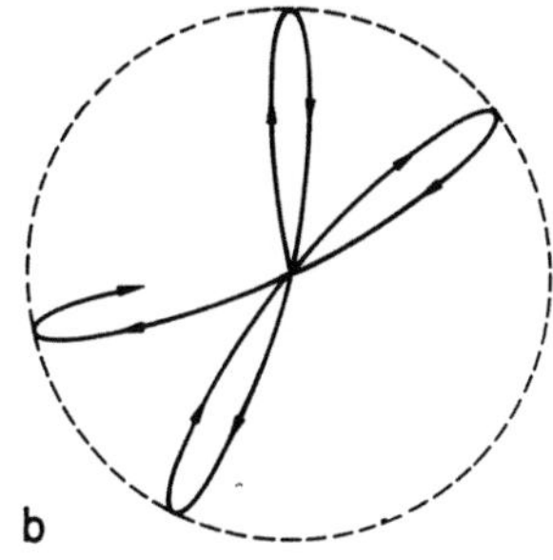

Fig. 5.2.2. The motion of Foucault's pendulum (shown schematically). In the northern hemisphere, the pendulum is always deflected to the right by the Coriolis force. (a) The shape of the trajectory if the pendulum is at rest at time $t = t_0$ and is then released from a point as shown, (b) The shape of the trajectory if the pendulum is pushed out of its equilibrium point at time $t = t_0$

6. Linear Oscillations

In Chaps. 3 and 4, we have seen that the Lagrangian for a holonomic, scleronomic system with f degrees of freedom has the form (with $q = q_1, \ldots, q_f$)

$$L = L(q, \dot{q}) = \frac{1}{2} \sum_{i,j=1}^{f} g_{ij}(q)\dot{q}_i\dot{q}_j - V(q)$$

in an inertial frame of reference. The equations of motion, i.e. Lagrange's equations derived from this Lagrangian, are, in general, very complicated non-linear differential equations, which can only be solved numerically.

Often, though, it is clear from the physical context that there exists a configuration of stable equilibrium for the system, that is, a state in which the system can remain for all times t and about which it can carry out small oscillations.

As an example, consider again the pendulum. Its state of rest must certainly correspond to a solution of the equation of motion, and it can make small oscillations around this equilibrium position. As a further example, a crystalline rigid body can be imagined as a lattice of atoms or molecules. Within the framework of classical physics, there is an equilibrium configuration of the crystal, in which all of its component parts are regularly ordered and at rest. If the crystal is excited by introducing a small amount of energy into the system, its components will perform small oscillations about their equilibrium points.

Many other examples of oscillation can be found in the most varied realms of nature, and indeed physically measurable quantities are often found to behave periodically in time. In technical fields, such as electronics, the "theory of oscillations" is given much consideration and there are correspondingly many textbooks in this area (see e.g. [Butenin, Nayfeh et al.]).

As long as the amplitude of the oscillations remains small enough, the behavior of the oscillating quantities in time can be described by linear equations of motion. In this case, these motions are referred to as *linear oscillations* or just as oscillations or vibrations. If, however, the non-linearity of the full equations of motion cannot be ignored, we must consider *non-linear oscillations*. Non-linear oscillations are especially significant with reference to periodic phenomena in chemistry and biology.

6.1   Linear Approximations About a Point of Equilibrium

In the following, we will be concerned with linear oscillations about states of equilibrium. On the one hand, these small oscillations describe to a good approximation a great number of physical phenomena. On the other hand, they are easy to treat mathematically.

We define the following:

A point q^0 is called a *point of equilibrium* if the trajectory $q(t) \equiv q^0$, $\dot{q}(t) \equiv 0$ is a solution of the equations of motion.

We have:

The point q^0 is a point of equilibrium if and only if

$$\left. \frac{\partial V}{\partial q_i} \right|_{q=q^0} = 0 \quad \text{for} \quad i = 1, \ldots, f \ . \tag{6.1.1}$$

This is clear since Lagrange's equations read:

$$\frac{d}{dt} \frac{\partial T}{\partial \dot{q}_i} - \frac{\partial T}{\partial q_i} + \frac{\partial V}{\partial q_i} = 0 \ , \quad i = 1, \ldots, f \ .$$

If $\dot{q} = 0$, the only term left is

$$\frac{\partial V}{\partial q_i} = 0$$

since $\partial T / \partial q_i$ and $\partial T / \partial \dot{q}_i$ still depend quadratically and linearly on $\dot{q}_i$ respectively. Thus this equation holds for a point of equilibium q^0.

If, conversely, (6.1.1) holds for a point q^0, then q^0 is also a point of equilibrium, since then $q(t) = q^0$ (which implies $\dot{q} = 0$) is a solution of the equation of motion.

Let $q^0 = (q_1^0, \ldots, q_f^0)$ be a point of equilibrium. In the neighborhood of this point, we set

$$q_i = q_i^0 + \eta_i \ . \tag{6.1.2}$$

Then,

$$V(q_1, \ldots, q_f) = V(q_1^0, \ldots, q_f^0) + \sum_{i=1}^{f} \left. \frac{\partial V}{\partial q_i} \right|_{q_0} \eta_i$$

$$+ \frac{1}{2} \sum_{i,j=1}^{f} \left. \frac{\partial^2 V}{\partial q_i \partial q_j} \right|_{q_0} \eta_i \eta_j + O(\eta^3)$$

$$= V(q^0) + \frac{1}{2} \sum_{i,j=1}^{f} K_{ij} \eta_i \eta_j + O(\eta^3) \tag{6.1.3}$$

with

$$K_{ij} = \frac{\partial^2 V}{\partial q_i \partial q_j}\bigg|_{q_0} = K_{ji} \ .$$

It is also true, since $\dot{q}_i = \dot{\eta}_i$, that

$$T(q, \dot{q}) = \frac{1}{2} \sum_{i,\,j=1}^{f} g_{ij}(q)\dot{q}_i\dot{q}_j$$

$$= \frac{1}{2} \sum_{i,\,j=1}^{f} g_{ij}(q^0)\dot{\eta}_i\dot{\eta}_j + \mathrm{O}(\eta^3) \ . \tag{6.1.4}$$

If we now ignore all terms of order $\mathrm{O}(\eta^3)$, we find a Lagrangian which is bilinear in the variables η_i and $\dot{\eta}_i$.

Using this approximate Lagrangian, we find Lagrange's equations to be differential equations linear in the η_i. This approximation is called a *linearization* of the system, and the corresponding system is called a *linear system*.

Thus, we have the linear system

$$L = \frac{1}{2} \sum_{i,\,j=1}^{f} (M_{ij}\dot{\eta}_i\dot{\eta}_j - K_{ij}\eta_i\eta_j) \tag{6.1.5}$$

with

$$M_{ij} = M_{ji} = g_{ij}(q^0) \tag{6.1.6}$$

and

$$K_{ij} = \frac{\partial^2 V}{\partial q_i \partial q_j}\bigg|_{q_0} \ . \tag{6.1.7}$$

The corresponding equation of motion

$$\sum_{j=1}^{f} (M_{ij}\ddot{\eta}_j + K_{ij}\eta_j) = 0 \ , \quad i = 1, \ldots, f \tag{6.1.8}$$

is a system of second-order linear differential equations.

Linear differential equations play an especially important role in theoretical physics.

We have already seen one reason why this is so: the behavior of a mechanical system in the neighborhood of a point of equilibrium can be approximately described by a linear equation of motion and the smaller the deviation from equilibrium, the better the approximation becomes. We also meet linear differential equations in many other areas of theoretical physics.

Fortunately, we can say much more about the solutions of linear differential equations than we can about the general case of non-linear equations of motion

– which explains why we attempt to reduce physical problems to *linear* equations of motion, whose solutions are a routine matter. We must keep in mind, though, that real problems are only more or less approximately linear and that non-linear features of these systems can also lead to qualitatively new properties.

6.2 A Few General Remarks About Linear Differential Equations

In the equations of motion produced in Sect. 6.1, the coefficients M_{ij} and K_{ij} were constant, i.e. not dependent on the time t. For now, we will drop this restriction and consider general linear differential equations. Later, we will return to equations with constant coefficients. We begin with the simplest case of the general linear differential equation:

i) Let us then consider a general homogeneous linear differential equation of second order for a system with one degree of freedom, which has the form

$$\ddot{x}(t) + a(t)\dot{x}(t) + b(t)x(t) = 0 \ . \tag{6.2.1}$$

Here, the adjective "homogeneous" means that the right side of the equation, i.e. the side independent of $x(t)$, vanishes.

The solutions of this equation satisfy the *superposition principle*[1].

If $x^{(1)}(t)$ and $x^{(2)}(t)$ are solutions of (6.2.1), then for all $\alpha, \beta \in \mathbb{R}$

$$\alpha x^{(1)}(t) + \beta x^{(2)}(t)$$

is also a solution of (6.2.1).

In other words, the set of solutions of (6.2.1) forms a vector space.
The dimension of this vector space is $d = 2$.
To prove this, we will produce a basis of the space of solutions: Let $x^{(1)}(t)$ be a solution of (6.2.1) with the initial values

$$x^{(1)}(0) = 1 \ , \quad \dot{x}^{(1)}(0) = 0$$

and $x^{(2)}(t)$ a solution of (6.2.1) with the initial values

$$x^{(2)}(0) = 0 \ , \quad \dot{x}^{(2)}(0) = 1 \ .$$

Obviously, $x^{(1)}(t)$ and $x^{(2)}(t)$ are linearly independent. Then, since any other solution $x(t)$ is uniquely determined by its initial values $x(0) = \alpha$, $\dot{x}(0) = \beta$, we

---

[1] Superposition (Latin) from *superponere*: superimpose.

must have

$$x(t) = \alpha x^{(1)}(t) + \beta x^{(2)}(t) \ .$$

Thus, $x^{(1)}$ and $x^{(2)}$ form a basis of the solution space. In particular, any solution can be expressed uniquely as a linear combination of $x^{(1)}$ and $x^{(2)}$.

ii) For a system with n degrees of freedom, a general homogeneous linear differential equation of second order reads:

$$\ddot{x}(t) + A(t)\dot{x}(t) + B(t)x(t) = 0 \ , \tag{6.2.2}$$

where now $x(t) \in \mathbb{R}^n$ for all t, and thus takes on values in an n-dimension vector space, and $A(t)$ and $B(t)$ are linear maps $\mathbb{R}^n \to \mathbb{R}^n$.

In terms of components, this equation has the form

$$\ddot{x}_i(t) + \sum_{j=1}^{n} A_{ij}(t)\dot{x}_j(t) + \sum_{j=1}^{n} B_{ij}(t)x_j(t) = 0 \ , \quad i = 1, \ldots, n \ .$$

The superposition principle still holds. The set of solutions is a $2n$-dimensional vector space.

A basis of the solution space is given by the solutions

$$u^{(1)}(t), \ldots, u^{(n)}(t) \ , \quad v^{(1)}(t), \ldots, v^{(n)}(t)$$

which have initial conditions

$$u^{(1)}(0) = \begin{pmatrix} 1 \\ 0 \\ 0 \\ \vdots \\ 0 \end{pmatrix} , \quad \dot{u}^{(1)}(0) = \begin{pmatrix} 0 \\ 0 \\ 0 \\ \vdots \\ 0 \end{pmatrix} ;$$

$$u^{(2)}(0) = \begin{pmatrix} 0 \\ 1 \\ 0 \\ \vdots \\ 0 \end{pmatrix} , \quad \dot{u}^{(2)}(0) = \begin{pmatrix} 0 \\ 0 \\ 0 \\ \vdots \\ 0 \end{pmatrix} ; \ldots ;$$

$$v^{(1)}(0) = \begin{pmatrix} 0 \\ 0 \\ 0 \\ \vdots \\ 0 \end{pmatrix} , \quad \dot{v}^{(1)}(0) = \begin{pmatrix} 1 \\ 0 \\ 0 \\ \vdots \\ 0 \end{pmatrix} ;$$

$$v^{(2)}(0) = \begin{pmatrix} 0 \\ 0 \\ 0 \\ \vdots \\ 0 \end{pmatrix} , \qquad \dot{v}^{(2)}(0) = \begin{pmatrix} 0 \\ 1 \\ 0 \\ \vdots \\ 0 \end{pmatrix} ; \ldots$$

The solution $x(t)$ with initial conditions

$$x(0) = \begin{pmatrix} \alpha_1 \\ \alpha_2 \\ \vdots \\ \alpha_n \end{pmatrix} , \qquad \dot{x}(0) = \begin{pmatrix} \beta_1 \\ \beta_2 \\ \vdots \\ \beta_n \end{pmatrix}$$

is then

$$x(t) = \sum_{i=1}^{n} \left[\alpha_i u^{(i)}(t) + \beta_i v^{(i)}(t) \right] . \tag{6.2.3}$$

Once again we have shown:

The fundamental characteristic of linear systems is the validity of the superposition principle. Every linear combination of any two solutions is again a solution of the equation. Further, every solution can be expressed as a linear superposition of a set of fundamental solutions (the basis of the solution space).

iii) For general considerations, it is useful to note that a system of second order differential equations can be rewritten as a system of first order equations with double the number of components. Thus the differential equation

$$\ddot{x} + A\dot{x} + Bx = 0 \tag{6.2.4}$$

is clearly equivalent to the system

$$\dot{x} = z , \quad \dot{z} + Az + Bx = 0 \quad \text{or}$$

$$\frac{d}{dt}\begin{pmatrix} x \\ z \end{pmatrix} + \begin{pmatrix} 0 & -1 \\ B & A \end{pmatrix}\begin{pmatrix} x \\ z \end{pmatrix} = 0 . \tag{6.2.5}$$

iv) In physics, we frequently encounter the problem of a linearly oscillating system which is also subject to an external time-dependent force.

Here are the equations of motion for a few such examples:

a) A pendulum with addition torque $d(t)$:

$$ml\ddot{\theta}(t) + mg\theta(t) = d(t) .$$

b) A one-dimensional system with external force $f(t)$:

$$m[\ddot{x}(t) + 2\varrho\dot{x}(t) + \omega_0^2 x(t)] = f(t) \ ,$$

c) An n-dimensional system with external forces $f_i(t)$:

$$\sum_{j=1}^{n} [M_{ij}\ddot{x}_j(t) + K_{ij}x_j(t)] = f_i(t) \ , \qquad i = 1, \ldots, n \ .$$

All of these systems are of the form

$$Lx(t) = f(t) \ , \tag{6.2.6}$$

where $x(t), f(t) \in \mathbb{R}^n$ and L is a linear differential operator.

They represent linear systems which are subject to external forces which depend only on t.

For these cases, the following important theorem holds:

If $x^{(0)}(t)$ is a solution of (6.2.6)

$$Lx^{(0)} = f \ , \tag{6.2.7}$$

then every other solution $x^{(1)}$ of (6.2.6) has the form

$$x^{(1)}(t) = x^{(0)}(t) + u(t) \ , \tag{6.2.8}$$

where $u(t)$ is a solution of the corresponding homogeneous equation

$$Lu = 0 \ . \tag{6.2.9}$$

Clearly, for any other solution $x^{(1)}(t)$ we have

$$x^{(1)}(t) = x^{(0)}(t) + [x^{(1)}(t) - x^{(0)}(t)] \quad \text{and}$$

$$L(x^{(1)} - x^{(0)}) = Lx^{(1)} - Lx^{(0)} = 0 \ ,$$

that is, $x^{(1)}(t) - x^{(0)}(t)$ is a solution of the homogeneous equation.

Conversely, if

$$Lu = 0 \quad \text{and} \quad Lx^{(0)} = f, \quad \text{then} \quad L(x^{(0)} + u) = f \ .$$

The inhomogeneous problem (6.2.6) is thus fully solved if we know the general solution of the homogeneous problem in addition to a single particular solution of the inhomogeneous problem.

v) Using the linearity of the equation, we can immediately construct solutions for the external force $c_1 f^{(1)}(t) + c_2 f^{(2)}(t)$ out of solutions for the external forces $f^{(1)}(t)$ and $f^{(2)}(t)$.

With

$$Lx^{(1)} = f^{(1)} \quad \text{and} \quad Lx^{(2)} = f^{(2)}$$

it is clear that

$$L(c_1 x^{(1)} + c_2 x^{(2)}) = c_1 Lx^{(1)} + c_2 Lx^{(2)}$$
$$= c_1 f^{(1)} + c_2 f^{(2)}$$

6.3 Homogeneous Linear Systems with One Degree of Freedom and Constant Coefficients

We now investigate the equation

$$\ddot{x}(t) + 2\varrho\dot{x}(t) + \omega_0^2 x(t) = 0 , \quad x(t) \in \mathbb{R} , \quad \varrho, \omega_0^2 > 0 , \tag{6.3.1}$$

which describes a harmonic oscillator with an additional frictional force $\sim -\varrho\dot{x}$. Only the case $\varrho \geq 0$ is physically meaningful, since the mechanical energy can only decrease as a result of the frictional force. We have,

$$\dot{E} = \frac{d}{dt}\left(\frac{1}{2}m\dot{x}^2 + \frac{1}{2}m\omega_0^2 x^2\right)$$
$$= m\dot{x}(\ddot{x} + \omega_0^2 x) = -2m\varrho\dot{x}^2 .$$

This equation (6.3.1) can be solved in the same way as all other linear differential equations with constant coefficients.

As a useful aid, we introduce the complex exponential function.

For complex $z = a + ib$, we define

$$e^z = \sum_{n=0}^{\infty} \frac{z^n}{n!} .$$

Then,

i) $e^{z_1+z_2} = e^{z_1}e^{z_2}$, in particular $e^{a+ib} = e^a e^{ib}$,

ii) $e^{ib} = \cos b + i\sin b$, and thus

iii) $(e^{ib})^* = e^{-ib}$, $\cos b = \frac{1}{2}(e^{ib} + e^{-ib}) = \text{Re}\{e^{ib}\}$,

$$\sin b = (1/2i)(e^{ib} - e^{-ib}) = \text{Im}\{e^{ib}\} ,$$

iv) $(d/dt)e^{zt} = ze^{zt}$ for $t \in \mathbb{R}$, $z \in \mathbb{C}$.

The solution of (6.3.1) is best found in two steps:

Step 1

We introduce complex variables: Consider (6.3.1) as a differential equation for a complex valued function $x(t)$ and look for all complex-valued solutions. In this case, naturally, there is a corresponding complex superposition principle: If

$$x^{(1)}(t) \quad \text{and} \quad x^{(2)}(t)$$

are solutions of (6.3.1), then for any $\alpha, \beta \in \mathbb{C}$

$$x(t) = \alpha x^{(1)}(t) + \beta x^{(2)}(t)$$

is also a solution of (6.3.1).

In addition, since $\varrho, \omega_0^2 \in \mathbb{R}$, if $x(t)$ is a solution of (6.3.1), its corresponding complex conjugate $x^*(t)$ is also a solution. Then

$$u(t) = \text{Re}\{x(t)\} = \tfrac{1}{2}[x(t) + x^*(t)]$$

is also real solution of (6.3.1) and we can find all the real solutions from the complex solutions by taking their real parts.

Step 2

We assume the solution has an exponential form: We look for solutions

$$x(t) = e^{\lambda t} , \quad (\lambda \in \mathbb{C}) ,$$

i.e. we try to find λ such that $\exp(\lambda t)$ is a solution of (6.3.1).

If we substitute $\exp(\lambda t)$ into (6.3.1), we are left with the following condition for λ:

$$\lambda^2 + 2\varrho\lambda + \omega_0^2 = 0 . \tag{6.3.2}$$

The exponential form we have assumed is a solution if and only if λ is a solution of the quadratic equation given by

$$\lambda_{1,2} = -\varrho \pm \sqrt{\varrho^2 - \omega_0^2} \tag{6.3.3}$$

We must differentiate three cases:

a) $\varrho^2 > \omega_0^2$: overdamping: there are two real, negative roots, the general solution

$$x(t) = e^{-\varrho t}(\alpha e^{\hat{\omega}t} + \beta e^{-\hat{\omega}t}) , \quad \hat{\omega} = \sqrt{\varrho^2 - \omega_0^2} \tag{6.3.4}$$

dies away as $t \to \infty$.

b) $\varrho^2 < \omega_0^2$: underdamping: there are two complex solutions, which are conjugates of each other, and thus the most general (complex) solution is given by

$$x(t) = e^{-\varrho t}(\alpha e^{i\omega t} + \beta e^{-i\omega t}) , \quad \omega = \sqrt{\omega_0^2 - \varrho^2} \tag{6.3.5}$$

or, written in real form

$$x(t) = e^{-\varrho t}(\alpha' \cos \omega t + \beta' \sin \omega t) \ .$$

For $\varrho > 0$, all solutions die away as $t \to \infty$.

c) $\varrho^2 = \omega_0^2$: Critical damping, aperiodic. The exponential assumption in this case yields only *one* linearly independent solution. The most general solution is of the form

$$x(t) = e^{-\varrho t}(\alpha + \beta t) \ . \tag{6.3.6}$$

Example. The Lagrangian for a planar pendulum is given by

$$L = \tfrac{1}{2}ml^2\dot\theta^2 - mgl(1 - \cos\theta)$$

thus

$$V(\theta) = mgl(1 - \cos\theta) \ ,$$

$$V(0) = 0 \ ,$$

$$V'(\theta) = mgl\sin\theta \quad (= 0 \quad \text{for} \quad \theta = 0) \ ,$$

i.e., θ^0 is a point of equilibrium.

We have $V''(0) = mgl > 0$, thus $\theta^0 = 0$ is a minimum of the potential. Expanding around $\theta^0 = 0$ yields

$$V(\theta) = mgl\left(1 - \left(1 - \frac{\theta^2}{2}\right) + O(\theta^4)\right)$$

$$= mgl\frac{\theta^2}{2} + O(\theta^4) \ ,$$

and thus

$$K = mgl \ , \quad M = ml^2 \ .$$

In linear approximation, the equation of motion is thus given by

$$ml^2\ddot\theta + mgl\theta = 0 \quad \text{or} \quad \ddot\theta + \frac{g}{l}\theta = 0 \ .$$

Solution: With the ansatz

$$\theta(t) = e^{i\omega t} \quad \text{it follows that}$$

$$-\omega^2 + \frac{g}{l} = 0 \quad \text{or} \quad \omega = \pm\sqrt{\frac{g}{l}} \ , \quad \text{real.}$$

Thus the most general real solution of the equation of motion is of the form

$$\theta(t) = \mathrm{Re}\{A_1 e^{i\omega t} + A_2 e^{-i\omega t}\} \ .$$

Note that we can satisfy any initial conditions $\theta(0) = \alpha$, $\dot{\theta}(0) = \beta$ by the proper choice of A_1 and A_2:

$$\theta(0) = \mathrm{Re}\{A_1 + A_2\} = \alpha \ , \qquad \dot{\theta}(0) = \mathrm{Re}\{(iA_1 - iA_2)\omega\} = \beta \ .$$

A solution is $A_1 = \alpha$, $A_2 = i\beta/\omega$ so that the solution for the initial conditions $\theta(0) = \alpha$, $\dot{\theta}(0) = \beta$ is given by

$$\theta(t) = \alpha \cos \omega t + (\beta/\omega)\sin \omega t \ ,$$

or, written another way,

$$\theta(t) = C \cos(\omega t - \delta) \quad \text{with}$$

$$C = \sqrt{\alpha^2 + \frac{\beta^2}{\omega^2}} \quad \text{and} \quad \tan \delta = \frac{\beta}{\alpha\omega} \ .$$

Remark. Note that $\theta^0 = \pi$ is also a point of equilibrium. At this point, though, $V''(\pi) = -mgl$ and thus the potential has a maximum. This point of equilibrium is therefore unstable, which is intuitively clear.

The expansion of $V(\theta)$ around $\theta = \pi$ yields

$$V(\theta) = V(\pi + \eta) = 2mgl - \tfrac{1}{2}mgl\eta^2 + \mathrm{O}(\eta^4) \ ,$$

and thus the equation of motion

$$ml^2\ddot{\eta} - mgl\eta = 0$$

follows with solutions

$$\eta(t) = e^{\pm \sqrt{(g/l)}t}$$

so that in general we wind up with an exponentially increasing motion. This indicates the instability of the point of equilibrium.

The stability of a point of equilibrium – or in general, of a solution of the equation of motion – can therefore be investigated by studying small displacements away from it and seeing if they increase with time. Stable, and thus realizable, solutions only allow oscillatory perturbations, which remain bounded for all time. This *linear analysis of stability* is a very important method in many areas of physics [Haken, Drazin et al.]

6.4 Homogeneous Linear Systems with n Degrees of Freedom and Constant Coefficients

6.4.1 Normal Modes and Eigenfrequencies

We now consider the equations produced in Sect. 6.1

$$\sum_{j=1}^{n} (M_{ij}\ddot{x}_j + K_{ij}x_j) = 0 , \qquad i = 1, \ldots, n .\tag{6.4.1}$$

The quantities M_{ij} and K_{ij} satisfy the conditions

$$M_{ij} = M_{ji} , \qquad K_{ij} = K_{ji} \qquad \text{as well as}$$

$$\sum_{i,j=1}^{n} M_{ij}a_i a_j > 0 \qquad \text{for} \qquad \sum_{i=1}^{n} a_i^2 \neq 0 ,\tag{6.4.2}$$

since the kinetic energy is strictly positive unless all the velocities vanish. The analogous inequality

$$\sum_{i,j=1}^{n} K_{ij}a_i a_j > 0 \qquad \text{for} \qquad \sum_{i=1}^{n} a_i^2 \neq 0 .\tag{6.4.3}$$

holds only if q^0 is a minimum of V.

Considering x as a vector in $\mathbb{R}^n$, the equation of motion can be written

$$M\ddot{x} + Kx = 0 ,\tag{6.4.4}$$

where M and K are linear mappings $\mathbb{R}^n \to \mathbb{R}^n$. Introducing a scalar product

$$x \cdot y = \sum_{i=1}^{n} x_i y_i\tag{6.4.5}$$

we can express the symmetry and positivity of M and K in the following way:

$$y \cdot (Mx) = (My) \cdot x , \qquad y \cdot (Kx) = (Ky) \cdot x$$

for all $x, y \in \mathbb{R}^n$ and

$$x \cdot Mx > 0 \qquad \text{for} \qquad x \neq 0 .$$

The solution of (6.4.4) again follows in two steps:

1) Introduction of complex variables: We interpret (6.4.4) as a differential equation for functions with values in $\mathbb{C}^n$ and obtain the real solutions as the real or imaginary parts of the complex solutions.

2) We assume an exponential form: We set

$$x(t) = ve^{i\omega t} \quad \text{with} \quad v \in \mathbb{C}^n ,$$

and look for v and ω such that $x(t)$ is a solution of the equation (6.4.4). This leads to the condition

$$(K - \omega^2 M)v = 0 .$$

This equation has a non-zero solution v only if $K - \omega^2 M$ is not a 1–1 mapping, that is, when

$$\det(K - \omega^2 M) = 0 . \tag{6.4.6}$$

This so-called *secular equation*[2] is an algebraic equation of n-th order in ω^2 (or $2n$-th order in ω). The possible frequency values ω are roots of the polynomial $\det(K - \omega^2 M)$.

Let ω_α^2 ($\alpha = 1, \ldots, n$) be the roots and $v^{(\alpha)}$ the corresponding "eigenvectors" with

$$(K - \omega_\alpha^2 M)v^{(\alpha)} = 0 , \quad \alpha = 1, \ldots, n .$$

We will show:

i) ω_α^2 *is real*
ii) *If* $x^* \cdot Kx \geq 0$ *for all* x, *then* $\omega_\alpha^2 \geq 0$.
iii) *If* $x^* \cdot Kx > 0$ *for* $x \neq 0$, *i.e. if* q^0 *is a minimum of* V, *then in fact* $\omega_\alpha^2 > 0$. ω_α *is real and* $x(t) = v^{(\alpha)}\exp(i\omega_\alpha t)$ *is bounded. Minima of* V *thus correspond to stable points of equilibrium.*
iv) *The eigenvectors* $v^{(\alpha)}$ *can be chosen to be real.*
v) *The real eigenvectors* $v^{(\alpha)}$ ($\alpha = 1, \ldots, n$) *are linearly independent and form a basis of* $\mathbb{R}^n$.

Proof. For (i): From

$$Kv^{(\alpha)} = \omega_\alpha^2 Mv^{(\alpha)} \quad \text{it follows} \quad v^{(\alpha)*} \cdot Kv^{(\alpha)} = \omega_\alpha^2 v^{(\alpha)*} \cdot Mv^{(\alpha)} .$$

Now, since M and K are symmetric and real-valued, $v^{(\alpha)*} \cdot Mv^{(\alpha)}$ and $v^{(\alpha)*} \cdot Kv^{(\alpha)}$ are real. Since in addition $v^{(\alpha)*} \cdot Mv^{(\alpha)} > 0$, we can solve for ω_α^2 and we find

$$\omega_\alpha^2 = \frac{v^{(\alpha)*} \cdot Kv^{(\alpha)}}{v^{(\alpha)*} \cdot Mv^{(\alpha)}} .$$

Therefore ω^2 is real. If, in addition, $v^{(\alpha)*} \cdot Kv^{(\alpha)} > 0$, then $\omega_\alpha^2 > 0$, so that (ii) has also been shown.

---

[2] It is called the secular equation because a similar equation appears in the calculation of long-term (secular) variations of planetary orbits caused by the mutual attraction of the planets.

For (iii): Since M and K are real matrices, from

$$(K - \omega_\alpha^2 M)v^{(\alpha)} = 0 \quad \text{it follows that}$$

$$[(K - \omega_\alpha^2 M)v^{(\alpha)}]^* = (K - \omega_\alpha^2 M)v^{(\alpha)*} = 0 \ .$$

Therefore both $\text{Re}\{v^{(\alpha)}\}$ and $\text{Im}\{v^{(\alpha)}\}$ are eigenvectors and they cannot both vanish.

For (iv): For real eigenvectors $v^{(\alpha)}$ and $v^{(\beta)}$, we have

$$v^{(\alpha)} \cdot Kv^{(\beta)} = \omega_\beta^2 v^{(\alpha)} \cdot Mv^{(\beta)} = (Kv^{(\alpha)}) \cdot v^{(\beta)}$$

$$= \omega_\alpha^2 (Mv^{(\alpha)}) \cdot v^{(\beta)} = \omega_\alpha^2 v^{(\alpha)} \cdot Mv^{(\beta)} \ .$$

It follows that

$$(\omega_\alpha^2 - \omega_\beta^2)v^{(\alpha)} \cdot Mv^{(\beta)} = 0$$

and thus

$$v^{(\alpha)} \cdot Mv^{(\beta)} = 0 \quad \text{for} \quad \omega_\alpha^2 \neq \omega_\beta^2 \ .$$

To prove the linear independence of the $v^{(\alpha)}$, we assume that all of the ω_α^2 are different (the case of degeneracy requires some small additional considerations).

From

$$\sum_{\alpha=1}^{n} \zeta_\alpha v^{(\alpha)} = 0$$

it follows then that

$$v^{(\beta)} \cdot M\left[\sum_{\alpha=1}^{n} \zeta_\alpha v^{(\alpha)}\right] = 0 = \sum_{\alpha=1}^{n} \zeta_\alpha v^{(\beta)} \cdot Mv^{(\alpha)}$$

$$= \zeta_\beta v^{(\beta)} \cdot Mv^{(\beta)} \ ,$$

so that $\zeta_\beta = 0$ for all β, and thus the vectors $v^{(\alpha)}$, $\alpha = 1, \ldots, n$ are linearly independent and must therefore form a basis of $\mathbb{R}^n$, since there are n of them.

If we substitute the inertia tensor I for K and the unity matrix $\mathbb{1}$ for M, we see that we have just given the missing proof from Sect. 4.2.1 of the existence of an orthonormal system of principle axes for a rigid body.

Since the above result is extremely important, we will reformulate it as the eigenvalue problem $(A - \lambda \mathbb{1})w = 0$.

We will use:

Theorem: Every real symmetric linear mapping $A: \mathbb{R}^n \to \mathbb{R}^n$ has a complete real orthonormal system of eigenvectors w_α ($\alpha = 1, \ldots, n$) such that $Aw_\alpha = \lambda_\alpha w_\alpha$, $w_\alpha \cdot w_\beta = \delta_{\alpha\beta}$. The corresponding eigenvalues are real.

Theorem: Every positive symmetrical real mapping $M: \mathbb{R}^n \to \mathbb{R}^n$ has a real symmetrical positive inverse M^{-1} and a real symmetric positive square root $M^{1/2}$ with $(M^{1/2})^2 = M$.

The following two equations are clearly equivalent:

$$(K - \omega_\alpha^2 M)v^{(\alpha)} = 0 \Leftrightarrow (M^{-1/2}KM^{-1/2} - \omega_\alpha^2 \mathbb{1})w_\alpha = 0$$

with $w_\alpha = M^{1/2}v^{(\alpha)}$. It is enough then to find an orthonormal system of eigenvectors w_α of $A = M^{-1/2}KM^{1/2}$. From $w_\alpha w_\beta = \delta_{\alpha\beta}$, it follows that

$$(M^{1/2}v^{(\alpha)}) \cdot (M^{1/2}v^{(\beta)}) = v^{(\alpha)} \cdot Mv^{(\beta)} = \delta_{\alpha\beta} \ .$$

The solutions $x^{(\alpha)} = v^{(\alpha)} \exp(i\omega_\alpha t)$ are called *normal modes*, the corresponding frequencies ω_α are called the *normal frequencies* of the system. In a normal mode, the harmonic time dependence of the motion is such that the proportions between the individual displacements of each coordinate from the equilibrium position are constant in time.

If now all the $\omega_\alpha^2 > 0$, then the most general solution of the linear equation is a linear superposition of normal modes

$$x(t) = \sum_{\alpha=1}^{n} (a_\alpha e^{i\omega_\alpha t} + b_\alpha e^{-i\omega_\alpha t})v^{(\alpha)} \quad \text{with} \quad a_\alpha, b_\alpha \in \mathbb{C} \ . \tag{6.4.7}$$

Since the $v^{(\alpha)}$ form a basis of $\mathbb{R}^n$, any set of initial conditions $x(0), \dot{x}(0)$ can be satisfied by a proper choice of the constants in the expansion.

If $\omega_\alpha = 0$, we obtain a translational solution instead of an oscillatory function proportional to $v^{(\alpha)}$, as we can see especially clearly using the so-called *normal coordinates*.

We introduce these new coordinates $Q_\alpha(t)$ using

$$x(t) = \sum_{\alpha=1}^{n} Q_\alpha(t)v^{(\alpha)} \ , \tag{6.4.8}$$

where the $v^{(\alpha)}$ are defined by

$$Kv^{(\alpha)} = \omega_\alpha^2 Mv^{(\alpha)}$$

and subject to the normalization

$$v^{(\alpha)} \cdot Mv^{(\beta)} = \delta_{\alpha\beta} \ . \tag{6.4.9}$$

Hence, we also have

$$Q_\alpha(t) = v^{(\alpha)} \cdot Mx(t) \ . \tag{6.4.10}$$

In the coordinates Q_α, the Lagrangian and the equations of motion read

$$L = \frac{1}{2}\dot{x} \cdot M\dot{x} - \frac{1}{2}x \cdot Kx = \frac{1}{2}\sum_{\alpha=1}^{n}(\dot{Q}_\alpha^2 - \omega_\alpha^2 Q_\alpha^2) \ , \tag{6.4.11}$$

$$\ddot{Q}_\alpha(t) + \omega_\alpha^2 Q_\alpha(t) = 0 \ , \quad \alpha = 1, \ldots, n \ . \tag{6.4.12}$$

The equations of motion are thus uncoupled when written in terms of the normal coordinates. For $\omega_\alpha^2 = 0$, we see that $Q_\alpha(t)$ is a linear function of t.

6.4.2 Examples of the Calculation of Normal Modes

i) Consider two coupled pendulums whose motion has small amplitudes (Fig. 6.4.1). Let k be the spring constant of the spring connecting the two pendulums. The Lagrangian is then given by

$$L = \tfrac{1}{2}ml^2(\dot{\varphi}_1^2 + \dot{\varphi}_2^2) - \tfrac{1}{2}mgl(\varphi_1^2 + \varphi_2^2) - \tfrac{1}{2}kl^2(\varphi_1 - \varphi_2)^2 \, , \tag{6.4.13}$$

are the equations of motion are given by

$$\ddot{\varphi}_1 + \frac{g}{l}\varphi_1 + \frac{k}{m}(\varphi_1 - \varphi_2) = 0 \; ;$$

$$\ddot{\varphi}_2 + \frac{g}{l}\varphi_2 + \frac{k}{m}(\varphi_2 - \varphi_1) = 0 \tag{6.4.14}$$

or, in the matrix form,

$$\frac{d^2}{dt^2}\begin{pmatrix} \varphi_1 \\ \varphi_2 \end{pmatrix} + \begin{pmatrix} \dfrac{g}{l} + \dfrac{k}{m} & -\dfrac{k}{m} \\ -\dfrac{k}{m} & \dfrac{g}{l} + \dfrac{k}{m} \end{pmatrix}\begin{pmatrix} \varphi_1 \\ \varphi_2 \end{pmatrix} = 0 \, . \tag{6.4.15}$$

Assuming an exponential form

$$\begin{pmatrix} \varphi_1 \\ \varphi_2 \end{pmatrix} = \begin{pmatrix} v_1 \\ v_2 \end{pmatrix} e^{i\omega t} \tag{6.4.16}$$

we find the eigenvalue equation

$$\begin{pmatrix} \dfrac{g}{l} + \dfrac{k}{m} - \omega^2 & -\dfrac{k}{m} \\ -\dfrac{k}{m} & \dfrac{g}{l} + \dfrac{k}{m} - \omega^2 \end{pmatrix}\begin{pmatrix} v_1 \\ v_2 \end{pmatrix} = 0 \tag{6.4.17}$$

(a)   (b)

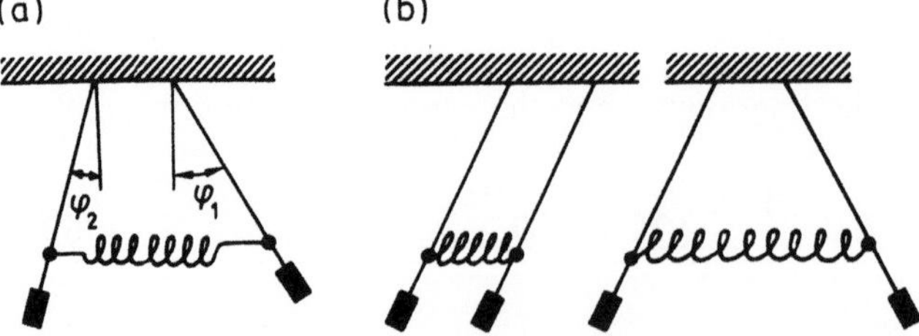

Fig. 6.4.1. Coupled pendulums (a) and their normal modes (b)

and the secular equation

$$\det\begin{pmatrix} \dfrac{g}{l}+\dfrac{k}{m}-\omega^2 & -\dfrac{k}{m} \\[2ex] -\dfrac{k}{m} & \dfrac{g}{l}+\dfrac{k}{m}-\omega^2 \end{pmatrix}$$

$$= \omega^4 - 2\omega^2\left(\frac{g}{l}+\frac{k}{m}\right) + \left(\frac{g}{l}+\frac{k}{m}\right)^2 - \frac{k^2}{m^2} = 0 \tag{6.4.18}$$

with solutions:

$$\omega_1^2 = \frac{g}{l}, \qquad \omega_2^2 = \frac{g}{l} + 2\frac{k}{m}. \tag{6.4.19}$$

For the eigenvectors, we find

$$\text{for } \omega_1^2: \quad v^{(1)} = \begin{pmatrix} v_1^{(1)} \\ v_2^{(1)} \end{pmatrix} = \begin{pmatrix} 1 \\ 1 \end{pmatrix}, \tag{6.4.20}$$

$$\text{for } \omega_2^2: \quad v^{(2)} = \begin{pmatrix} v_1^{(2)} \\ v_2^{(2)} \end{pmatrix} = \begin{pmatrix} 1 \\ -1 \end{pmatrix}. \tag{6.4.21}$$

Thus, the most general solution of the equation of motion is:

$$\begin{pmatrix} \varphi_1 \\ \varphi_2 \end{pmatrix} = \begin{pmatrix} a_1 e^{i\omega_1 t} + b_1 e^{-i\omega_1 t} + a_2 e^{i\omega_2 t} + b_2 e^{-i\omega_2 t} \\ a_1 e^{i\omega_1 t} + b_1 e^{-i\omega_1 t} - a_2 e^{i\omega_2 t} - b_2 e^{-i\omega_2 t} \end{pmatrix} \tag{6.4.22}$$

or, written in real form,

$$\begin{pmatrix} \varphi_1 \\ \varphi_2 \end{pmatrix} = \begin{pmatrix} a_1' \cos \omega_1 t + b_1' \sin \omega_1 t + a_2' \cos \omega_2 t + b_2' \sin \omega_2 t \\ a_1' \cos \omega_1 t + b_1' \sin \omega_1 t - a_2' \cos \omega_2 t - b_2' \sin \omega_2 t \end{pmatrix}. \tag{6.4.23}$$

We see that there are two normal modes, and the general solution is formed from the superposition of these two modes:

$$\begin{pmatrix} \varphi_1 \\ \varphi_2 \end{pmatrix} = \begin{pmatrix} 1 \\ 1 \end{pmatrix} e^{\pm i\omega_1 t}: \quad \text{The pendulums swing in the same direction}$$
$$(\varphi_1 = \varphi_2) \tag{6.4.24}$$

$$\begin{pmatrix} \varphi_1 \\ \varphi_2 \end{pmatrix} = \begin{pmatrix} 1 \\ -1 \end{pmatrix} e^{\pm i\omega_2 t}: \quad \text{The pendulums swing in the opposite directions}$$
$$(\varphi_1 = -\varphi_2). \tag{6.4.25}$$

In addition, the normal coordinates are given by (up to a normalization constant):

$$Q_1 = \varphi_1 + \varphi_2, \tag{6.4.26}$$

$$Q_2 = \varphi_1 - \varphi_2. \tag{6.4.27}$$

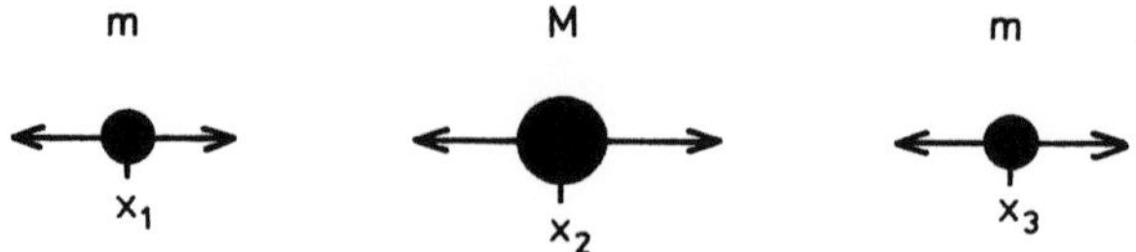

Fig. 6.4.2. A linear triatomic molecule

ii) We now consider a linear triatomic molecule. Let the mass of the center atom be M, the masses of the other two atoms, m (Fig. 6.4.2).

Let x_1, x_2, x_3 be the coordinates of the three atoms, and let

$$x_2 - x_1 = b , \quad x_3 - x_2 = b \tag{6.4.28}$$

be the equilibrium distances.

This means that the minimum values of the functions $V(x_2 - x_1)$ and $V(x_3 - x_2)$ are at $V(b)$.

Then, if we only take into account the interaction of neighboring particles, we have

$$
\begin{aligned}
V(x_1, x_2, x_3) &= V(x_2 - x_1) + V(x_3 - x_2) \\
&= 2V(b) + \tfrac{1}{2}V''(b)(x_2 - x_1 - b)^2 \\
&\quad + \tfrac{1}{2}V''(b)(x_3 - x_2 - b)^2 + \dots .
\end{aligned}
\tag{6.4.29}
$$

If x_i^0 are the coordinates in a state of equilibrium, then

$$\eta_i \quad \text{with} \quad x_i = x_i^0 + \eta_i$$

is the displacement of the atom i from its condition of equilibrium. Then,

$$x_2 - x_1 - b = x_2^0 - x_1^0 + \eta_2 - \eta_1 - b = \eta_2 - \eta_1 , \tag{6.4.30}$$

and analogously

$$x_3 - x_2 - b = \eta_3 - \eta_2 ,$$

and so we find that the potential to be considered is

$$V = \tfrac{1}{2}K(\eta_2 - \eta_1)^2 + \tfrac{1}{2}K(\eta_3 - \eta_2)^2 \quad \text{with} \quad K = V''(b) . \tag{6.4.31}$$

The Lagrangian is then given by

$$L = \tfrac{1}{2}m(\dot\eta_1^2 + \dot\eta_3^2) + \tfrac{1}{2}M\dot\eta_2^2 - \frac{K}{2}[(\eta_2 - \eta_1)^2 + (\eta_3 - \eta_2)^2] , \tag{6.4.32}$$

hence

$$M_{ij} = \begin{pmatrix} m & 0 & 0 \\ 0 & M & 0 \\ 0 & 0 & m \end{pmatrix}, \qquad K_{ij} = \begin{pmatrix} K & -K & 0 \\ -K & 2K & -K \\ 0 & -K & K \end{pmatrix}. \tag{6.4.33}$$

The secular equation is then

$$\det(K_{ij} - \omega^2 M_{ij}) = \det \begin{pmatrix} K - \omega^2 m & -K & 0 \\ -K & 2K - \omega^2 M & -K \\ 0 & -K & K - \omega^2 m \end{pmatrix} = 0 \tag{6.4.34}$$

or

$$(K - m\omega^2)^2(2K - M\omega^2) - 2K^2(K - m\omega^2) = 0 \;,$$

i.e.

$$(K - m\omega^2)[-2K^2 + (K - m\omega^2)(2K - M\omega^2)]$$

$$= (K - m\omega^2)(Mm\omega^4 - 2Km\omega^2 - KM\omega^2)$$

$$= (K - m\omega^2)\omega^2\left[\omega^2 - \frac{K(2m + M)}{mM}\right]mM = 0 \;. \tag{6.4.35}$$

We then find as solutions:

$$\omega_1^2 = 0 \;, \tag{6.4.36}$$

$$\omega_2^2 = \frac{K}{m} \;, \tag{6.4.37}$$

$$\omega_3^2 = K\left(\frac{1}{m} + \frac{2}{M}\right) > \omega_2^2 \;. \tag{6.4.38}$$

Since there is a solution $\omega_1 = 0$, $V(x_1, x_2, x_3)$ does not have a (true) minimum at $x_i = x_i^0$ and

$$\sum_{i,j=1}^{3} K_{ij} c_i c_j$$

is therefore not positive definite.

This was to be expected, since the sum of each row in the matrix K is zero, so that $\det(K_{ij})$ itself vanishes.

Then, the c_i, which satisfy

$$\sum_{i,j=1}^{3} K_{ij}c_i c_j = 0$$

are exactly given by $c_i = c$.

It is clear that the potential does not have a minimum at the equilibrium point with respect to these displacements

$$\eta_i = \eta \, , \quad i = 1, 2, 3 \, ,$$

since the potential is translation invariant and thus remains constant under these displacements, which are translations of the system as a whole.

The eigenvector corresponding to $\omega_1^2 = 0$ can be found from

$$\sum_{j=1}^{3} K_{ij}v_j^{(1)} = 0 = \begin{pmatrix} K & -K & 0 \\ -K & 2K & -K \\ 0 & -K & K \end{pmatrix} \begin{pmatrix} v_1^{(1)} \\ v_2^{(1)} \\ v_3^{(1)} \end{pmatrix}$$

which immediately yields

$$v_1^{(1)} = v_2^{(1)} = v_3^{(1)} = a \, . \tag{6.4.39}$$

The normalization condition is

$$\sum_{i,j=1}^{3} M_{ij}v_i^{(1)}v_j^{(1)} = (2m + M)a^2 = 1 \, , \tag{6.4.40}$$

so that

$$v^{(1)} = \frac{1}{\sqrt{2m + M}} \begin{pmatrix} 1 \\ 1 \\ 1 \end{pmatrix} . \tag{6.4.41}$$

For $\omega_2^2 = K/m$, we have

$$\begin{pmatrix} 0 & -K & 0 \\ -K & 2K - KM/m & -K \\ 0 & -K & 0 \end{pmatrix} \begin{pmatrix} v_1^{(2)} \\ v_2^{(2)} \\ v_3^{(2)} \end{pmatrix} = 0 \, , \tag{6.4.42}$$

so that

$$v^{(2)} = 0 \, , \quad v_1^{(2)} = -v_3^{(2)} \, . \tag{6.4.43}$$

The normalization condition in this case is

$$m[(v_1^{(2)})^2 + (v_3^{(2)})^2] + M(v_2^{(2)})^2 = 1 \tag{6.4.44}$$

so that

$$v^{(2)} = \frac{1}{\sqrt{2m}} \begin{pmatrix} 1 \\ 0 \\ -1 \end{pmatrix}. \tag{6.4.45}$$

Finally, for

$$\omega_3^2 = \frac{K}{m} \frac{M + 2m}{M} \tag{6.4.46}$$

we calculate

$$v^{(3)} = \frac{1}{\sqrt{M(1 + M/2m)}} \begin{pmatrix} M/2m \\ -1 \\ M/2m \end{pmatrix}. \tag{6.4.47}$$

We have now calculated the transformation matrix $(v_j^{(\alpha)})$, which, if we set

$$\eta_j = \sum_{\alpha=1}^{3} Q_\alpha v_j^{(\alpha)} \tag{6.4.48}$$

transforms the Lagrangian

$$L = \tfrac{1}{2}m(\dot\eta_1^2 + \dot\eta_3^2) + \tfrac{1}{2}M\dot\eta_2^2 - \tfrac{1}{2}K[(\eta_2 - \eta_1)^2 + (\eta_3 - \eta_2)^2] \tag{6.4.49}$$

into the normal coordinate form

$$L = \tfrac{1}{2}(\dot Q_1^2 + \dot Q_2^2 + \dot Q_3^2) - \tfrac{1}{2}\omega_2^2 Q_2^2 - \tfrac{1}{2}\omega_3^2 Q_3^2 \ . \tag{6.4.50}$$

In particular,

$$\eta_1 = \frac{1}{\sqrt{2m + M}} Q_1 + \frac{1}{\sqrt{2m}} Q_2 + \frac{M}{2m\sqrt{M[1 + (M/2m)]}} Q_3 \ , \tag{6.4.51}$$

$$\eta_2 = \frac{1}{\sqrt{2m + M}} Q_1 - \frac{1}{\sqrt{M[1 + (M/2m)]}} Q_3 \ , \tag{6.4.52}$$

$$\eta_3 = \frac{1}{\sqrt{2m + M}} Q_1 - \frac{1}{\sqrt{2m}} Q_2 + \frac{M}{2m\sqrt{M[1 + (M/2m)]}} Q_3 \ . \tag{6.4.53}$$

We see that it is always true that

$$m(\eta_1 + \eta_3) + M\eta_2 = \sqrt{2m + M}\, Q_1 \ . \tag{6.4.54}$$

On the other hand, it follows from L that

$$\ddot{Q}_1 = 0 , \quad \text{thus} \quad Q_1 = \alpha t + \beta \tag{6.4.55}$$

instead of

$$Q_1 = c_1 e^{i\omega_1 t} + d_1 e^{-i\omega_2 t} \tag{6.4.56}$$

but also

$$\ddot{Q}_2 + \omega_2^2 Q_2 = 0 , \tag{6.4.57}$$

and thus

$$Q_2 = c_2 e^{i\omega_2 t} + d_2 e^{-i\omega_2 t} \tag{6.4.58}$$

and, in the same way,

$$Q_3 = c_3 e^{i\omega_3 t} + d_3 e^{-i\omega_3 t} . \tag{6.4.59}$$

This means that while Q_2 and Q_3 have an oscillatory character, the coordinate Q_1, since $\ddot{Q}_1 = 0$, varies linearly. If only Q_1 and $\dot{Q}_1$ are nonzero, then

$$\eta_1 = \eta_2 = \eta_3 = \frac{1}{\sqrt{2m + M}}(\alpha t + \beta) , \tag{6.4.60}$$

that is, the complete molecule moves uniformly along its axis. Of course, the potential does not change in this motion; the potential is translation invariant.

As we see, this leads to the consequence that one root of the characteristic polynomial is exactly zero. Translational motion is free and unbounded and goes along the bottom of a potential energy valley.

If we do not allow this motion and set $Q_1 = \text{constant}$, we find

$$m(\eta_1 + \eta_3) + M\eta_2 = \text{constant} \tag{6.4.61}$$

so that the center of mass of the molecule is at rest.

The other degrees of freedom, represented by Q_2 and Q_3 are, however, oscillatory degrees of freedom.

If only $Q_2 \neq 0$, then

$$\eta_1 = -\eta_3 , \quad \eta_2 = 0 , \tag{6.4.62}$$

so that the atom with mass M remains at rest. Atoms 1 and 3 oscillate in opposite phase with the frequency $\omega_2^2 = K/m$.

If only $Q_3 \neq 0$, then at each point in time

$$\eta_1 = \eta_3 , \quad \eta_2 = -\frac{2m}{M}\eta_1 , \tag{6.4.63}$$

i.e., atoms 1 and 3 oscillate in the same direction, but opposite to the direction of motion of the heaveir atom 2, so that at each point in time,

$$m(\eta_1 + \eta_3) + M\eta_2 = 0 \qquad (6.4.64)$$

and the molecule as a whole remains at rest.

6.5 The Response of Linear Systems to External Forces

In many important applications in physics, a system will be stimulated from outside, that is, energy will be exchanged with the system. Since, in reality, there are always dissipative forces at work, such as friction (which transforms mechanical energy into heat), a linear system without an external source of energy will come to rest over a long period of time, that is, it will only undergo damped oscillations.

6.5.1 External Oscillating Forces

We will first investigate the inhomogeneous differential equation

$$\ddot{x}(t) + 2\varrho\dot{x}(t) + \omega_0^2 x(t) = f_0 \cos(\omega t) \qquad (6.5.1)$$

with harmonically varying external force. We reduce this problem to the simpler case of the complex differential equation

$$\ddot{x}(t) + 2\varrho\dot{x}(t) + \omega_0^2 x(t) = e^{i\omega t} \quad (\omega \geq 0) \ . \qquad (6.5.2)$$

In particular, if $x(t)$ is a solution of (6.5.2), then $f_0 \operatorname{Re}\{x(t)\}$ is a solution of (6.5.1).

We solve the differential equation (6.5.2) by assuming an exponential form for $x(t)$

$$x^{(0)}(t) = A e^{i\omega t}$$

with an amplitude A to be determined. If we substitute this expression for $x(t)$ in (6.5.2), we find

$$A(-\omega^2 + 2i\varrho\omega + \omega_0^2) = 1 \ .$$

We then obtain the special solution

$$x^{(0)}(t) = \frac{1}{\omega_0^2 - \omega^2 + 2i\varrho\omega} e^{i\omega t} \ , \qquad (6.5.3)$$

which yields the general solution if we add to it the general solution of the homogeneous equation (Sect. 6.2). Since the solutions of the homogeneous equation for $\varrho > 0$ always die away as $t \to \infty$, every solution of (6.5.2) approaches the special solution $x^{(0)}(t)$ as $t \to \infty$. The damping causes each solution of the harmonic equation with external force to "forget" its initial conditions after a sufficiently long period of time, and to turn into the particular fixed solution $x^{(0)}(t)$, which varies with the frequency of the applied force. The solution $x^{(0)}(t)$ is called the *steady state* solution, while the parts of the solution that die away are called the *transient* terms.

If we write

$$A = |A|e^{i\delta}$$

we see that (6.5.1) has the particular solution

$$f_0 \operatorname{Re}\{x^{(0)}(t)\} = f_0 |A| \cos(\omega t + \delta) \ .$$

Here

$$|A|^2 = \frac{1}{(\omega^2 - \omega_0^2)^2 + 4\varrho^2\omega^2} \ , \qquad \tan\delta = \frac{2\varrho\omega}{\omega^2 - \omega_0^2} \ . \tag{6.5.4}$$

$|A|^2$ has a minimum at $\omega^2 = \omega_0^2 - 2\varrho^2$ and δ varies between 0 and $-\pi$ as ω goes from 0 to ∞ (Fig. 6.5.1). For $\omega = \omega_0$, we have $\delta = -\frac{\pi}{2}$.

For weak damping $\varrho \ll \omega_0$ and $|\omega - \omega_0| \ll \omega_0$, we have, approximately,

$$|A|^2 = \frac{1}{4\omega_0^2} \frac{1}{(\omega - \omega_0)^2 + \varrho^2} \ , \qquad \tan\delta = \frac{\varrho}{\omega - \omega_0} \ . \tag{6.5.5}$$

If the applied force has a frequency near the eigenfrequency ω_0 of the undamped oscillator, the amplitude becomes particularly large (this phenomenon is called *resonance*). The smaller the damping ϱ, the narrower and higher

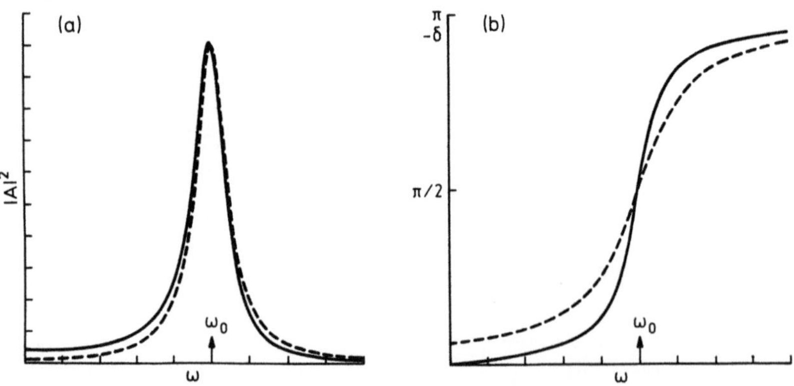

Fig. 6.5.1. The dependence of the square of the amplitude $|A|^2$ and the phase $-\delta$ on ω. The solid curve is the exact result (6.5.4), the dotted curve is the approximation for weak damping (6.5.5)

the maximum is. In terms of phase, the forced oscillation always lags behind the external force. For $\omega \to 0$, the phase difference approaches 0 and for $\omega \to \infty$ it approaches its maximum value of $-\pi$. The transition between phase difference 0 and $-\pi$ becomes more sudden as the damping decreases.

The behavior of a linear system under the influence of an external periodic force described above is of the greatest importance, because it can be observed in countless physical systems. For this reason, the dependence of the amplitude $A(\omega)$ and the phase $\delta(\omega)$ on the frequency ω should be carefully noted. It is especially important to keep in mind that the eigenfrequencies ω_α of the system are noticeable as resonance frequencies. This is to say, if the frequency of the external force is in the neighborhood of one of the resonance frequencies, the system responds with a particularly large amplitude of oscillation at this frequency.

The more general case of n degrees of freedom can be handled similarly. The inhomogeneous equation (without damping)

$$M\ddot{x}(t) + Kx(t) = f\cos\omega t \quad \text{with} \quad x(t), f \in \mathbb{R}^n$$

or, in coordinates

$$\sum_{j=1}^{n} M_{ij}\ddot{x}_j(t) + K_{ij}x_j(t) = f_i \cos\omega t , \quad i = 1, \ldots, n \tag{6.5.6}$$

can be solved as follows:

We consider first the equation

$$M\ddot{x} + Kx = f e^{i\omega t} , \tag{6.5.7}$$

and look for a solution of the form

$$x^{(0)}(t) = A e^{i\omega t}$$

for some $A \in \mathbb{C}^n$.

Substitution in (6.5.7) yields

$$(K - \omega^2 M)A = f , \quad \text{i.e.}$$

$$\sum_{j=1}^{n} (K_{ij} - \omega^2 M_{ij})A_j = f_i , \quad i = 1, \ldots, n ,$$

from which we obtain

$$A = (K - \omega^2 M)^{-1} f .$$

Here, $(K - \omega^2 M)^{-1}$ is the inverse matrix of $(K - \omega^2 M)$. Thus

$$x^{(0)}(t) = (K - \omega^2 M)^{-1} f e^{i\omega t} , \tag{6.5.8}$$

and $\text{Re}\{x^{(0)}(t)\}$ is then a solution of (6.5.6).

$(K - \omega^2 M)^{-1}$ must certainly exist, if $(K - \omega^2 M)$ is injective, that is, if ω^2 is different from all the resonance frequencies ω_α of the linear system. If ω^2 is in the neighborhood of one of the eigenfrequencies, then resonances, i.e. very large amplitudes of the forced motion, have to be dealt with.

The relationships are especially clear when we move to the normal coordinates $Q_\alpha = v^{(\alpha)} \cdot Mx$, where $v^{(\alpha)}$ is the eigenvector corresponding to the eigenvalue ω_α^2:

$$(K - \omega_\alpha^2 M)v^{(\alpha)} = 0 \ . \tag{6.5.9}$$

Scalar multiplication of

$$M\ddot{x}(t) + Kx(t) = f \cos \omega t$$

from the left with $v^{(\alpha)}$ yields

$$\ddot{Q}_\alpha + \omega_\alpha^2 Q_\alpha = f_\alpha \cos \omega t \quad \text{with}$$

$$f_\alpha = v^{(\alpha)} \cdot f \ . \tag{6.5.10}$$

We are therefore left with n uncoupled one-dimensional forced oscillations. Resonance phenomena appear if ω agrees with one of the eigenfrequencies ω_α and if, in this situation, f is not orthogonal to $v^{(\alpha)}$.

6.5.2 Superposition of External Harmonic Forces

We now consider again a system with one degree of freedom under the influence of an external force, which this time is not a single trigonometric function but rather a superposition of such periodic functions. The equation of motion is then given by

$$\ddot{x}(t) + 2\varrho\dot{x}(t) + \omega_0^2 x(t) = \sum_k c_k e^{i\omega_k t} \ , \tag{6.5.11}$$

where it is left open how far the index k in the summation runs.

Let us write

$$D^n := \frac{d^n}{dt^n} \ , \tag{6.5.12}$$

then the left-hand side can be conveniently expressed as

$$L(D)x = (D^2 + 2\varrho D + \omega_0^2)x \ .$$

Now, we have

$$L(D)e^{i\omega t} = L(i\omega)e^{i\omega t} \ .$$

If we substitute a real or complex number s for the operator D, we find

$$L(s) = s^2 + 2\varrho s + \omega_0^2 \;, \qquad \text{and} \tag{6.5.13}$$

$$Y(s) := \frac{1}{L(s)} = \frac{1}{s^2 + 2\varrho s + \omega_0^2} \tag{6.5.14}$$

is called the *transfer function*. It is a measure of the strength of the coupling of the external forces to the linear system.

Since now

$$L(D)e^{i\omega_k t} = L(i\omega_k)e^{i\omega_k t} \;, \qquad \text{we have}$$

$$x^{(0)}(t) = \sum_k c_k \, Y(i\omega_k)e^{i\omega_k t} \tag{6.5.15}$$

a solution of the inhomogeneous differential equation.

Thus, we can find the response of a linear system to any external force which can be written as a finite superposition of the periodic functions

$$e^{i\omega_k t} \;.$$

6.5.3   Periodic External Forces

We will now consider periodic functions which can be expressed as *infinite* linear combinations of the functions $\exp(i\omega_k t)$ with $\omega_k = k\omega_1$ (k integer). This class of periodic functions is extremely large, larger than the set of periodic functions which can be written in terms of a Taylor series. This class of functions is investigated in the theory *Fourier*[3] *series*, whose most important results are sketched briefly in Appendix D.

In this case, we can easily find a solution of the differential equation $Lx = f$, in which L is the general differential operator

$$L = L\left(\frac{d}{dt}\right) \equiv \sum_{r=0}^{N} L_r \frac{d^r}{dt^r} \tag{6.5.16}$$

and f is an arbitrary periodic function of period a

$$f(t) = \sum_{n=-\infty}^{\infty} f_n e^{2\pi i n t/a} \;.$$

---

[3] *Fourier, Joseph* (*1768 Auxerre, d. 1830 Paris). He made use of trigonometric series expansions in his work "Théorie analytique de la chaleur" in his treatment of heat conduction.

If we assume our solution has the form

$$x(t) = \sum_{n=-\infty}^{+\infty} x_n e^{2\pi i n t/a}$$
(6.5.17)

the equation for x_n reads:

$$L\left(\frac{2\pi i n}{a}\right) x_n = \frac{x_n}{Y\left(\dfrac{2\pi i n}{a}\right)} = f_n$$

with the transfer function

$$Y\left(\frac{2\pi i n}{a}\right) = \frac{1}{L\left(\dfrac{2\pi i n}{a}\right)} \ .$$

Thus,

$$x^{(0)}(t) = \sum_{n=-\infty}^{+\infty} Y\left(\frac{2\pi i n}{a}\right) f_n e^{2\pi i n t/a}$$
(6.5.18)

is a solution of the differential equation. The general solution is found by adding this particular solution to the most general solution of the homogeneous equation $Lx = 0$.

6.5.4 Arbitrary External Forces

Finally, with the help of the theory of Fourier transforms (Appendix D), we can find a solution of the inhomogeneous differential equation

$$\sum_{r=0}^{N} L_r \frac{d^r}{dt^r} x(t) =: L\left(\frac{d}{dt}\right) x(t) = f(t)$$
(6.5.19)

for an arbitrary function f, as long as f has a Fourier transform. Here, we proceed exactly as in Sect. 6.5.3, where we solved the corresponding problem for periodic f using a Fourier series expansion.

If $\tilde{f}(\omega)$ is the Fourier transform of $f(t)$, then a particular solution of the equation (6.5.19) is given by

$$x^{(0)}(t) = \frac{1}{\sqrt{2\pi}} \int_{-\infty}^{+\infty} d\omega\, Y(i\omega) \tilde{f}(\omega) e^{i\omega t} \ .$$
(6.5.20)

This can be seen immediately by differentiating inside the Fourier integral:

$$Lx^{(0)}(t) = \frac{1}{\sqrt{2\pi}} \int_{-\infty}^{+\infty} d\omega\, Y(i\omega) L(i\omega) e^{i\omega t} \tilde{f}(\omega)$$

$$= \frac{1}{\sqrt{2\pi}} \int_{-\infty}^{+\infty} d\omega\, e^{i\omega t} \tilde{f}(\omega) = f(t) \ .$$

Equation (6.5.20) also means

$$\tilde{x}^{(0)}(\omega) = Y(i\omega) \tilde{f}(\omega) \ , \tag{6.5.21}$$

and by the convolution theorem (see Appendix D) we can also write

$$x^{(0)}(t) = \sqrt{2\pi}(G*f)(t) = \int_{-\infty}^{+\infty} ds\, G(t-s) f(s) \tag{6.5.22}$$

with

$$G(t-s) = \frac{1}{2\pi} \int_{-\infty}^{+\infty} d\omega\, e^{i\omega(t-s)}\, Y(i\omega) \ . \tag{6.5.23}$$

The function $G(t-s)$, which yields a solution for $Lx = f$ for an arbitrary f by means of convolution, is called the *Green's*[4] *function* of the differential operator L.

The system

$$\sum_{j=1}^{n} L_{ij}\left(\frac{d}{dt}\right) x_j(t) = f_i(t) \ , \quad i = 1, \ldots, n \tag{6.5.24}$$

can be handled analogously

$$x_i^{(0)}(t) = \sum_{j=1}^{n} \int_{-\infty}^{+\infty} ds\, G_{ij}(t-s) f_j(s) \ . \tag{6.5.25}$$

is a solution of (6.5.24) if

$$G_{ij}(t-s) = \frac{1}{2\pi} \int_{-\infty}^{+\infty} d\omega\, e^{i\omega(t-s)}\, Y_{ij}(i\omega) \tag{6.5.26}$$

is the Green's function with Y_{ij} defined by

$$\sum_{j=1}^{n} L_{ij}(i\omega)\, Y_{jk}(i\omega) = \delta_{ik} \ . \tag{6.5.27}$$

---

[4] *Green, George* (*1793 Sneinton/Nottinghamshire, d. 1841 ibidem). Originally a baker, he taught himself mathematics and physics. His most important works concern the mathematical theory of electricity and magnetism. He coined the term "potential".

Remark. Note that

$$LG(t - s) = \frac{1}{2\pi} \int\limits_{-\infty}^{+\infty} d\omega \, e^{i\omega(t-s)} \equiv \delta(t - s) \,, \tag{6.5.28}$$

where the right side of the equation cannot be a function in the normal sense, since it diverges for $t = s$ and vanishes for $t \neq s$. The physicist P.A.M. Dirac[5] introduced thus "function" to simplify treatment of certain physical problems, and thus it is called the Dirac delta function. After Dirac introduced the delta function, mathematicians justified its use and developed rules of calculation for it in the theory of *generalized functions* or *distributions*[6]. In Appendix E, a short introduction to the theory of distributions is given.

Problems

6.1 Perturbations of Circular Motion. Discuss small deviations from circular motion for a particle in the following central fields:

a) a 3-dimensional harmonic oscillator $V = kr^2$
b) the Kepler problem

$$V = -\frac{A}{r} \,, \quad A > 0 \,.$$

Show that the deviation ϱ of the radial coordinate from the radius of the circle r_0 satisfies the equation of motion

$$\ddot{\varrho}(t) + \omega_0^2 \varrho(t) = 0$$

in linear approximation. Determine ω_0 in terms of the mass m of the particle, k or A and r_0 respectively, and interpret the result.

---

[5] *Dirac, Paul Adrien Maurice* (*1902 Bristol, d. 1984), one of the greatest English physicists, professor at Cambridge after 1932. Earned the Nobel Prize in physics in 1933 (with E. Schrödinger). He only turned to physics when he couldn't find a job after studying engineering. He is one of the founders of quantum mechanics and earned special praise for formulating it in a closed and elegant way (this is where he used the δ-function). He is also well-known as the co-founder of Fermi–Dirac statistics. His name is connected in particular with the Dirac equation, which gives a relativistic description of the electron, predicts the proper value for the gyromagnetic moment of the electron, and from which he predicted the existence of antimatter.

[6] Distributions are generalized functions. In 1945, L. Schwarz elaborated their theory. Heuristic methods of calculation used in physics (Dirac, Heaviside) were thus given a solid mathematical basis.

6.2 Coupled Differential Equations. a) Solve the equation of motion of a charged particle in a homogeneous magnetic field $\boldsymbol{B}_0$:

$$m\ddot{\boldsymbol{r}} = e\dot{\boldsymbol{r}} \times \boldsymbol{B}_0 \ .$$

b) Solve the equations of motion for small oscillations of the system shown here in its equilibrium state.

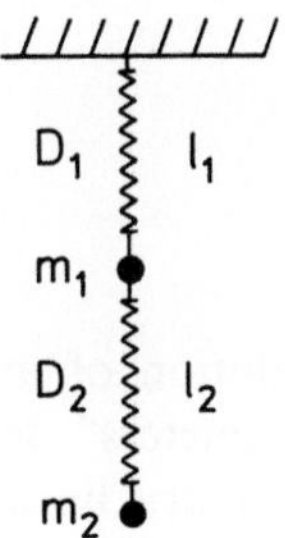

6.3 Linear Chains. We consider N particles of mass m each connected to its neighbors by springs with spring constant D which only undergo longitudinal oscillations; let the equilibrium distance between neighboring particles be d.

We choose the following possible boundary conditions:

– Periodic boundary conditions: Particles 1 and N are coupled to each other in the same way as the other neighboring particles.

(We can picture the particles as being arranged in a circle, where, if N is large enough, the curvature of the circle plays no role. Thus we identify particle N with particle 0, particle $N + 1$ with particle 1, etc.)

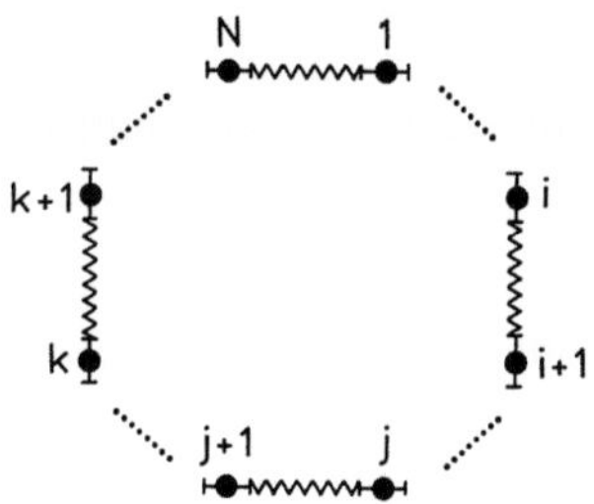

– Fixed endpoints: Particles 1 and N are attached on opposite sides to a fixed point.

a) Find the Lagrangian and the equations of motion of the system, and determine the normal frequencies using the assumptions

$$x_k(t) = r_k e^{-i\omega t} \quad \text{and} \quad r_k = a e^{i\lambda k} \ (\lambda \in \mathbb{R}) .$$

b) Determine also the normal coordinates, i.e., the possible values of λ which are consistent with the boundary conditions. In the case of fixed endpoints we should assume that the amplitudes satisfy

$$r_k = a \sin(\lambda k)$$

(why?)

c) How often are the possible normal frequencies degenerate?

d) Indicate the most general solution for $x_k(t)$.

6.4 Variation of Parameters. An important method in the solution of inhomogeneous differential equations is that of "variation of parameters" developed by Lagrange, which here is shown for the case of ordinary linear differential equations.

The equation to be solved is the following system of inhomogeneous differential equations of the first order:

$$\dot{y}(t) + A(t)y(t) = f(t) .$$

Here, $y(t)$ and $f(t)$ are n-component functions and $A(t)$ is an $n \times n$ matrix. $f(t)$ is a given external perturbation. Let y_i $(i = 1, \ldots, n)$ be a complete set of solutions of the corresponding homogeneous system of differential equations. Then, we can define the $n \times n$ matrix

$$Y(t) = \begin{pmatrix} y_1^1(t) \ \ldots \ y_n^1 \\ \vdots \qquad \vdots \\ y_1^n(t) \ \ldots \ y_n^n \end{pmatrix} \quad \text{so that} \quad Y_{ji}(t) = y_i^j(t)$$

which is a matrix solution of the homogeneous system of differential equations:

$$\dot{Y}(t) + A Y(t) = 0 .$$

The most general solution of the homogeneous system of differential equations is

$$y(t) = Y(t)c$$

with arbitrary constants c.

Our hypothesis for the solution of the inhomogeneous system of differential equations is

$$y(t) = Y(t)c(t) ,$$

i.e., the constants of the homogeneous equation can now change with time. The task is now to determine the function $c(t)$.

a) Show that

$$c(t) = \int_{t_0}^{t} dt' \, Y^{-1}(t') f(t') \quad \text{and thus}$$

$$y(t) = \int_{t_0}^{t} dt' \, Y(t') Y^{-1}(t') f(t')$$

is a solution of the inhomogeneous system of differential equations.

b) Use the method of "variation of parameters" to solve the following differential equation:

$$\ddot{x} + 2\gamma \dot{x} + \omega_0^2 x = f(t) \ .$$

7. Classical Statistical Mechanics

Matter on the macroscopic scale always consists of a very large number of particles (atoms or molecules). The number of particles in the macroscopic volume element of a cubic meter or a liter is of the order of magnitude of 10^{23}. It is self-evident that it makes no sense to try to write out and solve the equations of motion for this number of particles. Finding the explicit solution, i.e. the trajectories of all the particles, is not even desirable, and anyway it could not be verified experimentally.

Merely the process of gathering such data would far exceed the capabilities of a researcher, and even if the microscopic state of the system, that is the position and velocities of all 10^{23} particles, could be specified at one particular time, the smallest uncertainty in its determination would be magnified so quickly that just a little later the state of the system would be practically unknown. The knowledge of the microscopic state of a macroscopic system is thus neither possible nor useful. Because of the impossibility of such a description, other kinds of information are used to describe macroscopic systems.

We certainly know some of these concepts through intuition or experience, though they cannot be directly related to the mechanical quantities of the individual particles. For example, we can measure the temperature, the pressure, the volume, the heat capacity, etc. of a gas enclosed in a container. As we learn from experiment, these quantities cannot all be varied independently of each other in a system, and indeed there are laws which connect these (macroscopic) state variables with each other. An example is the ideal gas law.

In general, then, macroscopic systems are described by *state variables* and the *macroscopic state* of a system is determined by the values of a large enough set of these state variables.

Thermodynamics is a general theory of macroscopic systems, the description of their macroscopic states, and the interdependence of the state variables and of the possible changes in state.

It is the task of *statistical mechanics* to describe the connection between the micro- and the macro-system and to calculate the properties of macroscopic system from the microscopic interactions.

7.1   Thermodynamic Systems and Distribution Functions

Before we occupy ourselves with the larger problem, we will first explain a few basic concepts of thermodynamics which always appear in the description of macroscopic systems.

A *system* is an identifiable part of the world which can be theoretically, and in principle also operationally, separated from the rest of the world, which in general can be described by giving certain *boundary conditions* (for instance, a volume V to which the system is confined). Anything which acts on the system must be carefully recorded and is said to belong to the *environment* of the system. Several systems can be combined into a *total system.*

The state of a system is given by the values of all of its state variables (or of a complete set of independent state variables). The identification of the relevant variables involves a certain amount of abstraction. We do not consider the values of the irrelevant variables, and thus we identify the system with the totality of the possible values of its (relevant) state variables.

A system is called

closed,   if it does not exchange matter with its environment.
isolated, if it does not exchange either matter or energy with its environment
            and
open,   otherwise.

We can usually make nonisolated systems into isolated systems by enlarging them to include some of their environment.

A state variable of a system is called *extensive* ($=$ additive $=$ quantified), if its value doubles when the system is doubled (that is, when the system is joined to a perfect copy of itself), and *intensive* if its value remains unchanged.

Examples of extensive quantities are volume, energy, and particle number, while intensive quantities include pressure, temperature, and density. In general, the most important quantities in thermodynamics are either extensive or intensive.

From experience, we know that after a certain period of time (the *relaxation time*) an isolated macrosystem will reach a *state of equilibrium*, which is uniquely determined by the boundary conditions, and which cannot change spontaneously. A state of equilibrium can be described by a small number of independent state variables, whereas to specify a non-equilibrium state we may need a far greater number of variables.

Example. A stirred fluid in a container is at first in a state of non-equilibrium. After a certain relaxation time has passed, the fluid, because of the effects of friction, reaches the equilibrium state of a liquid at rest.

In this and the next chapter, we will only be concerned with states of equilibrium. The state variables are then all time independent.

We now turn to the relationship between macroscopic and microscopic descriptions of equilibrium states. In this context, we consider a macroscopic system to be a typical mechanical system with an unimaginably large number, say 10^{23}, degrees of freedom. This process may seem questionable at first, because classical mechanics is not valid at atomic dimensions and must be replaced by quantum mechanics. It turns out, though, that at atmospheric pressure and at temperatures which are not too small, the treatment of many systems using classical mechanics can be justified. In addition, "quantum statistics" is so similar to the theory of classical statistical mechanics presented here, that, given a knowledge of quantum mechanics, there are not too many additional difficulties in its formulation.

Let us then think of a macroscopic system, say a gas with $N \approx 10^{23}$ molecules, which is in a container of volume $V = 1$ liter. The microscopic states are characterized by points

$$(q, p) = (r_1, \ldots, r_N, p_1, \ldots, p_N)$$

in $6N$-dimensional phase space, and the trajectories can in principle be calculated if we know the Hamiltonian $H(q, p)$. Somehow the macroscopic characteristics of the system must also be determined by the Hamiltonian H.

The basic idea of statistical mechanics is to introduce probability statements about the microscopic state of a system. Since the number of microstates of a macroscopic system is far "larger" than the number of macroscopic states, there must be many microstates which correspond to one and the same macroscopic state.

If we could measure the microstate of a system with a given fixed macroscopic state, the results would turn out differently from measurement to measurement, and it is thus meaningful to speak of the probability that a particular microstate will be measured.

We will write the probability that, at time t_0, the microstate will lie in a volume element of size $d\tau = d^{3N}q\, d^{3N}p$ around a point (q, p) in phase space as

$$dw(q, p) = \varrho(q, p)d^{3N}q\, d^{3N}p \ . \tag{7.1.1}$$

The function $\varrho(q, p)$ is called the *probability density* in phase space or the *"distribution function."*

We begin by making the following claims about this probability density:

i) Since the microstate must certainly lie somewhere in phase space (the probability of this is 1), we have

$$\int \varrho(q, p)d^{3N}q\, d^{3N}p = 1 \tag{7.1.2}$$

integrating over all of phase space.

ii) The average measured value (statistical average) $\langle A \rangle$ of a quantity $A(p, q)$ at time t_0 is then calculated as follows:

$$\langle A \rangle = \int A(q, p)\varrho(q, p)d^{3N}q\,d^{3N}p \; . \tag{7.1.3}$$

In an equilibrium state, $\langle A \rangle$ is, of course, time-independent.

iii) If we also want to study the time behavior of a non-equilibrium state, we must introduce an explicitly time-dependent distribution function $\varrho(q, p, t)$. The equation in (i) also holds for this function, as does the procedure in (ii) for the calculation of time-dependent averages. In general, it is possible to show that *Liouville's equation*

$$\partial\varrho/\partial t + \{\varrho, H\} = 0 \tag{7.1.4}$$

holds for the distribution function (with the help of Liouville's theorem from Sect. 3.7, or see, for example, [McQuarrie]), where $\{\varrho, H\}$ represents the Poisson bracket. If the distribution function is time-independent, as in the case of equilibrium, it follows that the function $\varrho(q, p)$ is a conserved quantity in the sense of Sect. 3.7.1. We will further assume that the energy E, the momentum P, and the angular momentum L are the only conserved quantities of the system. Any other case, in which other independent conserved quantities would exist, is highly untypical, and certainly cannot happen for a complicated N-particle system. To simplify our calculations we further assume that $P = 0$ and $L = 0$. The system as a whole is then at rest. Then the function $\varrho(q, p)$ must be of the form

$$\varrho(q, p) = f(H(q, p)) \; , \tag{7.1.5}$$

that is, it is a presently undetermined function of the energy.

In order to determine the distribution function more exactly, we limit ourselves to an isolated system, so that the total energy of the system has a fixed value E. The totality of possible microstates for given values of the macroscopic variables E, N, and V in phase space is called a *microcanonical ensemble*. In this chapter, we will also become acquainted with other ensembles, for example in Sect. 7.5 we will see the canonical ensemble in which we are given the temperature T (which we still need to introduce) rather than the energy E.

From what we have said in (iii), it follows that $\varrho(q, p)$ must be constant on the energy surface $\{q, p \,|\, H(q, p) = E\}$ and that it must vanish outside of this surface. This means further that, for a given energy, all microscopic states compatible with this macroscopic state must have equal probability. We could have started with this assumption even without knowing what is above in (iii), as it seems quite plausible.

The condition $H(q, p) = E$ defines a $6N - 1$ dimensional surface in the $6N$ dimensional phase space. For the purposes of calculation, it is better (as well as fundamentally more appropriate) to smear the sharp energy value E a little, and demand only

$$\{E - \Delta \leq H(q, p) \leq E\}$$

with a very small and otherwise not more precisely specified value of Δ.

Then, the probability density $\varrho(q, p)$ for the microstate in equilibrium has the form

$$\varrho(q, p) = \begin{cases} c = \text{constant} & \text{for } E - \Delta \leq H(q, p) \leq E , \\ 0 & \text{otherwise .} \end{cases} \tag{7.1.6}$$

The constants c can be determined from the condition

$$\int \varrho(q, p) d^{3N}q \, d^{3N}p = c \int\limits_{E - \Delta \leq H \leq E} d^{3N}q \, d^{3N}p$$

$$=: c\tilde{\Omega}_\Delta = 1 . \tag{7.1.7}$$

It then follows that

$$c = \frac{1}{\tilde{\Omega}_\Delta} , \tag{7.1.8}$$

where $\tilde{\Omega}_\Delta$ is the *phase space volume* which is filled by all the states with energy between $E - \Delta$ and E.

The formula (7.1.6), together with (7.1.8), represents the probability density of the microcanonical ensemble. This result is plausible. The greater the volume in phase space, the more microstates there are compatible with the macrostate. The probability of a particular fixed microstate is correspondingly lower.

Example. We consider a gas of N particles of mass m in a volume V. We ignore the interaction of the particles with each other, i.e. we consider the gas to be ideal.

The Hamiltonian is then given by

$$H(q, p) = \sum_{i=1}^{N} \frac{p_i^2}{2m} + H_{\text{wall}}(r_1, \ldots, r_N) . \tag{7.1.9}$$

The component H_{wall} represents the influence of the walls of the container which ensure that none of the particles escapes the volume V. It is given by a strongly repulsive potential which attains very high values when a particle comes in the immediate area of the wall, but vanishes otherwise. For particles in the interior of the volume, the inequality

$$E - \Delta \leq H(q, p) \leq E$$

can then be satisfied.

For any N, the phase space volume is given by

$$\tilde{\Omega}_\Delta(E, V, N) = \int_{E - \Delta \le H \le E} d^{3N}r\, d^{3N}p =: V^N f(E) \ . \qquad (7.1.10)$$

$f(E)$ is the volume of a spherical shell in $3N$-dimensional space. The volume of a sphere with radius R in D-dimensional space is given by

$$V(R) = \alpha(D)R^D \quad \text{with}$$

$$\alpha(D) = \frac{\pi^{D/2}}{(D/2)!} = \frac{\pi^{D/2}}{\Gamma(D/2 + 1)} \ . \qquad (7.1.11)$$

Here, $\Gamma(z)$ is the gamma function (see Appendix A). $\Gamma(z)$ satisfies the identities $\Gamma(N + 1) = N!$, $\Gamma(1/2) = \sqrt{\pi}$ (for $D = 3$, we have $\Gamma(\frac{5}{2}) = \frac{3}{2}\cdot\frac{1}{2}\Gamma(\frac{1}{2}) = \frac{3}{4}\sqrt{\pi}$ and thus $V = (4\pi/3)R^3$). Then the volume of a spherical shell with radii R and $R - s$ is given by:

$$V(R) - V(R - s) = \alpha R^D\left[1 - \left(1 - \frac{s}{R}\right)^D\right] \ .$$

For large D, we have

$$\left(1 - \frac{s}{R}\right)^D = e^{[D\ln(1 - \frac{s}{R})]} \approx e^{-Ds/R} \xrightarrow[D\to\infty]{} 0 \ , \qquad (7.1.12)$$

and so for very large D, the volume of a very small spherical shell is approximately the volume of the full sphere. Since here, $R = \sqrt{2mE}$, we have

$$f(E) = \alpha R^{3N} = \frac{\pi^{3N/2}}{\Gamma(3N/2 + 1)}(\sqrt{2mE})^{3N} \sim E^{3N/2} \ . \qquad (7.1.13)$$

Thus, in the case of particles which do not interact,

$$\tilde{\Omega}_\Delta(E, V, N) = \tilde{\Omega}(E, V, N) = \alpha' V^N E^{3N/2} \qquad (7.1.14)$$

Note that the dimensions of $\tilde{\Omega}$ are

$$[\tilde{\Omega}] = [p\cdot q]^{3N} = (\text{Js})^{3N} \ .$$

In order to make a dimensionless quantity out of $\tilde{\Omega}$, we must therefore divide it by a quantity with the same dimensions. Such a quantity exists in quantum theory: h, Planck's constant. The quantity

$$\Omega(E, V, N) = \frac{\tilde{\Omega}(E, V, N)}{N! h^{3N}} \qquad (7.1.15)$$

can thus be viewed as the number of microstates. We must in addition divide by $N!$ from quantum mechanical considerations, since the particles are indistinguishable. Later, we will have more to say about the additional factor $1/N!$.

7.2 Entropy

A macroscopic system is determined by specifying the values of the state variables, while the corresponding microstates are unknown at every moment in time. We will now formulate a measure of the uncertainty of the microstate. This quantity is called entropy.

For the following considerations, it is convenient to divide phase space into small cells C_i of equal volume with midpoints (q_i, p_i). A natural scale for the volume of such cells is h^{3N}. This is borrowed from the uncertainty principle in quantum mechanics, which states that it is impossible to determine position and momentum, i.e. a position in phase space, more exactly than one of these cells. Thus, in principle, we can only determine which cell in phase space contains the state of the system, while the location within the cell remains entirely uncertain. We now represent by P_i the probability that the microstate is located in the i-th cell:

$$P_i = \int_{C_i} d^{3N}q \, d^{3N}p \, \varrho(q, p) \ . \tag{7.2.1}$$

For the microcanonical ensemble, the probabilities are all equal, as long as the cells C_i lie in the energy shell $\{q, p \,|\, E - \Delta \leq H(q, p) \leq E\}$. Otherwise, $P_i = 0$.

We now define the *entropy*[1] of a probability distribution defined by probabilities P_i as

$$\sigma = - \sum_i P_i \ln P_i \ . \tag{7.2.2}$$

We want to show that σ is really a good measure of the uncertainty of the microstate:

i) Since $0 \leq P_i \leq 1$, we always have $\sigma \geq 0$.

ii) We have $P_i \ln P_i = 0$ if and only if $P_i = 0$ or $P_i = 1$. Cells C_j with $P_j = 0$ or $P_j = 1$ do not contribute to the entropy, since in these cases the microstate is either certainly within or certainly outside of the cell.

---

[1] Entropy (Greek) *trepein* = turn, change. This term was coined by A. Clausius in 1865 in contrast to "energy". The farther the system is from its equilibrium state and the smaller its entropy is in comparison to the equilibrium entropy, the greater the set of attainable states is.

Further, we have

$$\sum_i P_i = 1 \ , \tag{7.2.3}$$

so that not all the P_i can vanish.

iii) The expression σ also attains its smallest possible value $\sigma = 0$ if and only if there is an i_0 with $P_{i_0} = 1$ (and thus $P_j = 0$ for $j \neq i_0$). In this case, it is certain that the microstate lies in C_{i_0} and the uncertainty of the microstate is minimal.

iv) Let the system consist of two independent (or approximately independent) systems of particles. Let the probability distribution in the partial systems be $P_i^{(1)}$ and $P_j^{(2)}$. The probability that system 1 is in the cell $C_i^{(1)}$ and at the same time system 2 is in the cell $C_j^{(2)}$ is then

$$P_{ij}^{(1,2)} = P_i^{(1)} P_j^{(2)} \ . \tag{7.2.4}$$

Thus the total entropy of the combined system is

$$\begin{aligned}
\sigma^{(1,2)} &= -\sum_{i,j} P_{ij}^{(1,2)} \ln P_{ij}^{(1,2)} \\[2mm]
&= -\sum_{i,j} P_i^{(1)} P_j^{(2)} (\ln P_i^{(1)} + \ln P_j^{(2)}) \\[2mm]
&= -\sum_{i,j} P_j^{(2)} P_i^{(1)} \ln P_i^{(1)} - \sum_{i,j} P_i^{(1)} P_j^{(2)} \ln P_j^{(2)} \\[2mm]
&= -\sum_i P_i^{(1)} \ln P_i^{(1)} - \sum_j P_j^{(2)} \ln P_j^{(2)} \\[2mm]
&= \sigma^{(1)} + \sigma^{(2)} \quad \text{since}
\end{aligned} \tag{7.2.5}$$

$$\sum_i P_i^{(1)} = \sum_j P_j^{(2)} = 1 \ .$$

We would certainly demand that any usable measurement of uncertainty satisfy properties (i–iv). With a few plausible assumptions, we can show that the entropy must be defined as above.

If we calculate the entropy

$$\sigma = -\sum_i P_i \ln P_i$$

of a system whose energy with certainty lies between $E - \varDelta$ and E, then the summation will only run over the cells which lie in the energy shell. If the distribution is the microcanonical one, then for all the cells in the energy shell

$$P_i = P_i^{\mathrm{MC}} = \frac{1}{M} \ , \tag{7.2.6}$$

where M is the number of cells in the energy shell. The entropy is then given by

$$\sigma^{\text{MC}} = - \sum_{i=1}^{M} \frac{1}{M} \ln \frac{1}{M} = \ln M \ . \qquad (7.2.7)$$

We will now calculate the entropy of an ideal gas of N particles in volume V.

The number of cells of size h^{3N} in the energy shell is, from the calculations of the preceding section,

$$M = \frac{\tilde{\Omega}(E, V, N)}{h^{3N}} = \frac{\pi^{3N/2}}{\Gamma(3N/2 + 1)} \left(\frac{2mE}{h^2} \right)^{3N/2} V^N \ , \qquad (7.2.8)$$

where M differs from Ω by the factor $1/N!$. In Sect. 7.1, we introduced $\Omega(E, V, N)$ as a measurement of the number of microstates and justified the factor $1/N!$ on quantum mechanical grounds. We will see now that there is also a purely classical argument for this factor.

For $\ln M$, we have

$$\ln M = N \ln \left[V \left(\frac{2\pi mE}{h^2} \right)^{3/2} \right] - \ln \Gamma \left(\frac{3N}{2} + 1 \right) \ .$$

Now for large arguments of the Γ-function we have (see Appendix A):

$$\ln \Gamma(x) = x(\ln x - 1) + O(\ln x) \ ,$$

so that we have

$$\ln M = N \left[\ln V + \frac{3}{2} \ln \left(\frac{2\pi mE}{h^2} \right) \right]$$

$$- \left(\frac{3N}{2} + 1 \right) \left[\ln \left(\frac{3N}{2} + 1 \right) - 1 \right] + O \left[\ln \left(\frac{3N}{2} + 1 \right) \right]$$

$$= N \left[\ln \left(\frac{V}{N} \right) + \frac{3}{2} \ln \left(\frac{4\pi mE}{3Nh^2} \right) + \frac{3}{2} \right] + N \ln N + O(\ln N) \ . \qquad (7.2.9)$$

This result cannot hold for the entropy, since it is an extensive quantity and must therefore be strictly proportional to N given a fixed volume per particle V/N and fixed energy per particle E/N. Here, classical mechanics alone leads to a paradoxical result, known as *Gibbs's*[2] *paradox*. Considering the factor $1/N!$

---

[2] *Gibbs, Josiah Willard* (*1839, New Haven, Connecticut, d. 1903 New Haven). The first great American theoretical physicist, at Yale after 1871. One of the founders of theoretical statistical mechanics and thermodynamics. Also known for his phase rules and the paradox named after him.

motivated by quantum mechanics, we must take away the additional quantity

$$\ln N! = N(\ln N - 1) + O(\ln N)$$

from $\ln M$. Then we find the correct relationship

$$\sigma^{MC} = \ln \Omega(E, V, N)$$

$$= N\left[\ln\left(\frac{V}{N}\right) + \frac{3}{2}\ln\left(\frac{4\pi mE}{3Nh^2}\right) + \frac{5}{2}\right], \tag{7.2.10}$$

where the terms $O(\ln N)$ have been ignored (consider that $\ln N = \ln 10^{23} \approx 55$). Note also that the extra term $\ln N!$ depends only on N and thus does not play a role if N does not change.

We therefore need to consider those microstates which are only distinguished by an exchange of particles as equivalent, since the particles are quantum mechanically indistinguishable. Then we only need to count $M/N! = \Omega$ cells.

For gases under normal conditions, $\sigma/N \approx 10$, up to an order of magnitude. We now have the following important theorem:

The entropy of an isolated system is maximal if and only if the distribution of microstates is microcanonical, that is, if and only if the system is in equilibrium.

The proof is simple: we have

$$\sigma^{MC} - \sigma = \sum_{i=1}^{\Omega} P_i \ln P_i + \ln \Omega$$

$$= \sum_{i=1}^{\Omega} P_i \ln P_i + \sum_{i=1}^{\Omega} P_i \ln \Omega$$

$$= \frac{1}{\Omega} \sum_{i=1}^{\Omega} (P_i \Omega) \ln (P_i \Omega)$$

$$= f(x_1, \ldots, x_\Omega) \tag{7.2.11}$$

with $x_i = P_i \Omega$. We will show that under the additional condition

$$\sum_{i=1}^{\Omega} P_i \Omega = \sum_{i=1}^{\Omega} x_i = \Omega \tag{7.2.12}$$

the function f has a minimum at $x_i = 1$ for all i.

This problem is solved in general using Lagrange multipliers. In this case, though, we can prove the desired result more rapidly using a trick. Add

$$0 = \frac{1}{\Omega} \sum_{i=1}^{\Omega} (1 - x_i) , \tag{7.2.13}$$

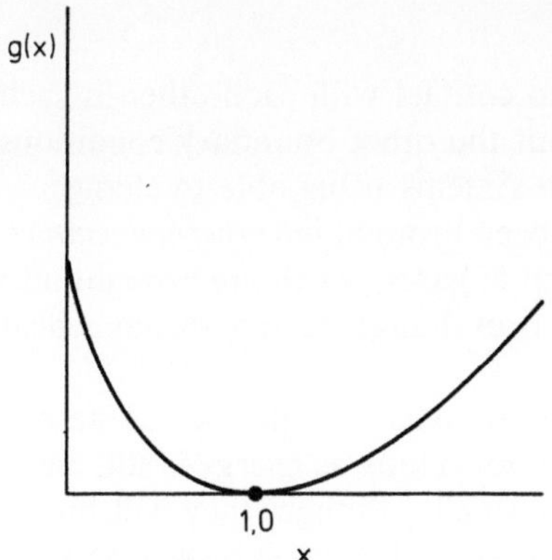

Fig. 7.2.1. Graph of the function $g(x) = x \ln x + 1 - x$

to f to obtain

$$\sigma^{\mathrm{MC}} - \sigma = \frac{1}{\Omega} \sum_{i=1}^{\Omega} (x_i \ln x_i + 1 - x_i) \tag{7.2.14}$$

for $\sigma^{\mathrm{MC}} - \sigma$. The function $g(x) = x \ln x + 1 - x$ is shown for $0 \le x < \infty$ in Fig. 7.2.1. It is never negative and has its sole minimum at $x = 1$. There, $g(1) = 0$. Thus the sum of such functions vanishes only if all the $x_i = 1$, otherwise this sum is positive.

Thus, we have $\sigma^{\mathrm{MC}} > \sigma$ and the equality holds if and only if $P_i \Omega = 1$ for all $i = 1, \ldots, \Omega$.

Of course, this result is plausible. Given that in the microcanonical distribution every state is equally likely, it stands to reason that this distribution has the maximal uncertainty as to the actual microstate of the system.

But under the given conditions, the microcanonical distribution is the uniquely determined distribution which is present at equilibrium. We thus see:

The entropy of an isolated system has its greatest possible value if the system finds itself in macroscopic equilibrium.

Entropy is the central concept of thermodynamics.

7.3   Temperature, Pressure, and Chemical Potential

We consider an isolated system consisting of two sub-systems. At first, let both of these sub-systems be isolated and let both be in states of equilibrium with values E_1, V_1, N_1 and E_2, V_2, N_2 respectively for energy, volume, and particle number.

7.3.1 Systems with Exchange of Energy

Let the two sub-systems now be brought into contact with each other in such a way that they can exchange energy, without the other boundary conditions determining the equilibrium states of the sub-systems being able to change.

In this case, we say that the systems have been brought into *thermal contact*. As an example, we consider two containers full of gases, which are brought into interaction with each other so that they can exchange energy through heat conduction.

Just after the contact has been made, the total system will not be in a state of equilibrium; indeed, it will not attain equilibrium so long as energy is still being exchanged between the two sub-systems. Eventually, though, they will finally reach a new state of equilibrium. In this new state, let the sub-systems have energies E_1' and E_2'. Then the total energy of the system is always given by $E = E_1 + E_2 = E_1' + E_2'$.

How can the value of E_1' (or, equivalently, $E_2' = E - E_1'$) now be determined?

One measure of the set of microstates of each of the two subsystems would be the number of cells in the energy shells $\Omega_i(E_i', N, V)$, $i = 1, 2$. For each E_1', the product

$$\Omega_{1,2} = \Omega_1(E_1', V_1, N_1)\Omega_2(E - E_1', V_2, N_2)$$

is a measure of the number of microstates of the combined system in equilibrium. Here, in principle, E_1' can lie anywhere in the interval between 0 and E. If, however, for a certain E_1' there are many more microstates than for any other value of E_1', the system would be more likely to reach the state in which system 1 has this value of energy E_1' and system 2 has energy $E_2' = E - E_1'$. We will thus attempt to find the maximum of $\Omega_{1,2}$, or, equivalently, the maximum of

$$\ln \Omega_{1,2} = \sigma_{1,2}(E_1') = \sigma_1(E_1') + \sigma_2(E_2')$$

as a function of E_1'. At a maximum, it must be true that

$$d\sigma_{1,2} = \frac{\partial \sigma_1}{\partial E_1'} dE_1' + \frac{\partial \sigma_2}{\partial E_2'} dE_2' = 0 \ . \tag{7.3.1}$$

To find the derivative of the entropy $\sigma(E, V, N)$ of a system with respect to its energy, we introduce the designation

$$\frac{\partial \sigma(E, V, N)}{\partial E} = \frac{1}{\tau(E, V, N)} \tag{7.3.2}$$

τ will turn out to be identical, up to a constant, with a quantity which we will call the *absolute temperature* of the system in equilibrium. Since for every mechanical system σ increases with E, it is always true that $\tau > 0$. Since σ and E are extensive variables, τ must be an intensive variable. The condition of equilibrium for energy exchange through thermal contact then reads, since dE_2 must equal

$-dE_1$:

$$\tau_1 = \tau_2 \ .$$

The sharper the maximum, the more unlikely it is that the combined system would *not* satisfy

$$\tau_1 = \tau_2 \ .$$

The systems therefore exchange energy until the energy of sub-system 1 changes from its initial value E_1 to a value E_1' which satisfies

$$\tau_1(E_1', V_1, N_1) = \tau_2(E_2', V_2, N_2) \ . \tag{7.3.3}$$

Here, τ_i, $i = 1, 2$ is a function which is calculated separately for each system. In equilibrium, the entropy cannot be further increased by an exchange of energy.

We note:

i) If equilibrium has not yet been reached, the entropy increases with a change dE_1 of energy. In this case, we have

$$d\sigma_{1,2} = \left(\frac{1}{\tau_1} - \frac{1}{\tau_2}\right) dE_1 > 0 \ , \qquad \text{that is, for}$$

$$\tau_2 > \tau_1 \ , \quad \text{we have} \quad dE_1 > 0 \ , \quad \text{and for}$$

$$\tau_1 > \tau_2 \ , \quad \text{we have} \quad dE_1 < 0 \ .$$

The system with the higher temperature gives energy to the system with lower temperature.

From our intuitive experience, we know that energy transferred by heat conduction always flows from warmer to cooler bodies and that this energy exchange ceases when the temperatures become equal. This does not only justify us, but rather obliges us to find a connection between the quantity τ and the intuitively given concept of temperature T, which is measured with a thermometer.

In any case, from the arguments thus far there could be a relationship of the form

$$T = h(\tau)$$

with some unspecified monotonically increasing function h. However, we will see soon that the temperature τ defined here actually has the relationship

$$\tau = kT \tag{7.3.4}$$

with the usual absolute temperature T with the dimension Kelvin (K). In this equation,

$$k = 1.38066 \times 10^{-23} \ \text{Nm K}^{-1}$$

is the Boltzmann constant. It is also conventional to define entropy not as the dimensionless quantity σ but as the quantity

$$S = k\sigma \tag{7.3.5}$$

Then it is true that

$$1/T = \partial S/\partial E \ . \tag{7.3.6}$$

In the following, we will always use the quantities T and S instead of τ and σ.

ii) If the two sub-systems can only exchange energy, we have

$$dE_i = T_i\, dS_i \ .$$

Since

$$dE_1 = -\, dE_2 \quad \text{we also have}$$

$$dS_2 = \frac{dE_2}{T_2} = -\,\frac{dE_1}{T_2} = -\,\frac{dS_1\, T_1}{T_2} \ .$$

If a condition of equilibrium has not yet been reached, we know $dS_{1,2} > 0$, so that

$$\begin{aligned}
dS_{1,2} &= dS_1 + dS_2 \\
&= dS_1\left(1 - \frac{T_1}{T_2}\right) \\
&= dS_2\left(1 - \frac{T_2}{T_1}\right) > 0 \ .
\end{aligned}$$

For $T_1 > T_2$ we then have

$$dS_1 < 0 \quad \text{and} \quad dS_2 > 0 \ ,$$

that is, the entropy of the warmer system (the system with higher temperature) decreases, while the entropy of the colder system increases upon contact until equilibrium is reached. We can thus say:

Entropy is exchanged (that is transported from 1 to 2), but also produced, since the total entropy does not remain constant, but increases.

iii) For an ideal gas, we found

$$\Omega(E, V, N) = \alpha V^N E^{3N/2} \quad \text{thus}$$

$$S(E, V, N) = Nk\ln V + \tfrac{3}{2}Nk\ln E \tag{7.3.7}$$

$$+ \text{ terms independent of } E \text{ and } V$$

and therefore

$$\frac{1}{T} = \frac{\partial S}{\partial E} = \frac{3}{2}\frac{Nk}{E} \ .$$

Hence we find

$$E = \tfrac{3}{2} NkT \tag{7.3.8}$$

as the relationship between the energy of the system and the temperature.

iv) We would like to demonstrate in an example that the maximum is very sharp. We consider two systems of ideal gases. Then

$$S_{1,2} = \tfrac{3}{2}k(N_1 \ln E_1 + N_2 \ln E_2) + \text{terms independent of energy} \ .$$

A maximum occurs if

$$\frac{3}{2}\frac{N_1}{E_1} = \frac{3}{2}\frac{N_2}{E_2} \quad \text{or} \quad E_i = \hat{E}_i \quad \text{with}$$

$$\hat{E}_i = N_i \frac{E}{N} \ , \qquad N = N_1 + N_2 \ . \tag{7.3.9}$$

(To see that there really is a maximum for $E_i = \hat{E}_i$, consider the second derivative

$$\frac{\partial^2 S_{1,2}}{\partial E_1^2} = -\frac{3}{2}k\left(\frac{N_1}{E_1^2} + \frac{N_2}{E_2^2}\right) < 0 \ .)$$

We would like to investigate how sharp the maximum is. With a deviation of Δ from the maximum, that is, with the values

$$E_1 = \hat{E}_1 + \Delta \ , \qquad E_2 = \hat{E}_2 - \Delta$$

we find for $S_{1,2}$

$$S_{1,2}(E_1 + \Delta) = \frac{3}{2}N_1 k \ln\left[\hat{E}_1\left(1 + \frac{\Delta}{\hat{E}_1}\right)\right] + \frac{3}{2}N_2 k \ln\left[\hat{E}_2\left(1 - \frac{\Delta}{\hat{E}_2}\right)\right] ,$$

and for small values of $\Delta/\hat{E}_1$ and $\Delta/\hat{E}_2$, we find for the right hand side

$$\frac{3}{2}N_1 k \ln \hat{E}_1 + \frac{3}{2}N_2 k \ln \hat{E}_2 - \frac{3}{2}N_1 k \frac{1}{2}\frac{\Delta^2}{\hat{E}_1^2} - \frac{3}{2}N_2 k \frac{1}{2}\frac{\Delta^2}{\hat{E}_2^2} + O\left(\frac{\Delta^3}{\hat{E}_1^3}\right) ,$$

so that with (7.3.9) we have

$$\Omega_{1,2} = (\Omega_{1,2})_{\max} \exp\left[-\frac{3}{4}\frac{\Delta^2}{E^2}N^2\left(\frac{1}{N_1} + \frac{1}{N_2}\right)\right] \ .$$

If now we have, say, $N_1 = N_2 = 10^{22}$, then

$$N^2 \left(\frac{1}{N_1} + \frac{1}{N_2} \right) = 8 \times 10^{22} \; ,$$

and thus already for, say, $\Delta/E = 10^{-10}$, i.e. a relatively small deviation from the mean, $\Omega_{1,2}$ is smaller than the maximum value by a factor of

$$e^{6 \times 10^{-20} \times 10^{22}} = e^{600} \approx 10^{260} \; .$$

This means that such deviations practically never occur, since they are so extremely unlikely. For $\Delta/E = 10^{-11}$, though, this factor is only about e^6 and for $\Delta/E = 10^{-12}$ only $\exp(6 \times 10^{-2}) \approx 1$.

According to this last calculation, though, all macrostates with

$$E_1 = \hat{E}_1(1 \pm 10^{-12}) \; , \qquad E_2 = \hat{E}_2(1 \mp 10^{-12})$$

are approximately equally probable, hence the system, with overwhelming probability, will be in a macrostate which is defined by these energy values of the individual systems, because the number of microstates with these energy values is incomparably greater than for other energies. We therefore expect that the energy of the subsystems will only fluctuate around the equilibrium point by a few parts in 10^{-12}.

7.3.2 Systems with an Exchange of Volume

We now consider two systems which are separated by a sliding wall. The two volumes will adjust in such a way that the entropy will be maximal, that is, the macroscopic state variables V_1' and V_2' will achieve exactly those values which correspond to the greatest number of microstates.

We thus consider again

$$S_{12} = S_1(E_1, N_1, V_1) + S_2(E_2, N_2, V_2) \; ,$$

and if E_i and V_i can both change,

$$dS_{1,2} = \frac{\partial S_1}{\partial V_1} dV_1 + \frac{\partial S_2}{\partial V_2} dV_2 + \frac{\partial S_1}{\partial E_1} dE_1 + \frac{\partial S_2}{\partial E_2} dE_2 \; .$$

Now,

$$dE_1 = - dE_2 \; , \qquad dV_1 = - dV_2 \; ,$$

and with $T_1 = T_2$, which we will assume to simplify our calculation, we have

$$dS_{1,2} = \left(\frac{\partial S_1}{\partial V_1} - \frac{\partial S_2}{\partial V_2} \right) dV_1 \; ,$$

and in equilibrium, i.e. if $S_{1,2}$ no longer changes,

$$\frac{\partial S_1}{\partial V_1} = \frac{\partial S_2}{\partial V_2} \ .$$

We now define the state variable $p(E, V, N)$ for a system by

$$\frac{p(E, V, N)}{T(E, V, N)} = \frac{\partial S(E, V, N)}{\partial V} \ . \tag{7.3.10}$$

Then this means: the combined system is in equilibrium, that is in a state which is overwhelmingly more likely than any other, if the volumes of the two subsystems have been arranged so that

$$p_1(E_1, V_1, N_1) = p_2(E_2, V_2, N_2) \ . \tag{7.3.11}$$

Before equilibrium is reached, let, say, $p_1 > p_2$, so that again $dS_{1,2} > 0$ and thus

$$dS_{1,2} = \frac{p_1 - p_2}{T} dV_1 > 0 \ .$$

Then, since $p_1 - p_2 > 0$, it is also true that $dV_1 > 0$, that is the volume of the system in which the variable p is greater will increase. We call the variable p the *pressure* of the system with volume V. p is therefore a state variable, but depends on E, N, and V. The exact form of this dependence is determined by the form of the entropy.

For a classical ideal gas, we have

$$\frac{S(E, V, N)}{k} = N \ln V + \tfrac{3}{2} N \ln E + \text{terms independent of } V \ .$$

Therefore,

$$\frac{p}{T} = \frac{\partial S}{\partial V} = \frac{kN}{V} \quad \text{and thus} \quad pV = NkT \ .$$

This is the well-known equation of state for classical ideal gases. This demonstrates that the temperature T introduced in (7.3.4) is identical to the temperature used in the gas law.

7.3.3   Systems with Exchanges of Energy and Particles

We now want to admit an additional exchange between two systems which exchange energy, namely the exchange of particles. We can imagine two gases which are separated by a thermally conducting permeable membrane. Then, since both N_i and E_i can now change,

$$S_{1,2} = S_1 + S_2$$

is maximal if

$$dS_{1,2} = \frac{\partial S_1}{\partial N_1} dN_1 + \frac{\partial S_2}{\partial N_2} dN_2 + \frac{\partial S_1}{\partial E_1} dE_1 + \frac{\partial S_2}{\partial E_2} dE_2 = 0 \ .$$

Since again

$$dN_1 = -dN_2 \ , \qquad dE_1 = -dE_2$$

it follows that in equilibrium

$$dS_{1,2} = \left(\frac{\partial S_1}{\partial N_1} - \frac{\partial S_2}{\partial N_2} \right) dN_1 + \left(\frac{\partial S_1}{\partial E_1} - \frac{\partial S_2}{\partial E_2} \right) dE_1 = 0 \ .$$

Since now dN_1 and dE_1 are independent, in equilibrium we must have

$$T_1 = T_2$$

and, if we define

$$-\frac{\mu(E, V, N)}{T(E, V, N)} = \frac{\partial S(E, V, N)}{\partial N} \ : \tag{7.3.12}$$

$$\mu_1(E_1, N_1, V_1) = \mu_2(E_2, N_2, V_2) \ . \tag{7.3.13}$$

$\mu(E, N, V)$ is called the *chemical potential*. This is another state variable like temperature and pressure.

If two systems with different chemical potentials are brought into contact in such a way that particles can be exchanged, then particles move from one system to another (that is, there is a net flow of particles) until the chemical potentials have become equal. We will again determine the direction of this flow.

Let $\mu_2 > \mu_1$, but $T_1 = T_2$ already. Then

$$dS_{1,2} = \frac{-\mu_1 + \mu_2}{T} dN_1 > 0 \ ,$$

and therefore, since $\mu_2 - \mu_1 > 0$, $dN_1 > 0$. This means that particles flow from the system with higher chemical potential to the system with lower chemical potential (analogous to the situation with temperature and pressure).

If two systems with different chemical potentials and temperatures are first brought only into thermal contact, they will exchange only energy at first, until the energies E_i have attained the values $\hat{E}_i$ with

$$T_1(\hat{E}_1, N_1, V_1) = T_2(\hat{E}_2, N_2, V_2) \ .$$

If we now make the wall between the two systems porous, particles will also flow from system 2 to system 1, if we have $\mu_2 > \mu_1$. Energy also flows with the particles from 2 to 1. This is plausible, since the values $\hat{E}_i$ which satisfy

$$T_1(\hat{E}_1, N_1, V_2) = T_2(\hat{E}_2, N_2, V_2)$$

certainly depend on N_1 and N_2. If the N_i change, then so do the $\hat{E}_i$.

For the classical ideal gas, we have (see (7.2.10))

$$\frac{S}{k} = N\left[\ln\left(\frac{V}{N}\right) + \frac{3}{2}\ln\left(\frac{4\pi mE}{3Nh^2}\right) + \frac{5}{2}\right]$$

and thus

$$\mu(E, V, N) = kT\ln\left(\frac{N}{V}\right) - \frac{3}{2}kT\ln\left(\frac{4\pi mE}{3Nh^2}\right) . \tag{7.3.14}$$

The chemical potential thus depends logarithmically on the particle density $n = N/V$.

7.4 The Gibbs Equation and the Forms of Energy Exchange

So far we have developed the following strategy for the calculation of macroscopic properties: We calculate the volume in phase space $\Omega(E, V, N)$ of the microcanonical ensemble, then the entropy is given by

$$S(E, V, N) = k\ln\Omega(E, V, N) ,$$

and the state variables T, p, μ satisfy

$$\frac{1}{T} = \frac{\partial S}{\partial E} , \qquad \frac{p}{T} = \frac{\partial S}{\partial V} , \qquad -\frac{\mu}{T} = \frac{\partial S}{\partial N}$$

so that

$$dS = \frac{1}{T}dE + \frac{p}{T}dV - \frac{\mu}{T}dN \tag{7.4.1}$$

or

$$dE = TdS - pdV + \mu dN , \tag{7.4.2}$$

that is, the differential form of the function $S(E, V, N)$ has as a consequence the differential form dE of the energy $E(S, V, N)$.

Then, it also follows that

$$T = \frac{\partial E}{\partial S}, \qquad p = -\frac{\partial E}{\partial V}, \qquad \mu = \frac{\partial E}{\partial N} . \tag{7.4.3}$$

We see the following:

i) The state variables can be divided into two categories:
a) E, S, V, N, and
b) T, p, μ.

Category (a) includes quantities which are extensive (additive). Energies, numbers of particles, volumes, and entropies add together, if two sub-systems are combined into a single system. By contrast, category (b) contains intensive, non-additive quantities, such as temperatures, pressures, and chemical potentials, which become equal when two systems are placed in the proper contact.

ii) If we know $S(E, V, N)$ or $E(S, V, N)$, we can calculate the other variables.
Functions from which all other state variables can be calculated are called *thermodynamic potentials*. $S(E, V, N)$ and $E(S, V, N)$ are therefore thermodynamic potentials.

iii) The differential form

$$dE = T dS - p dV + \mu dN \tag{7.4.4}$$

is called the Gibbs form. It indicates the various ways in which the system can exchange energy with the environment:

a) If only particles are exchanged, then the energy change is

$$dE = \mu dN .$$

The chemical potential can thus also be interpreted as the quantity which gives the energy change per particle exchange (with volume and entropy held fixed). In this case, we say that the energy is exchanged in the form of *chemical energy*.

b) The pressure p is the change in energy per change in volume. This corresponds to our intuition. Consider a container with sliding wall of area A. If we slowly push the wall towards the inside of the container, that is we move the wall using pressure $p = F/A$, the *work* done in this process $dW = F dh$ is given by

$$dW = dE = F dh = pA dh = - p dV .$$

This work appears in the system as an increase dE in energy.

Of course, this displacement can only occur if the pressure $p = F/A$ is somewhat greater than the pressure of the gas in the container. Here we have assumed that the difference in pressure is "very small" and that the wall slides "very slowly" so that we can assume that the system is in a state of equilibrium at each point in time (see also Sect. 7.8).

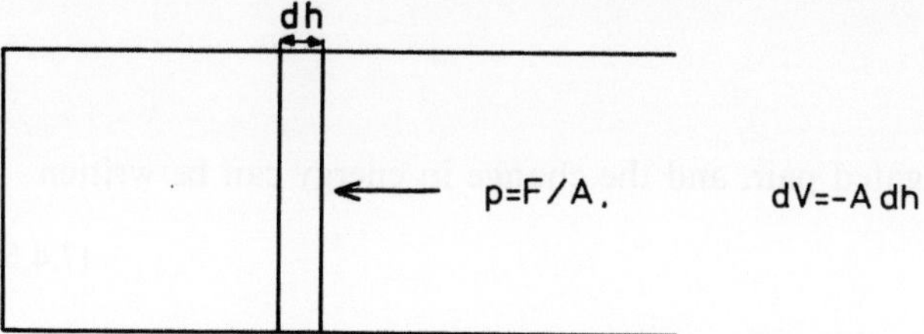

Fig. 7.4.1. The change in energy of a gas due to a change in volume

c) If neither the number of particles nor the volume is changed, energy can still be exchanged in such a way that the temperatures of the systems become equal (thermal contact). This is called an exchange of *heat*. This exchange is always connected with an exchange of entropy

$$dE = T\,dS \ .$$

This is evident, since as the energy E of subsystem 1 changes, so does its entropy (see Sect. 7.3.1). Of course, the entropy can also change if the volume or number of particles changes, since

$$dS = \frac{1}{T}dE - \frac{\mu}{T}dN + \frac{p}{T}dV \ .$$

In the present case, though, we are saying that if V and N are both held constant,

$$dS = \frac{1}{T}dE \quad \text{or} \quad dE = T\,dS$$

and the exchange of energy is then characterized as an exchange of heat.

iv) In the Gibbs form

$$dE = T\,dS - p\,dV + \mu\,dN$$

variables always appear in energy conjugated pairs

$$(T, S) \ , \quad (-p, V) \ , \quad (\mu, N) \ .$$

The product of each pair of variables has the dimensions of energy, and one of the two variables is always intensive, the other extensive.

In the framework of single particle mechanics, the two variables

$$(v, p)$$

also represent an energy conjugated pair of variables. Here the momentum p is an extensive quantity, the speed v, intensive. Similarly, if $F(r)$ represents the force on a particle,

$$(-\boldsymbol{F}, \boldsymbol{r})$$

is another energy conjugated pair, and the change in energy can be written

$$dE = \boldsymbol{v}\cdot d\boldsymbol{p} - \boldsymbol{F}\cdot d\boldsymbol{r} \ . \tag{7.4.5}$$

But here the situation is a bit different from thermodynamics.

Here, we have

$$\boldsymbol{v} = \frac{\boldsymbol{p}}{m} \quad \text{and thus} \quad \boldsymbol{v}\cdot d\boldsymbol{p} = d\left(\frac{\boldsymbol{p}^2}{2m}\right) \ .$$

Similarly, with

$$\boldsymbol{F} = -\nabla V(\boldsymbol{r}) \quad \text{it follows} \quad -\boldsymbol{F}\cdot d\boldsymbol{r} = dV(\boldsymbol{r})$$

hence,

$$dE = d\left(\frac{\boldsymbol{p}^2}{2m} + V(\boldsymbol{r})\right) = dE_{\text{kin}} + dE_{\text{pot}} \ , \tag{7.4.6}$$

i.e., here the individual expressions for the exchange of energy (either in the form of kinetic or potential energy) are themselves total differentials (of E_{kin} and E_{pot}), since

$$\boldsymbol{v} = \boldsymbol{v}(\boldsymbol{p}) \ , \quad \boldsymbol{F} = \boldsymbol{F}(\boldsymbol{r})$$

that is, the intensive variables depend only on their energy conjugate variables.

In thermodynamics, it turns out differently: $T\,dS$ is not a total differential, since $T = T(S, V, N)$. Thus we can indeed say that there is an exchange of energy in the form of heat if $dS \neq 0$, but it is impossible to talk of "the quantity of heat which a system possesses", as we can about its kinetic or potential energy.

The concepts *heat*, *work*, and *chemical energy* are thus only examples of possible forms of energy exchange.

We therefore write

$$\delta Q = T\,dS \tag{7.4.7}$$

$$\delta W = -p\,dV \tag{7.4.8}$$

$$\delta E_{\text{chem}} = \mu\,dN \tag{7.4.9}$$

where the δ indicates that here we mean only energy changes which arise from changes in extensive variables.

For systems with several types of particles, the Gibbs form is given by

$$dE = T\,dS - p\,dV + \sum_i \mu_i\,dN_i \; , \tag{7.4.10}$$

where μ_i is the chemical potential of the i-th kind of particle.

7.5   The Canonical Ensemble and the Free Energy

Up until now, we have always considered an isolated system with a thermodynamic potential given by the functions

$$S = S(E, V, N) \quad \text{or} \quad E = E(S, V, N) \quad \text{respectively} \; .$$

In many cases, it is not realistic to give the energy of a system directly or to measure it. In practice, it is often much easier to give the temperature of a system by bringing it into thermal contact with a large system called a *reservoir*. By means of an exchange of energy, the two systems achieve equal temperatures. If the reservoir is large enough so that the amount of energy coming into or leaving it can be neglected in comparison to its total energy, we can say: the temperature of the system is determined by the thermal contact, namely it is fixed at the value which it has for the reservoir.

We will now consider such thermal contact between a system 1 and a reservoir 2 (Fig. 7.5.1). In equilibrium, the temperatures of the two systems are equal. Let the total energy of the two systems be given by E_0.

We will pose the following question:

With what probability is system 1 in a given microscopic state which has energy E_1? Note that we are not asking about the energy of system 1. This can be calculated as in Sect. 7.3. Here, we are asking about the probability of a specific microscopic state of system 1.

Let us again think of phase space as divided into cells of size h^{3N}. We will identify any two microscopic states with values of (q, p) which lie in the same cell,

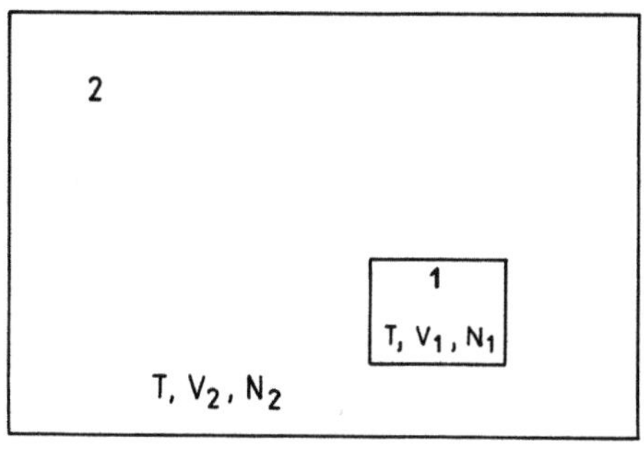

Fig. 7.5.1. System 1 in thermal contact with a reservoir

so that we can talk of a finite probability for a particular microscopic state. If $\varrho(q, p)$ is the probability density, then

$$\bar{\varrho}_i(q, p) = \frac{1}{h^{3N}} \int\limits_{Z_i} d^{3N}q' d^{3N}p' \varrho(q', p') , \qquad (q, p) \in Z_i$$

can be regarded as the probability of a microscopic state in the cell containing (q, p).

Since we are given a particular microscopic state of system 1, it follows that the probability of its realization must be proportional to the number of microscopic states of the reservoir which have the energy $E_2 = E_0 - E_1$, since the probability of each particular microscopic state of the combined system has the same value, as the combined system is isolated. If we sum over the possible states of the reservoir (the number of states we sum over is the number of microscopic states of the reservoir with energy E_2), we obtain the probability of the specific microscopic state of system 1. Thus, the probability of a microscopic state is given by

$$\bar{\varrho}^C(q, p) \sim \Omega_2(E_2, V_2, N_2) = e^{S_2(E_2, V_2, N_2)/k} . \tag{7.5.1}$$

Now $E_2 = E_0 - E_1$, and $E_1 \ll E_0$, so that the entropy of system 2 (the reservoir) can be expanded around the value E_0, to obtain the following approximation:

$$S_2(E_0 - E_1) = S_2(E_0) - E_1 \frac{\partial S_2}{\partial E_2}\bigg|_{E_0} + \frac{1}{2} E_1^2 \frac{\partial^2 S_2}{\partial E_2^2}\bigg|_{E_0} + \dots . \tag{7.5.2}$$

For the term of second order in E_1, we find

$$\frac{1}{2} E_1^2 \frac{\partial}{\partial E_2} \frac{1}{T} = -\frac{1}{2} \frac{E_1^2}{T^2} \frac{\partial T}{\partial E_2} . \tag{7.5.3}$$

But $E_2 = O(N_2)$, and therefore $\partial E_2/\partial T = O(N_2)$, and thus

$$\frac{\partial T}{\partial E_2} = O\left(\frac{1}{N_2}\right) ,$$

so that this term is smaller by a factor N_2 than the term of first order in E_1. It can then be neglected in the same way as the terms we already neglected in (7.5.2). The greater the number of particles in the reservoir, the closer this approximation will be.

Let us consider the term of first order in E_1. We have

$$\frac{\partial S_2}{\partial E_2}\bigg|_{E_0} \approx \frac{\partial S_2}{\partial E_2}\bigg|_{E_2} = \frac{1}{T_2} = \frac{1}{T} ,$$

and thus, with $\beta := 1/kT$:

$$\varrho^C(q, p) \sim e^{-E_1/kT} = e^{-\beta E_1} \ . \tag{7.5.4}$$

This exponential term is called the *Boltzmann*[3] *factor*.

We can substitute the Hamiltonian of system 1 for the energy E_1 and introduce suitable normalization factors. Then, with $N = N_1 =$ the number of particles in system 1,

$$\bar{\varrho}^C(q, p) = \frac{1}{ZN!h^{3N}} e^{-\beta H(q, p)} \tag{7.5.5}$$

is the probability (density), that in system 1 the microscopic state (q, p) with energy $H(q, p)$ occurs. The quantities Z and $N!$ represent normalization factors.

Since the system must be in some microscopic state, we have

$$\int d^{3N}q\, d^{3N}p\, \varrho^C(q, p) = 1 \ .$$

The the normalization factor Z is given by

$$Z(\beta, V, N) = \frac{1}{N!h^{3N}} \int d^{3N}q\, d^{3N}p\, e^{-\beta H(q, p)} \ . \tag{7.5.6}$$

Z is called the *partition function*. The collection of states which are given with probability density $\varrho^C(q, p)$ in system 1 with given values of T, V, and N is called the *canonical ensemble*, in contrast to the microcanonical ensemble of Sect. 7.1. In the latter, all microscopic states were considered equally; in the canonical ensemble, each microscopic state is given a weight, the Boltzmann factor. N and V are also given here, but not the energy. But the temperature is given by the thermal contact with the reservoir, and thus the microscopic states are no longer equally probable.

The factor $\exp[-\beta H(q, p)]$ means that the smaller the energy of a microscopic state of system 1, the more probable it is. If we ask, though, about the probability that system 1 has the energy E_1, we must consider all states with the energy E_1, weighted by the factor $\exp(-\beta E_1)$. This yields the probability

$$\frac{1}{Z} \Omega_1(E_1) e^{-\beta E_1} = \frac{1}{Z} e^{-\beta E_1 + S_1(E_1)/k} \ . \tag{7.5.7}$$

This is maximal, if

$$F = E_1 - TS_1(E_1, V_1, N_1) \tag{7.5.8}$$

---

[3] *Boltzmann, Ludwig* (*1844 Vienna, d. 1906 Duino near Trieste). His work ranged from experimental physics ($n = \sqrt{\varepsilon}$; early confirmation of the electrodynamic theory of light) to theoretical physics to philosophy. The central themes of his meditations were the kinetic theory of gases and statistical mechanics. He formulated the connection between entropy and probability. He is known, among other things, for the Stefan–Boltzmann law and the Boltzmann transport equation.

has a minimum, as a function of E_1. This is the case if

$$\frac{\partial F}{\partial E_1} = 1 - T\frac{\partial S_1(E_1, V_1, N_1)}{\partial E_1} = 0$$

or

$$\frac{1}{T} = \frac{\partial S_1(E_1, V_1, N_1)}{\partial E_1} = \frac{1}{T_1}$$

i.e., the temperature of system 1 agrees with the temperature of the reservoir, which was obviously what we expected.

Thus the most probable value of energy for system 1 is not the smallest value of energy, but rather the energy E_1 for which the expression (7.5.8) has a minimum.

The quantity

$$F(T, V_1, N_1) = E - TS(E, V_1, N_1) \tag{7.5.9}$$

with E fixed so that

$$\frac{\partial S(E, V_1, N_1)}{\partial E} = \frac{1}{T} \tag{7.5.10}$$

is called the *free energy*.

The smaller weighting factor that microscopic states with greater energy have due to the Boltzmann factor is compensated for by the fact that greater values of energy have more microscopic states. As a result of these two competing considerations, E_1 yields the most probable energy.

We can also interpret the construction of $F(T, V, N)$ from $S(E, V, N)$ in the following way:

Let $S(E, V, N)$ be given, and let a new variable T be introduced by

$$\frac{1}{T} = \frac{\partial S(E, V, N)}{\partial E} \ .$$

Now eliminate E in favor of T, that is, solve the above equation for E by finding $E = E(T, V, N)$ and then form

$$F(T, V, N) = E(T, V, N) - TS(E(T, V, N), V, N) \ . \tag{7.5.11}$$

Such a transformation from a function $S(E, V, N)$ to a function $F(T, V, N)$ is called a *Legendre*[4] *transformation*. We have already met such a transformation

---

[4] *Legendre, Adrien Marie* (*1752 Paris, d. 1833 Paris). French mathematician. Important work in number theory (quadratic residues), elliptical functions, geodesy, and celestial mechanics.

in going from the Lagrangian $L(q, \dot{q})$ to the Hamiltonian $H(q, p)$. We will return to this in Sect. 7.6.

We then obtain also

$$\frac{\partial F}{\partial T} = \frac{\partial E}{\partial T} - S(E(T, V, N), V, N) - T\frac{\partial S}{\partial E}\frac{\partial E}{\partial T}$$

$$= - S(E(T, V, N), V, N) \equiv - S(T, V, N) , \tag{7.5.12}$$

$$\frac{\partial F}{\partial V} = \frac{\partial E}{\partial V} - T\frac{\partial S}{\partial V} - T\frac{\partial S}{\partial E}\frac{\partial E}{\partial V}$$

$$= - T\frac{p}{T}$$

$$= - p(E(T, V, N), V, N) \equiv - p(T, V, N) , \tag{7.5.13}$$

$$\frac{\partial F}{\partial N} = \frac{\partial E}{\partial N} - T\frac{\partial S}{\partial N} - T\frac{\partial S}{\partial E}\frac{\partial E}{\partial N}$$

$$= T\frac{\mu}{T} = \mu(E(T, V, N), V, N) \equiv \mu(T, V, N) . \tag{7.5.14}$$

Partial differentiation with respect to the variables T, V, and N, yields (up to a sign) their energy conjugate variables S, p, and μ.

$F(T, V, N)$ is therefore another thermodynamic potential. If, based on the experimental situation the values of the variables T, V, and N are given, the free energy $F(T, V, N)$ is then the quantity naturally given as the thermodynamic potential.

Of course, F can be calculated by first finding $S(T, V, N)$ and then carrying out the Legendre transformation, as we have done. For this, though, the canonical ensemble would not be particularly necessary and the free energy $F(T, V, N)$ would not be a very interesting variable. However, there is a direct way to calculate the free energy.

We found that the probability that system 1 has the energy E which minimizes $E - TS(T, V, N)$ was

$$\frac{1}{Z}\mathrm{e}^{- \beta F(T, V, N)} ,$$

and this is just about 1, since system 1 has energy E with overwhelming probability.

Thus

$$Z = \mathrm{e}^{- \beta F(T, V, N)} , \tag{7.5.15}$$

and we need only calculate

$$Z(T, V, N) = \frac{1}{N! h^{3N}} \int d^{3N}q \, d^{3N}p \, e^{-\beta H(q, p)}$$

(7.5.16)

following (7.5.6). This is thus another, often simpler strategy to find a thermodynamic potential.

Applications. i) For an ideal gas,

$$
\begin{aligned}
Z &= \frac{1}{N! h^{3N}} \int d^{3N}q \, d^{3N}p \exp\left(-\beta \sum_{i=1}^{N} p_i^2 / 2m\right) \\
&= \frac{1}{N! h^{3N}} V^N \int d^{3N}p \exp\left(-\beta \sum_{i=1}^{N} p_i^2 / 2m\right) \\
&= \frac{1}{N! h^{3N}} V^N (2\pi m k T)^{3N/2} ,
\end{aligned}
$$

(7.5.17)

where we have used the identity

$$\int_{-\infty}^{+\infty} dx \, e^{-x^2} = \sqrt{\pi} .$$

It follows then that

$$
\begin{aligned}
-\beta F = \ln Z &= N\left[\ln V + \frac{3}{2}\ln\left(\frac{2\pi m k T}{h^2}\right)\right] - \ln N! \\
&= N\left[\ln\left(\frac{V}{N}\right) + \frac{3}{2}\ln\left(\frac{2\pi m k T}{h^2}\right) + 1\right] + O(\ln N) ,
\end{aligned}
$$

hence

$$-F = kTN\left[\ln\left(\frac{V}{N}\right) + \frac{3}{2}\ln\left(\frac{2\pi m k T}{h^2}\right) + 1\right] ,$$

(7.5.18)

and therefore

$$p = -\frac{\partial F}{\partial V} = kT\frac{N}{V} , \quad \text{thus} \quad pV = NkT ,$$

(7.5.19)

$$S = -\frac{\partial F}{\partial T} = kN\left[\ln\left(\frac{V}{N}\right) + \frac{3}{2}\ln\left(\frac{2\pi m k T}{h^2}\right) + \frac{5}{2}\right]$$

(7.5.20)

in agreement with Sect. 7.2, if we also use $E = 3NkT/2$.

We can now calculate $\langle H(q,p)\rangle$ very easily. We have

$$\langle H\rangle = \int d^{3N}q\, d^{3N}p\, H(q,p)\varrho^C(q,p)$$

$$\dot= \frac{1}{ZN!h^{3N}}\int d^{3N}q\, d^{3N}p\, H(q,p)\,e^{-\beta H(q,p)}$$

$$= \frac{1}{Z}\left(-\frac{\partial Z}{\partial\beta}\right)$$

$$= -\frac{\partial\ln Z}{\partial\beta} = \frac{3}{2}N\frac{1}{\beta} = \frac{3}{2}NkT\ , \tag{7.5.21}$$

which agrees with (7.3.8).

ii) The Maxwell[5] velocity distribution: We look for the probability that, in a system with the Hamiltonian

$$H(q,p) = \sum_{i=1}^{N}\frac{p_i^2}{2m} + V(q_1,\ldots,q_N)$$

some particle, say particle 1, will have momentum in the interval

$$p < |p_1| < p + dp\ .$$

The probability can be easily found from the canonical probability density $\varrho^C(q,p)$ by integrating over the quantities which we are not interested in, i.e., by integrating over $p_2,\ldots,p_N, q_1,\ldots,q_N$, yielding

$$dw(p) = \frac{1}{\tilde Z}4\pi\,e^{-\beta p^2/2m}p^2\,dp \tag{7.5.22}$$

with the normalization factor

$$\tilde Z = (2\pi mkT)^{3/2}\ ,\quad\text{so that}\quad \int dw(p) = 1$$

This distribution is the well-known *Maxwell velocity distribution* (Fig. 7.5.2). Note that this holds not only for ideal gases, but for arbitrary potentials $V(q)$.

It is also easy to calculate the average value

$$\int dw(p)\frac{p^2}{2m} = \frac{\langle p^2\rangle}{2m} = \frac{3}{2}kT\ , \tag{7.5.23}$$

---

[5] *Maxwell, James Clerk* (*1831 Edinburgh, d. 1879 Cambridge). Scottish physicist. Pioneering work in the area of kinetic gas theory (the Maxwell velocity distribution) which was an important influence on Boltzmann. His greatest accomplishment is his complete theory of classical electrodynamics (Maxwell's equations), which ranks him among the greatest physicists.

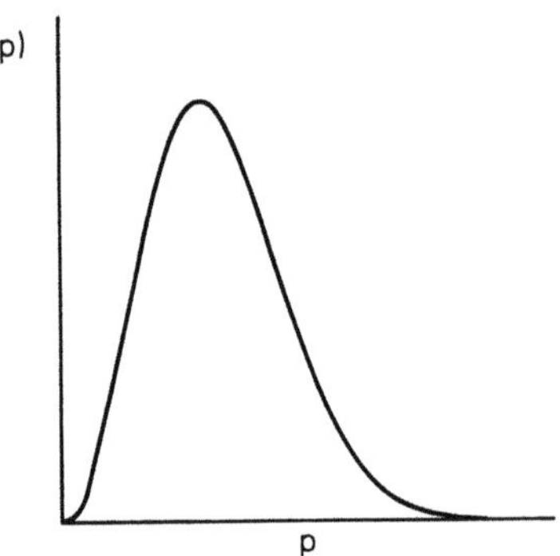

Fig. 7.5.2. The Maxwell velocity distribution $F(p)$ $\sim p^2 \exp(-\beta p^2/2m)$

and therefore for a monatomic ideal gas we have again

$$E = \langle H \rangle = N \frac{\langle p^2 \rangle}{2m} = \frac{3}{2} NkT \ .$$

iii) The barometric pressure equation: For an ideal gas in the earth's gravitational field, we have

$$H(q, p) = \sum_{i=1}^{N} \left(\frac{p_i^2}{2m} - m\boldsymbol{g} \cdot \boldsymbol{q}_i \right) \ .$$

If we integrate $\varrho^C(q, p)$ over $\boldsymbol{q}_2, \ldots, \boldsymbol{q}_N, \boldsymbol{p}_1, \ldots, \boldsymbol{p}_N$, we find the *barometric pressure equation*:

$$dw(h) \sim e^{-mgh/kT} dh \tag{7.5.24}$$

which is the probability of finding a particle in the altitude interval $[h, h + dh]$.

7.6 Thermodynamic Potentials

We have seen in Sect. 7.5 that a thermodynamic potential which depends on the variables T, V, N, namely the free energy, can be determined in either of two ways:

a) by a Legendre transformation from $S(E, V, N)$
b) by calculation from the partition function $Z(T, V, N)$ with the help of

$$-\beta F(T, V, N) = \ln Z(T, V, N) \ .$$

Here, we will discuss additional thermodynamic potentials, which depend on three other independent variables. We will begin by again deriving the free energy from the energy $E(S, V, N)$ by means of a Legendre transformation.

Starting with a given function $E(S, V, N)$ we set

$$T = \frac{\partial E(S, V, N)}{\partial S}$$

and solving this equation for S, we find

$$S = S(T, V, N) \ .$$

We then construct

$$F(T, V, N) = E(S(T, V, N), V, N) - TS(T, V, N) \ . \tag{7.6.1}$$

This is identical with the construction of Sect. 7.5, where we had

$$F(T, V, N) = E(T, V, N) - TS(E(T, V, N), V, N) \ . \tag{7.6.2}$$

For the total differential, we find

$$\begin{aligned}
dF &= dE - T\,dS - S\,dT \\
&= T\,dS - p\,dV + \mu\,dN - T\,dS - S\,dT \\
&= -S\,dT - p\,dV + \mu\,dN \ ,
\end{aligned} \tag{7.6.3}$$

from which we can immediately read off

$$\frac{\partial F}{\partial T} = -S \ , \qquad \frac{\partial F}{\partial V} = -p \ , \qquad \frac{\partial F}{\partial N} = \mu \tag{7.6.4}$$

as we found earlier [compare (7.5.12–14)].

Thus, the relations between the partial derivatives of a thermodynamic potential constructed in this manner and the corresponding conjugate variables can easily be discovered by considering the total differential.

In mechanics, where starting from the Lagrangian $L(q, \dot{q})$

$$p = \frac{\partial L}{\partial \dot{q}}$$

is defined and then

$$-H = L(\dot{q}(q, p), q) - p\dot{q}(q, p)$$

is introduced, we find

$$\begin{aligned}
-dH &= dL - p\,d\dot{q} - \dot{q}\,dp \\
&= \frac{\partial L}{\partial q}\,dq + p\,d\dot{q} - p\,d\dot{q} - \dot{q}\,dp = \frac{\partial L}{\partial q}\,dq - \dot{q}\,dp \ .
\end{aligned}$$

From this we see that

$$\frac{\partial H}{\partial p} = \dot{q} \quad \text{and} \quad \frac{\partial H}{\partial q} = -\frac{\partial L}{\partial \dot{q}}\left(= -\frac{d}{dt}p = -\dot{p}\right),$$

where the last equation in brackets on the right side follows from Lagrange's equation. Thus in mechanics a consideration of the total derivatives also yields the connection between the partial derivatives of the quantities obtained through a Legendre transformation and the other variables.

We now consider a few other Legendre transformations:

i) First, we will produce a thermodynamic potential which depends on the variables S, p, and N.
We will again start with

$$E(S, V, N) \quad \text{and construct} \quad -p(S, V, N) = \frac{\partial E(S, V, N)}{\partial V}.$$

We then solve for V to find

$$V = V(S, p, N).$$

If we now construct

$$H(S, p, N) = pV(S, p, N) + E(S, V(S, p, N), N) \tag{7.6.5}$$

then

$$dH = p\,dV + V\,dp + dE$$

$$= T\,dS + V\,dp + \mu\,dN, \tag{7.6.6}$$

and we have

$$\frac{\partial H(S, p, N)}{\partial S} = T, \quad \frac{\partial H(S, p, N)}{\partial p} = V,$$

$$\frac{\partial H(S, p, N)}{\partial N} = \mu. \tag{7.6.7}$$

We thus obtain the state variables T, V, and μ as functions of the quantities we decided to consider as variables S, p, N.
The quantity $H(S, p, N)$ is called the *enthalpy*[6]. For a process with $dp = dN = 0$, we have

$$dH = T\,dS = \delta Q.$$

---

[6] Enthalpy (Greek) from *thalpein*, to heat: thus it means approximately heat content. Named in analogy with energy and entropy because of its close connection to latent heat and heat of reaction.

ii) In order to obtain a thermodynamic potential which depends on the variables T, p, and N, we construct

$$G(T, p, N) = E - TS + pV \ . \tag{7.6.8}$$

Then we have

$$dG = dE - T\,dS - S\,dT + p\,dV + V\,dp$$

$$= -S\,dT + V\,dp + \mu\,dN \ , \tag{7.6.9}$$

and thus from $G(T, p, N)$, we can find

$$S = -\frac{\partial G(T, p, N)}{\partial T} \ , \qquad V = \frac{\partial G(T, p, N)}{\partial p} \ , \qquad \mu = \frac{\partial G(T, p, N)}{\partial N} \ . \tag{7.6.10}$$

$G(T, p, N)$ is called the *Gibbs free energy*. We will make abundant use of this thermodynamic potential in Chap. 8, when we discuss phase transformations.

iii) We construct

$$K(T, V, \mu) = E - TS - \mu N \ , \tag{7.6.11}$$

then,

$$dK = dE - T\,dS - S\,dT - \mu\,dN - N\,d\mu$$

$$= -S\,dT - p\,dV - N\,d\mu \ , \tag{7.6.12}$$

and it then follows that

$$S = -\frac{\partial K(T, V, \mu)}{\partial T} \ , \qquad p = -\frac{\partial K(T, V, \mu)}{\partial V} \ ,$$

$$\tag{7.6.13}$$

$$N = -\frac{\partial K(T, V, \mu)}{\partial \mu} \ .$$

The thermodynamic potential $K(T, V, \mu)$ has no special name. The fact that in this case T, V, and μ are the variables given indicates that the system observed here is one which is in contact with a reservoir with which it can exchange energy and particles, so that besides the volume, the temperature and the chemical potential of the system are given.

All of these further thermodynamic potentials can, like the free energy $F(T, V, N)$, be calculated from a generalized partition function. In particular, it can be shown that $K(T, V, \mu)$ can be determined from the partition function of the *"grand canonical ensemble"*. This is the collection of microscopic states in phase space with distributions compatible with the macroscopic state specified by the values of T, V, and μ. We have

$$-\beta K(T, V, \mu) = \ln Y(T, V, \mu) \quad \text{with} \tag{7.6.14}$$

$$Y(T, V, \mu) = \sum_{N=0}^{\infty} \frac{1}{N! h^{3N}} \int d^{3N}q \, d^{3N}p \, e^{-\beta(H(q, p) - \mu N)} . \tag{7.6.15}$$

7.7 Material Constants

In Sect. 7.6, we saw that by starting with $E(S, N, V)$ then using the appropriate Legendre transformation, we could find a thermodynamic potential for any choice of three macroscopic variables from each of the following pairs, (T, S), $(-p, V)$, (μ, N). However, we are interested not only in the state variables of a gas, a fluid, or a solid body, but also in a few of the quantities which characterize a material, which are called material constants. The most often used and most easily measured material constants are:

i) the *isothermal compressibility*

$$\kappa_T = -\frac{1}{V}\frac{\partial V(T, p, N)}{\partial p} . \tag{7.7.1}$$

If we use T, p, and N as the independent variables, the thermodynamic potential is given by $G(T, p, N)$ and we have

$$V(T, p, N) = \frac{\partial G(T, p, N)}{\partial p} ,$$

and therefore

$$\kappa_T = -\frac{1}{V}\frac{\partial^2 G(T, p, N)}{\partial p^2} . \tag{7.7.2}$$

ii) the *volume coefficient of expansion*:

$$\alpha = \frac{1}{V}\frac{\partial V(T, p, N)}{\partial T} = \frac{1}{V}\frac{\partial^2 G(T, p, N)}{\partial p \, \partial T} . \tag{7.7.3}$$

Of course,

$$\frac{\partial^2 G(T, p, N)}{\partial p \, \partial T}$$

can also be interpreted as

$$\frac{\partial}{\partial p}\frac{\partial G}{\partial T} = \frac{\partial}{\partial p}[-S(T, p, N)] \; ,$$

so that

$$-\frac{\partial S(T, p, N)}{\partial p} = \frac{\partial V(T, p, N)}{\partial T} \; . \tag{7.7.4}$$

Such relations between the first derivatives of macroscopic variables are called *Maxwell relations*. These relations can be derived from the mixed second derivatives of a thermodynamic potential. We will not list all these relations here, but only use them as the need requires.

iii) The most important material constants are the *specific heats*. These measure the amount of energy $T\,dS$ which we need to introduce into a system in order to raise its temperature dT. Of course, the amount of heat energy depends on which variables are held constant.

If both the number of particles and the volume V is held constant, we can define

$$C_V = T\frac{\partial S(T, V, N)}{\partial T} = -T\frac{\partial^2 F(T, V, N)}{\partial T^2} \tag{7.7.5}$$

and if the pressure p is held constant,

$$C_p = T\frac{\partial S(T, p, N)}{\partial T} = -T\frac{\partial^2 G(T, p, N)}{\partial T^2} \; . \tag{7.7.6}$$

We expect intuitively that $C_p > C_V$, since we would need to expend additional energy to increase the volume if we want to keep p constant.
We find:

a) We can also write

$$C_V \equiv T\frac{\partial S(T, V, N)}{\partial T} = \frac{\partial E(T, V, N)}{\partial T} \; , \tag{7.7.7}$$

since, starting with

$$dE = T\,dS + p\,dV - \mu\,dN$$

it follows, with $dN = dV = 0$, that

$$dE = T\,dS = T\frac{\partial S(T, V, N)}{\partial T}\,dT \; ,$$

i.e., the change in energy only occurs by means of an exchange of heat energy.

For an ideal gas, since $E = 3NkT/2$,

$$C_V = \tfrac{3}{2} Nk \ , \tag{7.7.8}$$

and thus the specific heat of a mole of ideal gas (which according to convention contains $N_0 = 6.022 \times 10^{23}$ particles) is given by:

$$C_V = \tfrac{3}{2} N_0 k = \tfrac{3}{2} R \ . \tag{7.7.9}$$

N_0 is called *Avogadro's*[7] *number* and R is called the *gas constant*. Its value is given by

$$R = 8.31441 \ \mathrm{J \, mol^{-1} K^{-1}} \ .$$

b) We also have

$$C_p \equiv T \frac{\partial S(T, p, N)}{\partial T} = \frac{\partial H(T, p, N)}{\partial T} \ , \tag{7.7.10}$$

since

$$dH = T\,dS + V\,dp + \mu\,dN$$

implies that for constant N and p, a change in enthalpy is also a pure exchange of heat.

For an ideal gas,

$$H = E + pV = \tfrac{3}{2} NkT + NkT = \tfrac{5}{2} NkT \tag{7.7.11}$$

and thus $H(T, p, N)$ is independent of p.

Note: $H(T, p, N)$ is not a thermodynamic potential for the variables (T, p, N). $H(T, p, N)$ is obtained from the thermodynamic potential $H(S, p, N)$ by calculating $S = S(T, p, N)$ from $G(T, p, N)$ and substituting this in $H(S, p, N)$.

But from this, of course, we find

$$\frac{\partial H}{\partial T} = \frac{\partial H}{\partial S} \frac{\partial S(T, p, N)}{\partial T} = T \frac{\partial S(T, p, N)}{\partial T} \ .$$

Thus for an ideal gas

$$C_p = \tfrac{5}{2} Nk \tag{7.7.12}$$

---

[7] *Avogadro, Amadeo* (*1776 Turin, d. 1856 Turin). In 1811, he formulated the law that different gases at the same pressure and temperature have the same number of molecules per unit volume. The first calculated estimate of this number was made in 1865 by Josef Loschmidt.

and for one mole

$$C_p = \tfrac{5}{2} R \ . \tag{7.7.13}$$

Thus, for one mole of an ideal gas

$$C_p - C_V = R \ . \tag{7.7.14}$$

In general, it can be shown

$$C_p - C_V = \frac{TV\alpha^2}{\kappa_T} \ . \tag{7.7.15}$$

c) For an ideal gas

$$\kappa_T = -\frac{1}{V}\frac{\partial V(T, p, N)}{\partial p}$$
$$= -\frac{1}{V}\frac{\partial(NkT/p)}{\partial p} = \frac{NkT}{Vp^2} = \frac{1}{p} \ , \tag{7.7.16}$$

$$\alpha = \frac{1}{V}\frac{\partial V(T, p, N)}{\partial T}$$
$$= \frac{1}{V}\frac{\partial(NkT/p)}{\partial T} = \frac{Nk}{pV} = \frac{1}{T} \ , \tag{7.7.17}$$

and thus

$$\frac{TV\alpha^2}{\kappa_T} = \frac{TVp}{T^2} = Nk = R \ .$$

iv) Finally we must mention the *isochoric*[8] *stress coefficient*

$$\beta = \frac{1}{p}\frac{\partial p(T, V, N)}{\partial T} = \frac{1}{p}\left(-\frac{\partial^2 F(T, V, N)}{\partial V \partial T} \right) \tag{7.7.18}$$

and the *adiabatic*[9] *compressibility*

$$\kappa_S = -\frac{1}{V}\frac{\partial V(S, p, N)}{\partial p} = -\frac{1}{V}\frac{\partial^2 H(S, p, N)}{\partial p^2} \ . \tag{7.7.19}$$

---

[8] Isobaric, isochoric (Greek) from *isos*: equal, *barys*: heavy, chora: *volume*, "equal pressure", "equal volume".

[9] Adiabatic (Greek) from *diabainein*: to go through. In an adiabatic process, heat energy is not allowed to go through the boundary of a system.

7.8 Changes of State

In this section we will discuss a series of especially important and typical processes and their various realizations.

Consider a system with three independent variables, one from each of the pairs

$$(T, S) , \quad (-p, V) , \quad (\mu, N) .$$

The thermodynamic potential corresponding to these variables then yields the corresponding energy-conjugate quantities.

Given the values of the independent variables, the values of all of the other quantities are determined. Thus, a state of the system is determined by the values of the independent variables.

A change in state is a transition from one state to another. If this transition occurs continuously, we call it a process.

Processes are called

isothermal	if $T = $ constant
isobaric	if $p = $ constant
isochoric	if $V = $ constant
isentropic	if $S = $ constant
isoenergetic	if $E = $ constant.

Here we always assume that N remains constant.

In special cases, for example an ideal gas, an isothermal process is also isoenergetic (since for an ideal gas, $E = \frac{3}{2} NkT$).

As a rule, though, the knowledge of which variables are held constant does not uniquely characterize a process. To demonstrate this, we consider an *isothermal expansion*.

7.8.1 Reversible and Irreversible Processes

a) We consider a gas in contact with a reservoir. Let the top of the container be weighted down with small masses (Fig. 7.8.1). In equilibrium, let the pressure,

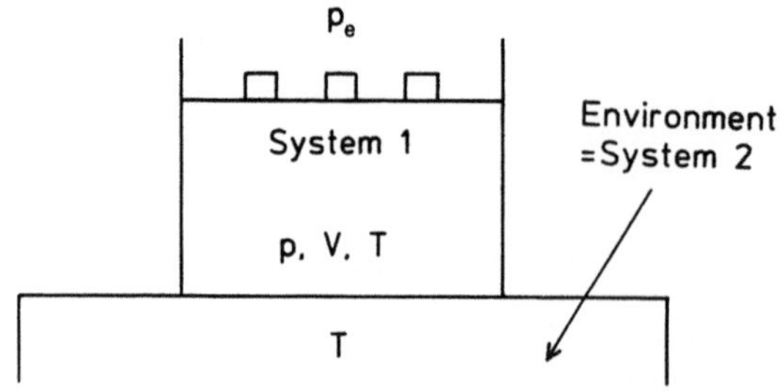

Fig. 7.8.1. An ideal gas in contact with a reservoir. After removing one of the weights from the cover, the gas expands isothermically

volume, and temperature be p, V, and T. If a small weight is taken away, p_e will be somewhat smaller than p, and the gas will expand because of the smaller force from above and will push the cover upwards until p has decreased to the point where $p = p_e$. We assume that the difference in pressure $p - p_e$ is so small that the gas is almost always in a state of equilibrium.

The change in energy of the gas in the form of work is given by:

$$\Delta W = - \int_{V_1}^{V_2} p \, dV = - \int_{V_1}^{V_2} dV \frac{NkT}{V} = - NkT \ln\left(\frac{V_2}{V_1}\right) . \tag{7.8.1}$$

We have $V_2 > V_1$ and $\Delta A < 0$, i.e., the gas performs work and thus gives up some of its energy to the outside. On the other side, the reservoir makes sure that the temperature T and the energy E of the system "gas" remains constant. This means that the work performed by the system is balanced by an influx of heat:

$$Q = \int_{S_1}^{S_2} T \, dS = T(S_2 - S_1)$$

$$= NkT \ln\left(\frac{V_2}{V_1}\right) = T \Delta S . \tag{7.8.2}$$

(Remember that for an ideal gas:

$$S = Nk \ln V + \tfrac{3}{2} Nk \ln E + f(N) .)$$

The energy balance of the process can be represented as in Fig. 7.8.2.

Volume and entropy are thus exchanged. Heat is introduced, and work is performed on the environment. (This is not a periodically functioning machine, and there is nothing to prevent all of the heat that is introduced from being changed into work.) This process can also be reversed. If the cover is again pushed down, work is performed on the system and then is given up in the form of heat to the reservoir.

A process realized this way is called *reversible*. Here, we have altered the variables p, V, S in a reversible manner

$$p_1 \rightarrow p_2 , \quad V_1 \rightarrow V_2 , \quad S_1 \rightarrow S_2 ,$$

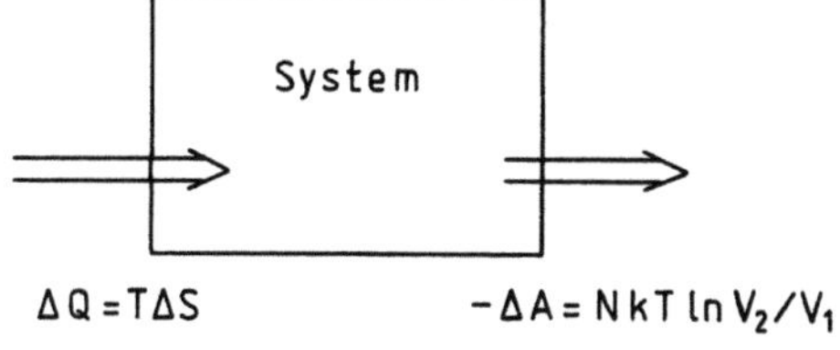

Fig. 7.8.2. During an isothermal expansion, energy will be introduced in the form of heat and energy will be lost in the form of work performed. For an ideal gas, this process is also isoenergetic

while holding T, N, and E constant. Energy in the form of heat and work has thus been exchanged in this system. This exchange is mediated by the exchange of entropy and volume.

b) Now consider the same process, but with an isolated system, so that there is no energy exchange of any kind with the environment. The volume V_1 is closed off by a cover which can be opened so that the gas "spontaneously" expands into the greater volume V_2 (Fig. 7.8.3).

After the expansion, when the system is again in a state of equilibrium, let it have entropy S_2. We then have

$$S = S_2 - S_1 = Nk\ln(V_2/V_1) \ .$$

The entropy S is produced in the system, so that the entropy increase of the gas does not occur through exchange, but rather through production. This additional contribution to the entropy can only be "removed" by "exporting" it outside the system, but then it is given to the environment. It is impossible to eliminate this additional entropy, as we will formulate in one of the central theorems of thermodynamics. Such an elimination would be equivalent to the system moving into an extremely unlikely state. In this sense, the production of entropy cannot be undone.

Since, to this point, we have only considered systems in equilibrium, we can say nothing about the gas during the free expansion. After the start of the free expansion, the gas flows, macroscopic vortices form and only after some time, the relaxation time, does the gas again achieve a state of rest and equilibrium. If the expansion is divided into many small steps, and after each step enough time is given so that the system again comes into equilibrium, we can determine the values of $T\Delta S_i$ and $-p(V)\Delta V_i$. Thus we can form the sum of all the quantities $T\Delta S_i$ and $-p(V)\Delta V_i$ and view these as approximations of the forms of energy

$$(\Delta Q)' = T\Delta S \quad \text{and} \quad (\Delta W)' = -\int_{V_1}^{V_2} p\,dV \ .$$

Thus the quantity $(\Delta Q)'$ is given by

$$(\Delta Q)' = T\Delta S = NkT\ln\left(\frac{V_2}{V_1}\right) \tag{7.8.3}$$

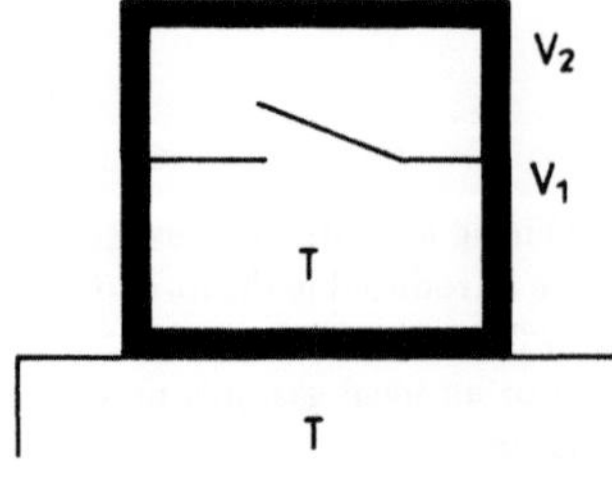

Fig. 7.8.3. The isothermal expansion of a gas realized irreversibly. The gas spontaneously expands into a greater volume. In this process, entropy is produced

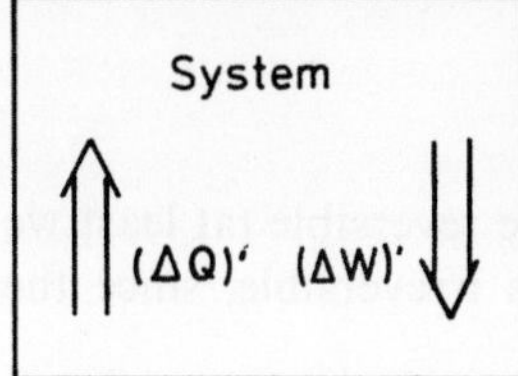

Fig. 7.8.4. The energy balance of spontaneous expansion. The quantities $(\Delta Q)'$ and $(\Delta W)'$ can be calculated using a special experimental set-up, but they represent at most expressions of energy which the system exchanges with itself

and the quantity $(\Delta W)'$:

$$(\Delta A)' = - \int_{V_1}^{V_2} p\, dV = - \int_{V_1}^{V_2} dV \frac{NkT}{V} = - NkT \ln\left(\frac{V_2}{V_1}\right) . \tag{7.8.4}$$

Note that the quantities $(\Delta Q)'$ and $(\Delta W)'$ do not represent energies which are exchanged with the environment.

We can represent the balance of energy graphically, as in Fig. 7.8.4. We see that the energy changes in the Gibbs equation

$$dE = T\, dS - p\, dV + \mu\, dN$$

express something about the forms of energy which can be exchanged with the environment, but that these changes in energy can also appear inside of the system, i.e. the system can, in a sense, exchange energy in various forms with itself. This occurs in the case of free expansion.

We have learned the following:

The process, in which

$$p_1 \rightarrow p_2 , \quad V_1 \rightarrow V_2 , \quad S_1 \rightarrow S_2 ,$$

where $T = $ constant and thus $E = $ constant can be realized in various ways. Let us write

$$\Delta S = \Delta_e S + \Delta_i S \quad \text{or} \quad dS = d_e S + d_i S \tag{7.8.5}$$

where here $d_e S$ signifies the change in the entropy which occurs due to exchange with the environment, and $d_i S$ is the amount due to production of entropy. In the first case,

$$d_e S \neq 0 , \quad d_i S = 0 ;$$

in the second case

$$d_e S = 0 , \quad d_i S > 0 .$$

A realization of a process in which

$$d_i S = 0$$

is called *reversible*, one for which

$$d_i S > 0$$

is called *irreversible*. The realization in (a) was therefore reversible (at least we neglected any production of entropy), that in (b) was irreversible, since the production of entropy cannot be undone.

7.8.2 Adiabatic and Non-adiabatic Processes

If in a realization of a process

$$d_e S = 0 \; ,$$

the realization is called *adiabatic*, otherwise it is called *nonadiabatic*.

In an adiabatically realized process, no entropy and thus no heat is exchanged with the environment. Inside the system itself, though, entropy can be produced. In order to have an adiabatic realization of a process, the system must be enclosed within so-called adiabatic walls, that is, walls which are heat-insulated. (A thermos bottle is a good example of an adiabatic enclosure.)

An adiabatic realization is not necessarily connected to heat insulation, though. The compressions and rarefactions associated with sound waves, for example, occur so rapidly that even without insulation there is no transfer of heat.

The realization (a) from Sect. 7.8.1 is thus reversible and non-adiabatic, since

$$d_i \dot{S} = 0 \; , \quad d_e S \neq 0$$

The realization (b) from Sect. 7.8.1 is irreversible and adiabatic since

$$d_i S > 0 \; , \quad d_e S = 0 \; .$$

Obviously, there are other possible types of realizations, such as a reversible adiabatic realization in which

$$d_i S = 0 \; , \quad d_e S = 0 \; .$$

In this case, the realization is also isentropic. An isothermal isentropic change of state is, in any case, impossible for an ideal gas, as we see from the formula

$$S = Nk \ln(VT^{3/2}) + f(N) \; . \tag{7.8.6}$$

If this were isothermal and isotropic, T as well as V would have to remain constant, and thus all the variables would remain unchanged.

Let us consider a reversible adiabatic expansion of an ideal gas.

Then, for any case,

$$dE = -p\,dV = -\frac{NkT}{V}\,dV \; ,$$

since $dS = dN = 0$, and thus, since also $E = \tfrac{3}{2} NkT$,

$$\frac{3}{2} Nk\, dT + \frac{NkT}{V} dV = 0 , \quad \text{or}$$

$$\frac{3}{2} \frac{dT}{T} + \frac{dV}{V} = 0 , \quad \text{i.e.}$$

$$d\ln(T^{3/2} V) = 0 ,$$

so that in this realization

$$T^{3/2} V = \text{constant} . \tag{7.8.7}$$

This result can also be seen directly from the form of $S(E(T, V, N), V, N)$ in (7.8.6).

Since $pV = NkT$, so that $T = pV/Nk$, we have

$$V^{5/2} p^{3/2} = \text{constant} \quad \text{or}$$

$$pV^{5/3} = \text{constant} \tag{7.8.8}$$

for an ideal gas undergoing a reversible adiabatic realization of a process. In a p–V diagram, this describes the "*adiabatic*" curves which are steeper than the "*isotherms*" for which $pV = \text{constant}$ (see Fig. 7.8.5).

Of course, we should say "isentropic" instead of "adiabatic." We tacitly assume here, when we use the word adiabatic, that $d_i S = 0$.

In an adiabatic reversible realization if we set $c = pV^{5/3}$, we then find:

$$\Delta E = \Delta A = - \int_{V_1}^{V_2} p\, dV = - c \int_{V_1}^{V_2} V^{-5/3}\, dV$$

$$= \frac{3c}{2} (V_2^{-2/3} - V_1^{-2/3})$$

$$= \tfrac{3}{2}(p_2 V_2^{5/3} V_2^{-2/3} - p_1 V_1^{5/3} V_1^{-2/3})$$

$$= \tfrac{3}{2}(p_2 V_2 - p_1 V_1)$$

$$= \tfrac{3}{2}(NkT_2 - NkT_1) ,$$

as we could have also calculated directly from $E = \tfrac{3}{2} NkT$. If we are given V_1 and T_1, T_2 can be found as a function of V_2 from the equation

$$V_2 T_2^{3/2} = V_1 T_1^{3/2} .$$

In a reversible adiabatic expansion, the temperature of the system drops, and thus the system gives energy to the outside in the form of work.

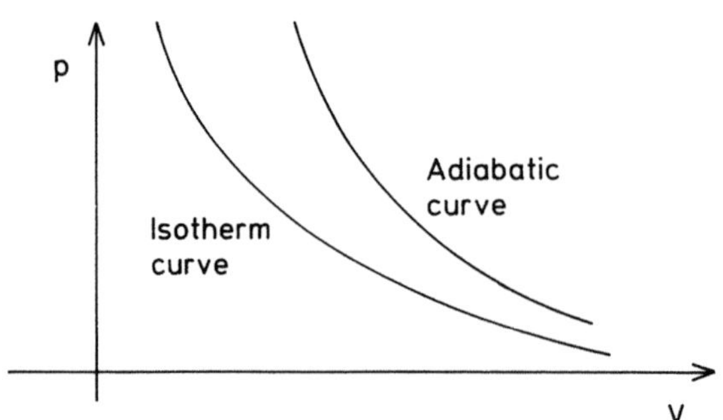

Fig. 7.8.5. Isotherms $(p \sim V^{-1})$ and adiabatic curves $(p \sim V^{-5/3})$ in a p–V diagram

In general, for polyatomic ideal gases in reversible adiabatic realizations,

$$pV^{\kappa} = \text{constant} \quad \text{with} \quad \kappa = C_p/C_V \tag{7.8.9}$$

(for a monatomic gas, $C_p = \frac{5}{2}Nk$, $C_V = \frac{3}{2}Nk$, $C_p/C_V = \frac{5}{3}$).

Finally, to consider an irreversible non-adiabatic realization, we need only find an "unsatisfactory" version of the realization (a) of Sect. 7.8.1, that is, one which does not guarantee reversibility. Let us thus consider this process more exactly (Fig. 7.8.1):

After removing the weight, we have $p_e < p$. System 1 and the environment (system 2) are no longer in complete equilibrium. After an exchange of volume, though, equilibrium is restored.

In this process, the total entropy increases, as we have already discussed in regard to the exchange of energy during the disturbance of the equilibrium. In this process, system 1 gains more entropy than system 2 loses, that is,

$$\Delta S_{1,2} = \Delta S_1 + \Delta S_2 > 0 \ , \quad \text{and}$$

$$\Delta S_1 > 0 \ , \quad \Delta S_2 < 0 \ .$$

Recall: At first, we have

$$S_{1,2} = S_1(E_1, V_1, N_1) + S_2(E_2, V_2, N_2) \ ,$$

after equilibrium has been restored

$$S_{1,2} = S_1(E_1, V_1', N_1) + S_2(E_2, V_2', N_2)$$

and during this process

$$dS_{1,2} = dS_1 + dS_2 = \frac{p}{T}dV_1 + \frac{p_u}{T}dV_2 \ .$$

Since $dV_1 > 0$, $dV_2 = -dV_1$, we have

$$dS_1 = \frac{p}{T}dV_1 > 0 \quad \text{and} \quad dS_2 = -\frac{p_u}{T}dV_1 \ ,$$

and thus, since $p > p_e$, also

$$dS_{1.2} > 0 \; .$$

The gain ΔS_1 in entropy is thus in principle always connected to a component $- \Delta S_2$ which is introduced to system 1, and a component $\Delta S_1 - (- \Delta S_2) = \Delta S_{1,2}$ which is produced within system 1.

Only when we can ignore the component $\Delta S_{1,2}$ can we speak of a reversible process. The smaller the difference in pressure is, the less entropy will be produced. If, then, the amount of heat removed from the environment is $- T\Delta S_2$, the environment can gain only the same amount of energy in the form of work, since the energy in both systems must remain fixed because $T = $ constant. This work is done by the gas, in that the volume of the gas is increased against the pressure p_e, i.e. against the force $A \cdot p_e$ (A is the area of the cover).

Thus, energy is transferred out of system 1 in the form of work, namely

$$(\Delta W)_1 = - \int p_u \, dV \; .$$

On the other hand, the total change in energy of the gas in the form of work is given by:

$$(\Delta W)_1' = - \int p \cdot dV \; .$$

Since $p > p_e$, and $dV > 0$, we then have $- (\Delta W)_1' > - (\Delta W)_1$, and thus the total change in energy of system 1 in the form of work $(\Delta W)_1'$ is not entirely delivered to system 2, just as the total change of energy of system 1 in the form of heat is not entirely taken from system 2. The component $- (\Delta W)_{1,2} = - (\Delta W)_1' + (\Delta W)_1$ remains in the system, corresponding to the quantity $\Delta S_{1,2}$.

This balance is indicated graphically in Fig. 7.8.6.

In the case of energy change in the form of heat, work, or chemical energy, we thus also need to distinguish whether these quantities of energy are completely exchanged or whether there is a transformation within the system, that is, whether some of the exchange occurs within the system itself. This possibility occurs precisely because entropy can be produced, and it is realized most clearly

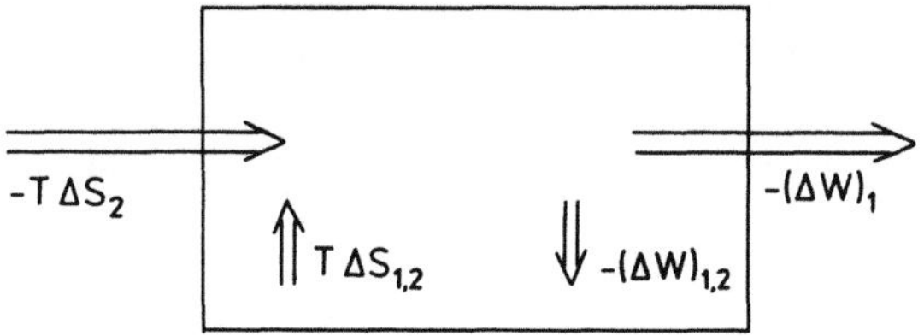

Fig. 7.8.6. An irreversible non-adiabatic expansion: The gain in entropy $T\Delta S_1$ consists of a component $- T\Delta S_2$ which is introduced and a component $T\Delta S_{1,2}$ which is produced. Analogously, the total change of energy in the form of work is composed of a component $- (\Delta W)_1$ which is exchanged and a component $- (\Delta W)_{1,2}$ which remains in the system

in realization (b) of Sect. 7.8.1, an irreversible adiabatic realization of a process in which the transformation occurs completely within the system.

Remarks. i) Often, in the literature, we find the convention that heat is understood only as the quantity $T d_e S$. Then, the exchanged heat is written

$$\delta Q = T d_e S \ .$$

Then, of course

$$dS = d_i S + d_e S = d_i S + (\delta Q / T) \ ,$$

and therefore it is always true that

$$dS \geq \delta Q / T$$

with the " $>$ " sign applies in the case of irreversible realizations and the " $=$ " sign applies when the realization is reversible. In this book, though, heat will always means the entire quantity $T\,dS$.

ii) In ordinary conversation, the meaning of heat is somewhat different. If we consider heat as a function of T, V, and N, we find

$$\delta Q(T, V, N) = T\,dS(T, V, N)$$

$$= T\frac{\partial S(T, V, N)}{\partial T}dT + T\frac{\partial S(T, V, N)}{\partial V}dV$$

$$+ \frac{\partial S(T, V, N)}{\partial N}dN \ . \tag{7.8.10}$$

In everyday usage, heat refers to only the first term on the right-hand side. This is exactly equal to $C_V\,dT$ and is always connected to a change in temperature.

Now, if $V = $ constant, so that $dV = 0$, and, in addition, $dN = 0$, this term corresponds to the complete exchange of heat. In general, though, a change in volume is also connected to the exchange of heat, as we have seen in the examples (a) and (b) in Sect. 7.8.1.

If, as in those examples, we hold T constant, then, considering T, N, and V as the independent variables:

$$\delta Q = T\,dS = T\frac{\partial S(T, V, N)}{\partial V}dV$$

$$= p\,dV + \frac{\partial E(T, V, N)}{\partial V}dV \ , \tag{7.8.11}$$

since

$$\frac{\partial S(E(T, V, N), V, N)}{\partial V} = \frac{p}{T} + \frac{1}{T}\frac{\partial E(T, V, N)}{\partial V}$$

and thus we again have

$$\delta A = -p\,dV \ .$$

If the gas is ideal, $\partial E(T, V, N)/\partial V = 0$. We can see immediately that for a fixed T the two quantities of exchanged energy δQ and δW are exactly the same in magnitude.

7.8.3 The Joule–Thomson Process

Finally, we consider another very instructive and also technically important process: the *Joule*[10]*–Thomson*[11] or *Joule–Kelvin* process.

In this process, a gas under constant pressure p_0 is forced through a narrow passage which works as a throttle into a region of constant lower pressure (Fig. 7.8.7). The gas is adiabatically isolated from the environment and thus only exchanges energy with the outside in the form of work. If a volume V_0 of gas has been forced through the throttle from the left side, then work $p_0 V_0$ has been performed on the gas, whereas on the right side the gas itself has performed work $p_1 V_1$. The gas changes its state from (p_0, V_0, E_0) into the state (p_1, V_1, E_1) and therefore experiences a change of energy equal to

$$A = p_0 V_0 - p_1 V_1 = E_1 - E_0 \ .$$

It therefore holds that

$$E_0 + p_0 V_0 = E_1 + p_1 V_1 \ . \tag{7.8.12}$$

The enthalpy thus remains constant during this process.

In order to calculate the temperature change of the gas in this process, we first consider the enthalpy as a function of T and p (and N). The fact that H remains constant can then be expressed in the form

$$0 = dH(T, p, N) = \frac{\partial H(T, p, N)}{\partial T}dT + \frac{\partial H(T, p, N)}{\partial p}dp \ . \tag{7.8.13}$$

---

[10] *Joule, James Prescott* (*1818 Salfort, Lancashire, d. 1889 Sale Cheshire). English physicist, brewery owner, privately educated. Most important work in the area of thermodynamics, 1840 heat conduction equation, 1843 value of the mechanical equivalent of heat, 1852 Joule–Thomson process.

[11] *Thomson, Sir William*, Lord Kelvin (*1824 Belfast, d. 1907 Netherhall, Largs/Scotland). 1846 Professor at Glasgow, close friendship and collaboration with J.P. Joule. Physicist and inventor with diverse interests. His principal areas of work were electrodynamics and thermodynamics.

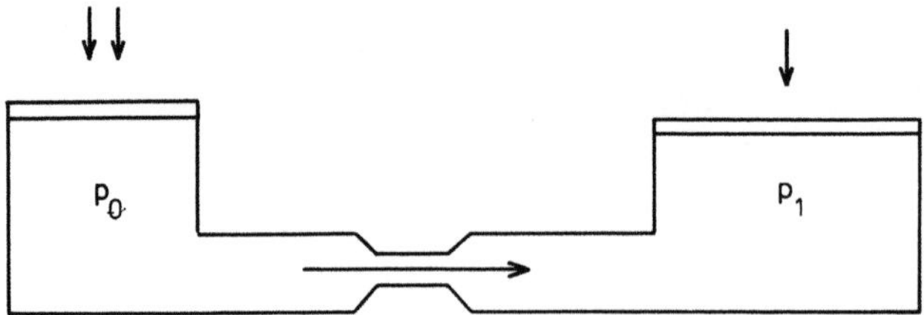

Fig. 7.8.7. The Joule–Thomson effect. Under adiabatic isolation, a real gas is forced through a throttle into a region of lower pressure

If we write $H = H(S, p, N)$, we have

$$0 = dH(S, p, N) = T\,dS + V\,dp \ .$$

Since $dp < 0$, $dH = 0$, it then follows that:

$$dS = -\frac{V}{T}dp > 0 \ ,$$

and therefore during the Joule–Thomson process entropy is created, and the process is always irreversible.

It follows further that

$$\frac{\partial H(T, p, N)}{\partial T} = \frac{\partial H(S(T, p, N), p, N)}{\partial T}$$

$$= T\frac{\partial S(T, p, N)}{\partial T} = C_p \ , \tag{7.8.14}$$

$$\frac{\partial H(T, p, N)}{\partial p} = \frac{\partial H(S(T, p, N), p, N)}{\partial p}$$

$$= V + T\frac{\partial S(T, p, N)}{\partial p}$$

$$= V - T\frac{\partial V(T, p, N)}{\partial T} \ . \tag{7.8.15}$$

Here, the last step follows from the Maxwell relation

$$\frac{\partial^2 G(T, p, N)}{\partial p\,\partial T} = \frac{\partial V(T, p, N)}{\partial T} = -\frac{\partial S(T, p, N)}{\partial p} \ .$$

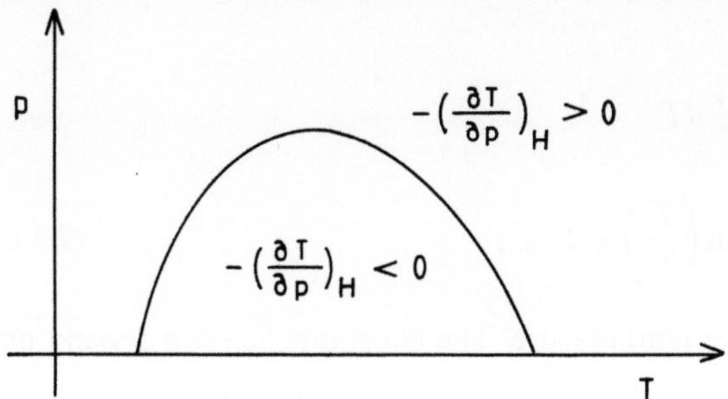

Fig. 7.8.8. A typical inversion curve for a real gas

Thus, we find

$$-\left(\frac{\partial T}{\partial p}\right)_H = \frac{\partial H(T, p, N)/\partial p}{\partial H(T, p, N)/\partial T}$$

$$= \frac{1}{C_p}\left(V - T\frac{\partial V(T, p, N)}{\partial T}\right). \tag{7.8.16}$$

A positive sign for $-(\partial T/\partial p)_H$ indicates a rise in temperature.

For an ideal gas, $H = \frac{5}{2} NkT$, and thus if H remains constant so does T, so that always $(\partial T/\partial p)_H = 0$. For a real gas, though, there is an "inversion curve" which, in a p–T diagram, separates from each other the two regions in which the process leads to a rise or a fall in the temperature of the gas (Fig. 7.8.8).

7.9   The Transformation of Heat into Work, the Carnot Efficiency

We consider two systems with temperatures T_1 and T_2, $T_1 > T_2$. Let the volumes of the two systems be held constant and, for simplicity, let the number of particles N_i in each of the two systems be equal. If the two systems are brought into contact with each other, heat flows from system 1 to system 2 (that is, energy in the form of heat), until the temperatures of the two systems become equal. We assume that the combined system is isolated.

We assume that the specific heats, which after all, are proportional to N_i, are equal and constant over the observed temperature range.

Then, since

$$C_V = T\frac{\partial S_i(T, V, N)}{\partial T}, \quad i = 1, 2 ,$$

it holds that

$$S_i(T) = S_i(T, V_i, N_i) = \int_{T_0}^{T} \frac{C_V}{T} dT$$

$$= C_V \ln\left(\frac{T}{T_0}\right) + S_{i,0} \, , \tag{7.9.1}$$

where T_0 is some fixed reference temperature. The constants $S_{i,0}$ can depend on T_0, N_i, and V_i.

For an ideal gas, we know that $C_V = \frac{3}{2} Nk$, so that it is constant. In this case, we see also that

$$S(T, V, N) = C_V \ln(T/T_0) + S_0 \, ,$$

which is also clear from (7.5.20).

Before the two systems are in contact with each other,

$$S_A = S_1(T_1) + S_2(T_2)$$

$$= C_V \ln\left(\frac{T_1 T_2}{T_0^2}\right) + S_{1,0} + S_{2,0} \, . \tag{7.9.2}$$

After thermal contact, a common temperature T_E has been established. The entropy is now given by

$$S_E = S_1(T_E) + S_2(T_E)$$

$$= 2C_V \ln(T_E/T_0) + S_{1,0} + S_{2,0} \, , \tag{7.9.3}$$

and thus the change in entropy is

$$\Delta S = S_E - S_A = 2C_V \ln\left(\frac{T_E}{\sqrt{T_1 T_2}}\right) . \tag{7.9.4}$$

The amount of entropy produced thus depends on the common temperature achieved by the two systems after thermal contact.

What is the value of T_E? In the case of thermal contact in which heat can be freely and directly exchanged, we have

$$T_E = \tfrac{1}{2}(T_1 + T_2) \, . \tag{7.9.5}$$

This is intuitively obvious, since both systems have the same number of particles ($N_1 = N_2$) with the same heat capacity. The final temperature must then be the average of the two starting temperatures. We can also show this using energy conservation:

For each system it is true that

$$E_i = C_V T_i \, ,$$

and therefore, before thermal contact,

$$E = E_1 + E_2 = C_V(T_1 + T_2)$$

and afterwards,

$$E = E_1' + E_2' = 2C_V T_E \, ,$$

Thus, $2T_E = T_1 + T_2$.

Then, the amount of entropy produced by this form of contact is given by

$$\Delta S = 2C_V \ln\left(\frac{T_1 + T_2}{2\sqrt{T_1 T_2}}\right) . \tag{7.9.6}$$

We have $\Delta S \geq 0$, since the arithmetic mean is always greater than or equal to the geometric mean (since from $(a - b)^2 \geq 0$, it follows that $a^2 + 2ab + b^2 \geq 4ab$, so that $(a + b)^2/4 \geq ab$).

The process is therefore irreversible, as entropy is produced, so that it cannot be run in reverse. Clearly, we cannot expect that the two systems will again attain their initial temperatures without interference from outside.

The process would be reversible, if we could arrange

$$T_E = \sqrt{T_1 T_2} \, . \tag{7.9.7}$$

But then, the amount of energy after thermal contact would be given by

$$E = 2C_V\sqrt{T_1 T_2} \, . \tag{7.9.8}$$

But since the energy before of the combined system was

$$E = C_V(T_1 + T_2) > 2C_V\sqrt{T_1 T_2} \tag{7.9.9}$$

that would mean that energy

$$E = 2C_V[\tfrac{1}{2}(T_1 + T_2) - \sqrt{T_1 T_2}] \tag{7.9.10}$$

would have to be given to the environment, for example, in the form of work. This means that we would have to partially transform the heat ΔQ_1, which is given up by system 1 with higher temperature, into work, while part of the heat would flow into system 2.

This would appear diagrammatically as in Fig. 7.9.1.

We therefore introduce a third system M between the two original systems, which can effect this transformation. We call this system the "machine." The machine should receive a flow of heat ΔQ_1, give up heat ΔQ_2, and also perform

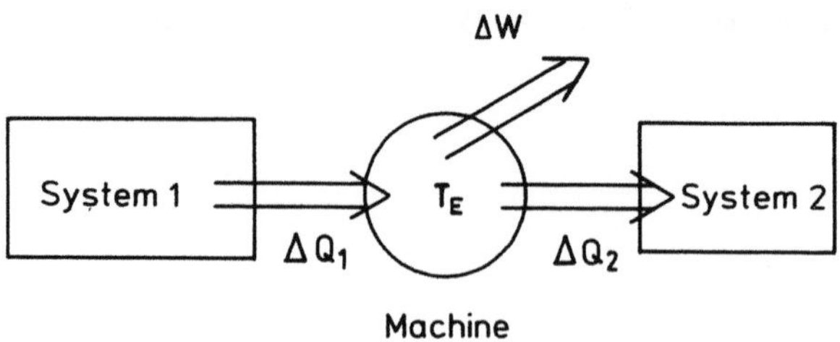

Fig. 7.9.1. A "machine" can moderate the equalization of temperature between two systems and thus perform work. Less entropy is then produced

work ΔW. If the machine receives heat ΔQ_1, then

$$\delta Q_1 = -T dS = -C_V dT ,$$

and therefore

$$\Delta Q_1 = -\int_{T_1}^{T_E} C_V dT = C_V(T_1 - T_E) \tag{7.9.11}$$

is the amount of heat which the machine receives from the cooling of system 1 to the temperature T_E.

On the other hand

$$\Delta Q_2 = \int_{T_2}^{T_E} C_V dT = C_V(T_E - T_2) \tag{7.9.12}$$

is the amount of energy that the machine must give up to system 2 in order to heat it to temperature T_E. (To simplify the situation, imagine that the machine has temperature T_E, so that it cools system 1 and heats system 2.)

The difference of these two, i.e. the quantity of energy

$$\Delta W = \Delta Q_1 - \Delta Q_2 = 2C_V[\tfrac{1}{2}(T_1 + T_2) - T_E] \tag{7.9.13}$$

can therefore be transformed into work.

If we note once again the expression for the amount of entropy produced

$$\Delta S = 2C_V \ln\left(\frac{T_E}{\sqrt{T_1 T_2}}\right) , \tag{7.9.14}$$

then we recognize that: for

$$\sqrt{T_1 T_2} \leq T_E \leq \tfrac{1}{2}(T_1 + T_2)$$

the amount of entropy produced decreases and the work ΔW increases, as T_E decreases.

The limiting cases are:

$$T_E = \tfrac{1}{2}(T_1 + T_2) \ , \qquad \text{maximal production of entropy, no work} \ ,$$

$$T_E = \sqrt{T_1 T_2} \ , \qquad \text{no entropy produced, maximal work} \ .$$

Each time entropy is produced, then, work is being "wasted," i.e., we lose the chance to perform work.

If we define the efficiency of the machine as

$$\eta = \frac{\Delta A}{\Delta Q_1} = \frac{\text{work performed by the machine}}{\text{heat introduced into the machine}} \tag{7.9.15}$$

we have

$$\eta = \frac{\Delta Q_1 - \Delta Q_2}{\Delta Q_1} = 1 - \frac{\Delta Q_2}{\Delta Q_1} \ , \tag{7.9.16}$$

and thus

$$\eta = 1 - \frac{(T_E - T_2)}{(T_1 - T_E)} \ . \tag{7.9.17}$$

The energy and entropy balance of the system "machine" are shown in Fig. 7.9.2.

The process involving this machine, which slows down the equalization of temperatures between two systems and thereby produces work, stops when the two temperatures have been equalized.

Of course, it is more interesting to consider machines which work periodically, that is, systems in which cyclical processes occur. In these systems, after a given amount of time, called the period, the final state is identical to the initial state. In a state diagram – a diagram in which the changes in the state variables are portrayed – a cyclical process is indicated by a closed curve (Fig. 7.9.3).

Circular processes in which a certain amount of heat ΔQ_1 is taken from a reservoir at the fixed temperature T_1 in order to convert it to work are especially important as applications. Besides the work, the energy ΔQ_2 in the form of heat will also be given to reservoir 2 at constant temperature. The energy balance then appears again as in Fig. 7.9.2a.

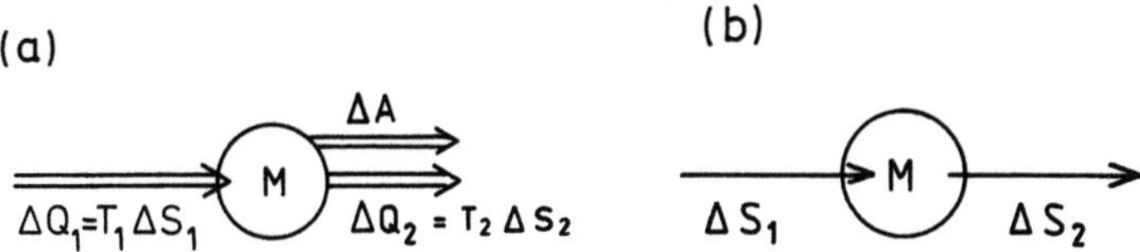

Fig. 7.9.2. Energy balance (a) and entropy balance (b) for the machine

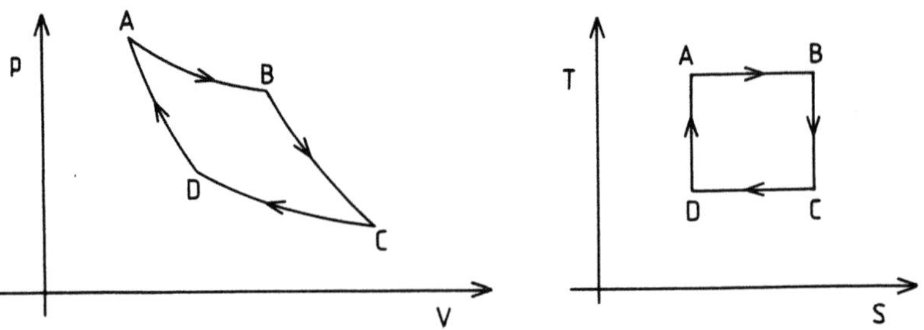

Fig. 7.9.3. In a cyclical process, the state variables run through a closed curve in a diagram. Here, the process consists of two isothermal and two isentropic subprocesses. This is the Carnot cycle

From energy conservation, it is clear that

$$\Delta Q_1 = \Delta W + \Delta Q_2 \ .$$

With the exchange ΔQ_i of heat, there is also an associated entropy exchange. The entropy balance is again as in Fig. 7.9.2b, with $\Delta Q_1 = T_1 \Delta S_1$, $\Delta Q_2 = T_2 \Delta S_2$.

It now follows:

If after one period the machine has exactly the same value of entropy (which must be true for a cyclical process), then all the entropy introduced with ΔQ_1 must be removed with ΔQ_2, and the any entropy which might still be produced, would also have to be removed with ΔQ_2. That is,

$$\Delta S_2 \geq \Delta S_1$$

and in the best case $\Delta S_2 = \Delta S_1$ if the process in the machine is reversible.

The efficiency η is thus given by

$$\eta = \frac{\Delta A}{\Delta Q_1} = \frac{(\Delta Q_1 - \Delta Q_2)}{\Delta Q_1} = 1 - \frac{T_2 \, \Delta S_2}{T_1 \, \Delta S_1} \ . \tag{7.9.18}$$

The efficiency is maximal if

$$\frac{\Delta S_2}{\Delta S_1} = 1 \ .$$

Thus, the highest possible efficiency for a periodically working machine, which takes heat from a heat reservoir with temperature T_1 and gives it up to a reservoir with temperature T_2 to do work, is given by

$$\eta_{\text{max}} = 1 - \frac{T_2}{T_1} \ . \tag{7.9.19}$$

This is the Carnot[12] efficiency. The Carnot cycle is an idealized process in which this efficiency is attained.

Consequences. a) In the cyclical process just described, heat is transformed into work. The machine works as a heat engine. The cycle can also be run in the other direction. This means:

With the help of a system which contributes energy (not in the form of heat), energy can be removed from a reservoir at lower temperature and delivered to a reservoir at higher temperature. The machine thus functions as a heat pump. If the process is irreversible, that is, the entropy of the environment is increased, the process can still be run in reverse. In this case, too, the entropy of the environment increases.

Thus, we need to be very careful in talking about what precisely is irreversible in an irreversible process. Here, the cyclical process is reversible, but not the balance of entropy between system and environment. If, for example, in one cycle the entropy of the system and the environment grows from S_0 to $S_0 + \Delta S$, this entropy can never be reduced back to S_0, because entropy can never be destroyed.

b) The result

$$\frac{[(\text{Heat given up by reservoir 1}) - (\text{heat acquired by reservoir 2})]}{(\text{heat given up by reservoir 1})}$$

$$= 1 - \frac{T_2}{T_1}$$

for reversible processes is independent of the substance involved.

This is valid not only for cyclical processes, but also for engines using any reversible processes which take energy in the form of heat at temperature T_1 and give it up at temperature T_2.

Since we must have $\Delta S_1 + \Delta S_2 = 0$, the difference of the amounts of energy which can be transformed into work in proportion to the energy received is always given by

$$\frac{(T_1 \Delta S_1 - T_2 \Delta S_1)}{T_1 \Delta S_1} = 1 - \frac{T_2}{T_1} \ .$$

---

[12] *Carnot, Sadi Nicolas Léonard* (*1796 Paris, d. 1832 Paris). Engineer and officer. He became famous for his publication "Réflexions sur la puissance motrice du feu et sur les machines propres à developper cette puissance" about the theory of heat engines, written in 1821 in Magdeburg (where Carnot had fled from the Bourbons with his family). He started with the impossibility of a perpetual motion machine of the second kind. His considerations are partially contradictory, since he assumed the existence of a heat substance, which would perform work as it flowed, but which itself was not a form of energy. In his posthumous work, though, we find the currently accepted formulation of heat.

Measuring temperatures can thus be reduced to measuring heat or energy. However, this measurement cannot produce a unit of temperature, since here only proportions are measured.

With the help of the ideal gas law,

$$pV = NkT$$

we could also measure temperature proportions: if we hold N and p constant, T is proportional to V, or holding N and V constant, T is proportional to p. But in this case, the gas would have to be quite rarefied so that we could treat it as ideal.

In any case, if a temperature T_0 is assigned to some arbitrarily chosen state of a system, the temperatures of all the other states of the system are uniquely determined.

In 1954, a standard temperature convention was established, using as reference point the triple point of water, which is the unique temperature at which ice, water, and water vapor are in phase equilibrium in a container closed on all sides.

The pressure at this point is given by 6.1×10^2 Pa $= 4.58$ Torr $= 6.1 \times 10^{-3}$ bar. The temperature at this point is defined as $T_0 = 273.16$ K (for historical reasons).

Using these definitions, the proportionality constant k is also determined, since if an ideal gas is brought into thermal contact with such a container at the triple point, we must have

$$pV = NkT_0 \ .$$

Measuring p, V, and N, we can then determine k and we obtain the value of the Boltzmann constant k quoted in Sect. 7.3.

7.10 The Laws of Thermodynamics

Laws formulate principles motivated by experimental results, but they can never be proved in the strong sense of the word. The reliability of physical laws lies in the number of times they have been confirmed, the lack of results which deviate from them, and in the number of successful predictions which have been made based on them.

a) The first law of thermodynamics is essentially the law of conservation of energy:

"Energy can be neither created nor destroyed."

A system can exchange energy with its environment, i.e., the energy loss of the system is equal to the energy gain of the environment or vice versa. Energy is therefore only a conserved quantity when we consider all the systems which take part in a certain process. Thus the statement of the conservation of energy is a statement about the possibility of realizing a process.

Historically, the development of this conservation law came at the same time as the discovery that it was possible to transform mechanical energy into heat and vice versa, i.e. with the determination of the mechanical equivalent of heat.

This determination was first made by *R. Mayer*[13], and later more exactly by J. Joule. Today, we write

$$1 \text{ Nm} = 1 \text{ Ws} = 1 \text{ Joule} = 0.238845 \text{ cal}$$

or

$$1 \text{ kWh} = 859.8 \text{ kcal} ,$$

where 1 cal is defined as the amount of heat that must be introduced to 1 g of water in order to raise its temperature from $14.5°C$ to $15.5°C$.

The experiments which were used to determine the mechanical equivalent of heat do not, of course, prove the conservation of energy. They only make clear the fact that different types of energy can be freely exchanged.

b) The second law states:

"Entropy can be created, but not destroyed."

We have seen that entropy is created in the process in which two bodies of different temperature come to thermal equilibrium with each other. The destruction of entropy would mean, for example, that the process of temperature equalization could be reversed, that is, that a system in a highly probable state could spontaneously transform itself into a state of extremely small probability, and then remain in this second state long enough for us to talk of a condition of equilibrium.

There are two formulations of the second law, which now are only of historical interest, but which are often found in textbooks. These two formulations follow from the above statement:

---

[13] *Mayer, Julius Robert* (*1814 Heilbronn, d. 1878 Heilbronn). A doctor by profession and encouraged by his observations of tropical diseases, he formulated a principle of the conservation of energy in 1841. Energy can be transferred among different forms (including biological energy), but energy can be neither created nor destroyed. He calculated an accurate value for the mechanical equivalent of heat from a comparison of C_p and C_V. As an outsider, it was difficult for him to achieve recognition from people in the field. His conclusions were independently discovered and expanded upon by Helmholtz, Joule, and others.

a) The formulation of *R. Clausius*[14]:

It is impossible to deliver heat from a colder to a warmer reservoir without somehow changing the environment.

In particular, if we take heat $T_1 \Delta S_1 = \Delta E$ from the colder reservoir and introduce it into the warmer reservoir as $\Delta E = T_2 \Delta S_2$, then the entropy surplus is

$$\Delta S = \Delta S_1 - \Delta S_2 = \Delta E \left(\frac{1}{T_1} - \frac{1}{T_2} \right) > 0$$

since $T_1 < T_2$. This entropy cannot be destroyed, thus the state of the environment has been changed.

b) Thomson's formulation:

It is impossible to construct a periodically working engine which does nothing but performing work and cooling off a reservoir of heat.

Because: The cooling of the heat reservoir means removing from it energy in the form of heat: $\Delta E = T \Delta S$. Since entropy cannot be destroyed, the engine must also give off heat in addition to work in order to accommodate the entropy somewhere.

In case of the expansion of a gas in thermal contact with a reservoir, it is true that the entire heat energy is transformed into work, but the entropy removed from the reservoir remains in the gas, which is shown in the greater volume occupied by the gas. Entropy here is not destroyed, but remains constant, i.e. it is exchanged.

c) The third law states that entropy is a quantity with a naturally given zero-point. One formulation reads:

"When a system is in its ground state, its entropy is zero."

In this formulation, the ground state means the state of least energy. If the ground state is not degenerate, this law follows directly from the definition

$$S = k \ln \Omega(E, V, N) \ ,$$

since then in the ground state $\Omega = 1$ and thus $S = 0$.

A more general formulation of this law reads:

As $T \to 0$, the entropy becomes independent of external parameters like pressure, volume, etc. The value of entropy can thus be defined as zero.

It is often said that absolute zero cannot be reached. By this, we mean that it cannot be reached in a finite number of isothermal and isentropic processes.

---

[14] *Clausius, Rudolf* (*1822 Köslin, d. 1888 Bonn). One of the founders of thermodynamics. In 1865, he discovered and named entropy and formulated the second law of thermodynamics. He also made important contributions to kinetic gas theory.

Here, we will not go into the behavior of macroscopic bodies at low temperatures, where the quantum mechanical description of microscopic interactions must be used, and we will not consider the question of entropy for degenerate ground states. These topics arise in a course on statistical mechanics.

Consequences

a) Every kind of heat capacity goes to zero as $T \to 0$.

Proof. We have

$$\frac{C_R}{T} = \frac{\partial S(T, R, N)}{\partial T}, \quad R = p, V, \ldots \tag{7.10.1}$$

and thus

$$S(T, R, N) = \int_0^T dT' \frac{C_R(T')}{T'} . \tag{7.10.2}$$

Therefore, since the integral must exist, it follows that

$$C_R(T) \to 0 \quad \text{for} \quad T \to 0 . \tag{7.10.3}$$

b) We found

$$\alpha = \frac{1}{V} \frac{\partial V(T, p, N)}{\partial T} = -\frac{1}{V} \frac{\partial S(T, p, N)}{\partial p} . \tag{7.10.4}$$

But now, as $T \to 0$, S is independent of T, p, and N, and therefore

$$\frac{\partial S(T, p, N)}{\partial p} = 0 .$$

Hence, it follows

$$\alpha \to 0 \quad \text{for} \quad T \to 0 . \tag{7.10.5}$$

c) Correspondingly, for the isochoric pressure coefficient

$$\beta = \frac{1}{p} \frac{\partial p(T, V, N)}{\partial T}$$

$$= \frac{1}{p} \frac{\partial}{\partial V} S(T, V, N) \to 0 \quad \text{for} \quad T \to 0 , \tag{7.10.6}$$

where again as in (b) we have used a Maxwell relation.

7.11 The Phenomenological Basis of Thermodynamics

7.11.1 Thermodynamics and Statistical Mechanics

In the previous sections, we have seen that in order to describe macroscopic physical systems we needed to use new concepts such as temperature and entropy which are not present in microphysics, even though in principle macroscopic systems are only much larger versions of microscopic systems. To justify this, we have recognized the unimaginable complexity of a detailed microscopic description of a macroscopic system, and thus realized that very different problems arise in macroscopic systems along with new phenomena such as irreversibility.

Classical statistical mechanics relates the microscopic and macroscopic descriptions of a macroscopic system under the assumption that the microscopic theory is given by classical mechanics. Microscopic states of an N-particle system correspond to points in a $6N$-dimensional phase space $\mathbb{R}^{6N}$, whereas macroscopic states are desribed by certain ensembles with probability distribution $\varrho(q, p)$ on $\mathbb{R}^{6N}$. The calculation of the thermodynamic characteristics of the macroscopic system is reduced, for example, to the calculation of the Gibbs function

$$\Omega(E, V, N) = \frac{1}{N! h^{3N}} \int_{H(q,p) \leq E} d^{3N}q \, d^{3N}p$$

$$= \exp\left[\frac{S(E, V, N)}{k} \right] \tag{7.11.1}$$

or

$$Z(T, V, N) = \frac{1}{N! h^{3N}} \int d^{3N}q \, d^{3N}p \exp\left[-\frac{H(q, p)}{kT} \right]$$

$$= \exp\left[-\frac{F(T, V, N)}{kT} \right]. \tag{7.11.2}$$

Thermodynamics, however, presents a general theory of macroscopic systems, independent of any assumptions about the underlying microscopic theory. We will now describe the structure of this theory.

The basis of this theoretical construction consists of the three laws of Sect. 7.10, at least with slightly different formulations. This will take into account the fact that we must now introduce the concept of entropy in a different form.

Within the framework of classical statistical mechanics, these laws summarize the results of investigation and emphasize the essential properties of energy and entropy. However, in a purely phenomenological theory of thermodynamics, which does not consider the microscopic structure of matter, these laws represent basic postulates from which we can derive an abundance of consequences.

There are good reasons to build a phenomenological thermodynamics independent of microscopic physics.

– All statements of a phenomenological thermodynamics hold in general, independent of any assumptions about physics at the microscopic level, as long as the fundamental laws are correct. These laws can be obtained relatively easily by induction from intuitive and accessible facts of experience.
– The meaning of microscopic physics for thermodynamics is limited to the calculation of a single Gibbs function. All further conclusions are fully within the framework of thermodynamics and thus rely precisely on those phenomena which are important for practical applications. It is much more comfortable and economical for an applied scientist, like a chemical engineer, to ignore the microscopic background of thermodynamics.
– We should not underestimate the increased conceptual clarity, which comes from using the general results of thermodynamics rather than results derived from the microscopic properties of systems. There is also a peculiar charm in pursuing the consequences of a few basic laws.
– Historically, thermodynamics has been derived from a theory of heat and not been treated as a subfield of mechanics.
– A fully complete derivation of the laws of thermodynamics from equations of microscopic physics does not yet exist.

On the other hand, there are also limitations to the approach of thermodynamics, i.e. ignoring microscopic characteristics.

– Thermodynamics is limited to systems in (local) equilibrium.
– Thermodynamics only yields *relations* between state variables. The Gibbs function, which is essential in working out the characteristics of a system, must be found in a different way.
– Thermodynamics makes statements about the characteristics of (local) states of equilibrium and places limits on the set of attainable states of a closed system using the laws of energy conservation and increase of entropy. The exact physical course of the changes of macroscopic variables cannot be calculated in thermodynamics.

For these reasons, we also need additional theories, namely:

a) The thermodynamics of irreversible processes, a macroscopic-phenomenological theory whose basis will be described in Chap. 9, or
b) Kinetics, a subdiscipline of the statistical mechanics of non-equilibrium states.

He will now restate the laws of phenomenological thermodynamics in a formulation which is suited to their role as fundamental postulates.

7.11.2 The First Law of Thermodynamics

The *first law* of thermodynamics reads:

For every system, the total energy E is an extensive state variable. For an isolated system, the value of E remains constant in time.

The first law was first expressed by Robert Mayer in 1841.
In general, the change in an extensive quantity X can be expressed as follows

$$dX = \delta_e X + \delta_i X \tag{7.11.3}$$

i.e., the change dX in X is made up of a change $\delta_e X$ of X due to an influx from outside the system and a change $\delta_i X$ from production within the system. The first law then states

$$dE = \delta_e E \ , \quad \delta_i E = 0 \ . \tag{7.11.4}$$

The introduction of energy into a system can be accomplished in many ways by changing its state variables. We now indicate for several types of changes of state variables X the corresponding amount $\delta_X E$ of energy introduced.

– If the volume of a system is changed, work must generally be done against forces due to pressure. In a homogeneous isotropic state, pressure has a constant value p. The pressure is defined as the force per unit area. If by displacing the cover of a system its volume is changed by an amount dV, then in the process the amount of work (Fig. 7.4.1)

$$\delta_V E = - A\,dh\,p = - p\,dV \tag{7.11.5}$$

is introduced into the system. The sign is explained by the fact that pressure pushes outwards, so that in a compression ($dV < 0$), the energy is increased.
Thus we have

$$\delta_V E = - p\,dV \ . \tag{7.11.6}$$

– In other physical situations, other possible ways of introducing energy become important, e.g. through a:

change in momentum dp:	$\delta_p E = \boldsymbol{v}\cdot d\boldsymbol{p}$	(v: velocity)
change in angular momentum dL:	$\delta_L E = \boldsymbol{\omega}\cdot d\boldsymbol{L}$	(ω: angular velocity)
change in electric charge dQ^e:	$\delta_{Q^e} E = \phi\,dQ^e$	(ϕ: electric potential)
change in electrical polarization dq:	$\delta_q E = \boldsymbol{E}\cdot d\boldsymbol{q}$	(E: electric field strength)
change in position dx:	$\delta_x E = - \boldsymbol{K}\cdot d\boldsymbol{x}$	(K: force)
change in the number of particles dN:	$\delta_N E = \mu\,dN$	(μ: chemical potential)

The total change in energy from these and other mechanisms is always has the form

$$\delta_{\{X\}} E = \sum \xi_\alpha dX_\alpha \ . \tag{7.11.7}$$

Here, ξ_α is never extensive and X_α is never intensive. The quantities X_α and ξ_α are said to be *energy-conjugate* to each other.

It is important to note that *no state variables* correspond to the individual possibilities of introducing energy. For example, there is no state variable "volume-energy". We can explain this by noting that we can easily find examples of processes where energy is introduced by changing volume in such a way that the system is always in a state of equilibrium, so that the amount of energy thus introduced can be calculated:

$$\int_{①}^{②} \delta_V E = - \int_{①}^{②} p\, dV \ , \tag{7.11.8}$$

but for which the amount of energy so introduced depends on the path chosen between states ① and ②. On the other hand, the value of a state variable, by definition, can depend only on the state, not on the way the state has been produced. (This is analogous to the situation of a conservative field, $K = -\nabla U$. It is certainly true that $\oint (K_x dx + K_y dy + K_z dz) = 0$ along every closed path, but, in general, $\oint K_x dx \neq 0$. Thus, U_x, U_y, and U_z are not state variables, only the single potential energy U is.)

7.11.3 The Second and Third Laws

In a thermodynamic system, there is another typical method of energy change, namely through the *introduction of heat*.

Instead of $\delta_Q E$, we normally write δQ. We then have:

$$dE = \delta Q + \sum_\alpha \xi_\alpha dX_\alpha := \delta Q + \delta A \ . \tag{7.11.9}$$

The quantity

$$\delta W = \sum_\alpha \xi_\alpha dX_\alpha \tag{7.11.10}$$

is often called the *work* done on the system (it would be better to describe it as the "non-heat energy", since δW can also include chemical energy). In general, for a cyclical process with the same beginning and end state

$$\oint dE = 0 \tag{7.11.11}$$

but most generally

$$\oint \delta Q = -\oint \delta A \neq 0 \ . \tag{7.11.12}$$

Thus "heat" and "work" are not state variables.

The possibility of another method of energy transfer as "non-work" is especially clear when thermal equilibrium is disturbed. Two systems which cannot exchange energy in the form of work are, in general, not in equilibrium with each other and they attempt to find a new equilibrium state when they come into thermal contact with each other, i.e. a state of thermal equilibrium. For example, two vessels containing gas which are put into thermal contact will bring themselves to a state of equilibrium, as is evident as the pressure in each of the gases changes during a relaxation time. The relaxation times for thermal equilibrium are often quite long. The property "two systems A and B in the states a and b are in thermal equilibrium" defines an equivalence relation. Intuitively, two systems, between which this equivalence relation holds, prove to be equally warm and, as the systems come to equilibrium, energy flows from the warmer to the colder system. This fact is often formulated as the "*Zeroth law of thermodynamics*":

There is an intensive state variable ϑ in every thermodynamic system, named the empirical temperature, such that two systems are in thermal equilibrium with each other if and only if they are in states with the same value of ϑ. Greater values of ϑ correspond to warmer temperatures.

Clearly any monotonically increasing function $f(\vartheta)$ of ϑ is also an empirical temperature. As we have seen, the quantity τ defined in Chap. 7.3.1 is an empirical temperature.

In order to measure an empirical temperature, we can choose some fixed system in which the values of all the independent variables save one are held constant. We then observe the behavior of the changing variable as the system is put into thermal equilibrium with another system. If this value is monotonically dependent on the temperature, this system is suited to measure empirical temperature. Examples are the gas thermometer ($\vartheta = pV$) and the mercury thermometer. It is then true, exactly as for the other forms of energy transfer, that there exists a pair of conjugate variables T and S such that $\delta Q = TdS$. More exactly, the *second law* reads:

There is an intensive variable T (the absolute temperature) and an extensive (additive) variable S (the entropy) such that for a homogeneous system in equilibrium we have:

$$dE = TdS + \sum_{\alpha} \xi_{\alpha} dX_{\alpha} \ . \tag{7.11.3}$$

The entropy of a closed system never decreases and achieves its maximum in the state of equilibrium (which is determined by the given boundary conditions).

In Sect. 7.2, we showed how the second law in this formulation could be understood in the framework of classical statistical mechanics. The second law of thermodynamics was formulated for the first time by Rudolf Clausius in the year 1850, who coined the word "entropy" in the year 1865.

Here are a few immediate consequences of the second law:

a) As is the case for every extensive quantity, $dS = \delta_e S + \delta_i S$, where $\delta_e S$ is the change due to introduction of entropy and $\delta_i S$ is the change through entropy production. The second law then says that $\delta_i S \geq 0$.

b) T has the properties of an empirical temperature. Consider two systems which are both isolated from the environment and which can exchange energy with each other only in the form of heat after they have been brought into thermal contact. Then, $S = S_1 + S_2$ and

$$dS = \frac{dE_1}{T_1} + \frac{dE_2}{T_2} \ .$$

Since the systems are isolated, $dE_1 = dE_2$, so that $dS = [(1/T_1) - (1/T_2)]dE_1$.

If we have thermal equilibrium, entropy can no longer be increased through an exchange of energy, and we have $dS = 0$, so that $T_1 = T_2$. Otherwise $dS > 0$, and for $dE_1 > 0$, we must have

$$\frac{1}{T_1} - \frac{1}{T_2} > 0 \ , \quad T_2 > T_1 \ .$$

Thus energy flows from the warmer to the colder system (Sect. 7.3.1).

In the Gibbs form:

$$dE = TdS + \sum \xi_\alpha dX_\alpha$$

only differences in entropy appear. A zero-point of entropy, as opposed to the temperature, is not defined. This is done in the *third law*:

At absolute zero, $T = 0$, the entropy of a system in equilibrium nears its smallest possible value $S_0 = 0$, which is independent of volume, pressure, phase, etc.

Thus, we can speak of the entropy of a system in the same way that we speak of its volume or number of particles. The third law was formulated by *W. Nernst*[15]. In the framework of statistical mechanics, the third law means that at absolute zero the uncertainty of the microscopic state of a macroscopic system vanishes. Now, at absolute zero a system has its smallest possible energy and the set of possible microscopic states decreases along with the energy. However the assertion that the state with the lowest energy, called the ground

---

[15] *Nernst, Walter* (*1864 Briesen/West Prussia, d. 1941 Gut Zibelle/Oberlausitz). One of the founders of physical chemistry. He discovered this law around 1905. Nobel prize in 1920.

state, can have so few microscopic states that its entropy vanishes can only be correctly understood with the help of quantum statistics. It is precisely in the vicinity of this ground state, where the energies of all particles become very small, that the applicability of classical mechanics must be questioned.

Remarks. i) T and S are essentially uniquely determined by (7.11.13). To see this, let $\hat{T}$ and $\hat{S}$ be quantities defined differently, such that $\hat{T}$ is intensive, $\hat{S}$ is extensive, and $T\,dS = \hat{T}\,d\hat{S}$.
   Then,

$$\frac{d\hat{S}}{dS} = \frac{T}{\hat{T}} \quad \text{and} \quad \hat{S} = g(S) \quad \text{with} \quad g'(S) = \frac{T}{\hat{T}} \ .$$

Since S and $\hat{S}$ are extensive, g must be linear: thus $\hat{S} = \alpha S + \beta$. Here, $\alpha > 0$, since $dS > 0$ implies $d\hat{S} > 0$. From this we see that the zero point $T = 0$ and the sign of T have an absolute meaning, thus the absolute temperature is uniquely defined after we have fixed a unit of T using a conventional temperature determination for some system in a reproducible state of equilibrium.
   For a mechanical system, as we have already remarked in Sect. 7.3.1, we always have $T \geq 0$. Temporarily, we will always assume that $T \geq 0$.

ii) If a procedure to measure T is known, entropy differences can then be determined by energy measurements. In particular, for fixed values of the extensive variables X_α, we have $dE = T\,dS$, so that $dS = dE/T$.

iii) Normal mechanical systems can be understood as special thermodynamic systems for which no entropy exchange and production is possible.

iv) The form $dE = T\,dS + \sum \xi_\alpha\,dX^\alpha$ (the *Gibbs equation*) describes all of the properties of a system.

All of the (relevant) properties of a system are therefore known if the function $E(S, X_1, \ldots, X_n) = E(S, X)$ is known, or, equivalently, if the functions $T(S, X)$ and $\xi_\alpha(S, X)\,(\alpha = 1, \ldots, n)$ are known. The function $E(S, X)$ is a *Gibbs function* or a *thermodynamic potential*, since all of the quantities of the system can be derived from it, e.g.

$$T(S, X) = \frac{\partial E(S, X)}{\partial S} \ .$$

Already as a consequence of the existence of the Gibbs equation, there are restrictions on the functions

$$T(S, X) = \frac{\partial E(S, X)}{\partial S} \quad \text{and} \quad \xi_\alpha(S, X) = \frac{\partial E(S, X)}{\partial X_\alpha} \ . \tag{7.11.14}$$

By differentiating and using the symmetry of second order derivatives, we can show immediately that

$$\frac{\partial^2 E}{\partial S \partial X_\alpha} = \frac{\partial^2 E}{\partial X_\alpha \partial S} , \quad \text{thus} \quad \frac{\partial T(S, X)}{\partial X_\alpha} = \frac{\partial \xi_\alpha(S, X)}{\partial S} \tag{7.11.15}$$

and

$$\frac{\partial \xi_\alpha(S, X)}{\partial X_\beta} = \frac{\partial \xi_\beta(S, X)}{\partial X_\alpha}. \tag{7.11.16}$$

These are the Maxwell relations (Sect. 7.7). Note the analogy to the equation $\partial F_i/\partial x_j = \partial F_j/\partial x_i$ for a conservative force field, which justifies the use of the term "thermodynamic potential".

7.11.4   The Thermal and Caloric Equations of State

Since in phenomenological thermodynamics we cannot calculate a thermodynamic potential by recourse to microscopic interactions, we must start from definite measured quantities. The following functions can be measured:

The *thermal equation of state*:

$$p = p(T, V) \tag{7.11.17a}$$

and the so-called *caloric equation of state*:

$$E = E(T, V) \tag{7.11.17b}$$

(in the following, we always assume that the number of particles N is held constant).

Together, the thermal and caloric equations of state determine the Gibbs function $E = E(S, V)$ and thus all of the properties of the system.

First, we have

$$S(T, V) = \int_{T_0, V_0}^{T, V} \frac{dE + p\, dV'}{T'}$$

$$= \int_{T_0, V_0}^{T, V} \frac{\dfrac{\partial E(T', V')}{\partial T'} dT' + \left(\dfrac{\partial E(T', V')}{\partial V'} + p \right) dV'}{T'} ,$$

so that $S(T, V)$ is determined by the thermal and caloric equations of state. By solving for T, we can find $T(S, v)$ and from the caloric equation of state we can calculate $E(S, V) = E(T(S, V), V)$.

The thermal and caloric equations of state are as little independent of each other as the functions $T(S, V)$ and $p(S, V)$, since

$$dS = \frac{dE + pdV}{T} = \frac{1}{T}\frac{\partial E(T, V)}{\partial T}dT + \frac{1}{T}\left[\frac{\partial E(T, V)}{\partial V} + p(T, V)\right]dV$$

thus

$$\frac{\partial S(T, V)}{\partial T} = \frac{1}{T}\frac{\partial E(T, V)}{\partial T}\,,$$

$$\frac{\partial S(T, V)}{\partial V} = \frac{1}{T}\left[\frac{\partial E(T, V)}{\partial V} + p(T, V)\right]\,,$$

therefore

$$\frac{\partial}{\partial V}\left(\frac{1}{T}\frac{\partial E(T, V)}{\partial T}\right) = \frac{\partial}{\partial T}\left\{\frac{1}{T}\left[\frac{\partial E(T, V)}{\partial V} + p(T, V)\right]\right\}\,,$$

i.e.

$$\frac{1}{T}\frac{\partial^2 E}{\partial T\partial V} = -\frac{1}{T^2}\left[\frac{\partial E(T, V)}{\partial V} + p(T, V)\right] + \frac{1}{T}\frac{\partial^2 E(T, V)}{\partial T\partial V} + \frac{1}{T}\frac{\partial p(T, V)}{\partial T}\,,$$

thus

$$\frac{\partial E(T, V)}{\partial V} + p(T, V) = T\frac{\partial p(T, V)}{\partial T}$$

or, writing this a different way,

$$\left(\frac{\partial E}{\partial V}\right)_T + p = T\left(\frac{\partial p}{\partial T}\right)_V\,, \tag{7.11.18}$$

so that the V-dependence of $E(T, V)$ is determined by the thermal equation of state.

For several variables, each of the following sets (A–D) fully determines the thermodynamic system:

A) the Gibbs function $E(S, X)$,
B) the Gibbs function $S(E, X)$,
C) the functions

$$T(S, X) = \frac{\partial E(S, X)}{\partial S}\,, \qquad \xi_\alpha(S, X) = \frac{\partial E(S, X)}{\partial X_\alpha}$$

with the restrictions (Maxwell relations)

$$\frac{\partial T(S, X)}{\partial X_\alpha} = \frac{\partial \xi_\alpha(S, X)}{\partial S}, \qquad \frac{\partial \xi_\alpha(S, X)}{\partial X_\beta} = \frac{\partial \xi_\beta(S, X)}{\partial X_\alpha}, \qquad (7.11.19)$$

D) the functions $E(T, X)$ (the caloric equation of state) and $\xi_\alpha(T, X)$ (the thermal equation of state) with the restrictions (Maxwell relations)

$$\xi_\alpha(T, X) - \frac{\partial E(T, X)}{\partial X_\alpha} = T\frac{\partial \xi_\alpha(T, X)}{T}$$

and

$$\frac{\partial \xi_\alpha(T, X)}{\partial X_\beta} = \frac{\partial \xi_\beta(T, X)}{\partial X_\alpha} .$$

For high enough temperatures and/or low enough density, matter behaves like an ideal gas (i.e., like a collection of free noninteracting particles). For an ideal gas, we find that $pV = $ constant for constant temperature, so that pV is an empirical temperature and must be a function of the absolute temperature:

$$pV = f(T) .$$

Further, we find, as we might expect from a model based on a collection of free particles, that the energy of an ideal gas does not depend on volume:

$$E = E(T) .$$

This is the result of a well-known experiment of *Gay-Lussac*[16]. The function $f(T)$ is further determined by (7.11.18):

$$\left(\frac{\partial E}{\partial V}\right)_T = 0 = T\left(\frac{\partial p}{\partial T}\right)_V - p = \frac{Tf'(T)}{V} - \frac{f(T)}{V}$$

so that

$$Tf'(T) = f(T) \quad \text{and thus} \quad f(T) = \text{constant} \cdot T .$$

The empirical temperature measured with an ideal gas is thus proportional to the absolute temperature. Since obviously $f(T)$ must be proportional to the number of particles N, we have $f(T) = NkT$ and the constant k can be experimentally determined.

---

[16] *Gay-Lussac, Joseph-Louis* (*1778 St Léonhard, d. 1850 Paris). French chemist and physicist. Work on gas theory, 1804 balloon flights, co-discoverer of the element boron.

7.12 Equilibrium and Stability Conditions

7.12.1 Equilibrium and Stability in Exchange Processes

We consider again, as in Sect. 7.4, as isolated system consisting of two subsystems which can exchange energy with each other (Fig. 7.12.1 – the subsystems are only in thermal contact).

We have then have

$$dS = dS_1 + dS_2 = \frac{dE_1}{T_1} + \frac{dE_2}{T_2} = \left(\frac{1}{T_1} - \frac{1}{T_2}\right)dE_1 \geq 0 , \qquad (7.12.1)$$

so that in the case of equilibrium, $T_1 = T_2$. If this equilibrium is stable, the total entropy must have a maximum, so that to second order in $dE_1 = - dE_2$, we must have

$$dS = \left(\frac{\partial S_1}{\partial E_1}\right)_{X_1} dE_1 + \left(\frac{\partial S_2}{\partial E_2}\right)_{X_2} dE_2$$
$$+ \frac{1}{2}\left(\frac{\partial^2 S_1}{\partial E_1^2}\right)_{X_1} dE_1^2 + \frac{1}{2}\left(\frac{\partial^2 S_2}{\partial E_2^2}\right)_{X_2} dE_2^2 \leq 0 \qquad (7.12.2)$$

and in equilibrium

$$\frac{1}{2}\left(\left(\frac{\partial^2 S_1}{\partial E_1^2}\right)_{X_1} + \left(\frac{\partial^2 S_2}{\partial E_2^2}\right)_{X_2}\right) \leq 0 .$$

Now the characteristics of the two subsystems can be varied independently of each other, in particular both subsystems can be the same so that in stable equilibrium

$$\left(\frac{\partial^2 S_1}{\partial E_1^2}\right)_{X_1} \leq 0$$

or, if we eliminate the unnecessary index "1":

$$\left(\frac{\partial^2 S}{\partial E^2}\right)_X \leq 0 . \qquad (7.12.3)$$

This is the condition for stability of equilibrium when energy exchange is allowed.

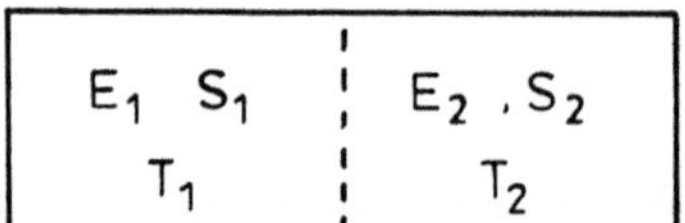

$$\boxed{\begin{array}{c|c} E_1 \;\; S_1 & E_2 \; , S_2 \\ T_1 & T_2 \end{array}}$$

Fig. 7.12.1. Two systems which are only in thermal contact

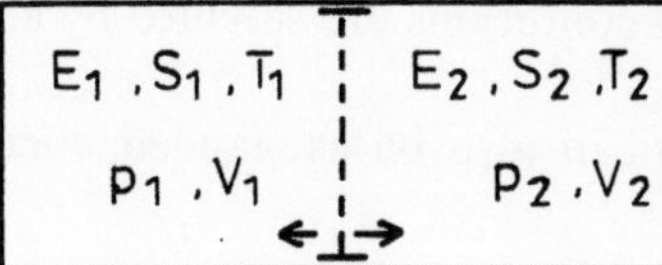

Fig. 7.12.2. Two systems which can exchange of thermal and volume energy

Since $(\partial S/\partial E)_X = 1/T$, this means that $(\partial T/\partial E)_X \geq 0$ or

$$\left(\frac{\partial E}{\partial T}\right)_X \geq 0 \; , \tag{7.12.4}$$

so that E must increase with T.

If besides energy, volume can also be exchanged, say through a movable, heat conducting wall (Fig. 7.12.2), then since

$$dS_\alpha = \frac{1}{T_\alpha}\left(dE_\alpha + p_\alpha dV_\alpha - \sum_i \xi_i^{(\alpha)} dX_i^{(\alpha)} \right) \; , \quad \alpha = 1,2 \; ;$$

$$dE_1 + dE_2 = dV_1 + dV_2 = 0 \; , \quad dX_i^{(\alpha)} = 0 \; ,$$

$$\left(\frac{1}{T_1} - \frac{1}{T_2}\right)dE_1 + \left(\frac{p_1}{T_1} - \frac{p_2}{T_2}\right)dV_1 = 0 \; , \tag{7.12.5}$$

the conditions for equilibrium and stability must read

$$\left(\frac{\partial^2 S}{\partial E^2}\right)_{V,X} dE^2 + 2\left(\frac{\partial^2 S}{\partial E \partial V}\right)_X dE dV + \left(\frac{\partial^2 S}{\partial V^2}\right)_{E,X} dV^2 \leq 0 \; . \tag{7.12.6}$$

In stable equilibrium with both energy and volume exchange possible, we then have

$$T_1 = T_2 \; , \quad p_1 = p_2 \quad \text{and}$$

$$\left(\frac{\partial^2 S}{\partial E^2}\right)_{V,X} \leq 0 \; , \quad \left(\frac{\partial^2 S}{\partial V^2}\right)_{E,X} \leq 0 \; ,$$

$$\det\begin{pmatrix} \left(\dfrac{\partial^2 S}{\partial E^2}\right)_{V,X} & \left(\dfrac{\partial^2 S}{\partial E \partial V}\right)_X \\[2ex] \left(\dfrac{\partial^2 S}{\partial E \partial V}\right)_X & \left(\dfrac{\partial^2 S}{\partial V^2}\right)_{E,X} \end{pmatrix} \geq 0 \; . \tag{7.12.7}$$

We already found the first stability condition above. The second means in particular

$$\left(\frac{\partial}{\partial V}\frac{p}{T}\right)_{E,X} \leq 0 \; . \tag{7.12.8}$$

We can immediately verify that all these stability conditions are satisfied by an ideal gas.

If, finally, momentum (say through friction) can also be exchanged, then since

$$dS_\alpha = \frac{1}{T_\alpha}(dE_\alpha + p_\alpha dV_\alpha - \boldsymbol{v}_\alpha \cdot d\boldsymbol{p}_\alpha - \sum \xi_i^{(\alpha)} dX_i^{(\alpha)}) \ , \quad \alpha = 1, 2 \ ,$$

$$dV_1 + dV_2 = dE_1 + dE_2 = 0 \ , \quad d\boldsymbol{p}_1 + d\boldsymbol{p}_2 = \boldsymbol{0}$$

we have the additional equilibrium and stability condition

$$\boldsymbol{v}_1 = \boldsymbol{v}_2 \ ; \tag{7.12.9}$$

i.e., in equilibrium, no relative motion is possible and

$$-\frac{\partial}{\partial p_i}\left(\frac{v_i}{T}\right) \leq 0 \tag{7.12.10}$$

(p_i, v_i are the i-components of $\boldsymbol{p}$ and $\boldsymbol{v}$).

Since $\boldsymbol{v} = \boldsymbol{p}/M$ and T is independent of $\boldsymbol{p}$ (that is, the temperature is independent of the state of motion of the reference system) the stability condition becomes $T/M \geq 0$, or just $T \geq 0$.

A system which possesses translational energy must therefore have a non-negative absolute temperature at all times. A more direct way of seeing this is as follows:

The Gibbs function $S(E, \boldsymbol{p}, X)$ depends on E and $\boldsymbol{p}$ in a very special way; namely, it is possible to slow down the system in a reversible way, that is, to make the momentum $\boldsymbol{p}$ vanish without changing the values of the variables X. Because of the kinetic component, the energy decreases during the breaking.

We thus have

$$S(E, \boldsymbol{p}, X) = S\left(E - \frac{\boldsymbol{p}^2}{2M}, \boldsymbol{0}, X\right) , \tag{7.12.11}$$

and S depends only on the internal energy $U = E - (\boldsymbol{p}^2/2M)$.

If we could have $T < 0$, then we would have $(\partial S/\partial E)_X < 0$, so that S would be a decreasing function of the internal energy and thus the entropy of a system could spontaneously increase if the internal energy were reduced and the kinetic energy increased, i.e. if the system "burst" into several subsystems moving with respect to each other.

The stability condition $(\partial^2 S/\partial E^2)_X \leq 0$ means that the function $S(E, X)$ is always curved towards the E-axis (convex) as E is varied.

In view of the third law, there are the following possibilities (Fig. 7.12.3):

a) The energy can take on arbitrarily large values. The temperature $T = 1/(\partial S/\partial E)_X$ is always positive and the minimum of energy and entropy is at $T = 0$.

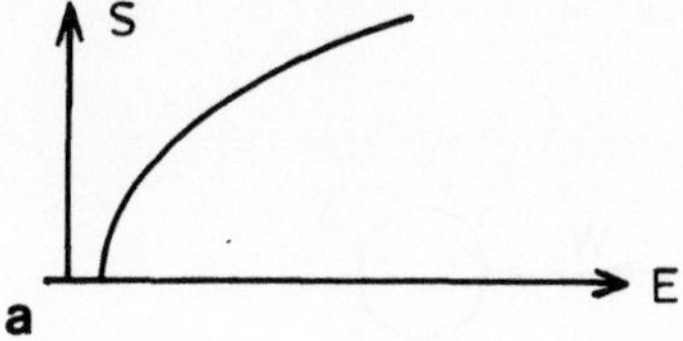

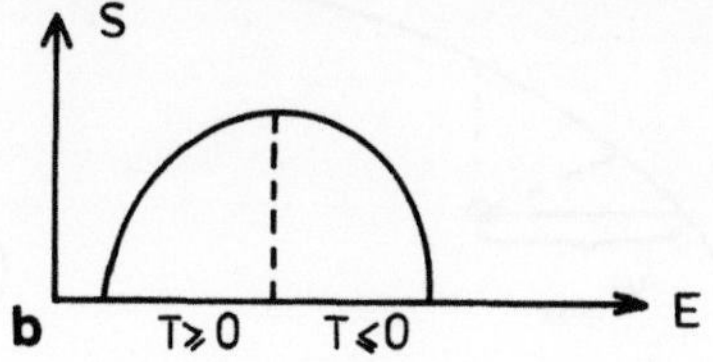

Fig. 7.12.3. The function $S(E)$ for fixed extensive variables X (**a**) if there is no maximum of E and S, (**b**) if there are maximal values of E and S

$\vec{B}$	↑ ↑ ↑ ↑ ↑	$T = 0_+$
	↗ ↖ ↗ ↖ ↗	$T > 0$
	↙ ↑ ↘ ↗ ↖	$T \to \infty$
	↓ ↘ ↙ ↘ ↙	$T < 0$
	↓ ↓ ↓ ↓ ↓	$T = 0_-$

Fig. 7.12.4. The ordering of a spin system in an external magnetic field. The most likely ordering occurs at the maximum as well as the minimum energy

b) There are maximal values of energy and entropy, the temperature can take on positive and negative values, and the entropy is minimal at $T = 0_+$ and $T = 0_-$, maximal as $T \to \infty$. Negative temperatures correspond to *higher energies* than positive temperatures.

Systems which can have negative temperature can only be realized if the energy remains limited, in particular, if no translational energy can be introduced (in which case the above stability argument for $T \geq 0$ is not applicable). An example is a system of spins in an external magnetic field. The smallest possible energy corresponds to a parallel lineup of all the magnetic moments with respect to the magnetic field, and the highest possible energy to an antiparallel lineup (Fig. 7.12.4).

We see that the state of maximal energy has the same degree of order (the same entropy) as the state of least energy.

From now on, we will only consider systems whose temperature is always positive. For an isolated system, the entropy has its greatest value $S(E)$ at a given energy E when the system is in equilibrium. Equilibrium states therefore lie on the curve $S(E)$ in the diagram (Fig. 7.12.5) and nonequilibrium states are represented by points under the curve. We are now interested in finding the largest possible amount of energy which a system Σ can deliver to another system Z if Σ is not allowed to exchange entropy with the environment (Fig. 7.12.6). Under these conditions, Σ is adiabatically closed and the entropy of Σ cannot decrease. We see immediately from the diagram in Fig. 7.12.5 that the maximum amount of work can be given up in a reversible process (compare Sect. 7.9).

As an example, consider the expansion of an ideal gas. In a free expansion, we clearly have

$$W = 0 , \quad S_1 - S_0 = Nk \ln \frac{V_1}{V_0} > 0 .$$

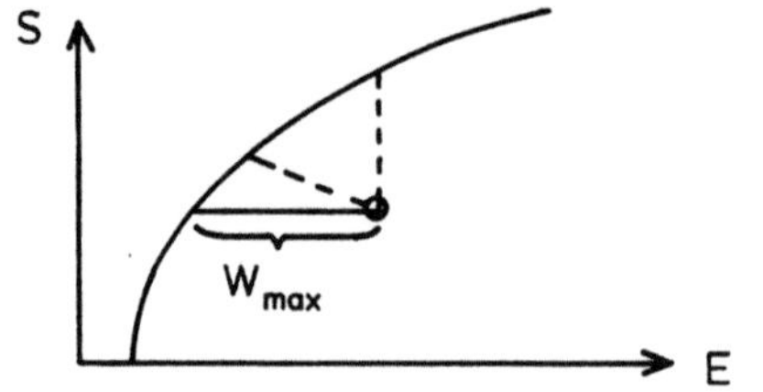

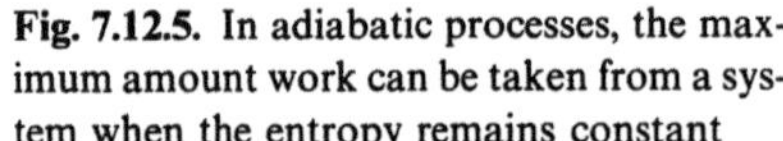

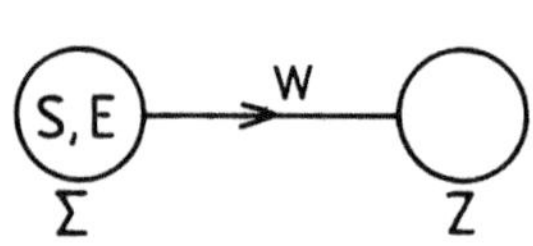

Fig. 7.12.5. In adiabatic processes, the maximum amount work can be taken from a system when the entropy remains constant

Fig. 7.12.6. A system Σ delivers work to a mechanical system Z whose entropy cannot be altered

By contrast, in an isentropic expansion since $dS = 0$,

$$W = \int p\,dV = \int (p\,dV - T\,dS) = -\int dE = E_0 - E_1 \ . \tag{7.12.12}$$

Another important (and clearly equivalent) formulation of the equilibrium condition for a closed system is the following:

A closed system in equilibrium has the least possible energy corresponding to a given value of entropy.

7.12.2 Equilibrium, Stability, and Thermodynamic Potentials

As a special case in which the physical meaning of the thermodynamic potentials F, G, and H can be demonstrated, let us consider the following situation: Let a system Σ be such that it can exchange entropy with a heat reservoir $R_{\tilde{T}}$ at temperature $\tilde{T}$ and volume with a "volume reservoir" $R_{\tilde{p}}$ with pressure $\tilde{p}$ (Fig. 7.12.7).

(A volume reservoir is a system with Gibbs form $d\tilde{E} = -\tilde{p}\,d\tilde{V}$, thus a system which can exchange energy only in the form of volume energy at constant pressure $\tilde{p}$. For example, a volume reservoir can be made using a movable piston which keeps the pressure constant. In many concrete cases, the atmosphere works as a heat and volume reservoir.) Let the assembled system $(R_{\tilde{T}}, R_{\tilde{p}}, \Sigma)$ deliver energy to a system Z only in the form of work. Z and $R_{\tilde{p}}$ thus have fixed entropy.

The balance of energy for any variation with fixed $\tilde{T}$ and $\tilde{p}$ reads:

$$\Delta E + \Delta E_{R_{\tilde{T}}} + \Delta_{R_{\tilde{p}}} + \Delta E_Z = 0 \ ,$$

thus

$$\Delta E + \tilde{T}\Delta\tilde{S} - \tilde{p}\Delta\tilde{V} + A = 0 \ . \tag{7.12.13}$$

The entropy balance is $\Delta S + \Delta\tilde{S} \geq 0$, so that, if we note that $\Delta V = -\Delta\tilde{V}$:

$$\Delta E - \tilde{T}\Delta S + \tilde{p}\Delta V = \Delta(E - \tilde{T}S + \tilde{p}V) \leq -A \ . \tag{7.12.14}$$

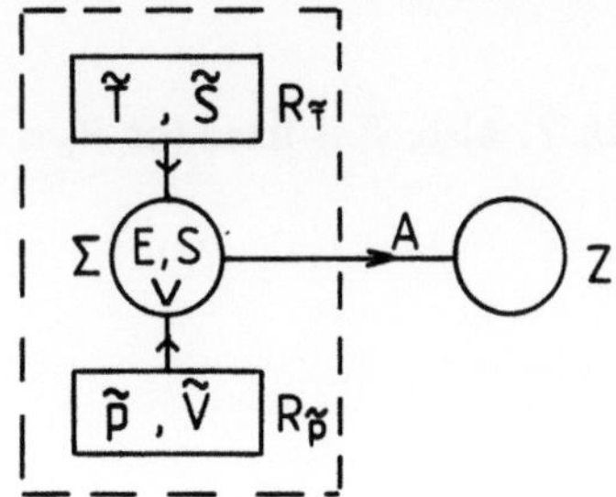

Fig. 7.12.7. A system $(\Sigma, R_{\tilde{T}}, R_{\tilde{p}})$ composed of (a) a work system, (b) a heat reservoir at temperature $\tilde{T}$, (c) a volume reservoir with pressure $\tilde{p}$ exchanges energy in the form of work with a mechanical system Z

The maximum work performed is determined by the change in the quantity $E - \tilde{T}S + \tilde{p}V$ and occurs during a reversible process. If, in particular, the system Z is disconnected, i.e. if $W = 0$, then

$$\Delta(E - \tilde{T}S + \tilde{p}V) \leq 0 \ . \tag{7.12.15}$$

The following special cases are important:

i) $\Delta S = 0 \ , \qquad \Delta V = 0 \ .$

The entropy produced in Σ is conducted to $R_{\tilde{T}}$ and the volume V does not change.

Alternatively, we can think of the volume reservoir as absent or disconnected.

Then,

$$\Delta E \leq - A \ . \tag{7.12.16}$$

If Z is disconnected then $\Delta E \leq 0$, so that in this case E does not spontaneously increase and it is minimal in stable equilibrium.

ii) $\Delta S = 0 \quad$ and $\quad p = \tilde{p} = $ constant .

There is then a uniform pressure in Σ which agrees with p.

Then,

$$\Delta(E + pV) = \Delta H \leq - A$$

so that at constant entropy and constant pressure for Σ, the work done is not greater than the magnitude of the change in enthalpy and thus

$$\Delta H \leq 0 \ ,$$

if Z is disconnected. The enthalpy does not spontaneously increase if the entropy is constant and the pressure of Σ is constant and it is minimal in stable equilibrium.

iii) $T = \tilde{T} = \text{constant}$, $\Delta V = 0$.

In Σ the temperature is uniform and it agrees with $\tilde{T}$. Also, V is fixed (or $R_{\tilde{p}}$ is disconnected).

Then,

$$\Delta(E - TS) = \Delta F \leq - A \ ,$$

so that when Σ is at constant temperature, the work done is no greater than the magnitude of the change of the free energy.

If Z is disconnected, then

$$\Delta F \leq 0 \ .$$

The free energy does not spontaneously increase at constant temperature and is minimal in stable equilibrium.

iv) $T = \tilde{T}$, $p = \tilde{p}$.

Then

$$\Delta(E - TS + pV) = \Delta G \leq - A \ ,$$

so that when Σ has constant temperature and pressure, the work done is not greater than the magnitude of the change in the Gibbs free energy, so that

$$\Delta G \leq 0 \ .$$

The Gibbs free energy therefore does not increase spontaneously at constant T and p and its minimum occurs at stable equilibrium.

As an example of (iii), we consider the free and the isothermal expansion of an ideal gas.

Free expansion:

$$E_1 - E_0 = \Delta E = 0 \ , \quad A = 0 \ ,$$

$$\Delta S = Nk \ln\left(\frac{V_1}{V_0}\right) > 0 \ , \tag{7.12.17}$$

$$\Delta F = - NkT \ln\left(\frac{V_1}{V_0}\right) < 0 = - A \ .$$

Isothermal expansion:

$$\Delta E = 0 \ , \qquad \Delta S = Nk\ln\!\left(\frac{V_1}{V_0}\right) > 0 \ ,$$

$$\Delta F = -NkT\ln\!\left(\frac{V_1}{V_0}\right) \ ,$$

$$(7.12.18)$$

$$A = \int_{V_0}^{V_1} p\,dV = NkT\int_{V_0}^{V_1}\frac{dV}{V} = NkT\ln\!\left(\frac{V_1}{V_0}\right)$$

$$= -\Delta F = \Delta E_{R_T} \ . \tag{7.12.19}$$

We again obtain equilibrium and stability conditions for the cases (i–iii) if we think of Σ as being divided into two subsystems and allow the exchange of extensive variables between the subsystems.

For (i), entropy and volume exchange:

$$dE_\alpha = TdS_\alpha - p_\alpha dV_\alpha \ , \qquad \alpha = 1, 2 \ ,$$

$$dS_1 + dS_2 = 0 \ , \qquad dV_1 + dV_2 = 0 \ .$$

Equilibrium:

$$T_1 = T_2 \ , \qquad p_1 = p_2 \ .$$

Stability:

$$\left(\frac{\partial^2 E}{\partial S^2}\right)_V \geq 0 \ , \qquad \text{i.e.} \qquad \left(\frac{\partial T}{\partial S}\right)_V \geq 0 \ ,$$

thus

$$C_V = T\left(\frac{\partial S}{\partial T}\right)_V \geq 0 \ ,$$

$$\left(\frac{\partial^2 E}{\partial V^2}\right)_S \geq 0 \ , \qquad \text{i.e.} \qquad -\left(\frac{\partial p}{\partial V}\right)_S \geq 0 \ ,$$

thus

$$\kappa_S = -\frac{1}{V}\left(\frac{\partial V}{\partial p}\right)_S \geq 0 \ . \tag{7.12.20}$$

For (ii), entropy exchange:

$$dH_\alpha = T_\alpha dS_\alpha + V_\alpha dp_\alpha = T_\alpha dS_\alpha \ ,$$

$$\alpha = 1,2 \ , \quad dS_1 + dS_2 = 0 \ .$$

Equilibrium:

$$T_1 = T_2 \ .$$

Stability:

$$\left(\frac{\partial^2 H}{\partial S^2}\right)_p \geq 0 \ , \quad \text{i.e.} \quad \left(\frac{\partial T}{\partial S}\right)_p \geq 0 \ ,$$

thus

$$C_p = T\left(\frac{\partial S}{\partial T}\right)_p \geq 0 \ . \tag{7.12.21}$$

For (iii), volume exchange:

$$dF_\alpha = -S_\alpha dT_\alpha - p_\alpha dV_\alpha = -p_\alpha dV_\alpha \ ,$$

$$\alpha = 1,2 \ , \quad dV_1 + dV_2 = 0 \ .$$

Equilibrium:

$$p_1 = p_2 \ .$$

Stability:

$$\left(\frac{\partial^2 F}{\partial V^2}\right)_T = -\left(\frac{\partial p}{\partial V}\right)_T \geq 0 \ ,$$

thus

$$\kappa_T = -\frac{1}{V}\left(\frac{\partial V}{\partial p}\right)_T \geq 0 \ . \tag{7.12.22}$$

Since [compare (7.7.15)]

$$C_p - C_V = -T\left(\frac{\partial V}{\partial T}\right)_p^2 \bigg/ \left(\frac{\partial V}{\partial p}\right)_T \tag{7.12.23}$$

it follows that

$$C_p \geq C_V \ .$$

Problems

7.1   A Gas Made up of Bi-atomic Molecules. Consider a gas made up of N bi-atomic molecules with the following Hamiltonian:

$$H(q, p) = \sum_{i=1}^{N} \left\{ \frac{1}{2m_1}(p_i^{(1)})^2 + \frac{1}{2m_2}(p_i^{(2)})^2 + \frac{D}{2}(q_i^{(1)} - q_i^{(2)})^2 \right\} .$$

a) Through Gaussian integration, find the canonical partition function

$$Z = \frac{1}{N! h^{6N}} \int d^{3N} p^{(1)} d^{3N} p^{(2)} d^{3N} q^{(1)} d^{3N} q^{(2)} e^{-H/kT} .$$

Hint: Transform the position variables into center of mass and relative coordinates. For the new limits of integration, choose the entire volume for the center of mass coordinates and an indefinite integral for the relative coordinates. Justify why – and under which conditions on β and D – the error made in this choice of the limits of integration is negligible.

b) Determine the free energy $F = -kT \ln Z$, the entropy $S = -(\partial F/\partial T)_V$ and the specific heat $C_V = T(\partial S/\partial T)_V$ at constant volume, as well as the thermal equation of state $p = p(T, V)$ and the caloric equation of state $E = E(T, V)$.

7.2   The Specific Heat of a Gas of Bi-atomic Molecules. Consider again an ideal gas made up of N bi-atomic molecules. In the previous exercise, the internal energy was given by $E = \frac{9}{2} NkT$ and the specific heat at constant volume was $C_V = \frac{9}{2} Nk$. Experimentally, however, it is found that for nitrogen, for example, the specific heat increases in steps from $C_V = \frac{3}{2}$ to $\frac{5}{2}$ to $C_V = \frac{7}{2}$ as the temperature is increased. This discrepancy occurs because too imprecise a model of a bi-atomic gas was used.

a) Repeat the calculation from the previous exercise with the new Hamiltonian:

$$H(q, p) = \sum_{i=1}^{N} \left\{ \frac{1}{2m_1}(p_i^{(1)})^2 + \frac{1}{2m_2}(p_i^{(2)})^2 + V(|q_i^{(1)} - q_i^{(2)}|) \right\} ,$$

where

$$V(r) = \frac{D}{2}(r - a)^2 \quad \text{with} \quad a \gg \sqrt{\frac{kT}{D}} \, (*) .$$

(Consider suitable approximations for the q-integration which follow from the condition $(*)$.)

Determine the partition function Z, the internal energy E, and the specific heat C_V. How does this Hamiltonian differ from the one in the previous exercise?

b) In light of these two models, discuss the thermodynamic rule of thumb: "Every thermodynamic degree of freedom increases the internal energy". Make this claim more precise by listing the mechanical degrees of freedom (as different possible types of motion) and their contributions to the kinetic and potential energies. Interpret the empirical behavior of $E/NkT = C_V/Nk$ as it changes with T by identifying the different levels as successive "freezing" and "thawing" of rotational and vibrational degrees of freedom. (A consistent description of the discrete rotational and vibrational energy levels underlying this phenomena is only possible using quantum theory.)

7.3   The Canonical Ensemble. The canonical distribution is given by

$$\varrho(p, q) = \frac{1}{Z} e^{-\beta H(q, p)}$$

$$Z = \frac{1}{Z_0} \int dq\, dp\, e^{-\beta H(q, p)} \quad \left(\beta = \frac{1}{kT} \right) ,$$

where Z_0 is a normalization constant which depends only on the number n of degrees of freedom and which has dimensions (action)^n.

Show that the mean value $\langle H \rangle$ and the standard deviation

$$(\Delta H)^2 = \langle (H - \langle H \rangle)^2 \rangle$$

of the Hamiltonian satisfy the following relationships

$$E = \langle H \rangle = -\frac{\partial}{\partial \beta} \ln Z$$

$$(\Delta H)^2 = +\frac{\partial^2}{\partial \beta^2} \ln Z = kT^2 C_V .$$

Use this result in order to show that for an ideal gas the relative width $\Delta H/\langle H \rangle$ of the energy in the canonical distribution changes as $1/\sqrt{N}$ with the number of particles N.

7.4   Heat Engine. We consider a periodically working heat engine such that the closed curve in state space traced out over one complete cycle has the shape sketched here when projected onto the T–S plane; here, $S_3 = \frac{1}{2}(S_1 + S_2)$.

Sketch the corresponding curve in the p–V plane for an ideal gas, and calculate the efficiency $\eta = A/Q$ of this engine for a reversible process ($Q =$ the heat absorbed at T_1 and T_3). Compare this with the efficiency of a Carnot engine between the temperatures T_1 and T_2.

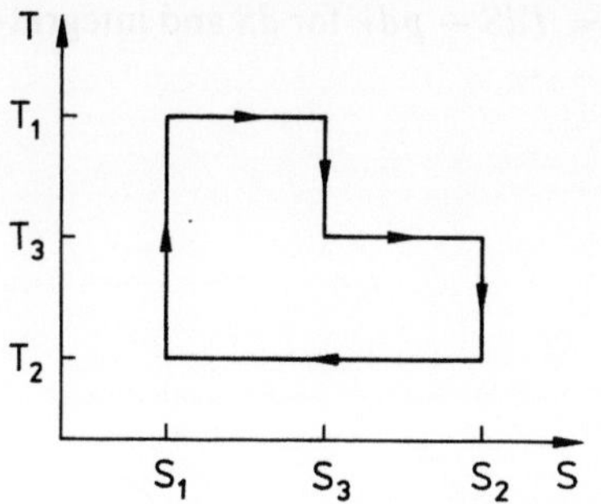

7.5 Stirling and Ericsson Engines. We consider periodically operating heat engines whose closed curves traced out in state space over one complete cycle have the following shapes when projected onto the p–V plane:

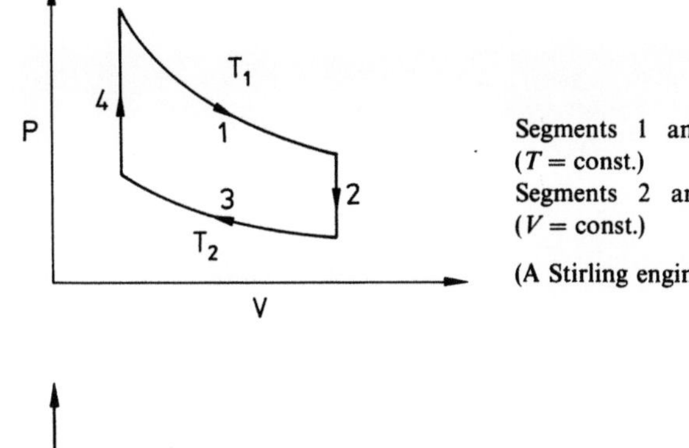

Segments 1 and 3 of the curve are isotherms $(T = \text{const.})$
Segments 2 and 4 of the curve are isochors $(V = \text{const.})$

(A Stirling engine between T_1 and T_2)

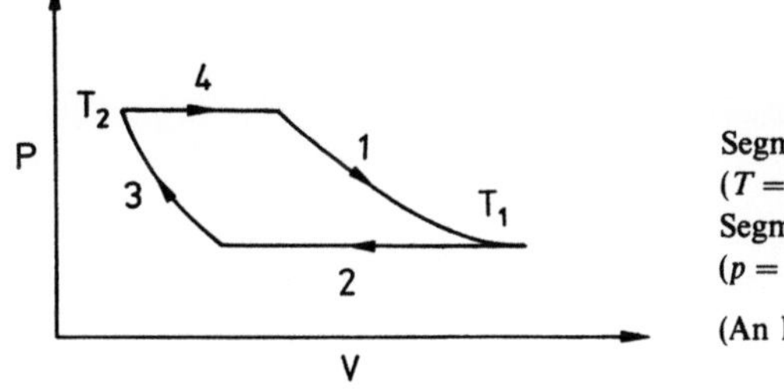

Segments 1 and 3 of the curve are isotherms $(T = \text{const.})$
Segments 2 and 4 of the curve are isobars $(p = \text{const.})$

(An Ericsson engine between T_1 and T_2)

Determine the corresponding curves in the T–S plane for an ideal gas and calculate the efficiency $\eta = A/Q$ for a reversible process ($Q = $ the heat absorbed at T_1). Show that in both cases the efficiency of a Carnot engine between T_1 and T_2 is obtained.

Hint: For one mole of an ideal gas, the thermal equation of state $p = p(T, V)$ and the caloric equation of state $E = E(T, V)$ read

$$pV = RT \quad \text{and} \quad E = C_V T + E_0$$

$(R, C_V, E_0 = \text{const.})$. Solving the equation $dE = TdS - pdV$ for dS and integrating yields

$$S = C_V \ln T + R \ln V + S_0$$

which, since

$$V = \frac{RT}{p} \quad \Rightarrow \quad \ln V = \ln T - \ln p + \text{const.}$$

can also be written in the form

$$S = C_p \ln T - R \ln p + S_0'$$

$$(C_p = C_V + R = \text{constant}) \ .$$

8. Applications of Thermodynamics

In Chap. 7, we attempted to explain the behavior and properties of macroscopic bodies with the assistance of our knowledge of microscopic interactions. We introduced terms such as entropy S, pressure p, temperature T and chemical potential μ. We called these quantities state variables and the set of state variables was structured by the Gibbs form

$$dE = T\,dS - p\,dV + \mu\,dN \ .$$

This differential form reveals not only the different forms in which energy can be exchanged but also allows the state variables to be paired into energy-conjugate quantities

$$(T, S), \quad (-p, V), \quad (\mu, N) \ .$$

Another important concept was the thermodynamic potential. These were functions with the dimensions of energy which depend on one variable from each pair. We saw that from such thermodynamic potentials, we could derive all the relationships between the thermodynamic state variables and we have learned how such thermodynamic potentials could be calculated, at least in principle.

So far the only application we have discussed is that to the ideal gas, because in this case it is easy to actually calculate the thermodynamic potential, and using this approach we derived well-known laws which are known empirically.

However, we often deal with gases which are not ideal, that is, gases in which we cannot neglect the interactions between the molecules or the atoms of the gas. In addition, there are also solids and liquids. For all these materials, we use the following strategy: Based on a knowledge (or a supposition) of the form of the microscopic interaction, we try to calculate a thermodynamic potential from which we can easily derive the properties of the material.

Of course, there are no substances which are always in solid, liquid, or gaseous forms. We know from experience that a single substance, such as water, can exist in all of these *phases*[1]. In these different phases, the substance has different physical properties. The density, compressibility, etc. of the substance in these different phases depend in entirely different ways on the state variables.

---

[1] Phase, originally from the Greek, *phásis*: the rising of a star. Appearance, form of appearance.

More generally, a system in a state of equilibrium is no longer necessarily homogeneous, rather it can consist of several phases, each of which is homogeneous in itself down to molecular dimensions and which are separated from each other by boundary surfaces.

We thus have found an additional task for the formalism developed in Chap. 7: We must be able to demonstrate the existence of different phases with this formalism, as well as describe the behavior during a transformation from one phase to another.

The formalism can accomplish this, but it is evident that given the abundance of phenomena and of possible microscopic interactions, the *"theory of phase transformations"* is a large subject which, in many areas, is still a subject of research.

In this and the following sections, we will not attempt to calculate the thermodynamic potentials for realistic cases, rather we will use the terms we have introduced thus far in order to describe phenomenologically what can happen in a phase transition. In addition, we consider only the solid, liquid, and gaseous phases. It should not be forgotten, though, that there are many other phases. In the case of solid bodies, there are various crystalline structures of the same substance; for magnetic systems, there are for example, the ferromagnetic and paramagnetic phases, there is the superconducting phase, etc.

8.1   Phase Transformations and Phase Diagrams

A phase is homogeneously down to the molecular realm. *Homogeneous mixtures* of two different chemical substances also represent a phase. For example, in a system composed of a variety of substances there is only one gaseous phase, since gases are perfectly miscible. A sugar solution (sugar and water) also represents a phase. If this solution is cooled to a low enough temperature, the sugar begins to precipitate. Thus, two phases are obtained, a solid pure sugar phase, and a sugar solution with lower concentration of sugar.

An inhomogeneous system with several phases is called a *multiple-phase system*. Let us consider first a two-phase system composed of a single material, e.g. the system (water, water vapor). *"Vapor"* means the gaseous phase of a substance when this phase exchanges energy or mass with the liquid or solid phase, or when it can be transformed into another phase without too great a change in volume or temperature. Vapor is a real gas, so that relationships between state variables, i.e. state equations, will differ strongly from the equations of an ideal gas. However, if the vapor is separated from the other phases and progressively heated, the resulting superheated vapor will behave more like an ideal gas.

In the two-phase system we are considering, vapor and another phase are in contact, which means that substance is continuously exchanged between the two phases. Molecules emerge out of their bound states in the solid or liquid phase into vapor, and in the other direction molecules of vapor will be absorbed into the solid or liquid phase. Thermodynamic equilibrium is established in a closed two-phase system if both phases are at the same temperature and pressure and if the number of particles of vapor and of the other phase have a constant statistical average, i.e. when the chemical potentials of both phases are the same. The vapor is then called *saturated*.

Let p and T be the common pressure and temperature, $\mu(p, T)$, the chemical potential of water, and $\mu'(p, T)$, the chemical potential of water vapor. Then, in thermodynamic equilibrium

$$\mu'(p, T) = \mu(p, T) \ .$$

Let us assume these functions are given or have been calculated. In the p–T plane the equation

$$\mu' = \mu$$

(with $\mu' \not\equiv \mu$) defines a curve which contains all the points (p, T) at which water and water vapor can exist in equilibrium (if, in any case, $\mu' \equiv \mu$, this equation is satisfied for all p, T and it is then no longer meaningful to talk of two distinguishable phases).

If there are three phases of a substance (e.g., if there is ice in addition to the water and water vapor), they can be in equilibrium if

$$\mu = \mu' = \mu''$$

where $\mu''(p, T)$ represents the chemical potential of the third phase. Since we have two equations for two variables, they will normally determine a single point in the p–T diagram, the so-called *triple point*.

In Fig. 8.1.1 we show a typical phase diagram, namely that of water. We see that, at $T = 100\,°\text{C}$ and $p = 1$ atm, water and water vapor are in thermodynamic equilibrium, while at lower pressures the equilibrium temperature has lower values, as we know from experience. (On a mountain, due to the lower pressure, water boils at temperatures under $100\,°\text{C}$.) We also see that the solid phase, ice, is in equilibrium with the liquid phase at $0\,°\text{C}$ and the normal pressure 1 atm. If the pressure is raised, the temperature at which the two phases can be at equilibrium then drops. Ice melts under the blade of an ice skate; glaciers can move slowly into valleys.

This behavior, that the melting point decreases with an increase in pressure, is rare. It occurs again in bismuth, but other materials always demonstrate the reversed behavior, that the melting temperature increases with growing pressure. The triple point of water is at $0.0098\,°\text{C}$ and 4.58 Torr $= 6.11$ mbar. Since the triple point of water is uniquely defined and can be easily reproduced, it is used as a reference point for the temperature scale. Below 6.11 mbar there is no longer a liquid phase. Ice transforms directly to vapor when the temperature is raised. This process is called *sublimation*. This process is much better

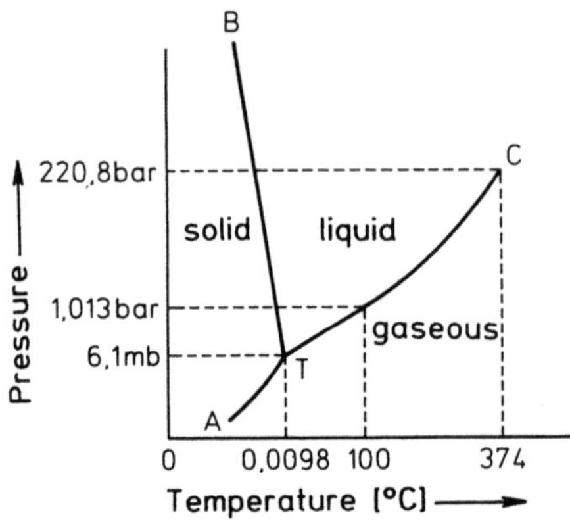

Fig. 8.1.1. Phase diagram of H$_2$O at lower pressures

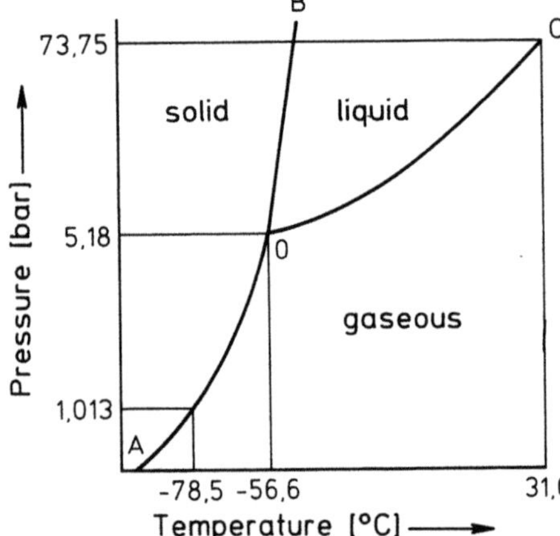

Fig. 8.1.2. Phase diagram of CO$_2$ (not to scale)

known in the case of dry ice (solid CO$_2$). If dry ice is heated at the pressure 1.013 bar = 1 atm (Fig. 8.1.2), then at $-78.5\,°$C it changes directly into a gaseous state.

The last characteristic point in our phase diagram Fig 8.1.1 is the critical point C at $T_{\mathrm{crit}} = 374\,°$C. Above this temperature, the chemical potentials of water and water vapor become identical as functions, that is, there is no longer any difference between the phases, and it no longer makes any sense to distinguish between them. This means the following: whereas for $T < 374\,°$C, $\mu(p, T)$ is defined only above the curve TC and to the right of TB, and $\mu'(p, T)$, the chemical potential of water vapor, is defined only under the curve TC and to the right of TA, for $T > 374\,°$C we have $\mu \equiv \mu'$ and they are both defined for all p.

This fact can also be formulated another way. It is possible to define a chemical potential $\mu(p, T)$ for the substance H$_2$O for all physical values of p and T. This function of two variables is then continuous on the lines TA, TB, and TC, but its derivatives can be discontinuous there.

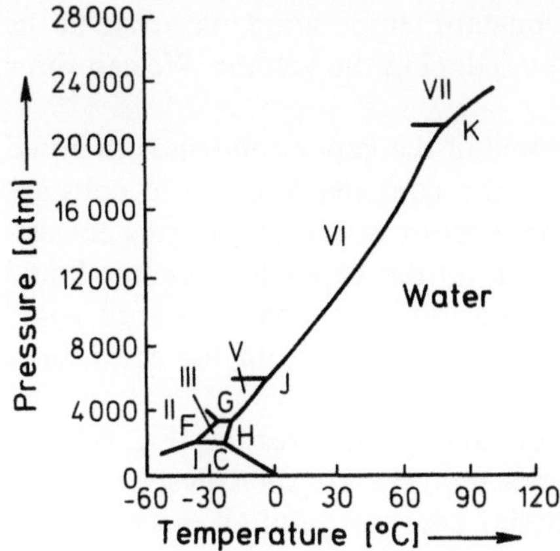

Fig. 8.1.3. Phase diagram of H_2O at higher pressures

The curve TC in Fig. 8.1.1 is called the *vapor pressure curve*. The pressure $p(T)$ on this curve is called the vapor pressure, because it is the pressure that vapor has at temperature T when it is in equilibrium with the liquid phase. Figure 8.1.3 shows a phase diagram for H_2O at higher pressures. We see that there can be several triple points; the different phases here are distinguished by their crystalline structures.

We ascertain from these diagrams that in one-phase regions there are two variables which can be varied independently, while in a two-phase region there is only one free variable. In a three-phase region, no variables are free. In an m-phase region there are thus $(3 - m)$ free variables.

We can generalize this statement for mixtures of n substances. It can be shown that if m phases are in equilibrium, there are

$$f = n + 2 - m$$

free variables. This is the *Gibbs phase rule*. This rule is discussed in courses in thermodynamics or statistical mechanics.

Example. Consider the two materials H_2O and $NaCl$ and their phases: ice, solid $NaCl$, saturated $NaCl$-solution, vapor. If all four phases are in equilibrium, there are no longer any free variables.

8.2   The Latent Heat of Phase Transitions

We consider first a point 1 in the p–T plane which is in the vicinity of the coexistence curve for liquid and gaseous phases and lies in the gaseous phase (Fig. 8.2.1).

If we increase the pressure (1 → 2) at constant temperature, we arrive at the coexistence curve. The pressure is raised by reducing the volume. We can show this in a p–V diagram (Fig. 8.2.2).

If we come to the coexistence curve, some of the vapor condenses to liquid form. Since the pressure remains fixed on the coexistence curve at constant temperature – independent of volume – the pressure is constant during coexistence as the volume is further decreased until all the vapor has become liquid (2 → 3). Only then does the pressure of the fully liquified substance again increase and the point in the p–T diagram move away from the coexistence curve (3 → 4).

On the other hand, we can heat the liquid at constant pressure (Fig. 8.2.3). If the temperature has reached the coexistence temperature then, despite further introduction of heat, the temperature remains constant until all the liquid has been vaporized. Only then will the additional heat energy introduced be used to increase the temperature of the vapor.

The line (3 → 4) in Fig. 8.2.4, like the line (1 → 2), cuts across various isotherms; the introduction of heat produces a rise in temperature. In the coexistence phase before all of the liquid has been vaporized (2 → 3), the heat is used to break up the bonds between the molecules in the liquid and to allow their expansion into the greater volume of the gas. We call this heat the *heat of*

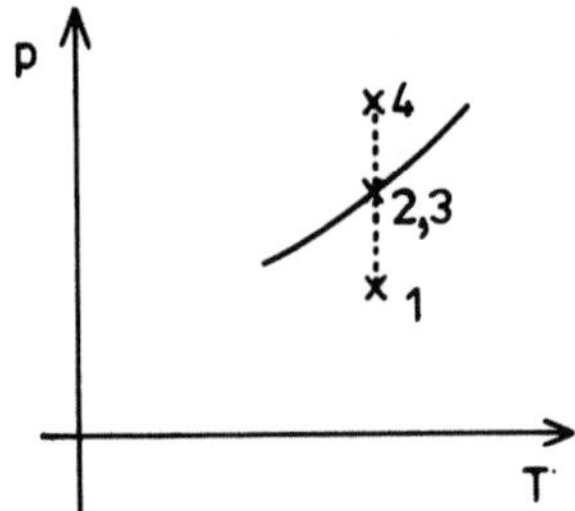

Fig. 8.2.1. Condensation at constant temperature in a p–T diagram

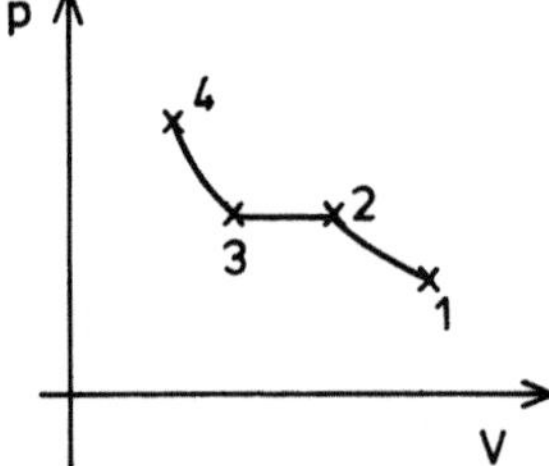

Fig. 8.2.2. Condensation at constant temperature in a p–V diagram

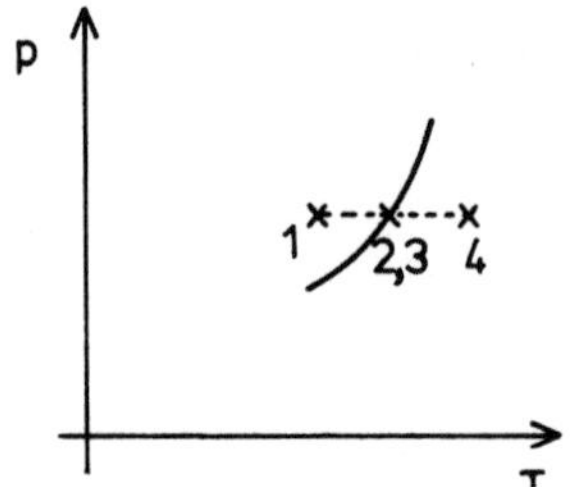

Fig. 8.2.3. Vaporization at constant pressure in a p–T diagram

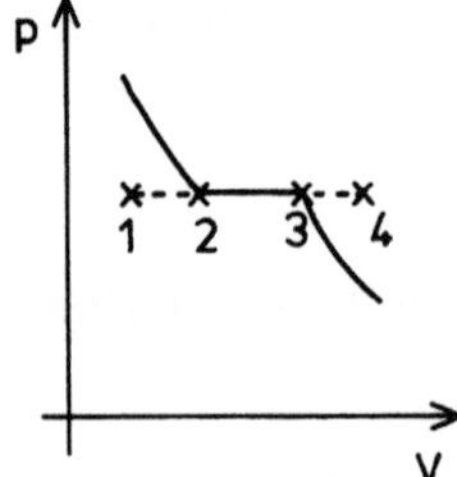

Fig. 8.2.4. Vaporization at constant pressure in a p–V diagram

transformation or the "*latent*" heat. The word "latent" comes from historical usage.

We will now calculate the latent heat.

Let $S_i(T, p, N)$ be the entropy of phase i. If dn particles go from phase 2 to phase 1 then $dN_1 = dn$, $dN_2 = - dn$ and the heat required is given by

$$\delta Q = T dS_1 + T dS_2$$

$$= T\left[\left.\frac{\partial S_1(T, p, N)}{\partial N}\right|_{N_1} - \left.\frac{\partial S_2(T, p, N)}{\partial N}\right|_{N_2}\right] dn \ . \tag{8.2.1}$$

Here, $p = p(T)$, since both phases coexist. Now, we have

$$dG = - S dT + V dp + \mu dN \ , \quad \text{thus} \quad \frac{\partial G(T, p, N)}{\partial T} = - S(T, p, N) \ .$$

In addition,

$$G(T, p, N) = N\mu(T, p) \ , \tag{8.2.2}$$

as we will show more generally in the next section. The intensive quantity $\mu(T, p)$ as a function of the intensive quantities T and p can no longer depend on the extensive quantity N.

Further, we have the Maxwell relation

$$\frac{\partial S(T, p, N)}{\partial N} = - \frac{\partial^2 G(T, p, N)}{\partial N \partial T} = - \frac{\partial \mu(T, p)}{\partial T} \ .$$

Thus for the energy of transformation q, the energy which we must introduce to the medium in order to bring a particle in this case from the liquid phase to the gaseous phase, we obtain:

$$q = \frac{\delta Q}{dn} = - T\left[\frac{\partial \mu_1(T, p)}{\partial T} - \frac{\partial \mu_2(T, p)}{\partial T}\right]\Bigg|_{p = p(T)} \tag{8.2.3}$$

From $\mu_1(T, p(T)) - \mu_2(T, p(T)) = 0$, it follows upon differentiation with respect to T that

$$\frac{\partial \mu_1}{\partial T} + \frac{\partial \mu_1}{\partial p}\frac{dp}{dT} - \frac{\partial \mu_2}{\partial T} - \frac{\partial \mu_2}{\partial p}\frac{dp}{dT} = 0$$

or

$$\left(\frac{\partial \mu_1}{\partial T} - \frac{\partial \mu_2}{\partial T}\right)\Bigg|_{p = p(T)} = - \left(\frac{\partial \mu_1}{\partial p} - \frac{\partial \mu_2}{\partial p}\right)\frac{dp}{dT} \ . \tag{8.2.4}$$

With

$$\frac{\partial \mu_i}{\partial p} = \frac{1}{N_i} \frac{\partial G_i(T, p, N_i)}{\partial p} = \frac{V_i}{N_i} =: v_i(T, p) , \qquad i = 1, 2 , \tag{8.2.5}$$

we then find the *Clausius–Clapeyron*[2] *equation*:

$$q = T[v_1(T, p) - v_2(T, p)] \frac{dp}{dT} . \tag{8.2.6}$$

Special Case. Let $v_2(p, T)$ be the volume per particle in a liquid phase, $v_1(p, T)$ the volume per particle in a gaseous phase. Then normally $v_1 \gg v_2$, and if we use $pv_1 = kT$ for the gas phase, we find

$$q = Tv_1 \frac{dp(T)}{dT} = \frac{kT^2}{p} \frac{dp(T)}{dT} \tag{8.2.7}$$

or

$$\frac{dp}{dT} = q \frac{p}{kT^2} . \tag{8.2.8}$$

If we assume that q does not depend on T, it follows that

$$p(T) = p_0 \, e^{-q/kT} \quad \text{or} \quad \ln p(T) = -\left(\frac{q}{kT}\right) + \ln p_0 , \tag{8.2.9}$$

that is, if $\ln p$ is plotted against $1/T$, we obtain a straight line. The validity of this approximation is shown in Fig. 8.2.5 for the region

$$10^{-2} \text{ Torr} \quad \text{to} \quad 10^6 \text{ Torr} \approx p_{\text{crit}} .$$

In this diagram, the vapor pressure of water and ice is shown in logarithmic scale.

Remarks. i) In (8.2.3), we have calculated the latent heat per particle. In applications, the latent heat per mole is frequently used. This is clearly

$$\bar{q} = Lq , \tag{8.2.10}$$

where $L = 6.02 \times 10^{23}$ is the number of particles per mole, Avogadro's number.
Let n be the number of moles, given N particles. We then have

$$n = \frac{N}{L} , \quad \text{and} \quad \bar{G} = \frac{G}{n} = L\frac{G}{N} = L\mu(T, p) \tag{8.2.11}$$

---

[2] *Clapeyron, Benoit Pierre Emile* (*1799 Paris, d. 1864 Paris). Engineer and mathematician. In 1835, took part in building the first French railroad. His equation for latent heat was generalized in 1850 by R. Clausius.

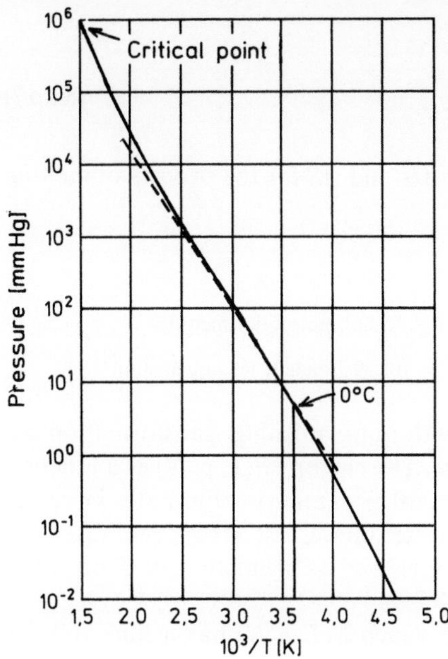

Fig. 8.2.5. The vapor pressure curve of H_2O with respect to $1/T$, using a logarithmic scale for the pressure. The dotted line is straight

is the Gibbs free energy per mole. For the enthalpy $\bar{H}$ per mole and entropy $\bar{S}$ per mole we find

$$\bar{S} = \frac{S}{n} = -\frac{1}{n}\frac{\partial G}{\partial T} = -\frac{\partial \bar{G}}{\partial T} = -L\frac{\partial \mu}{\partial T}\,, \tag{8.2.12}$$

$$\bar{H} = T\bar{S} + \bar{G}\,, \quad \text{so that} \tag{8.2.13}$$

$$\bar{q} = Lq = T[\bar{S}_1(p,\,T) - \bar{S}_2(p,\,T)] = T\,\Delta\bar{S} \tag{8.2.14}$$

or also, since on the coexistence curve

$$\bar{G}_1 = L\mu_1(p,\,T) = \bar{G}_2 = L\mu_2(p,\,T)$$

we have

$$\bar{q} = \Delta\bar{H}\,. \tag{8.2.15}$$

The latent heat per mole thus agrees with the change in enthalpy per mole or the product of the transition temperature with the change in entropy per mole. This is called the latent transformation enthalpy or transformation entropy.

We can obtain another frequently used form of the equation if, with

$$L[v_1(T,\,p) - v_2(T,\,p)] = \Delta\bar{V} \tag{8.2.16}$$

we write

$$\frac{dp}{dT} = \frac{\Delta \bar{H}}{T \Delta \bar{V}} = \frac{\Delta \bar{S}}{\Delta \bar{V}} \ . \tag{8.2.17}$$

There are tables with measured values of $\Delta \bar{H}$ and $\Delta \bar{S}$ for the process of melting or liquefaction at a fixed pressure.

For H_2O at 1 bar (normal pressure):

$T_M = 273.15 \text{ K}$, $\Delta \bar{H} = 6.03 \text{ kJ/mole}$, $\Delta \bar{S} = 22.02 \text{ J/mole K}$ for melting

$T_L = 373.15 \text{ K}$, $\Delta \bar{H} = 40.67 \text{ kJ/mole}$, $\Delta \bar{S} = 108.89 \text{ J/mole K}$ for liquefaction

ii) The phase transitions associated with non-vanishing transformation enthalpy are called *transitions of the first kind*. The entropy $S(T, p, N)$ as a function of T jumps at $T = T(p)$; for example, the number of microscopic states increases discontinuously from the ice phase to the water phase, just as between water and vapor. The same is true for $\bar{V}$. $\bar{G}(T, p, N)$, viewed as a function of T or p, has a kink at $T(p)$ or $p(T)$ respectively (Fig. 8.2.6).

There are other phase transitions for which $S(T, p, N)$ has a kink only as a function of T, but then, for example,

$$c_p = T \frac{\partial \bar{S}(T, p, N)}{\partial T}$$

jumps. Such transformations are called *phase transitions of the second kind*.

The following division of phase transitions comes from *Ehrenfest*[3]:

He calls a phase transition a transition of n-th order if the n-th derivative of $\mu(T, p)$ jumps, i.e. has a point of discontinuity, but the derivatives of lower order are continuous at that point. We can derive equations analogous to the Clausius–Clapeyron equation for these points. For a phase transition of second order, we have

$$c_p = Tv \frac{dp}{dT} \Delta \alpha \quad \text{(Ehrenfest's first equation)} \tag{8.2.18}$$

and

$$\Delta \alpha = \frac{dp}{dT} \Delta \kappa \quad \text{(Ehrenfest's second equation)} \ . \tag{8.2.19}$$

Not all phase transitions fit Ehrenfest's schema, though. The transition from Helium I to Helium II is very likely not a phase transition of first order, since the

---

[3] *Ehrenfest, Paul* (*1880 Vienna, d. 1933 Amsterdam). Fundamental work on statistical mechanics and quantum theory.

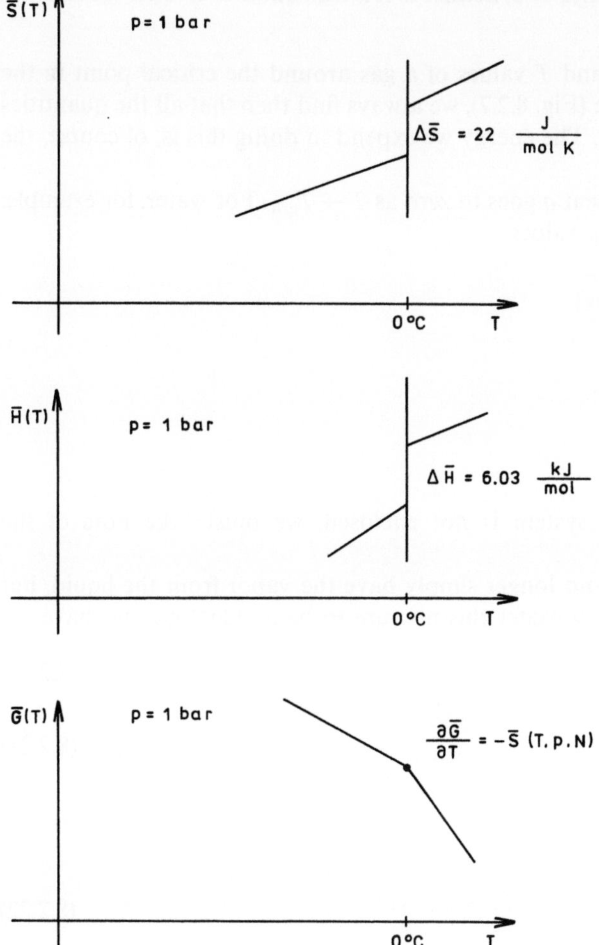

Fig. 8.2.6. A phase transition of first order is characterized by a kink in $\bar{G}(T)$ and a jump in the first derivative of $\bar{G}(p, T)$

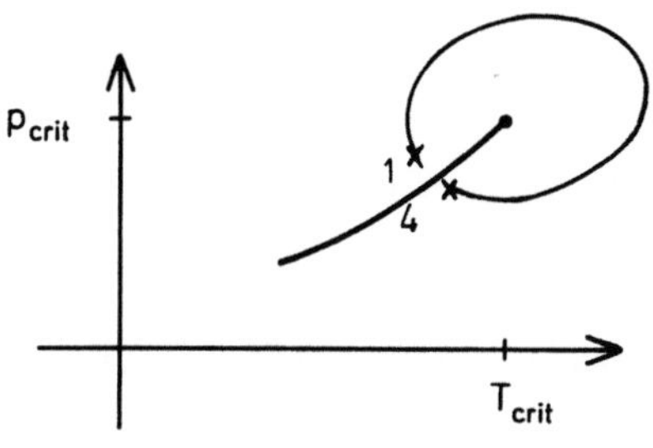

Fig. 8.2.7. In the transition from 1 to 4, all of the state variables change continuously

transformation enthalpy $\Delta\bar{H} = 0$, neither is it a transition of second order, since $c_p \to \infty$ as $T \to T_{\text{crit}}$.

iii) If we vary the p and T values of a gas around the critical point in the region of the liquid phase (Fig. 8.2.7), we always find then that all the quantities S, H, etc. are continuous. The energy we expend in doing this is, of course, the latent heat.

The transformation heat q goes to zero as $T \to T_{\text{crit}}$. For water, for example, we measure the following values:

T [K]	q [cal/g]
273	538
484.4	452.8
582	311.8
637.2	147.0
647	0

iv) If the two-phase system is not enclosed, we must take note of the following.

Above the liquid, we no longer simply have the vapor from the liquid, but a mixture of gases. If we consider this mixture to be an ideal gas, we have

$$pV = NkT \quad \text{or} \tag{8.2.20}$$

$$p = \frac{N}{V}kT \, , \tag{8.2.21}$$

where the particle number N is the sum of the particle numbers N_i of the individual types of gas:

$$N = \sum_i N_i \, . \tag{8.2.22}$$

The gas of type i thus has the partial pressure

$$p_i = \frac{N_i}{V}kT \, , \tag{8.2.23}$$

and the total pressure is the sum of the partial pressures:

$$p = \sum_i p_i \, . \tag{8.2.24}$$

Each component of the system of gases behaves as if it alone could fill the available volume. This is *Dalton's*[4] *Law*.

---

[4] *Dalton, John* (*1799 Eaglesfield/Cumberland, d. 1844 Manchester). One of the founders of atomic theory in chemistry.

In an open container, then, the partial pressure of the water vapor above the water at equilibrium would be $p(T)$, where $p(T)$ can be found from the vapor pressure diagram. At normal pressure, the other gases also contribute their partial pressures. But since the container is open, the volume is infinite and the water vapor molecules can diffuse out more or less rapidly so that the partial pressure of the water vapor never achieves its equilibrium value, the saturation point. The water thus continues to lose molecules in the form of vapor; it evaporates, and the more quickly the evaporated molecules are transported away, say by wind, the faster it evaporates.

To keep the temperature constant, the heat of vaporization would have to be introduced into the evaporating liquid. If this does not happen, the liquid then takes this energy from its own supply, the temperature sinks, and we speak of evaporation cooling. We feel this, for example, when our skin gets wet. There are countless examples of the use of evaporation cooling (see, for example, [Barrow]).

v) Air in the atmosphere possesses a certain concentration of water vapor, which is created by evaporation from the large water surfaces. The air pressure thus contains a contribution from the partial pressure of the water vapor. This is usually smaller than the vapor pressure $p(T)$, at which water and water vapor are at equilibrium. Since $p(T)$ sinks with T, it can happen that if the temperature suddenly drops, the partial pressure of the water vapor in the air becomes equal to the vapor pressure, and then the air becomes saturated with water vapor. The vapor condenses into water, and dew is formed. When a pair of glasses is cold, it "fogs up" because in the environment of the glasses the partial pressure of the water vapor is above the vapor pressure $p(T)$.

The partial pressure of the water vapor is also a measure of the number of water molecules in the air or the humidity. The humidity can be given as the partial pressure of the water vapor, the absolute humidity, or as the relative humidity. The relative humidity is the proportion of the given humidity to the maximum possible humidity at the given temperature.

8.3 Solutions

Dilute solutions are physical mixtures of different particles in the liquid phase. The solution contains the dissolved material in molecular distribution, so that we can speak of a homogeneous mixture of matter, i.e., the solvent is present in such great surplus that the finely distributed particles of the dissolved substance are no longer bound to each other and we can thus ignore their interactions. At given temperatures, a solvent can only absorb a certain maximal value of the substance to be dissolved. In this state, we speak of a *saturated solution.* Oversaturated solutions are unstable. The dissolved substance precipitates and the system is no longer homogeneous. The solubility of a substance is generally temperature dependent, and it usually increases as the temperature increases.

Let r types of particle be present in a solution, let N_i be the number of particles of each type, then

$$N = \sum_{i=1}^{r} N_i \tag{8.3.1}$$

is the total number of particles. Let us designate by $c_i = N_i/N$ the concentration of substance i, then, of course,

$$\sum_{i=1}^{r} c_i = 1 \; , \tag{8.3.2}$$

and thus there are exactly $r - 1$ independent concentrations.

In the following, c_1 will signify the concentration of the solvent.

Claim 1

The chemical potentials μ_i are functions of T, p, and of the $(r - 1)$ independent concentrations $c_1, \ldots, c_{r-1}$, or $c_2, \ldots, c_r$, that is, we have

$$\mu_i = \mu_i(T, p, c_1, \ldots, c_{r-1}) \quad or \tag{8.3.3}$$

$$\mu_i = \mu_i(T, p, c_2, \ldots, c_r) \; . \tag{8.3.4}$$

Proof. The Gibbs free energy $G(T, p, N_1, \ldots, N_r)$ is an extensive quantity, that is, if the number of particles N_i is increased by the factor v then the Gibbs free energy also increases by the factor v. We then have

$$G(T, p, vN_1, \ldots, vN_r) = vG(T, p, N_1, \ldots, N_r) \; . \tag{8.3.5}$$

Differentiating this equation with respect to v and then setting $v = 1$, we find

$$\sum_{i=1}^{r} N_i \frac{\partial G(T, p, vN_1, \ldots, vN_r)}{\partial vN_i}\bigg|_{v=1} = \sum_{i=1}^{r} N_i \mu_i(T, p, N_1, \ldots, N_r)$$

$$= G(T, p, N_1, \ldots, N_r) \; . \tag{8.3.6}$$

The relation

$$G = \sum_{i=1}^{r} \mu_i N_i \tag{8.3.7}$$

is called the *Duhem[5]–Gibbs relation*. We have already used this relation in Sect. 8.2 in the case of a single type of particle.

---

[5] *Duhem, Pierre-Maurice-Marie* (*1861 Paris, d. 1916 Cabrespine (Aude)). French physicist and philosopher, professor at Bourdeaux.

From this it follows that $\mu_i(T, p, N_1, \ldots, N_r)$ are homogeneous functions of degree 0, i.e., that

$$\mu_i(T, p, \nu N_1, \ldots, \nu N_r) = \mu_i(T, p, N_1, \ldots, N_r) , \quad i = 1, \ldots, r . \tag{8.3.8}$$

Thus, since the c_i are also homogeneous functions of degree 0, it follows, for example:

$$\mu_i = \mu_i(T, p, c_2, \ldots, c_r) , \quad i = 1, \ldots, r . \tag{8.3.9}$$

The equation

$$G = \sum_{i=1}^{r} \mu_i N_i$$

might at first be irritating, since we also have $\partial G/\partial N_i = \mu_i$ and it seems then like the μ_i can no longer be explicitly dependent on the N_j.

But we find explicitly

$$\frac{\partial G}{\partial N_j} = \mu_j + \sum_{i=1}^{r} N_i \frac{\partial \mu_i}{\partial N_j} = \mu_j + \sum_{i=1}^{r} N_i \frac{\partial \mu_j}{\partial N_i} , \tag{8.3.10}$$

where we have used the Maxwell relation

$$\frac{\partial \mu_i}{\partial N_j} = \frac{\partial \mu_j}{\partial N_i} . \tag{8.3.11}$$

Now since the μ_j are homogeneous of degree 0,

$$\sum_{i=1}^{r} N_i \frac{\partial \mu_j}{\partial N_i} = 0 . \tag{8.3.12}$$

Thus the second term vanishes in (8.3.10).

Claim 2

For small enough c_i ($i = 2, \ldots, r$) the chemical potential of the solvent satisfies

$$\mu_1(T, p, c_2, \ldots, c_r) = \mu_0(T, p) - kT \sum_{i=2}^{r} c_i , \tag{8.3.13}$$

and the chemical potentials of the dissolved materials satisfy

$$\mu_i(T, p, c_2, \ldots, c_r) = \hat{\mu}_i(T, p) + kT \ln c_i , \quad i = 2, \ldots, r . \tag{8.3.14}$$

$\hat{\mu}_i$ here is an unspecified function of T and p, $\mu_0(T, p)$ is the chemical potential of the pure solvent if the concentrations $c_2, \ldots, c_r$ all go to 0.

The dissolved substances thus lower the chemical potential of the solvent, and the chemical potential of a dissolved substance is logarithmically dependent on its concentration.

This claim is proved in statistical mechanics. We will not give a proof here.

Remarks. i) The form here of the chemical potential for a single dissolved substance seems justified, since we can treat it as an ideal gas. We therefore consider r different ideal gases which do not interact with each other.

If the gas of type i has partial pressure p_i, then from Sect. 7.3 the chemical potential is given by

$$
\begin{aligned}
\mu_i(T, p_i) &= kT \ln p_i + f(T) \\
&= kT \ln [p_i/p] + g(p, T) \\
&= kT \ln [n_i/n] + g(p, T) \\
&= kT \ln c_i + g(p, T) , \quad i = 2, \ldots, r .
\end{aligned}
\tag{8.3.15}
$$

For a mixture of ideal gases, then, the chemical potentials of the individual types of particles also depend logarithmically on their concentrations. In this way, we can imagine the dissolved substances as a mixture of ideal gases. The solvent then takes over the role of the vacuum.

ii) In concentrated solutions, we write

$$
\mu_i = \hat{\mu}_i(T, p) + kT \ln (f_i c_i) , \quad i = 2, \ldots, r ,
\tag{8.3.16}
$$

and we call $f_i(c_i)$ the activity coefficient, and the product $f_i c_i$ is called the activity of the dissolved substance. As $c_i \to 0$, we have, naturally

$$
f_i(c_i) \to 1 , \quad i = 2, \ldots, r .
\tag{8.3.17}
$$

We can also give the chemical potential of the solvent in the form

$$
\mu_1 = \mu_0(p, T) + kT \ln (f_1 c_1)
\tag{8.3.18}
$$

with $f_1(c_1) \to 1$ as $c_1 \to 1$. Since for $c_1 \approx 1$, $c_i \ll 1$ we then have again, to a good approximation,

$$
\mu_1 = \mu_0(T, p) + kT \ln \left(1 - \sum_{i=2}^{r} c_i \right) = \mu_0(T, p) - kT \sum_{i=2}^{r} c_i .
\tag{8.3.19}
$$

Of course, the Maxwell relations must also be satisfied under such generalized assumptions about the chemical potentials. We then obtain relationships between the activities.

8.4 Henry's Law, Osmosis

Since we are now familiar with the chemical potentials of dissolved substances and solvents, we can study a few important applications.

8.4.1 Henry's Law

We consider a two-phase system, say a gas and a liquid or two liquids, like water and benzene, which do not mix with each other (the benzene always lies on top of the water). In addition, let a substance A be dissolved in both phases, and let the concentrations of A be c_A^I and c_A^{II}. Then, it is clear from Sect. 8.3 that the following equations hold for the chemical potential of A in the two phases:

$$\mu_A^I = \hat{\mu}_A^I + kT \ln c_A^I \quad \text{and} \tag{8.4.1}$$

$$\mu_A^{II} = \hat{\mu}_A^{II} + kT \ln c_A^{II} \, . \tag{8.4.2}$$

In equilibrium the substance A divides between the phases in such a way that

$$\mu_A^I = \mu_A^{II} \quad \text{or} \tag{8.4.3}$$

$$\hat{\mu}_A^I - \hat{\mu}_A^{II} = kT \ln \left(\frac{c_A^{II}}{c_A^I} \right) \quad \text{or} \tag{8.4.4}$$

$$\frac{c_A^{II}}{c_A^I} = \exp \left[(\hat{\mu}_A^I - \hat{\mu}_A^{II})/kT \right] = \gamma(T, p) \, . \tag{8.4.5}$$

The concentrations of substance A in the two phases are thus proportional to each other. γ is called the distribution coefficient.

If phase II is a mixture of gases, in which gas A has partial pressure p_A, then using the ideal gas law for gas A,

$$p_A = n_A \frac{RT}{V} \quad \text{or} \tag{8.4.6}$$

$$c_A^{II} = \frac{n_A}{n} = \frac{p_A V}{nRT} \, , \tag{8.4.7}$$

so that the concentration of gas A in phase I is given by

$$c_A^I = \frac{c_A^{II}}{\gamma} = \frac{p_A V}{nRT\gamma} = K(T, p) p_A \, . \tag{8.4.8}$$

This means the following: in a two-phase system at fixed temperature and pressure, the concentration of gas A in phase I (e.g., in the liquid) is proportional to the partial pressure of A in phase II, the gas.

This is *Henry's*[6] *Law*. This relationship between the partial pressure of a gas over a liquid and the concentration of the gas in the liquid is seen frequently in everyday life:

---

[6] *Henry, William* (*1774 Manchester, d. 1836 Manchester). Not to be confused with Joseph Henry after whom the unit of inductance is named. W. Henry published his law in 1803.

– When a bottle of soda water is opened, the liquid bubbles. The partial pressure of the carbon dioxide in the top cf the bottle is greater before the bottle is opened than after. Then, after the decrease in partial pressure, carbon dioxide escapes from the liquid until an equilibrium has been established. If the bottle is not closed again, or if the soda is poured into a glass, the equilibrium is never reached. The volume for the gas to fill is then infinite and the partial pressure of the carbon dioxide above the water never achieves its equilibrium value. Thus carbon dioxide will continuously escape from the water.

– The coefficient $K(p, T)$ decreases with increasing temperature. By heating a liquid it is possible to force out most of the gas dissolved in it.

– A diver who goes 20 meters underwater with a compressed air device will breathe air of about three times atmospheric pressure (since 760 mm-Hg = 1 atm and since the density of water is about 13 times smaller as that of mercury, a 10 meter high column of water corresponds approximately to a pressure of 1 atmosphere – and thus the pressure increases by about 1 atmosphere for every additional 10 meters underwater). In the diver's blood, then, there is three times the normal amount of nitrogen dissolved. If the diver comes up too quickly, this dissolved nitrogen escapes quickly, producing gas bubbles in the bloodstream which can cause tissue damage and necrosis and ultimately paralysis, collapse, and, in severe cases, even death (this is caisson disease or the bends). The same dangers are faced by pilots who climb too quickly to great heights without pressure equalization devices.

8.4.2 Osmosis

We consider two solutions of the same substance, but with different concentrations c^1 and c^2. Let them be separated by a boundary wall which can let through the molecules of the solvent but not those of the dissolved substance.

Such semipermeable membranes actually exist. Cellophane is permeable to water, but not to sugar, rubber is permeable to carbon dioxide but not hydrogen. Many biological barriers, e.g. cell walls, have precisely this property.

Let us now consider such a rigid membrane, permeable only to the solvent (Fig. 8.4.1).

Since particle exchange is possible for the solvent, but the pressures cannot become equal by moving the wall, we have as a condition of equilibrium:

$$\mu_0(T, p_1) - c^1 kT = \mu_0(T, p_2) - c^2 kT \ . \tag{8.4.9}$$

We see that different concentrations cause different pressures.

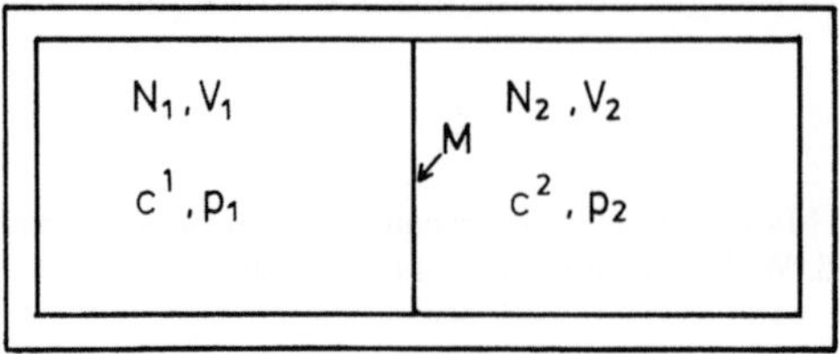

Fig. 8.4.1. A membrane M, permeable only to the solvent, separates the two solutions with concentrations c^1, c^2

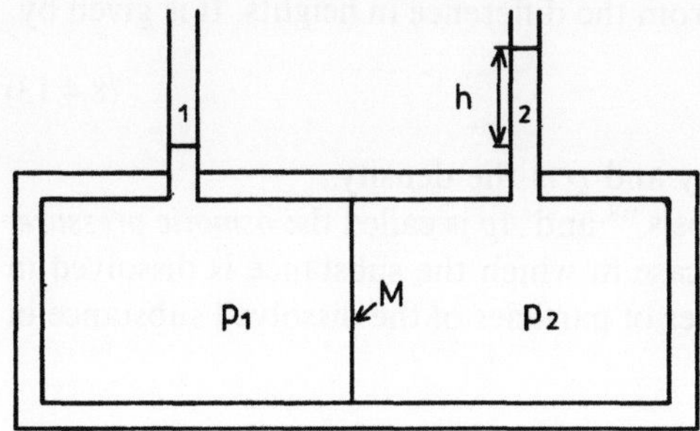

Fig. 8.4.2. We can read off the osmotic pressure $\Delta p = p_2 - p_1$ from the height difference in the overflow pipes

Since for small values of c^1, $\Delta p = p_2 - p_1$ will also be small, we write

$$\mu_0(T, p_2) = \mu_0(T, p_1) + \Delta p \frac{\partial \mu_0(T, p)}{\partial p}\bigg|_{p=p_1} , \qquad (8.4.10)$$

and we have

$$\frac{\partial \mu_0(T, p_1)}{\partial p_1} = \frac{1}{N_1} \frac{\partial G_1(T, p_1, N_1)}{\partial p_1} = \frac{V_1}{N_1} = v_1(T, p_1) ,$$

since $G_1(T, p, N) = \mu_0 N_1$ in system 1 when $c^1 = 0$. Here, N_1 is the number of particles of solvent in system 1, V_1 is the volume for $c^1 = 0$, and thus in good approximation also for $c^1 \neq 0$. $v_1(T, p_1) = v$ is the volume per particle which is equal in both systems. Then, we have

$$(p_2 - p_1)v = \Delta p v = (c^2 - c^1)kT . \qquad (8.4.11)$$

This is known as *van't Hoff's*[7] *law*. The side with the higher concentration, has the higher pressure.

Let us consider the arrangement in Fig. 8.4.2. In the beginning, let there be pure water in both containers, so that $p_1 = p_2$.

We then introduce a small amount of sugar into the container on the right hand side. In doing this, we lower the chemical potential of system 2, and water flows from 1 to 2 until

$$\Delta p = p_2 - p_1 = \frac{c^2 kT}{v} . \qquad (8.4.12)$$

In column 2, the solution rises. In column 1, the water level falls, and the

---

[7] *van't Hoff, Jacobus Henricus* (*1852 Rotterdam, d. 1911 Berlin). Physicist and chemist. Founder of stereochemistry. 1885 "Theory of Solutions".

pressure differences can be read off from the difference in heights. It is given by

$$\Delta p = \varrho g h \ , \tag{8.4.13}$$

where g is the acceleration of gravity and ϱ is the density.

This phenomenon is called "osmosis,"[8] and Δp is called the *osmotic pressure*. Let us consider more generally the case in which the substance is dissolved in both containers. Let $\bar{N}_i$ be the number of particles of the dissolved substance in system i, then

$$c^i = \frac{\bar{N}_i}{(N_i + \bar{N}_i)} \approx \frac{\bar{N}_i}{N_i} \ , \tag{8.4.14}$$

and therefore also

$$\frac{c^i}{v} = \frac{c^i N_i}{V_i} \approx \frac{\bar{N}_i}{V_i} \ , \tag{8.4.15}$$

so that we can write van't Hoff's law in the form

$$p_2 - p_1 = \left(\frac{\bar{N}_2}{V_2} - \frac{\bar{N}_1}{V_1}\right) kT \ . \tag{8.4.16}$$

On both sides of the membrane, we can consider the dissolved substance to be an ideal gas. According to the quantity present, it exerts a partial pressure on the dividing wall and the difference of these partial pressures is the osmotic pressure. The membrane must withstand this pressure.

If the solution contains several components for which the membrane is impermeable, and if $c_2, \ldots, c_r$ are the concentrations of these components and in system 1 there is only the pure solvent,

$$p_2 - p_1 = \frac{kT}{v} \sum_{i=2}^{r} c_i \ . \tag{8.4.17}$$

The quantity $\sum_{i=2}^{r} c_i$ is called the "*osmolarity*" of the solution.

Examples from Biophysics. i) If a biological cell is put into a watery NaCl solution with concentration 0.15 mole/l, then the cell (whose membrane is nearly impermeable for NaCl) retains its shape. This means that the osmotic pressure is exactly zero. If the concentration of the solution is increased, water escapes from the cell and it shrivels. If the concentration is decreased, water flows into the cell and it swells.

ii) Blood is in osmotic equilibrium with a salt solution if the salt concentration is 9 g/l (i.e. 9 grams of salt per liter of water). The injection of such a salt

---

[8] Osmosis (Greek) from *ozein*: to push.

solution into the bloodstream does not cause any additional osmotic pressure on the blood vessel walls so that no water (solvent) enters into or leaves the tissue from the veins.

8.5 Phase Transitions in Solutions

Consider a binary solution, that is a solution with a solvent A and only a single kind of dissolved substance, which we will call B. Lowering the temperature of the solution can give rise to several different phenomena.

1) It can lead to a *separation* of the two materials. The two substances, the solvent and the dissolved material, are then said to be only partially miscible beneath a critical solution temperature which also depends on the concentration of the dissolved substance. This case will not be discussed here.

2) The two substances might always be miscible in phase I (for example, the liquid phase) but never in phase II (the solid or gas phase). This could mean, for example, that dropping the temperature "freezes out" the solvent into the solid phase. As long as this has not been carried out to its conclusion, we still have a solid A and a concentrated solution in coexistence.

3) The substances can be fully miscible in both phases.

Here, we will consider the second and third cases.

8.5.1 Case (2): Miscibility in Only One Phase

If the solvent were pure, the equation

$$\mu_{A,0}^{I}(T, p) = \mu_{A,0}^{II}(T, p) \tag{8.5.1}$$

would give us the pressure for which both phases could coexist in terms of the temperature. For the solution, let

$$\mu_{A}^{I}(T, p, c_{B}^{I}) = \mu_{A,0}^{I}(T, p) - c_{B}^{I} kT \tag{8.5.2}$$

be the chemical potential of the solvent, and let $\mu_{A,0}^{II}(T, p)$ be the chemical potential for the pure phase II. The requirement that the solution coexist with the phase II material then gives us the equation

$$\mu_{A}^{I}(T, p, c_{B}^{I}) = \mu_{A,0}^{I}(T, p) - c_{B}^{I} kT = \mu_{A,0}^{II}(T, p) \ , \tag{8.5.3}$$

which yields a vapor pressure curve $p'(T)$ which approximates the vapor pressure curve $p(T)$ of the pure solvent, since c_{B}^{I} is small (Fig. 8.5.1).

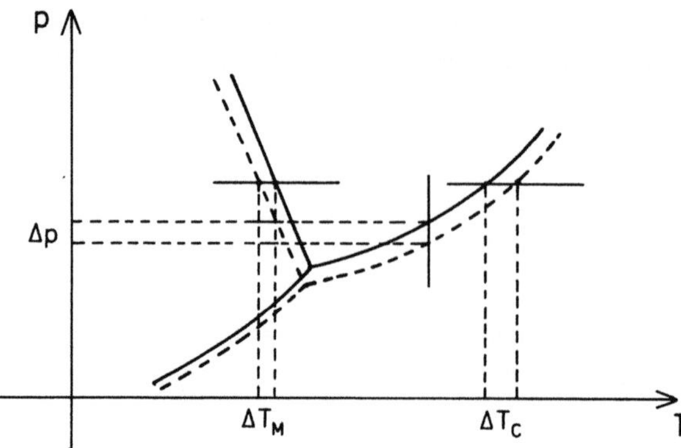

Fig. 8.5.1. The displacement of the phase boundary curve of the solution (–––) with respect to the pure solvent (——) and thus associated depression of vapor pressure (Δp), depression of the freezing point ($\Delta T_{\rm M}$), and elevation of the boiling point

Let (p_0, T_0) be a point on the curve $p(T)$, (p', T') a neighboring point on $p'(T)$, so that

$$p' = p_0 + \Delta p , \qquad T' = T_0 + \Delta T . \tag{8.5.4}$$

Then, if the solution coexists with pure A in phase II:

$$\mu_{\rm A}^{\rm I}(T', p', c_{\rm B}^{\rm I}) = \mu_{\rm A,0}^{\rm II}(T', p') , \tag{8.5.5}$$

i.e., if we ignore the term $c_{\rm B}^{\rm I}\,\Delta T$:

$$\mu_{\rm A,0}^{\rm I}(T_0, p_0) + \Delta p\,\frac{\partial \mu_{\rm A,0}^{\rm I}(T_0, p_0)}{\partial p_0} + \Delta T\,\frac{\partial \mu_{\rm A,0}^{\rm I}(T_0, p_0)}{\partial T_0} - c_{\rm B}^{\rm I}\,kT_0$$

$$= \mu_{\rm A,0}^{\rm II}(T_0, p_0) + \Delta p\,\frac{\partial \mu_{\rm A,0}^{\rm II}(T_0, p_0)}{\partial p_0} + \Delta T\,\frac{\partial \mu_{\rm A,0}^{\rm II}(T_0, p_0)}{\partial T_0} . \tag{8.5.6}$$

With

$$\frac{\partial \mu_{\rm A}}{\partial p} = v_{\rm A} , \qquad \frac{\partial \mu_{\rm A}}{\partial T} = -\frac{S_{\rm A}}{N} = -s_{\rm A}$$

we then obtain

$$\Delta p\,[v_{\rm A}^{\rm I}(T_0, p_0) - v_{\rm A}^{\rm II}(T_0, p_0)] - \Delta T[s_{\rm A}^{\rm I}(T_0, p_0) - s_{\rm A}^{\rm II}(T_0, p_0)] = c_{\rm B}^{\rm I}\,kT_0 . \tag{8.5.7}$$

Applications. i) Let $T' = T_0$, i.e. $\Delta T = 0$. Let I be the liquid phase, II the gas phase. Δp then signifies the difference in pressure between the partial pressure of the solvent vapor in equilibrium with the solvent alone and the partial pressure of the solvent vapor in equilibrium with the solution. Δp thus reveals how the vapor pressure changes at a given temperature T if another substance is dissolved in the solvent. The assumption that there is no dissolved substance in the gas phase means that the partial pressure of the dissolved substance is negligible. We say that substance B is "not volatile".

We then obtain

$$\Delta p = kT_0 \frac{c_B^I}{(v_A^I - v_A^{II})} \; , \tag{8.5.8}$$

and with $v_A^{II} \gg v_A^I$ (in the gas phase, a solvent molecule has a much greater volume at its disposal than in the liquid phase) we find, using

$$p v_A^{II} = kT_0 \tag{8.5.9}$$

the following:

$$\Delta p = kT_0 \frac{c_B^I}{- v_A^{II}} = - p c_B^I \quad \text{or}$$

$$- \frac{\Delta p}{p} = c_B^I \; . \tag{8.5.10}$$

This is *Raoult's*[9] *law*.

The pressure of the saturated vapor of the solvent is lowered if another substance is dissolved in it, and the relative change is exactly equal to the concentration of the dissolved substance independent of what the substance is.

ii) We set $\Delta p = 0$. Then ΔT is the change of the phase transition temperature for the solvent if another substance is dissolved in it. We find

$$\Delta T = - kT_0 \frac{c_B^I}{s_A^I - s_A^{II}} = - kT_0^2 \frac{c_B^I}{q} \; , \tag{8.5.11}$$

since $q = T_0(s_A^I - s_A^{II})$ is the latent heat in the transition from II to I.

a) Let phase II be the solid phase, phase I the liquid phase. Substance B is therefore insoluble in the solid phase of A. Then, since here $q > 0$,

$$\Delta T = - kT_0^2 \frac{c_B^I}{q} < 0 \; . \tag{8.5.12}$$

Here, T_0 is the melting point. The solvent freezes out from the solution at lower temperatures. By adding salt to water, the freezing point can be lowered.

b) Let phase II be the vapor phase, phase I the liquid phase. Substance B is therefore not "volatile." Then $q < 0$, and from (8.5.11) we find

$$\Delta T = kT_0^2 \frac{c_B^I}{|q|} > 0 \; , \tag{8.5.13}$$

---

[9] *Raoult, Francois Marie* (*1830, d. 1901). French chemist, most important work on solutions.

where now T_0 is the boiling point. By dissolving substances in water, the boiling point can be raised.

The concentration

$$c_{\text{B}} = \frac{N_{\text{B}}}{(N_{\text{A}} + N_{\text{B}})} \approx \frac{N_{\text{B}}}{N_{\text{A}}} = \frac{n_{\text{B}}}{n_{\text{A}}} \ , \qquad n = \frac{N}{L} \ ,$$

of substance B can be expressed in terms of a number which shows how many moles of the dissolved substance have been added to 1 kg of solvent. This quantity is called the *molality*. If we indicate it by m_{B}, we have

$$c_{\text{B}} = \frac{M_{\text{A}} m_{\text{B}}}{1000} \ , \tag{8.5.14}$$

since 1 kg of solvent contains $1000/M_{\text{A}}$ moles, where M_{A} is the molecular weight of substance A in grams.

We thus find

$$\Delta T = K_{\text{f}} m_{\text{B}} \ , \tag{8.5.15}$$

and K_{f} is called the cryoscopic[10] constant. It depends on the transition temperature, the latent heat, and on the molecular weight of the solvent.

We find, for example, for

Water, $K_{\text{f}} = 1.855$ degree kg/mole
Benzene, $K_{\text{f}} = 5.2$ degree kg/mole
Camphor, $K_{\text{f}} = 40.0$ degree kg/mole

Using camphor as a solvent, we can obtain a considerable depression of the freezing point.

Measuring freezing point depression can also help us to accurately determine relative molecular masses. If a_1 grams of the known substance B is added to the solvent, then

$$\Delta T_1 = K_{\text{f}} m_{\text{B}} = \frac{K_{\text{f}}}{M_{\text{B}}} a_1 \ . \tag{8.5.16}$$

Now add a_1 grams of an unknown substance B$'$ to the solvent. If this causes a freezing point depression of

$$\Delta T_2 = K_{\text{f}} m_{\text{B}'} = \frac{K_{\text{f}}}{M_{\text{B}'}} a_1 \ , \tag{8.5.17}$$

we can then determine the molecular weight of substance B$'$ from the equation

$$\frac{\Delta T_1}{\Delta T_2} = \frac{M_{\text{B}'}}{M_{\text{B}}} \ . \tag{8.5.18}$$

Of course, we can only use this method in cases where the approximation of an ideal solution is valid. The amount of deviation from an ideal solution is given by the factor

$$g = \frac{\Delta T}{\Delta T_{\text{ideal}}} \ . \tag{8.5.19}$$

---

[10] Cryoscopy (Greek) *kryos*: cold, frost, *skopein*: to observe. Method of chemical investigation by measuring melting points.

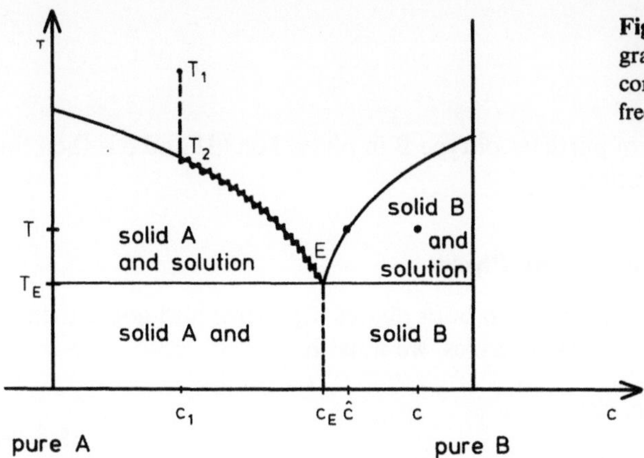

Fig. 8.5.2. Melting point diagram. For a solution with the concentration c_1 the solvent freezes out at $T = T_2$

The greater the concentration, the greater the deviation of g from 1. In tables, for example, we find values for NaCl from $g = 0.98$ to 0.91 for $c = 0.001$ to 0.7, and for p-glucose from $g = 1.0005$ to 1.03. It is reasonable to assume that substances which dissociate strongly when they dissolve have a larger deviation due to the Coulomb interaction.

As the concentration increases, the temperature at which the solid and liquid phases of the solvent coexist therefore decreases. If we plot the concentration against the coexistence temperature at fixed pressure, we obtain a *"melting point diagram"* (Fig. 8.5.2).

If the temperature is reduced from T_1 to T_2 at a given concentration c_1, the solvent then freezes out if more heat is taken away. This causes the concentration of the solution to increase. If even more heat is removed, we move along the melting point diagram as indicated in Fig. 8.5.2, until we reach a concentration c_E and a temperature T_E. At this point, there are three phases in coexistence, the solution, the solid solvent, and the solid dissolved substance, since for the inverted solution in which the solvent and the dissolved substance have switched their roles there is generally a similar freezing point depression. E is called the *eutectic*[11] point. Removing more heat at T_E will cause both the solvent and the solute to freeze in a fixed proportion c_E. Only when all the solution has frozen can the temperature decrease again.

In this diagram (Fig. 8.5.2), we also see the solubility boundary of the solute. If we try to reach a concentration c' at temperature T', we obtain a solid precipitate of the solute and a saturated solution of concentration $\hat{c}$. This diagram is thus also called a *"solubility diagram."* Note here that a point, e.g. (T', c') signifies:

$$T' = \text{temperature} , \quad c' = \text{net concentration} ,$$

---

[11] Eutectic (Greek) *eutektos*: easily melted. This property is usually applied to alloys of metals.

i.e.

$$c' = N_B/(N_A + N_B) \; ,$$

where N_B is the number of particles of type B in phase I *and* II and $\hat{c}$ is then the concentration of the solution.

8.5.2 Case (3): Miscibility in Two Phases

If two substances are fully miscible in both phases (e.g. silver and gold), then in the coexistence region of the two phases, we must have

$$\mu_A^I(T, p, c_B^I) = \mu_A^{II}(T, p, c_B^{II}) \; ,$$

$$\mu_B^I(T, p, c_B^I) = \mu_A^{II}(T, p, c_B^{II}) \; .$$

(8.5.20)

These are two equations for the two unknowns c_B^I and c_B^{II} with given T and p.

If we were to have $c_B^{II} = 0$, then from the first equation we would again find for $c_B^I \ll 1$ an equation as in (8.5.3), while the second equation would not have to be satisfied since the dissolved substance cannot exist in both phases.

We now assume that for a given p, T we can find a solution for c_B^I and c_B^{II}. Figure 8.5.3 shows a typical diagram of such a solution. Here we have fixed the pressure, say, to normal pressure, and shown the solutions of the equations c_B^I and c_B^{II} as functions of T.

If we start with such a solution with concentration c_2, and then cool it to $T_2 - \varepsilon$, a solid phase of concentration c_2' freezes out, and further cooling to the point T_3 leaves the remaining liquid phase with concentration c_2^I and the solid frozen out phase with concentration c_2^{II}. At T_4, we leave the two-phase region, which we entered at T_2, and then beneath T_4 the solid phase exists at concentration c_2. Note that we can read the values of c_2^I and c_2^{II} from the diagram only in the coexistence phase of the two substances, that is, not both values c_2^I and c_2^{II} are allowed to lie to the right or left of c_2 at the same time.

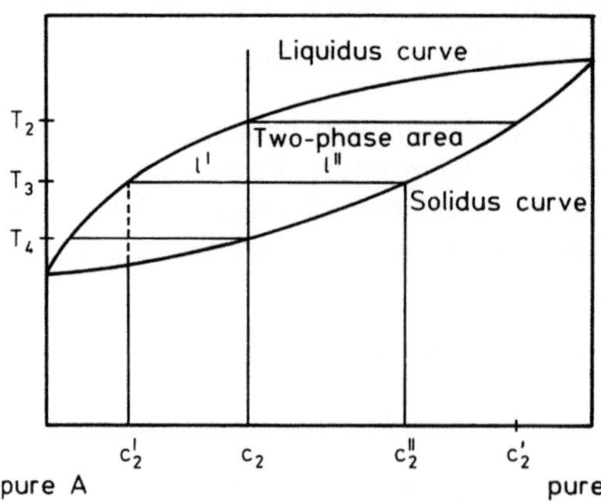

Fig. 8.5.3. Melting point diagram of a two component system in which both phases are fully miscible

It can be shown that for the amount of the substances

$$n^{\mathrm{I}} = n_A^{\mathrm{I}} + n_B^{\mathrm{I}} , \qquad n^{\mathrm{II}} = n_A^{\mathrm{II}} + n_B^{\mathrm{II}}$$

we have

$$n^{\mathrm{I}}/n^{\mathrm{II}} = l^{\mathrm{II}}/l^{\mathrm{I}} ,$$

where $l^{\mathrm{I}}, l^{\mathrm{II}}$ are the distances according to Fig. 8.5.3. At T_2, $l^{\mathrm{I}} = 0$, thus $n^{\mathrm{II}} = 0$. At T_4, $l^{\mathrm{II}} = 0$ thus $n^{\mathrm{I}} = 0$ (see [Barrow]).

Problem

8.1 The Law of Mass Action. Consider a chemical reaction

$$\sum_{i=1}^{a} |v_i| A_i = \sum_{i=a+1}^{a+e} |v_i| A_i$$

with reactants $A_1, \ldots, A_a$ and e final products $A_{a+1}, \ldots, A_{a+e}$ (where the $|v_i|$ are the numbers of molecules of each substance participating in a single reaction) for these reactants, the numbers are counted as negative. Example:

$$2H_2 + O_2 = 2H_2O$$

$$A_1 = H_2, \quad A_2 = O_2, \quad A_3 = H_2O, \quad v_1 = -2, \quad v_2 = -1, \quad v_3 = +2 .$$

Let the reaction take place in a homogeneous phase (e.g., in the gaseous phase). Let us assume that the chemical potentials depend on the concentrations

$$c_i = [A^i] := \frac{N_i}{N} \left(N = \sum_{i=1}^{a+e} N_i \right)$$

according to the following relationship

$$\mu_i = \mu_i^0 + kT \ln c_i .$$

a) Derive the law of mass action

$$\prod_{i=1}^{a+e} c_i^{v_i} = \frac{\prod_{i=a+1}^{a+e} [A_i]^{|v_i|}}{\prod_{i=1}^{a} [A_i]^{|v_i|}} =: K(p, T)$$

from the equilibrium condition $\sum_{i=1}^{a+e} v_i \mu_i = 0$.

b) Calculate $K(p, T)$ explicitly under the additional hypothesis (which is valid, e.g., for an ideal gas):

$$\mu_i^0(p, T) = kT \ln \frac{p}{p_0} - c_p^i T \ln \frac{T}{T_0} + S_0 T \ .$$

c) Discuss the displacement from equilibrium resulting from changes in temperature and pressure as it depends on the reaction enthalpy $\Delta h = \sum_{i=1}^{a+e} \nu_i h_i$, as well as on the change in volume (per reaction) $\Delta v = \sum_{i=1}^{a+e} \nu_i v_i$ where h_i and v_i are the enthalpy and volume, respectively, per particle of substance A_i.

9. Elements of Fluid Mechanics

In the previous two chapters, we have been concerned primarily with the thermodynamics of systems which are in a state of equilibrium. In particular, we have seen that we can calculate all of the state variables of a system in equilibrium from a thermodynamic potential. Further, we have formulated conditions for the existence and stability of equilibrium states.

However, we have found few important statements thus far about how equilibrium states are actually achieved or, more generally, about the behavior in time of thermodynamic systems which are not in equilibrium. We merely know that depending on the boundary conditions certain thermodynamic potentials can change in only one direction. For example, the entropy of a closed system never decreases and the Gibbs free energy of a system at given p and T never increases.

We now set ourselves the task of describing and calculating the time evolution of thermodynamic systems.

9.1   A Few Introductory Remarks About Fluid Mechanics

The global state quantities of equilibrium thermodynamics are no longer sufficient to provide a spatial and temporal description of the behavior of processes in systems which are not in a state of equilibrium. If, for example, we bring the two ends of a rod to different temperatures, the rod will not have a single unique temperature, rather the temperature T will be function of position, and if we then isolate the rod from the heat reservoirs at each end, it will also be a function of time, i.e.

$$T = T(r, t) \ .$$

Instead of working with a global temperature T, we are working with a temperature field $T(r, t)$ which gives the temperature at each point r and for each time t. Such fields arise quite naturally in the description of the state of a system in *local equilibrium.*

This means the following: The smaller the dimensions of a system are, the faster it will come into equilibrium. Small subsystems of a larger macroscopic system which is not in equilibrium will find themselves in a state of equilibrium more rapidly than the system as a whole. They will change their states then only because of their interactions with the neighboring subsystems. Such a state in a macroscopic system is called a state of local equilibrium. If the macroscopic system itself is isolated, eventually it will come into a state of global equilibrium in which all the subsystems are in equilibrium not only with themselves, but also with each other.

Let us think, in this sense, of a macroscopic system divided up into small but still macroscopic subsystems which are individually in equilibrium. Then the state of each subsystem is described by the values of a few thermodynamic variables and the state of the total system can be characterized by the states of the subsystems. If one of the subsystems is located at time t in the neighborhood of the point r, then the state of the whole system at time t is clearly described by functions such as

$n(r, t)$: the local particle number density,
$\varrho(r, t)$: the local mass density,
$e(r, t)$: the local energy density,
$s(r, t)$: the local entropy density,
$v(r, t)$: the local specific volume,
$T(r, t)$: the local temperature,
$p(r, t)$: the local pressure,
$\mu(r, t)$: the local chemical potential.

From the assumption that the small but still macroscopic subsystems are in equilibrium (assumption of local equilibrium) it follows that the same relationships hold between the various state fields as for the corresponding quantities in equilibrium thermodynamics. Thus, for example, an ideal gas in local equilibrium satisfies

$$p(r, t)v(r, t) = kT(r, t) \ .$$

If there is motion inside of the system, we also need a *velocity field*

$$v(r, t) \ ,$$

which indicates the velocity with which a subsystem located at point r moves at time t, in order to describe the state of the system. As opposed to the fields named above, this is a vector field, that is, a vector is assigned to each point of time and space. We have already met vector fields in Chap. 2 in the form of force fields $F(r, t)$. We will meet another important class of vector fields, the *current density fields* below.

In Appendix F, a few fundamental properties of vector fields are described as well as several basic operations on them, such as integration, divergence, and curl are defined.

We call a description of a system which changes in space and time based on the assumption of local equilibrium a *hydrodynamic description*. This method has applications in many areas of classical physics. In the following sections, we will discuss the fundamental equations of one of these areas, *fluid mechanics*.

Fluid mechanics, for its part, embraces many fields of physics (Fig. 9.1.1). The term "fluid dynamics" contains the two disciplines "hydrodynamics" and "rheology."[1] In these two areas of fluid mechanics we consider the behavior of fluids of *constant density* in space and time. While hydrodynamics is concerned with simple fluids such as water, rheology works with macromolecular fluids such as liquid polymers or blood which, because of their complicated molecular structures, behave quite differently from simple fluids in certain situations.

If the change of density plays a large part in the behavior of the fluid, as for gases at high speeds, we enter the field of gas dynamics which also contains many application oriented subfields.

On the other hand, fluid mechanics is a subdiscipline of continuum mechanics, the mechanics of deformable media, which treats deformable material in any state (solid, liquid, or gaseous) idealized as a continuum.

Continuum mechanics thus represents an important collection of disciplines of "applied physics". Within this framework, only mechanical properties are examined. If electromagnetic fields come into the picture, we must turn to the "electrodynamics of continuous media". In this field there are subdisciplines such as "magnetohydrodynamics" or "electromagnetic waves in anisotropic media".

Finally, the mechanical and electrical properties of substances can be studied not only by considering these substances as continuous media, but also considering their molecular structure in order to demonstrate the interactions of molecules as the source of special macroscopic behavior, as is done in statistical mechanics for equilibrium phenomena. This occurs in fields such as "kinetic gas theory" which is valid for gases of very low density or in "molecular hydrodynamics".

In these kinetic theories, the molecules or, more generally, the constituents of matter are treated as classical particles, or, when this approximation is no longer valid, as quantum systems. Molecular systems, in particular quantum systems, are of course very difficult to treat if the goal is to describe irreversible processes, and in this area much intensive research is being pursued. But complexity is not a measure of inherent interest or meaning and there is also much current research being done in the discipline of continuum mechanics. In many branches of science, the mechanical properties of raw materials, building materials, biological substances, etc. play an important role and it is often a good enough approximation to view these substances as continua. A consideration of the molecular structure is often impossible because of its complexity, and would in addition be unsuitable for certain questions.

---

[1] Rheology, (Greek) from *rhein*: to flow, science of fluid motion.

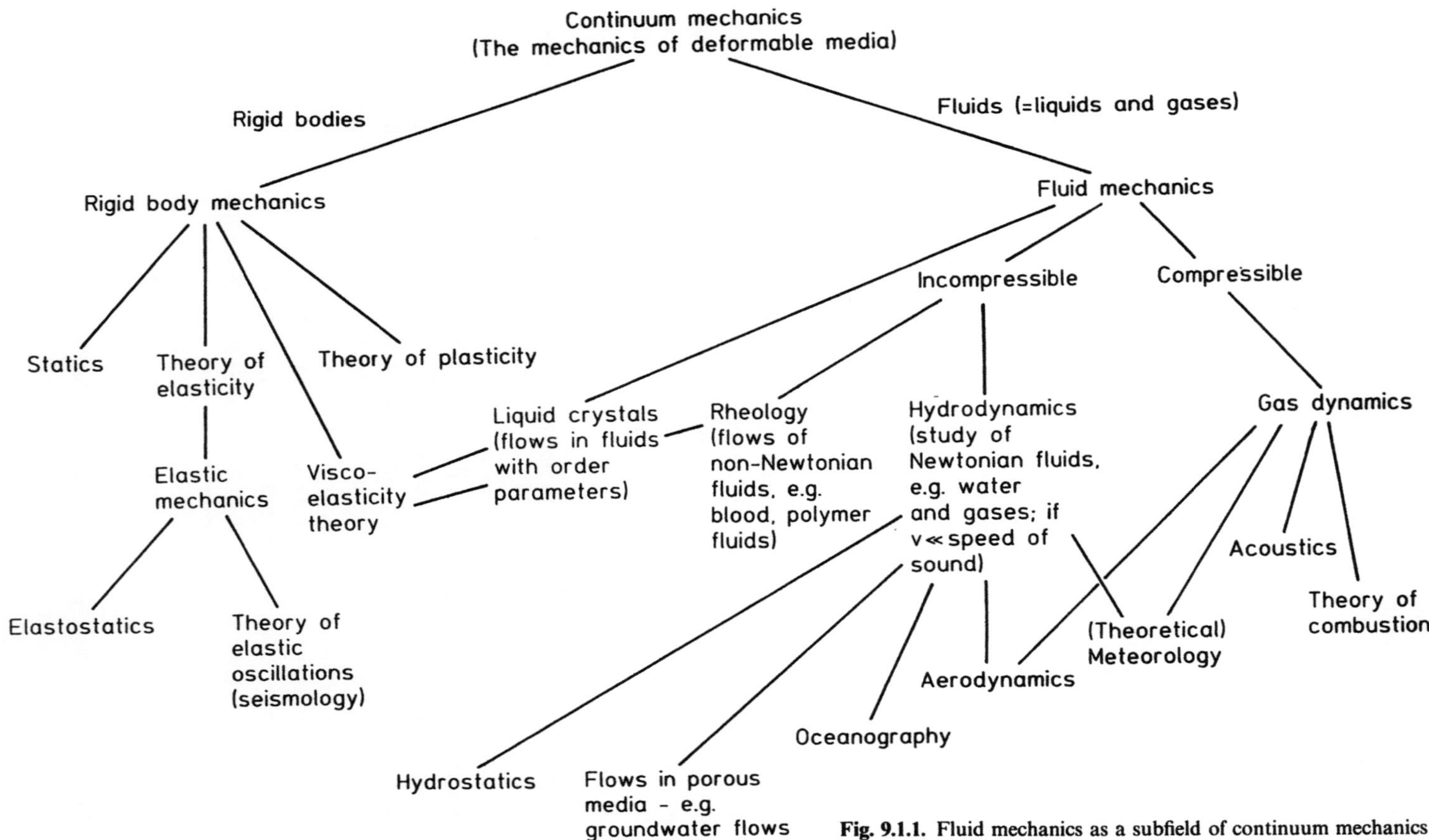

Fig. 9.1.1. Fluid mechanics as a subfield of continuum mechanics

9.2 The General Balance Equation

We consider state fields which are densities of extensive quantities like

$$\varrho(r, t) \; , \quad s(r,t) \; , \quad \text{or} \quad e(r, t) \; .$$

In general, let $a(r, t)$ be such a density, then the amount A of this extensive quantity in a volume V is given by

$$A = \int_V d^3r \, a(r, t) \; . \tag{9.2.1}$$

The change in A per unit time then consists of the amount $d_e A/dt$, which is exchanged through the surface of V per unit time, and the amount $d_i A/dt$ created or destroyed in V per unit time. Let $q_a(r, t)$ be the local *source strength* of A per unit volume, then the second contribution is

$$\frac{d_i A}{dt} = \int_V d^3r \, q_a(r, t) \; . \tag{9.2.2}$$

To describe the first term, we must introduce the *current density* $j_a(r, t)$ of the quantity A, for example, a particle current density or an energy current density. Then, by definition,

$$j_a(r, t) \cdot dS \tag{9.2.3}$$

is the amount of A which passes through the surface element dS per unit time. The field $j_a(r, t)$ is a vector field and is called the current density of the extensive quantity A.

Then,

$$\frac{dA}{dt} = \frac{d_e A}{dt} + \frac{d_i A}{dt}$$

$$= \int_V d^3r \frac{\partial}{\partial t} a(r, t)$$

$$= - \int_{\partial V} dS \cdot j_a(r, t) + \int_V d^3r \, q_a(r, t) \; , \tag{9.2.4}$$

where the minus sign in the current term ensures that the term is positive if j_a is antiparallel to dS, that is, if the quantity A flows into V.

From Gauss's[2] divergence theorem (see Appendix F)

$$\int_{\partial V} dS \cdot C = \int_V d^3r \nabla \cdot C \tag{9.2.5}$$

---

[2] *Gauss, Carl Frederich* (*1777 Braunschweig, d. 1855 Göttingen). "The prince of mathematics", he did pioneering work in algebra, number theory, geometry, theory of error, astronomy, celestial mechanics (orbital paths of asteroids), electricity, and magnetism (together with W. Weber).

so that

$$\int_V d^3r \left(\frac{\partial}{\partial t} a(r, t) + \nabla \cdot j_a(r, t) \right) = \int_V d^3r \, q_a(r, t) \tag{9.2.6}$$

and, since V is arbitrary,

$$\frac{\partial}{\partial t} a(r, t) + \nabla \cdot j_a(r, t) = q_a(r, t) \ . \tag{9.2.7}$$

This is the general balance equation. Since j_a and q_a are unknown, this equation is not a further assertion about A. In order to calculate the current of A, the current density $j_a(r, t)$ and the local source strength q_a must be determined.

In the following, we consider a system of B materials which are in gaseous or liquid states. No chemical reactions are allowed between the different materials.

Let $a(r, t)$ be the mass density $\varrho_\alpha(r, t) = m_\alpha n_\alpha(r, t)$, i.e. the local mass density of molecules of type α with mass m_α. Let v_α be the velocity of the molecules of type α in the mass element.

Then (Fig. 9.2.1)

$$\varrho_\alpha dS v_\alpha dt \tag{9.2.8}$$

is the mass which flows through the area dS in time dt if dS is parallel to v_α. More generally, this mass is

$$\varrho_\alpha v_\alpha \cdot dS \, dt \ , \tag{9.2.9}$$

and thus

$$j_\alpha = \varrho_\alpha v_\alpha \tag{9.2.10}$$

is the mass curent density, that is, the mass per unit time of molecules of type α which flow through a unit area perpendicular to j_α.

A nonzero source strength q_a of masses of type α could only arise through chemical reactions, so that here $q_a = 0$. We then have

$$\frac{\partial}{\partial t} \varrho_\alpha + \nabla \cdot (\varrho_\alpha v_\alpha) = 0 \ . \tag{9.2.11}$$

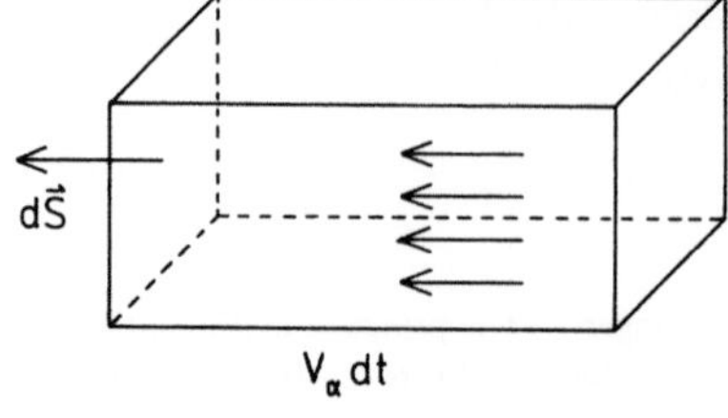

Fig. 9.2.1. A box of length $v_\alpha dt$ with a side of area dS. The particles in the box move through dS in time dt. Here, v_α is parallel to dS

If

$$\varrho = \sum_{\alpha} \varrho_{\alpha} \tag{9.2.12}$$

is the total mass density, then v, defined by

$$\varrho v := \sum_{\alpha} \varrho_{\alpha} v_{\alpha} \tag{9.2.13}$$

is the center of mass velocity of the mass elements. Summing equation (9.2.11) over all α then yields

$$\frac{\partial}{\partial t} \varrho(r, t) + \nabla \cdot [\varrho(r, t)v(r, t)] = 0 \ . \tag{9.2.14}$$

This is the *continuity equation* for the entire mass density $\varrho(r, t)$.

Now we want to express the general balance equation in a different form. We define the *substantial derivative* as

$$\frac{D}{Dt} a(r, t) = \frac{\partial}{\partial t} a(r, t) + v(r, t) \cdot \nabla a(r, t) \ . \tag{9.2.15}$$

The motivation for this is the following:

Consider a mass element at time t with position vector r, velocity $v(r, t)$ and value $a(r, t)$ for the quantity a at time t. At time $t + dt$, this mass element is at location $r + v\,dt$ and a has the value

$$a(r + v\,dt, t + dt) = a(r, t) + dt[(\partial/\partial t)a(r, t) + v \cdot \nabla a(r, t)] + \dots \ . \tag{9.2.16}$$

Thus the substantial derivative represents the change in a if at time $t + dt$ we measure the quantity a not at point r, but rather at the point where the mass element which was at point r at time t is now at time $t + dt$.

We introduce now the general mass-specific quantity

$$\hat{a}(r, t) = a(r, t)/\varrho(r, t) \tag{9.2.17}$$

so that $\hat{a}(r, t)$ is the amount of A per unit mass. Then,

$$\varrho \frac{D}{Dt} \hat{a} = \varrho \frac{\partial}{\partial t} \frac{a}{\varrho} + \varrho v \cdot \nabla \frac{a}{\varrho}$$

$$= \frac{\partial}{\partial t} a - \frac{a}{\varrho} \frac{\partial \varrho}{\partial t} + v \cdot \nabla a - \frac{a}{\varrho} v \cdot \nabla \varrho \ , \tag{9.2.18}$$

and with the continuity equation in the form

$$\frac{\partial \varrho}{\partial t} + v \cdot \nabla \varrho = - \varrho \nabla \cdot v \tag{9.2.19}$$

we have

$$\varrho \frac{D}{Dt}\hat{a} = \frac{\partial}{\partial t}a + a\mathbf{V}\cdot\mathbf{v} + \mathbf{v}\cdot\mathbf{V}a$$

$$= \frac{\partial}{\partial t}a + \mathbf{V}\cdot(a\mathbf{v}) = -\mathbf{V}\cdot\mathbf{j}_a + q_a + \mathbf{V}\cdot(a\mathbf{v}) \ . \tag{9.2.20}$$

Thus we obtain the balance equation for the specific quantity in the form

$$\varrho \frac{D}{Dt}\hat{a}(\mathbf{r},\ t) + \mathbf{V}\cdot[\mathbf{j}_a(\mathbf{r},\ t) - a(\mathbf{r},\ t)\mathbf{v}(\mathbf{r},\ t)] = q_a(\mathbf{r},\ t) \ . \tag{9.2.21}$$

In the following, we will frequently use the general balance equation in this form.
The term

$$\mathbf{J}_a(\mathbf{r},\ t) := \mathbf{j}_a(\mathbf{r},\ t) - a(\mathbf{r},\ t)\mathbf{v}(\mathbf{r},\ t) \tag{9.2.22}$$

is called the *conductive current density*; it is the current density of a relative to the velocity $\mathbf{v}$, i.e. the amount of A which per unit time passes through a unit area (perpendicular to $\mathbf{J}_a$) moving with velocity $\mathbf{v}$. The quantity

$$a(\mathbf{r},\ t)\mathbf{v}(\mathbf{r},\ t) \tag{9.2.23}$$

is called the *convective*[3] *current density*. It describes the flow of A as it is carried along in the moving medium. Obviously,

$$\mathbf{j}_a = (\mathbf{j}_a - a\mathbf{v}) + a\mathbf{v} = \mathbf{J}_a + a\mathbf{v} \ . \tag{9.2.24}$$

The total flow is thus the sum of its conductive and convective components.

Remark. Using the expression

$$\hat{v}(\mathbf{r},\ t) = 1/\varrho(\mathbf{r},\ t) \tag{9.2.25}$$

we can also write the continuity equation for ϱ in the form

$$\varrho \frac{D}{Dt}\hat{v}(\mathbf{r},\ t) - \mathbf{V}\cdot\mathbf{v}(\mathbf{r},\ t) = 0 \tag{9.2.26}$$

since

$$\varrho\frac{D\hat{v}}{Dt} = -\varrho\frac{1}{\varrho^2}\left(\frac{\partial\varrho}{\partial t} + \mathbf{v}\cdot\mathbf{V}\varrho\right) \equiv -\frac{1}{\varrho}\frac{D\varrho}{Dt} = \mathbf{V}\cdot\mathbf{v} \tag{9.2.27}$$

$\hat{v} = 1/\varrho$ is the volume per unit mass, that is, the specific volume, but this time

---

[3] Convective (Latin) from *convehere*: by carrying along.

with respect to mass and not to particle number. If we write the above continuity equation in the form

$$\frac{1}{\hat{v}}\frac{D\hat{v}}{Dt} = \nabla \cdot \boldsymbol{v}$$

(9.2.28)

we see that the velocity divergence $\nabla \cdot \boldsymbol{v}$ describes the relative change in volume.

9.3 Particular Balance Equations

We will now study the balance equation

$$\varrho \frac{D}{Dt}\hat{a}(\boldsymbol{r},\,t) + \nabla \cdot \boldsymbol{J}_a(\boldsymbol{r},\,t) = q_a(\boldsymbol{r},\,t)$$

(9.3.1)

for various quantities $\hat{a}(\boldsymbol{r},\,t)$.

i) Let $a = \varrho_\alpha$, the mass density of particles of type α. This case is simple. We have already found the mass current density and the source strength in Sect. 9.2. We have

$$\hat{a} = \frac{a}{\varrho} = \frac{\varrho_\alpha}{\varrho} = \frac{(m_\alpha n_\alpha)}{\left(\sum\limits_\alpha m_\alpha n_\alpha\right)} =: \hat{c}_\alpha \ .$$

(9.3.2)

We call $\hat{c}_\alpha$ the specific concentration of the substance α. In contrast to the concentration c_α of Chap. 8, this is not the particle concentration but rather the mass concentration of substance α.
Then

$$\boldsymbol{J}_\alpha := \varrho_\alpha(\boldsymbol{v}_\alpha - \boldsymbol{v})$$

(9.3.3)

is the conductive mass current density, also called the *diffusion*[4] *current density*. It describes the flow of particles of type α against the total flow, for example the diffusion of a dissolved substance in a solution. Of course, we have

$$\sum\limits_{\alpha=1}^{B} \boldsymbol{J}_\alpha = \boldsymbol{0} \ .$$

(9.3.4)

In any case, if there is only one type of particle then $\boldsymbol{J}_\alpha = \boldsymbol{0}$ and only the

---

[4] Diffusion (Latin) from *diffundere*: to scatter, spill. Dilution through equalization of concentration.

conservation of mass is expressed in the continuity equation

$$\frac{\partial \varrho}{\partial t} + \nabla \cdot (\varrho \boldsymbol{v}) = 0 \tag{9.3.5}$$

which was already derived in Sect. 9.2.

The balance equation for the specific concentration $\hat{c}_\alpha$ thus reads,

$$\varrho \frac{D\hat{c}_\alpha}{Dt} + \nabla \cdot \boldsymbol{J}_\alpha = 0 \quad \text{with} \tag{9.3.6}$$

$$\boldsymbol{J}_\alpha = \varrho_\alpha (\boldsymbol{v}_\alpha - \boldsymbol{v}) \ .$$

ii) Let

$$a = \varrho v_i \tag{9.3.7}$$

be the i component of the momentum density of the mass element.

If we call

$$\boldsymbol{J}_i := -(\tau_{1i}, \tau_{2i}, \tau_{3i}) \tag{9.3.8}$$

the conductive current density of the i-th component of momentum and we call the source term f_i, the transport equation reads

$$\varrho \frac{Dv_i}{Dt} - \nabla_j \tau_{ji} = f_i \tag{9.3.9}$$

or, in vector notation,

$$\varrho \frac{D\boldsymbol{v}}{Dt} - \nabla \cdot \tau = f \ . \tag{9.3.10}$$

Note that the current densities $\boldsymbol{J}_i$ of the various components of the momentum form a tensor field $- \tau_{ji}$, i.e. a map $\mathbb{R}^3 \rightarrow \mathbb{R}^3 \otimes \mathbb{R}^3$ (see Appendix C). Here the first index indicates the component of the flow, while the second index indicates which momentum component is being observed.

The quantity τ_{ji} thus indicates the flow of the i component of momentum per unit time and area through a plane normal to the e_j-direction.

Thus,

$$dF_i = -dS_j \tau_{ji} \tag{9.3.11}$$

is the flow of the i-component of momentum through the surface dS in the direction of dS. But this is precisely the i-component of the force which is exerted by the negative side of the surface element on the positive side (i.e. the side in the direction of dS).

This force is not always parallel to dS, but sometimes contains forces parallel to the area element, indicated by the non-diagonal elements of τ_{jk}. These are called *shearing forces* because they can cause a *shear*, i.e. a tangential displacement.

One component of the momentum flow, or the force per unit area, comes from the pressure p inside the mass element. The corresponding force is, in this case, always perpendicular to the area dF:

$$dF_i = p\, dS_i = p\, \delta_{ij}\, dS_j \ . \tag{9.3.12}$$

There is thus always a term of the form $-p\delta_{ji}$ in the tensor τ_{ji}:

$$\tau_{ji} = -p\delta_{ji} + \tau'_{ji} \ . \tag{9.3.13}$$

The component τ'_{ji} is not always diagonal and as we will soon see it arises in liquids and gases from the effects of friction. The tensor τ is thus a generalization of pressure and is called the *pressure tensor*.

The total momentum which flows through the surface of a volume V per unit time, i.e. the total force exerted on the environment arising from the flow of momentum, is clearly

$$F_i = -\int_{\partial V} dS_j \tau_{ji} \ . \tag{9.3.14}$$

From Gauss' theorem (the index i can be fixed and it plays a fully passive role here), we have

$$F_i = -\int_V d^3r\, \nabla_j \tau_{ji} =: -\int_V d^3r f_i \ . \tag{9.3.15}$$

The total force which is exerted by the environment on the volume V is the opposite of this force and can therefore be written as a volume integral of a force density (force per volume)

$$f_i = \nabla_j \tau_{ji} \ . \tag{9.3.16}$$

(This relation also motivates the choice of a minus sign in the definition of τ_{ji}.)

We want to show that the tensor field τ_{ji} is symmetric, if the mass element cannot have any internal angular momentum:

$$\tau_{ij} = \tau_{ji} \ . \tag{9.3.17}$$

To prove this, we consider the current density of the angular momentum. Since

$$l_i = \varepsilon_{ijk} x_j p_k \tag{9.3.18}$$

it is given by

$$m_{ji} = -\varepsilon_{irs} x_r \tau_{js} \ . \tag{9.3.19}$$

Here we have assumed that the total angular momentum of each mass element derives from its velocity v. If the mass elements can also have internal angular momentum, the following argument is not valid.

From Gauss' theorem, the torque per unit volume is given by

$$m_i = \nabla_j m_{ji} = - \nabla_j \varepsilon_{irs} x_r \tau_{js}$$

$$= - \varepsilon_{irs} \delta_{rj} \tau_{js} - \varepsilon_{irs} x_r \nabla_j \tau_{js}$$

$$= - \varepsilon_{irs} \tau_{rs} - \varepsilon_{irs} x_r f_s \; . \tag{9.3.20}$$

Since the angular momentum density must be given as a vector product of r and the force density f, it follows that $\varepsilon_{irs} \tau_{rs} = 0$ and thus from the antisymmetry of ε_{irs}

$$\tau_{rs} = \tau_{sr} \; .$$

The convective component of the momentum flow tensor is also symmetric

$$(\varrho v_j) v_i = \varrho v_j v_i \; , \tag{9.3.21}$$

so that the total momentum flow

$$j_{ik} = - \tau_{ik} + \varrho v_i v_k \tag{9.3.22}$$

must also have this property

$$j_{ik} = j_{ki} \; . \tag{9.3.23}$$

Examples of momentum sources are external forces, e.g. gravitational fields. In this case, the force per unit volume on a particle of type α is simply $f_\alpha = \varrho_\alpha g$. In general, with

$$\hat{f}_\alpha := \frac{f_\alpha}{\varrho_\alpha} \tag{9.3.24}$$

the total force per volume is given by

$$f = \sum_\alpha f_\alpha = \sum \varrho_\alpha \hat{f}_\alpha \; . \tag{9.3.25}$$

The balance equation for momentum in the form

$$\varrho \frac{Dv}{Dt} - \nabla \cdot \tau = f$$

can also be considered as a continuum version of Newton's second law.

iii) Let us consider the energy density of the system, $e(r, t)$: This contains one component for the particles' kinetic energy from the velocity $v(r, t)$, which is the center of mass velocity of a particle in a volume element. This term is $\frac{1}{2}\varrho v^2$, and

$$u(r,\, t) = e(r,\, t) - \tfrac{1}{2}\varrho v^2 \tag{9.3.26}$$

can thus be considered to be the *internal energy density*.

In the balance of energy we know the source density $q_e(r,\, t)$ which is brought about by the external forces f. On a particle of type α with velocity v_α, these forces perform work

$$v_\alpha \cdot f_\alpha$$

per unit time, and thus

$$q_e = \sum_\alpha v_\alpha \cdot f_\alpha = \sum_\alpha v_\alpha \cdot \varrho_\alpha \hat{f}_\alpha \; . \tag{9.3.27}$$

Then we can formulate the balance equation for $e(r,\, t)$ as

$$\varrho \frac{D\hat{e}}{Dt} + \nabla \cdot J_e = q_e \tag{9.3.28}$$

where J_e is the conductive energy current density.

In order to produce the balance equation for $\hat{u}(r,\, t)$, we need only calculate

$$\varrho \frac{D\hat{u}}{Dt} \quad \text{from} \quad \varrho \frac{D\hat{e}}{Dt}$$

i.e., we need to know

$$\varrho \frac{D}{Dt}\left(\frac{1}{2}v^2\right) = \varrho v \cdot \frac{Dv}{Dt} \; . \tag{9.3.29}$$

From the momentum balance equation (9.3.10), we obtain

$$\varrho v \cdot \frac{Dv}{Dt} = v \cdot (\nabla \cdot \tau) + v \cdot f \; . \tag{9.3.30}$$

Now,

$$v \cdot (\nabla \cdot \tau) = v_i \nabla_j \tau_{ji} = \nabla_j(\tau_{ji} v_i) - \tau_{ji} \nabla_j v_i$$

$$= \nabla_j(\tau_{ji} v_i) - \tau_{ji} V_{ji} \tag{9.3.31}$$

with

$$V_{ji} = \tfrac{1}{2}(\nabla_j v_i + \nabla_i v_j) \; . \tag{9.3.32}$$

Here we have used the fact that $\tau_{ij} = \tau_{ji}$. Thus we obtain

$$\varrho \frac{D}{Dt}\left(\frac{1}{2}v^2\right) = \nabla \cdot (\tau \cdot v) - \tau_{ji} V_{ij} + v \cdot f \tag{9.3.33}$$

as the balance equation for the energy of the center of mass motion of the mass element.

Then, we have

$$\varrho \frac{D\hat{u}}{Dt} = \varrho \frac{D\hat{e}}{Dt} - \varrho v \cdot \frac{Dv}{Dt}$$

$$= -\nabla \cdot J_e + q_e - \nabla \cdot (\tau \cdot v) + \tau_{ij} V_{ji} - v \cdot f \,, \tag{9.3.34}$$

and it follows that

$$\varrho \frac{D\hat{u}}{Dt} + \nabla \cdot (J_e + \tau \cdot v) = \tau_{ij} V_{ji} + \sum_\alpha \varrho_\alpha (v_\alpha - v) \cdot \hat{f}_\alpha \tag{9.3.35}$$

or

$$\varrho \frac{D\hat{u}}{Dt} + \nabla \cdot Q = \tau_{ij} V_{ji} + \sum_\alpha \varrho_\alpha (v_\alpha - v) \cdot \hat{f}_\alpha$$

$$= \tau_{ij} V_{ji} + \sum_\alpha J_\alpha \cdot \hat{f}_\alpha \tag{9.3.36}$$

with

$$Q = J_e + \tau \cdot v \,. \tag{9.3.37}$$

The quantity Q can be interpreted as the conductive current density of the internal energy. The conductive current density J_e contains, besides Q, the contribution $-\tau \cdot v$. However, this contribution is exactly the flow of energy corresponding to the work which must be performed against the molecular interaction during the flow, since this interaction is the reason for the momentum current density τ on the molecular level and for the macroscopic phenomenon "friction". Thus Q is an energy flow which can be interpreted as heat flow.

The term

$$\tau_{ij} V_{ji} \tag{9.3.38}$$

which, as can be shown, is always positive, represents a further source for the internal energy density. As we see from (9.3.33), this term appears with opposite sign as a sink for the energy of the center of mass motion. This term thus represents a transport of energy from the center of mass motion into molecular motion. This is a dissipation of energy: the term τ_{ij} describes, as we will see later, friction and "friction makes heat".

iv) If $\hat{a} = \hat{s}(r, t)$, the specific entropy density, we have

$$\varrho \frac{D\hat{s}}{Dt} + \nabla \cdot J_s = q_s \tag{9.3.39}$$

with the conductive entropy current density J_s and the local entropy production density q_s. The second law then asserts

$$q_s \geq 0 \ . \tag{9.3.40}$$

Remark. So far, the fact that we do not know the quantities

$$\tau_{ij}, \ Q_i$$

has prevented us from producing a closed system of differential equations from these transport equations which we could solve for the fields $v(r, t)$, $T(r, t)$, ... This would only be possible if we knew these current densities in terms of $v(r, t)$ and $T(r, t)$. The goal of the next section will be to find plausible assumptions for this dependence.

However, we can already say the following from intuitive considerations:

A heat flow will certainly occur if the temperature field is not homogeneous, i.e. when there are temperature gradients. We expect then, to lowest order, that

$$Q(r, t) = - \kappa \nabla T(r, t) \ , \tag{9.3.41}$$

where the minus sign means that the heat flow is in the opposite direction of the temperature gradient (from warmer to colder).

In order to find the easiest assumption for the momentum flow, let us consider three neighboring parallel layers of a fluid with somewhat different velocities in the x-direction (Fig. 9.3.1). Because of friction, whose molecular origin will not be discussed here, level 2 will slow down level 3, but it will accelerate level 1. This means that there will be a flow of momentum from level 2 to level 3 that only has a negative x-component, while in the negative z-

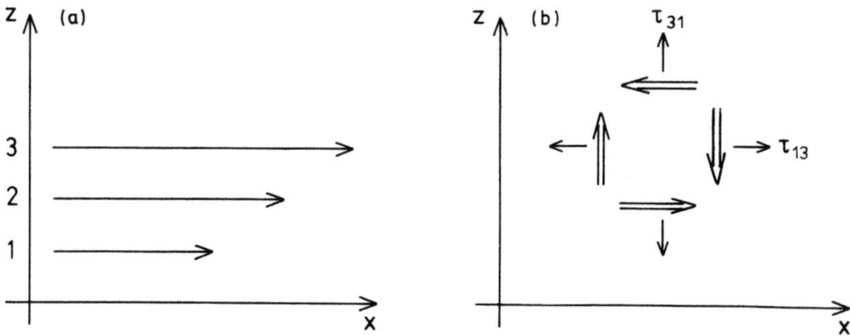

Fig. 9.3.1. (a) Three layers of a fluid which flow with different speeds in the x-direction. Through friction, level 2 slows down level 3, so that there is a momentum flow in the positive z-direction which has only a negative x-component. (b) Forces ($\Rightarrow$) and directions of the momentum flow ($\rightarrow$) which cause these forces. From the symmetry of τ_{ij}, $\tau_{13} = \tau_{31}$ and is thus nonzero. These forces prevent mass elements from collecting angular momentum

direction (towards level 1) there will be a positive x-component of the momentum flow. The degree of the slowing or acceleration, respectively, is dependent on the material properties of the fluid and on the gradients of the x-components of the velocity in the z-direction

$$\partial v_x/\partial z \ . \tag{9.3.42}$$

We thus expect that the momentum current density will be given by

$$J_1 = \left(0, 0, -\eta \frac{\partial v_x}{\partial z}\right) \tag{9.3.43}$$

so that in this case we would have

$$\tau_{31} = \eta \frac{\partial v_x}{\partial z} \tag{9.3.44}$$

where η is a material quantity which describes the viscosity of the fluid.

In fact, these equations have actually been established. The equation (9.3.41) for the heat flow is called *Fourier's Law*, and equation (9.3.44) for the momentum flow is attributed to Newton. In the next section we will recognize these equations as special cases of a more general relation for current densities.

9.4   Entropy Production, Generalized Forces, and Fluids

To find the entropy balance equation, we will calculate the quantities

$$\frac{D\hat{s}}{Dt}, \quad J_s \quad \text{and} \quad q_s$$

with the aid of the equation of state and the balance equations for $\hat{u}$, $\hat{v}$, and $\hat{c}_\alpha$.

In equilibrium thermodynamics, the entropy of a fluid or gas satisfies the equation

$$S = S(E, V, N_1, \ldots, N_B)$$

$$= Ns\left(\frac{E}{N}, \frac{V}{N}, \frac{N_1}{N}, \ldots, \frac{N_B}{N}\right) \quad \text{with} \quad N = \sum_{\alpha=1}^{B} N_\alpha \ .$$

From the assumption of local equilibrium, we can consider the local specific entropy as a function of the specific quantities

$\hat{u}$: the specific energy
$\hat{v}$: the specific volume and
$\hat{c}_\alpha$: the specific mass concentration of type α,

so that

$$\hat{s} = \hat{s}(\hat{u}(\mathbf{r}, t), \hat{v}(\mathbf{r}, t), \hat{c}_1(\mathbf{r}, t), \ldots, \hat{c}_B(\mathbf{r}, t)) \ . \tag{9.4.1}$$

Here, the extensive quantities such as energy, volume, and particle number of type α are taken with respect to the total mass rather than the total number of particles. The following relations also hold locally:

$$\frac{\partial \hat{s}}{\partial \hat{u}} = \frac{1}{T(\mathbf{r}, t)} \ , \qquad \frac{\partial \hat{s}}{\partial \hat{v}} = \frac{p(\mathbf{r}, t)}{T(\mathbf{r}, t)} \ , \qquad \frac{\partial \hat{s}}{\partial \hat{c}_\alpha} = -\frac{\hat{\mu}_\alpha(\mathbf{r}, t)}{T(\mathbf{r}, t)} \ . \tag{9.4.2}$$

Thus we obtain

$$\frac{D\hat{s}}{Dt} = \frac{\partial \hat{s}}{\partial \hat{u}} \frac{D\hat{u}}{Dt} + \frac{\partial \hat{s}}{\partial \hat{v}} \frac{D\hat{v}}{Dt} + \sum_{\alpha=1}^{B} \frac{\partial \hat{s}}{\partial \hat{c}_\alpha} \frac{D\hat{c}_\alpha}{Dt}$$

$$= \frac{1}{T} \frac{D\hat{u}}{Dt} + \frac{p}{T} \frac{D\hat{v}}{Dt} - \sum_{\alpha=1}^{B} \frac{\hat{\mu}_\alpha}{T} \frac{D\hat{c}_\alpha}{Dt} \ . \tag{9.4.3}$$

Since we produced equations in Sects. 9.2 and 3 for

$$\frac{D\hat{u}}{Dt} \ , \qquad \frac{D\hat{v}}{Dt} \quad \text{and} \quad \frac{D\hat{c}_\alpha}{Dt}$$

containing the corresponding current and source terms, we can express the entropy flow and the entropy production here in terms of these current and source terms. We obtain, with (9.3.36), (9.2.27), and (9.3.6):

$$\varrho \frac{D\hat{s}}{Dt} = \frac{1}{T}\left(-\nabla \cdot \mathbf{Q} + \tau_{ij} V_{ji} + \sum_{\alpha=1}^{B} \mathbf{J}_\alpha \cdot \hat{\mathbf{f}}_\alpha \right)$$

$$+ \frac{p}{T} \nabla \cdot \mathbf{v} - \frac{1}{T} \sum_\alpha \hat{\mu}_\alpha (-\nabla \cdot \mathbf{J}_\alpha) \ . \tag{9.4.4}$$

Now, we have

   i) $p\nabla \cdot \mathbf{v} = p\delta_{ij}\frac{1}{2}(\nabla_j v_i + \nabla_i v_j) = (\mathbf{p})_{ij} V_{ji}$ (9.4.5)

with the tensor

$$(\mathbf{p})_{ij} = p\delta_{ij} \ .$$

   ii) $-\dfrac{1}{T}\left(\nabla \cdot \mathbf{Q} - \displaystyle\sum_{\alpha=1}^{B} \hat{\mu}_\alpha \nabla \cdot \mathbf{J}_\alpha \right)$

$$= -\nabla \cdot \left(\frac{\mathbf{Q}}{T} - \sum_{\alpha=1}^{B} \hat{\mu}_\alpha \frac{\mathbf{J}_\alpha}{T} \right) + \mathbf{Q} \cdot \nabla \frac{1}{T} - \sum_{\alpha=1}^{B} \mathbf{J}_\alpha \cdot \nabla \frac{\mu_\alpha}{T} \ . \tag{9.4.6}$$

It then follows that

$$\varrho \frac{D\hat{s}}{Dt} = -\nabla\cdot\left(\frac{Q}{T} - \sum_{\alpha=1}^{B} \hat{\mu}_\alpha \frac{J_\alpha}{T}\right) + Q\cdot\nabla\frac{1}{T}$$

$$- \sum_{\alpha=1}^{B} J_\alpha\cdot\left(\nabla\frac{\hat{\mu}_\alpha}{T} - \frac{\hat{f}_\alpha}{T}\right) + \frac{1}{T}(\tau + p)_{ij}V_{ji} \ . \tag{9.4.7}$$

We thus find, for the entropy production

$$q_s = Q\cdot\nabla\frac{1}{T} - \sum_{\alpha=1}^{B} J_\alpha\cdot\left(\nabla\frac{\hat{\mu}_\alpha}{T} - \frac{\hat{f}_\alpha}{T}\right) + (\tau + p)_{ij}\frac{V_{ji}}{T}$$

$$= Q\cdot\nabla\frac{1}{T} - \sum_{\alpha=1}^{B-1} J_\alpha\cdot\left(\nabla\frac{\hat{\mu}_\alpha - \hat{\mu}_B}{T} - \frac{\hat{f}_\alpha - \hat{f}_B}{T}\right) + (\tau + p)_{ij}\frac{V_{ji}}{T} \ , \tag{9.4.8}$$

where we have eliminated J_B with the help of

$$\sum_{\alpha=1}^{B} J_\alpha = 0 \ .$$

The entropy production is caused here by

 i) a temperature gradient $\nabla\,1/T$
 ii) a gradient in chemical potential and an external force,
 iii) a gradient in the velocity field.

If the gradients of T, $\hat{\mu}_\alpha$, and v all vanish, there are indeed no flows of heat, diffusion, or momentum and no entropy will be produced. In this sense, we can consider the gradients to be "forces" which cause and propel the flow.

In order to formulate a hypothesis for the relation between forces and flows, let the components

$$Q, \ J_\alpha, \ \tau'_{ij} = \tau_{ij} + p\delta_{ij}$$

of the flow be indicated all together by J_A, where the index A runs through exactly $3 + 3(B-1) + 6 = 3B + 6$ values, and the components of the "forces"

$$\nabla\frac{1}{T}, \qquad -\nabla\frac{\hat{\mu}_\alpha - \hat{\mu}_B}{T} + \frac{\hat{f}_\alpha - \hat{f}_B}{T}, \qquad \frac{V_{ij}}{T}$$

be indicated by X_A. Then the entropy production density can be written as

$$q_s = \sum_A J_A X_A \ . \tag{9.4.9}$$

It is now seems reasonable for us to assume, for forces which are not too large, that the flows are proportional to the forces.

We therefore set

$$J_A = \sum_B L_{AB} X_B \tag{9.4.10}$$

with certain phenomenological coefficients L_{AB}. Then,

$$q_s = \sum_{A,B} L_{AB} X_A X_B \ . \tag{9.4.11}$$

Since the entropy production cannot be negative, the matrix L_{AB} must in any case be positive (semi)definite. In particular, we must have

$$L_{AA} \geq 0 \ . \tag{9.4.12}$$

It is the task of *transport theory*, a subdiscipline of non equilibrium statistical mechanics, to calculate the coefficients L_{AB} from the microphysical characteristics of the system. Within our framework, L_{AB} are material constants which can be found from experiment. *Onsager's*[5] *symmetry relations*:

$$L_{AB} = L_{BA} \ , \tag{9.4.13}$$

follow from microscopic theory. Thus, the matrix L_{AB} is symmetric. (More exactly, as Casimir has shown, $L_{AB} = \varepsilon_A \varepsilon_B L_{BA}$, where $\varepsilon_C = \pm 1$ depending on whether or not the force X_C changes its sign under time reversal, $t \to -t$.) The coefficient matrix L_{AB} is further limited by *Curie's*[6] *principle*:

In isotropic systems, only forces and flows with the same transformation properties under rotations are connected by equations

$$J_A = \sum_B L_{AB} X_B$$

thus, only vector fields wth vector fields, tensor fields with tensor fields, and scalar fields with scalar fields.

Scalar fields and flows appear because it is possible to split off a scalar field from a symmetric field by "constructing the trace".

$$A_{ij} = (A_{ij} - \tfrac{1}{3}\delta_{ij}A_{kk}) + \tfrac{1}{3}\delta_{ij}A_{kk} \ . \tag{9.4.14}$$

---

[5] *Onsager, Lars* (*1903 Oslo, d. 1976 Miami, Florida). Became professor at Yale University in 1934. Fundamental work on statistical mechanics and irreversible thermodynamics. Nobel prize in chemistry in 1968.

[6] *Curie, Pierre* (*1859 Paris, d. 1906 Paris). French physicist, husband of Marie Curie, known especially for his work on radioactivity, but also made important contributions to the theory of magnetic materials. Discovered piezoelectricity.

We thus have the following forces and flows:

$$\text{vectors: } \mathbf{Q},\ \mathbf{J}_\alpha;\ \nabla\frac{1}{T},\ -\nabla\frac{\hat{\mu}_\alpha - \hat{\mu}_B}{T} + \frac{\hat{f}_\alpha - \hat{f}_B}{T}\ ;$$

$$\text{tensors: } \tau_{ij} - \frac{1}{3}\delta_{ij}\tau_{kk};\ \frac{1}{T}\left(\nabla_i v_j + \nabla_j v_i - \frac{2}{3}\delta_{ij}\nabla\cdot\mathbf{v}\right)\ ;$$

$$\text{scalars: } \tau_{kk} + 3p;\ \frac{1}{T}\nabla\cdot\mathbf{v}\ .$$

If chemical reactions can also occur, then for each possible reaction there is another scalar field and scalar force.

Considering the symmetry relations and Curie's principle, the linear form we have assumed, $J_A = \sum_B L_{AB} X_B$, then reads:

$$\text{i)}\quad \mathbf{Q} = \tilde{\lambda}\nabla\frac{1}{T} + \sum_{\alpha=1}^{B-1} \tilde{\lambda}_\alpha\left(\frac{\hat{f}_\alpha - \hat{f}_B}{T} - \nabla\frac{\hat{\mu}_\alpha - \hat{\mu}_B}{T}\right),\tag{9.4.15}$$

$$\text{ii)}\quad \mathbf{J}_\alpha = \sum_{\beta=1}^{B-1} \tilde{\lambda}_{\alpha\beta}\left(\frac{\hat{f}_\beta - \hat{f}_B}{T} - \nabla\frac{\hat{\mu}_\beta - \hat{\mu}_B}{T}\right) + \tilde{\lambda}_\alpha\nabla\frac{1}{T}\tag{9.4.16}$$

with $\tilde{\lambda} \geq 0$, $\tilde{\lambda}_{\alpha\beta}$ positive semidefinite.

$$\text{iii)}\quad \tau_{ij} - \frac{1}{3}\delta_{ij}\tau_{kk} = \frac{\Lambda}{T}\left(\nabla_i v_j + \nabla_j v_i - \frac{2}{3}\delta_{ij}\nabla\cdot\mathbf{v}\right)\tag{9.4.17}$$

with $\Lambda \geq 0$,

$$\text{iv)}\quad \tau_{kk} + 3p = \frac{\tilde{\Lambda}}{T}\nabla\cdot\mathbf{v}\ ,\qquad \tilde{\Lambda}\geq 0\ .\tag{9.4.18}$$

The relationships between τ_{ij} and V_{ji} can also be written in a single formula as

$$\tau_{ij} = -\delta_{ij}p + \eta(\nabla_i v_j + \nabla_j v_i - \tfrac{2}{3}\delta_{ij}\nabla\cdot\mathbf{v}) + \zeta\delta_{ij}\nabla\cdot\mathbf{v}\ .\tag{9.4.19}$$

Here, $\eta = \Lambda/T$ is called the *viscosity* or *shear viscosity* and $\zeta = \tilde{\Lambda}/3T$ is called the *volume viscosity*.

If we substitute our linear hypothesis $J_A = \sum_B L_{AB} X_B$ into the transport equations for

$$\frac{D\hat{v}}{Dt},\quad \frac{D\hat{u}}{Dt},\quad \frac{D\hat{c}_\alpha}{Dt}$$

we obtain a closed system of partial differential equations for the desired field functions

$$\varrho(r, t) , \quad p(r, t) , \quad u(r, t) , \quad c_\alpha(r, t) , \quad \mu_\alpha(r, t) \quad \text{and} \quad v(r, t) .$$

We will study this in Sect. 9.5.

Remarks. i) For a one-component system ($B = 1$), or if diffusion effects can be neglected, the heat flow is given by

$$Q = \tilde{\lambda} \nabla \frac{1}{T} . \tag{9.4.20}$$

If the temperature does not deviate strongly from a reference value T_0, the relation simplifies to

$$Q = - \frac{\tilde{\lambda}}{T_0^2} \nabla T = - \kappa \nabla T . \tag{9.4.21}$$

This is the special case already discussed in Sect. 9.3, *Fourier's law of heat conduction,* κ is called the *heat conductivity.*

ii) For a system with two kinds of molecules ($B = 2$), the diffusion flow in the absence of external forces at constant temperature is given by:

$$J_1 = - \tilde{\lambda}_{11} \nabla \frac{\hat{\mu}_1 - \hat{\mu}_2}{T}$$

$$= - \frac{\tilde{\lambda}_{11}}{T} \frac{\partial(\hat{\mu}_1 - \hat{\mu}_2)}{\partial \hat{c}_1} \nabla \hat{c}_1 , \qquad \text{that is,} \tag{9.4.22}$$

$$J_1 = - D \nabla \hat{c}_1 \quad \text{with} \tag{9.4.23}$$

$$D = \frac{- \tilde{\lambda}_{11}}{T} \frac{\partial}{\partial \hat{c}_1} (\hat{\mu}_1 - \hat{\mu}_2) . \tag{9.4.24}$$

This relationship between the diffusion current density and the concentration gradient is called *Fick's*[7] *first law.* The quantity D is called the *diffusion coefficient.*

iii) The coefficients η and ζ describe the effects of friction during the deformation (with and without change in volume) which an element of the fluid experiences because of the inhomogeneity of the velocity field. The volume changes in liquids are generally small, so that for liquids the effect of it is less noticeable.

---

[7] *Fick, Adolf* (*1829 Kassel, d. 1901 Blankenberge (Belgium)). German medical researcher and physiologist, originally mathematician. Work in countless areas of physiology: the cardiovascular system, biomechanics, muscle and respiratory physiology. He formulated his law of diffusion in the year 1855.

iv) Our linear hypothesis shows us that, in general, a temperature gradient leads not only to a heat flow, but also to a diffusion flow, and that chemical potential gradients not only cause diffusion flow but also heat flow. These two effects, which are related by Onsager's symmetry relations, have actually been observed. They are called *thermodiffusion* and the *diffusion thermoeffect*.

9.5 The Differential Equations of Fluid Mechanics

We can substitute the hypotheses we have formulated about the flows into the equations for $\hat{c}_\alpha$, v, and $\hat{u}$:

i) Starting with the equation for v

$$\varrho \frac{Dv}{Dt} = \nabla \cdot \tau + f \,, \tag{9.5.1}$$

we find, if we assume additionally that η and ζ are constant

$$\varrho \frac{Dv_i}{Dt} = -\nabla_j \delta_{ji} p + \zeta \nabla_j \delta_{ji} \nabla \cdot v + \eta \nabla_j (\nabla_j v_i + \nabla_i v_j - \tfrac{2}{3} \delta_{ij} \nabla \cdot v) + f_i$$

or

$$\varrho \frac{Dv}{Dt} = -\nabla p + (\zeta + \tfrac{1}{3}\eta) \nabla (\nabla \cdot v) + \eta \Delta v + f \,. \tag{9.5.2}$$

The differential operator

$$\Delta := \nabla \cdot \nabla = \nabla^2 = \nabla_i \nabla_i \tag{9.5.3}$$

is called the *Laplacian*[8].

The equation (9.5.2) is called the *Navier*[9]*–Stokes*[10] *equation*. It is of extreme importance in all fields of fluid mechanics.

---

[8] *Laplace, Pierre Simon* (*1749 Beaumont-en-Auge (Normandy), d. 1827 Arcueil near Paris). Great mathematician, astronomer, and physicist. Principle work on partial differential equations, probability theory, and above all great achievements in celestial mechanics. At the same time he excelled as a scientific writer. His name is also connected to the Kant–Laplacian model of the world and the idea of Laplace's demon.

[9] *Navier, Claude Louis Marie Henri* (*1785 Dijon, d. 1836 Paris). French engineer and physicist, professor at the Ecole Polytechnique. Important works on elasticity theory and hydrodynamics.

[10] *Stokes, George Gabriel* (*1819 Skreen/Ireland, d. 1903 Cambridge). Professor at Cambridge in Newton's former chair from 1849 to 1903. Especially important work on hydrodynamics, optics, and geodesy.

ii) From the equation for $\hat{u}$:

$$\varrho\,\frac{D\hat{u}}{Dt} + \mathbf{V}\cdot\mathbf{Q} = \tau_{ij}V_{ij} + \sum_{\alpha=1}^{B} \mathbf{J}_\alpha\cdot\hat{\mathbf{f}}_\alpha$$

we obtain

$$\varrho\,\frac{D\hat{u}}{Dt} + \mathbf{V}\cdot(-\kappa\mathbf{V}T) = [-p\delta_{ij} + \zeta\mathbf{V}\cdot\mathbf{v}\delta_{ij} + \eta(2V_{ij} - \tfrac{2}{3}\delta_{ij}\mathbf{V}\cdot\mathbf{v})]\,V_{ij}$$

$$+ \sum_{\alpha=1}^{B} \mathbf{J}_\alpha\cdot\hat{\mathbf{f}}_\alpha\,, \tag{9.5.4}$$

or, if κ is considered to be constant and external forces can be neglected

$$\varrho\,\frac{D\hat{u}}{Dt} - \kappa\Delta T = -p\mathbf{V}\cdot\mathbf{v} + 2\eta V_{ij}V_{ji} + (\zeta - \tfrac{2}{3}\eta)(\mathbf{V}\cdot\mathbf{v})^2\,. \tag{9.5.5}$$

This is the *generalized heat conduction equation*.

So far we have four equations for the unknowns v, $\hat{u}$, p, T, and ϱ. If there is only one type of particle, these are all the unknown functions. In order to obtain a fully determined system of differential equations for these seven unknowns, we thus need three more equations. The continuity equation

$$\frac{D\varrho}{Dt} + \varrho\mathbf{V}\cdot\mathbf{v} = 0 \tag{9.5.6}$$

is certainly one, and if we now use the local state equations

$$p = p(T,\varrho)\,, \tag{9.5.7}$$

$$\hat{u} = \hat{u}(T,\varrho) \tag{9.5.8}$$

for the individual volume elements, we can finally determine the unknown state fields

$$v(r,t)\,, \quad T(r,t)\,, \quad \varrho(r,t) \tag{9.5.9}$$

from the differential equations, and then also p and $\hat{u}$.

Special Cases. i) Consider an incompressible fluid, that is, a substance with

$$\dot{\varrho} = 0\,, \quad \mathbf{V}\varrho = \mathbf{0}\,. \tag{9.5.10}$$

Then the equation of continuity implies

$$\mathbf{V}\cdot\mathbf{v} = 0\,. \tag{9.5.11}$$

The branch of fluid mechanics in which incompressibility is always assumed is called *fluid dynamics*, since in fluids the changes in density are negligible if the flow velocity is not too high.

The Navier–Stokes equation then reads

$$\frac{\partial v}{\partial t} + (v \cdot \nabla)\, v = -\frac{1}{\varrho}\nabla p + \frac{\eta}{\varrho}\Delta v + \frac{f}{\varrho}\,. \tag{9.5.12}$$

ii) In contrast, *gas dynamics* takes into account compressibility but sets the pressure tensor equal to

$$\tau_{ij} = -p\delta_{ij}$$

unless the friction present at the boundary layers in the vicinity of the walls cannot be ignored.

iii) A fluid is called *ideal* if viscosity, heat conduction, and diffusion (for systems with several materials) can be ignored. Then the Navier–Stokes equation becomes

$$\varrho\frac{Dv}{Dt} = \varrho\left[\frac{\partial v}{\partial t} + (v \cdot \nabla)\, v\right] = -\nabla p + f\,. \tag{9.5.13}$$

This is *Euler's equation* (discovered by Euler in 1755).

For an ideal fluid, the currents Q, J_α, and τ'_{ji} vanish, and thus also q_s and the entropy current J_s. The entropy transport equation is then simply

$$\frac{D\hat{s}}{Dt} = 0\,. \tag{9.5.14}$$

The change of state of each mass element is adiabatic and reversible. The relation $p = p(\varrho)$ is then simply given by the adiabatic equation. For an ideal fluid, the flow equation together with the equation of continuity and the adiabatic equation constitute a fully determined system.

iv) With the help of the identity

$$(v \cdot \nabla)\, v = \nabla \tfrac{1}{2} v^2 - v \times (\nabla \times v) \tag{9.5.15}$$

we obtain for an ideal, incompressible fluid, from Euler's equation

$$\frac{\partial v}{\partial t} + \frac{1}{2}\nabla v^2 - v \times (\nabla \times v) = -\nabla\frac{p}{\varrho} + \frac{f}{\varrho}\,. \tag{9.5.16}$$

If the external force f consists of the gravitational force of the earth, then $f/\varrho = g$, and if we take the z-axis to be in the direction of $-g$, we have

$$g = -\nabla(gz)\quad\text{and thus} \tag{9.5.17}$$

$$-\frac{\partial v}{\partial t} + v \times (\nabla \times v) = \nabla\left(\frac{1}{2}v^2 + \frac{p}{\varrho} + gz\right)\,. \tag{9.5.18}$$

A flow is said to be *stationary* if $\partial v/\partial t \equiv 0$. Further, a *streamline* is a line in the fluid to which $v(r, t)$ is always tangential. Streamlines are thus the field lines of the vector field $v(r, t)$. In the stationary case, streamlines are also the trajectories of the particles of the fluid.

Now, since $v \times (\nabla \times v)$ is perpendicular to v, in the stationary case the derivative in the direction v of

$$\frac{1}{2}v^2 + \frac{p}{\varrho} + gz \tag{9.5.19}$$

is equal to zero and thus along a streamline

$$\frac{1}{2}v^2 + \frac{p}{\varrho} + gz = \text{constant} , \tag{9.5.20}$$

where the constant can vary from streamline to streamline. This is *Bernoulli's*[11] *law*.

If in addition

$$\nabla \times v = 0 \tag{9.5.21}$$

is valid everywhere in the fluid, the constant in (9.5.20) in the stationary case is the same for all streamlines.

v) In an ideal incompressible liquid, no vortices can form. To see this, we construct the curl $(\nabla \times v)$ from Euler's equation in the form (9.5.18). For the vortex density,

$$\omega = \nabla \times v \tag{9.5.22}$$

we then find

$$\dot{\omega} = \nabla \times (v \times \omega) . \tag{9.5.23}$$

If at any point in time, say for $t = 0$, we have $\omega(r, 0) = 0$, then from continued differentiation with respect to t, we find

$$\frac{d^n}{dt^n}\, \omega(r, 0) = 0$$

for all n, which implies that $\omega(r, t) = 0$ for all t and r.

A flow for which

$$\omega = \nabla \times v \equiv 0$$

---

[11] *Bernoulli, Daniel I.* (*1700 Groningen, d. 1782 Basel). Scion of a well known family of scholars in Basel. Important work in continuum mechanics, physiology, pure mathematics. Anticipated the kinetic theory of gases by explaining gas pressure kinetically.

is called a *potential flow*, since a potential ϕ can be found for $\boldsymbol{v}$ with

$$\boldsymbol{v} = -\nabla\phi \ . \tag{9.5.24}$$

The equation of continuity in the form $\nabla\cdot\boldsymbol{v}$ for incompressible fluids then reads

$$\Delta\phi = 0 \ . \tag{9.5.25}$$

This is called *Laplace's equation*. This equation arises in many areas of physics. We will consider it further in Chap. 10.

Often, we can consider a fluid to be ideal over large regions, and thus we can ignore the vortex density in these regions. Then the velocity field can be determined from the Laplace equation. If the flow is also two-dimensional, i.e.

$$\boldsymbol{v} = (v_x(x, y), v_y(x, y), 0) \ ,$$

then $\phi(x, y)$ can be written as the real part of an analytic function $f(z)$, $z = x + \mathrm{i}y$, and with the help of methods from function theory, $f(z)$ can be determined in such a way as to satisfy the physical boundary conditions (see [Landau–Lifschitz]).

vi) We will now investigate heat conduction in the special case $\boldsymbol{v} = 0$ of a fluid at rest.

From the equation of continuity, it then follows that

$$\frac{\partial\varrho}{\partial t} + \nabla\cdot(\varrho\boldsymbol{v}) = \frac{\partial\varrho}{\partial t} = 0 \ ,$$

thus ϱ is constant in time.

The heat conduction equation then reads

$$\varrho\,\frac{\partial\hat{u}}{\partial t} - \kappa\Delta T = 0 \ . \tag{9.5.26}$$

The change in $\hat{u}(T, \varrho)$ over time now arises only from the temperature change

$$\frac{\partial\hat{u}}{\partial t} = \frac{\partial\hat{u}}{\partial T}\frac{\partial T}{\partial t} \ .$$

As we know, $\partial\hat{u}/\partial T = \hat{c}_v$, the specific heat per unit mass.

Thus it follows

$$\varrho\hat{c}_v\frac{\partial T(\boldsymbol{r}, t)}{\partial t} - \kappa\Delta T(\boldsymbol{r}, t) = 0 \tag{9.5.27}$$

or

$$\frac{\partial T(\boldsymbol{r}, t)}{\partial t} - \lambda\Delta T(\boldsymbol{r}, t) = 0 \ , \quad \text{with} \quad \lambda = \frac{\kappa}{\varrho\hat{c}_v} \ . \tag{9.5.28}$$

This equation is called the *heat conduction equation* and λ is called the *thermal conductivity*.

vii) Let us now examine diffusion effects, limiting ourselves to the case of constant temperature, $T =$ constant, and negligible flow ($v = 0$).

For a system with two components ($B = 2$), the assumed linear form for the diffusion current in the absence of external forces (Sect. 9.4) is then

$$J_1 = -D\nabla \hat{c}_1$$

with diffusion constant D. From the balance equation

$$\frac{\partial \hat{c}_1}{\partial t} + \nabla \cdot J_1 = 0$$

it follows, if we take $D =$ constant, that

$$\frac{\partial c(r, t)}{\partial t} - D\Delta c(r, t) = 0 \ . \tag{9.5.29}$$

where we have written c instead of $\hat{c}_1$. The diffusion equation in this approximation has the same form as the heat conduction equation.

9.6   A Few Elementary Applications of the Navier–Stokes Equations

An important theme of hydrodynamics is the investigation of the Navier–Stokes equation as well as the discussion of physically important solutions, and there is a full literature in this area. Here, we will discuss only the most elementary applications.

i) From Euler's equation with

$$f = -\nabla(\varrho g z) \tag{9.6.1}$$

for a fluid at rest we obtain

$$\nabla(p + \varrho g z) = 0 \quad \text{or}$$
$$p = -\varrho g z + \text{constant} \ . \tag{9.6.2}$$

This is the hydrostatic pressure of a liquid at rest. The force on a body immersed in the liquid is then equal to the momentum per unit time lost by the fluid on the boundary corresponding to the surface of the body:

$$F_i = \int_K d^3 r \nabla_j \tau_{ji} = -\int_K d^3 r \nabla_j (p\delta_{ji}) = -\int_K d^3 r \nabla_i p \ . \tag{9.6.3}$$

Using (9.6.2), we find

$$\boldsymbol{F} = (0, 0, F); \qquad F = g\varrho V , \tag{9.6.4}$$

where V is the volume of the body. The *buoyancy force* exerted on the body is thus equal to the weight of the fluid which is displaced by the body. This is *Archimedes'*[12] *principle*.

ii) Consider a gas which as a whole is at rest and in which local density and pressure waves are produced (sound waves) due to external influences. Then $v(r, t)$ is small and $(v \cdot \boldsymbol{V})v$ can be neglected in comparison to $\dot{v}$ and we can write

$$p(\boldsymbol{r}, t) = p_0 + \tilde{p}(\boldsymbol{r}, t) , \qquad \varrho(\boldsymbol{r}, t) = \varrho_0 + \tilde{\varrho}(\boldsymbol{r}, t) \tag{9.6.5}$$

with similarly small values for $\tilde{p}$ and $\tilde{\varrho}$. Let p_0 and ϱ_0 be the equilibrium values. Then the equation of continuity is approximately given by

$$\frac{\partial \tilde{\varrho}}{\partial t} + \varrho_0 \boldsymbol{V} \cdot \boldsymbol{v} = 0 , \tag{9.6.6}$$

and from Euler's equation we find

$$\varrho_0 \frac{\partial \boldsymbol{v}}{\partial t} = - \boldsymbol{V}\tilde{p} . \tag{9.6.7}$$

The equations (9.6.6 and 7) represent four equations for the five unknowns v, $\tilde{p}$, and $\tilde{\varrho}$. We therefore need one more equation. If we assume that the energy transfer occurs so quickly in the pressure and density waves that no heat is transferred, we can use the state equation for an adiabatic process. Then

$$p \cdot \varrho^{-\kappa} = \text{constant} ,, \quad \kappa = c_p/c_v \tag{9.6.8}$$

or

$$(p_0 + \tilde{p}) (\varrho_0 + \tilde{\varrho})^{-\kappa} = p_0 \varrho_0^{-\kappa} ,$$

and thus

$$p_0 \varrho_0^{-\kappa} \left(1 + \frac{\tilde{p}}{p_0}\right)\left(1 + \frac{\tilde{\varrho}}{\varrho_0}\right)^{-\kappa} = p_0 \varrho_0^{-\kappa}\left(1 + \frac{\tilde{p}}{p_0} - \frac{\kappa\tilde{\varrho}}{\varrho_0} + \dots\right)$$

or, to a good approximation,

$$\frac{\tilde{p}}{p_0} = \kappa \frac{\tilde{\varrho}}{\varrho_0} . \tag{9.6.9}$$

---

[12] *Archimedes of Syracuse* (ca. 287–212 B.C.). Universal thinker, mathematician, physicist, and inventor.

Thus we finally obtain from (9.6.6 and 7)

$$0 = \frac{\partial^2 \tilde{\varrho}}{\partial t^2} + \varrho_0 \nabla \cdot \frac{\partial \boldsymbol{v}}{\partial t} = \frac{\partial^2 \tilde{\varrho}}{\partial t^2} + \nabla^2 \left(-\frac{\kappa p_0}{\varrho_0} \tilde{\varrho} \right)$$

or

$$\left(\frac{1}{c^2} \frac{\partial^2}{\partial t^2} - \Delta \right) \tilde{\varrho}(r, t) = 0 \tag{9.6.10}$$

with

$$c^2 = \kappa \frac{p_0}{\varrho_0} . \tag{9.6.11}$$

This is the wave equation for the propagation of sound waves in a medium at rest in which viscosity can be neglected. The quantity c is the speed of sound.

We still would like to justify neglecting $(\boldsymbol{v} \cdot \nabla)\boldsymbol{v}$ in comparison to $\dot{\boldsymbol{v}}$. In a wave

$$\boldsymbol{v} = \boldsymbol{v}_0 e^{i(\omega t - \boldsymbol{k} \cdot \boldsymbol{r})}$$

with $|\boldsymbol{k}| = \omega/c$, we have, to an order of magnitude: $|(\boldsymbol{v} \cdot \nabla)\boldsymbol{v}| \approx |\boldsymbol{k}| v_0^2$ and $|\dot{\boldsymbol{v}}| \approx |\boldsymbol{v}_0| \omega$; $|(\boldsymbol{v} \cdot \nabla)\boldsymbol{v}| \ll |\dot{\boldsymbol{v}}|$ thus means that $|k v_0| \ll \omega$ or $|\boldsymbol{v}_0| \ll \omega/k = c$. For a sound wave, the "sound particle velocity" $|\boldsymbol{v}_0|$ is of the order of magnitude 10^{-2} cm/s, so that indeed $|\boldsymbol{v}_0| \ll c$.

iii) We consider a cylindrical pipe in the z-direction through which an incompressible fluid is flowing (Fig. 9.6.1). In a stationary state, the Navier–Stokes equation holds in the form

$$(\boldsymbol{v} \cdot \nabla)\boldsymbol{v} = -\frac{1}{\varrho} \nabla p + \frac{\eta}{\varrho} \Delta \boldsymbol{v} , \tag{9.6.12}$$

where we have neglected external forces.

We are interested in the velocity profile. We look for a solution for which

$$v_x = v_y = 0$$

and since $\nabla \cdot \boldsymbol{v} = 0$, also

$$\partial v_z / \partial z = 0 . \tag{9.6.13}$$

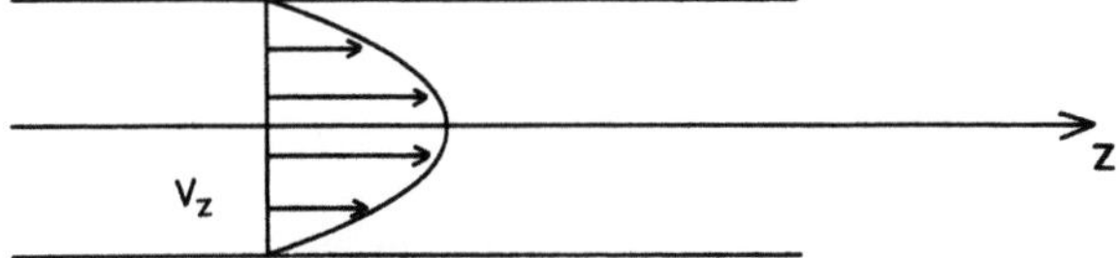

Fig. 9.6.1. Flow in a pipe with axial symmetry. A parabolic velocity profile results. This is called a Poiseuille flow

This means: v must always point in the z-direction and the v_z components must be independent of z.

Then

$$(v \cdot \nabla)v = v_z \frac{\partial}{\partial z} (0, 0, v_z) = 0 ,$$

and from (9.6.12) we find the equation

$$\eta \Delta v_z = \partial p/\partial z . \tag{9.6.14}$$

$\partial p/\partial z$ is the pressure gradient along the pipe. From (9.6.13), this gradient is constant and can be expressed in terms of the pressures p_1 and p_2 at the two ends of the pipe:

$$\partial p/\partial z = -(p_1 - p_2)/l , \tag{9.6.15}$$

where l is the length of the pipe.

If we introduce cylindrical coordinates

$$x = r \sin \theta ,$$

$$y = r \cos \theta ,$$

$$z = z , \quad 0 \le \theta \le 2\pi , \quad 0 \le r \le R , \tag{9.6.16}$$

then the Laplacian is given by (as shown in Appendix F)

$$\Delta = \frac{\partial^2}{\partial r^2} + \frac{1}{r} \frac{\partial}{\partial r} + \frac{1}{r^2} \frac{\partial^2}{\partial \theta^2} + \frac{\partial^2}{\partial z^2} . \tag{9.6.17}$$

Then, since $v_z = v_z(r)$, the differential equation (9.6.14) reads

$$\Delta v_z(r) = \left(\frac{\partial^2}{\partial r^2} + \frac{1}{r} \frac{\partial}{\partial r} \right) v_z(r) = -\frac{1}{\eta} \frac{(p_1 - p_2)}{l} , \tag{9.6.18}$$

and the solution which vanishes at the boundaries is

$$v_z(r) = \frac{1}{4\eta} \frac{p_1 - p_2}{l} (R^2 - r^2) , \tag{9.6.19}$$

as can easily be verified. The amount V of liquid which flows per unit time through the cross section is given by

$$V = 2\pi \int_0^R dr\, r v_z(r) = \frac{\pi R^4}{8\eta l}\,(p_1 - p_2)\ . \tag{9.6.20}$$

This is the *Hagen*[13]*–Poiseuille*[14] *equation.*

iv) Because of the large scale in the motion of air in the atmosphere, gravitation and the Coriolis force have a very important influence on the weather. The Navier–Stokes equation in a rotating coordinate system is given by

$$\varrho\left(\frac{D\boldsymbol{v}}{Dt} + 2\boldsymbol{\Omega} \times \boldsymbol{v}\right) = -\nabla p + \varrho \boldsymbol{g} + \eta \Delta \boldsymbol{v} \tag{9.6.21}$$

where $\boldsymbol{\Omega}$ here represents the angular velocity vector of the Earth's rotation. At altitudes above ≈ 1500 m above the earth's surface, the friction term $\Delta \boldsymbol{v}$ can be ignored.

If we again introduce a coordinate system on the earth in which $\boldsymbol{e}_3$ points upwards, $\boldsymbol{e}_1$ points to the east, and $\boldsymbol{e}_2$ points north, then for large-scale motions we find the approximations

$$dp(z)/dz = -\varrho g \tag{9.6.22}$$

from equation (9.6.21), and for the horizontal wind velocity, $\boldsymbol{v} = v_1 \boldsymbol{e}_1 + v_2 \boldsymbol{e}_2$:

$$f\boldsymbol{v} \times \boldsymbol{e}_3 = \frac{1}{\varrho}\nabla p \tag{9.6.23}$$

with $f = 2|\boldsymbol{\Omega}|\sin\phi$, $\phi = $ geographic latitude (see [Houghton, Pedlosky]). This means that for horizontal motion, the Coriolis term $2\boldsymbol{\Omega} \times \boldsymbol{v}$ and the pressure term ∇p are dominant, while the vertical term vanishes in this very rough first approximation.

With the equation of state

$$p = \varrho R T$$

for example, the equation (9.6.22) yields the barometric altitude formula:

$$p = p_0\, e^{-gz/RT}\ , \tag{9.6.24}$$

---

[13] *Hagen, Gotthilf* (*1793 Königsberg, d. 1884 Berlin). Hydraulic engineer, worked at the Engineering Academy, the Artillery and Engineering School, and at the Prussian Ministry of Commerce in Berlin. He was involved in the construction of Wilhelmshaven.

[14] *Poiseuille, Jean Louis Marie* (*1799 Paris, d. 1869 Paris). French medical researcher, investigated the flow of liquids through pipes as a physiologist.

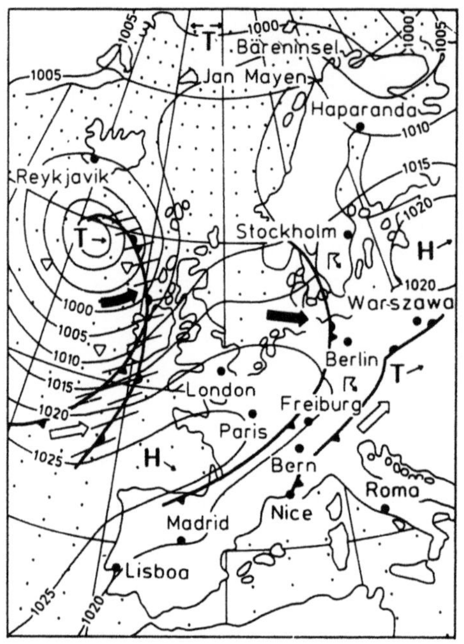

Fig. 9.6.2. Typical weather map. The dominant wind direction is parallel to the isobars

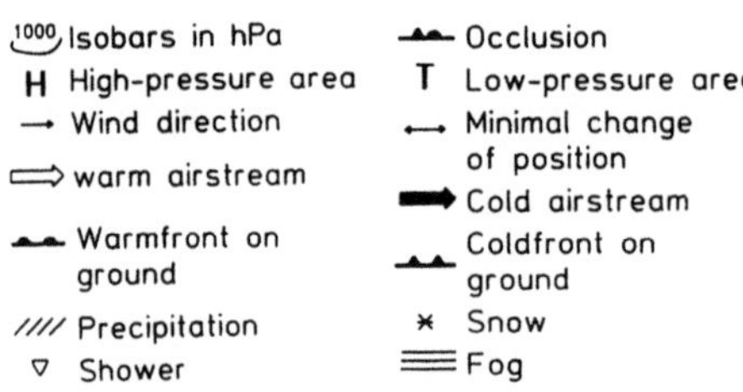

while equation (9.6.23) has the solution

$$v = \frac{1}{f\varrho}\, e_3 \times \nabla p \ . \tag{9.6.25}$$

The direction of the horizontal wind is thus always perpendicular to the pressure gradients, i.e. parallel to the isobars. This dominant wind component is called the *geostrophic wind*. A look at an everyday weather map confirms this effect (Fig. 9.6.2). Around a region of low pressure, the winds move counterclockwise; around a region of high pressure, they move clockwise.

Remarks. i) Flows in which fluids slide in layers next to or on top of each other are called *laminar*[15]. In contrast, for *turbulent flow* there are strong fluctuations

---

[15] Laminar (Latin) layered.

in the velocity field over time and space. Energy, momentum, and mass flow, as well as such quantities as the resistance to flow, are strongly dependent on the form of the flow. Understanding turbulence is one of the most difficult problems of classical physics and is a goal of current research. Seen mathematically, this is the problem of obtaining information about the solutions of nonlinear partial differential equations.

ii) Fluids whose behavior is described well by the Navier–Stokes equation are called Newtonian fluids, because then they have the Newtonian hypothesis (for $\nabla \cdot \boldsymbol{v} = 0$)

$$\tau_{ij} = \eta(\nabla_i v_j + \nabla_j v_i)$$

as their *constitutive relation*, i.e., this relation between the pressure tensor and the velocity gradients is valid for the fluid. However, as soon as the molecular structure of the constituents of a fluid becomes complicated, as in the case of polymer fluids, entirely new effects appear, which can no longer be explained in terms of this simple hypothesis and the Navier–Stokes equation. In a more general constitutive equation, non-linear terms in the velocity gradients and also "memory effects" (as in the case of so-called viscoelastic behavior) must be included (see [Bird et al.]).

Problem

9.1   The Production of Entropy During Diffusion. We consider solutions of the heat conduction equation $\dot{u} - \lambda \Delta u = 0$ $(\lambda > 0)$ in a compact region V with boundary ∂V, subject to either homogeneous Neumann boundary conditions or time-independent Dirichlet boundary conditions.

Show that under these assumptions, the total production of entropy

$$\dot{S} = \int\limits_V d^3x (\nabla u)^2$$

is a monotonically decreasing function of time, i.e. that $d\dot{S}/dt \leq 0$.

10. The Most Important Linear Partial Differential Equations of Physics

Of all the partial differential equations we encountered last chapter, we will consider three in some detail because of their particular importance. These three equations are the *wave equation*

$$\left(\frac{1}{c^2}\frac{\partial^2}{\partial t^2} - \Delta\right)u(t, r) = 0 \; ,$$

the *heat conduction equation*

$$\left(\frac{\partial}{\partial t} - \lambda\Delta\right)u(t, r) = 0$$

and *Laplace's equation*

$$\Delta u(r) = 0 \; .$$

These three equations represent the principle types of partial differential equations in theoretical physics. In this chapter we will discuss solutions of these equations as well as the conditions needed to uniquely characterize these solutions.

10.1 General Considerations

10.1.1 Types of Linear Partial Differential Equations, the Formulation of Boundary and Initial Value Problems

The wave equation is seen in many fields of physics. In electrodynamics, we will find that wave equations are valid in a vacuum for the electromagnetic fields $E(t, r)$ and $B(t, r)$.

However, such equations are also found in continuum mechanics. Consider a taut string of length l and let us examine the transverse displacement $u(t, x)$ $(0 \leq x \leq l)$ of an element of string from equilibrium. This also satisfies the wave equation

$$\left(\frac{1}{c^2}\frac{\partial^2}{\partial t^2} - \frac{\partial^2}{\partial x^2}\right)u(t, x) = 0 \; , \tag{10.1.1}$$

where now, we have

$$c^2 = \frac{S}{\varrho} \tag{10.1.2}$$

where ϱ is the mass per unit length, S the tension in the string. Clearly, we can consider $\partial^2/\partial x^2$ to be the Laplacian in one dimension.

On the other hand, consider a taut membrane, say the top of a drum, then the transverse displacement $u(t, x, y)$ satisfies the wave equation

$$\left(\frac{1}{c^2} \frac{\partial^2}{\partial t^2} - \Delta \right) u(t, x, y) = 0 \ , \tag{10.1.3}$$

where now

$$\Delta = \frac{\partial^2}{\partial x^2} + \frac{\partial^2}{\partial y^2}$$

is the Laplacian in two dimensions and

$$c^2 = \frac{\sigma}{\varrho} \ . \tag{10.1.4}$$

Here, σ is the tension per unit length on the edge of the membrane, and ϱ is the mass per unit area.

Following these considerations, we will define the wave equation in D dimensions as

$$\left(\frac{1}{c^2} \frac{\partial^2}{\partial t^2} - \Delta \right) u(t, x) = 0 \tag{10.1.5}$$

where $x = (x_1, \ldots, x_D)$ is a D-dimensional vector and

$$\Delta = \frac{\partial^2}{\partial x_1^2} + \cdots + \frac{\partial^2}{\partial x_D^2} \tag{10.1.6}$$

is the Laplacian in D dimensions. u here is a scalar field or one component of a vector field.

The heat conduction equation, as we have seen, can also describe the change in concentration of a substance during diffusion. In D spatial dimensions, the equation reads

$$\left(\frac{\partial}{\partial t} - \lambda \Delta \right) u(t, x) = 0 \ . \tag{10.1.7}$$

Despite formal similarities, the wave equation and the heat conduction equation differ in the order of the time derivative:

The wave equation, like Newton's equation of motion, contains second-order time derivatives. It remains invariant under the substitution $t \to -t$, that is, if $u(t, x)$ is a solution of the wave equation then so is $u(-t, x)$. This time reversal symmetry is not valid for the diffusion equation, which has a time derivative of first order. This corresponds to the fact that the wave equation describes *oscillatory processes*, while the heat conduction equation applies to *relaxation processes* such as heat conduction and diffusion, for which one direction in time is privileged because of the monotonic increase in entropy.

The difference in the order of time derivative also effects the number of initial conditions needed to uniquely determine a solution of the equation.

It is intuitively clear that the temperature distribution in a heat conduction problem is uniquely determined for all later times when at some initial point of time, say $t = 0$, the temperature distribution $u(0, x) = a(x)$ is given for all space.

In contrast, for the wave equation we need to have two functions at $t = 0$ in order to uniquely characterize a solution for all later times, namely the amplitude

$$u(0, x) = a(x)$$

and the first derivative of the amplitude

$$\dot{u}(0, x) = b(x) \ .$$

This is analogous to Newton's equation of motion for a system with N degrees of freedom, for which we need $2N$ initial values. The wave equation can be understood as Newton's equation of motion for a system with infinitely many degrees of freedom (one for every point in space). Seen this way, two functions must be given as the initial values.

In stationary, i.e. time-independent cases, the wave and heat conduction equations reduce to Laplace's equation

$$\Delta u(x) = 0 \ . \tag{10.1.8}$$

This equation no longer describes a process but rather a *state of equilibrium*. In this situation, there is clearly no more reason to give initial conditions.

The equations (10.1.5, 7, and 8) are the basic forms of the most important linear differential equations of physics. They are, however, suited to different types of physical problems. Mathematically, the wave equation is classified as a *hyperbolic*, the heat conduction equation as a *parabolic*, and Laplace's equation as an *elliptical* partial differential equation. In general, all linear partial differential equation can be classified in one of these categories, and the equations (10.1.5, 7, and 8) are the most elementary representatives from each category.

In many cases, the physical problem is set up in such a way that (10.1.5, 7, and 8) are only considered in a region V and not in all of space, e.g. sound waves, diffusion, or heat conduction in a container V. Of course, in this case it is no longer meaningful to give initial conditions outside of V. Instead, we are often given the behavior at all times t on the boundary ∂V of V. These boundary conditions usually occur in one of two forms: first, *Dirichlet*[1] *boundary conditions*:

$u(t, \mathbf{x})$ *is known for all t and* $\mathbf{x} \in \partial V$,

or second, *Neumann*[2] *boundary conditions*:

The normal derivative $(\mathbf{n} \cdot \mathbf{V}) u(t, \mathbf{x})$ *is known for all t and all* $\mathbf{x} \in \partial V$ *where* $\mathbf{n}$ *is the normal to* ∂V *at the point* $\mathbf{x} \in \partial V$.

Here are a few examples:

In the case of the heat conduction equation, the temperature of the boundary ∂V of V can be fixed from outside, so that $u(t, \mathbf{x}) = T(\mathbf{x})$ for $\mathbf{x} \in \partial V$ (Dirichlet boundary conditions).

Heat conduction can also occur in a vessel with insulated walls. In this case, the heat flow $\mathbf{Q} = -\kappa \mathbf{V} T$ cannot have any component perpendicular to the surface. Then, we must have

$$(\mathbf{n} \cdot \mathbf{V}) u(t, \mathbf{x}) = 0 \quad \text{for} \quad \mathbf{x} \in \partial V$$

(homogeneous Neumann boundary conditions).

The same boundary condition occurs for sound waves of diffusion in a rigid container V.

In these cases, the particle velocities

$$\dot{\mathbf{v}} = -\frac{1}{\varrho} \mathbf{V} p$$

or the diffusion current

$$\mathbf{J} = -D \mathbf{V} c$$

respectively can have no normal component.

It can be shown that Dirichlet and Neumann boundary conditions cannot both be given.

For the heat conduction and wave equations, we need additional initial conditions inside of V if the solution is to be characterized uniquely. The boundary conditions and the initial conditions must of course be compatible with each other.

The solutions of Laplace's equation (10.1.8) are already characterized by the boundary conditions. This agrees with the physical fact that equilibrium states are globally determined by boundary conditions.

---

[1] *Dirichlet, Peter Gustav Lejeune* (*1805 Düren, d. 1839 Göttingen). German mathematician, follower of Gauss in Göttingen. Important works on number theory, analysis, and mechanics.

[2] *Neumann, Franz Ernst* (*1798 Joachimsthal, d. 1895 Königsberg). German mathematician, physicist and mineralogist. Works on specific heats, optics, and electricity.

The equations (10.1.5, 7, and 8) have in common their *linearity*. Therefore, the superposition principle is valid:

If u_1 and u_2 are solutions of (10.1.5, 7, or 8) then so is $\alpha_1 u_1 + \alpha_2 u_2$ for any $\alpha_1, \alpha_2 \in \mathbb{R}$ or $\mathbb{C}$.

This principle considerably simplifies the solution of initial value problems for the heat conduction and wave equations. As in the case of linear equations for harmonic motion with a finite number of degrees of freedom, we can find the solution of the general initial value problem as a linear superposition of several basic solutions.

10.1.2   Initial Value Problems in $\mathbb{R}^D$

We now will discuss a few special solutions.

i) We claim:

Let $D(t, x)$ be a special solution of the heat conduction equation

$$\left(\frac{\partial}{\partial t} - \lambda \Delta\right) D(t, x) = 0$$

with initial values $D(0, x) = \delta(x)$ (see Appendix E). Then the solution of the initial value problem in $\mathbb{R}^D$,

$$\left(\frac{\partial}{\partial t} - \lambda \Delta\right) u_a(t, x) = 0 \; ; \quad u_a(0, x) = a(x) \tag{10.1.9}$$

is given by

$$u_a(t, x) = \int d^D x' \, a(x') D(t, x - x') \; . \tag{10.1.10}$$

As proof, we simply evaluate

$$\left(\frac{\partial}{\partial t} - \lambda \Delta\right) u_a(t, x) = \int d^D x' a(x') \left(\frac{\partial}{\partial t} - \lambda \Delta\right) D(t, x - x') = 0 \tag{10.1.11}$$

and

$$u_a(0, x) = \int d^D x' \, a(x') D(0, x - x')$$

$$= \int d^D x' \, a(x') \delta(x - x') = a(x) \; . \tag{10.1.12}$$

The special solution $D(t, x)$ defined in this way is called a *heat conduction kernel*.

We want to calculate $D(t, x)$ explicitly. Let us write

$$D(t, x) = \int d^D k \, \tilde{D}(t, k) e^{i k \cdot x} \; . \tag{10.1.13}$$

(Every function $D(t, x)$ can be written in this way with a suitable $\tilde{D}$ which can easily be found using the Fourier transform as in Appendix D.)

Then,

$$\left(\frac{\partial}{\partial t} - \lambda\Delta\right) D(t, x) = \int d^D k \left(\frac{\partial}{\partial t} - \lambda\Delta\right) \tilde{D}(t, k) e^{ik\cdot x}$$

$$= \int d^D k [\dot{\tilde{D}}(t, k) + \lambda k^2 \tilde{D}(t, k)] e^{ik\cdot x} = 0 \ . \tag{10.1.14}$$

From this, it follows that

$$\dot{\tilde{D}}(t, k) + \lambda k^2 \tilde{D}(t, k) = 0 \ , \tag{10.1.15}$$

so that $\tilde{D}(t, k)$ must be of the form

$$\tilde{D}(t, k) = C(k) e^{-\lambda k^2 t} \tag{10.1.16}$$

for some function $C(k)$. $C(k)$ can then be determined by the initial conditions. We have

$$D(t, x) = \int d^D k \, e^{-\lambda k^2 t} C(k) e^{ik\cdot x} \ . \tag{10.1.17}$$

Since (see Appendix D, E)

$$D(0, x) = \delta(x) = \frac{1}{(2\pi)^D} \int d^D k \, e^{ik\cdot x}$$

if follows that

$$C(k) = \frac{1}{(2\pi)^D} \ ,$$

and finally

$$D(t, x) = \frac{1}{(2\pi)^D} \int d^D k \, \exp(-\lambda k^2 t + ik\cdot x)$$

$$= \frac{1}{(2\pi)^D} \int d^D k \, \exp\{-\lambda t[k^2 - (ik\cdot x/\lambda t) - (x^2/4\lambda^2 t^2) + (x^2/4\lambda^2 t^2)]\}$$

$$= \frac{1}{(2\pi)^D} \exp(-x^2/4\lambda t) \int d^D k \, \exp\{-\lambda t[k - (ix/2\lambda t)]^2\}$$

$$= \frac{1}{(4\pi\lambda t)^{D/2}} \exp(-x^2/4\lambda t) \tag{10.1.18}$$

(since $\int dr \exp[-(r + is)^2] = \sqrt{\pi}$).

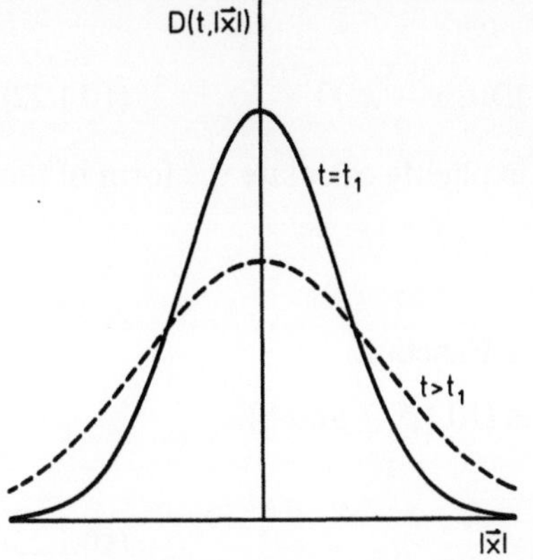

Fig. 10.1.1. The heat conduction kernel $D(t, |x|)$ for various values of t

Obviously, the solution is only meaningful for $t \geq 0$. $D(t, x)$ describes the change in temperature through heat conduction if there is an instantaneous heat pulse at time $t = 0$ at the point $x = 0$.

Figure 10.1.1 shows how the effects of the heat pulse spread out more and more from the single point $x = 0$ as the time t increases. At time t, the temperature distribution, which is always in the shape of a bell curve, has width

$$2\sqrt{\lambda t} \ .$$

We see further that, for $t > 0$, $D(t, x)$ is a series of smooth functions of x which approach the δ-distribution as $t \to 0$.

The solution of the initial value problem then reads explicitly

$$u_a(t, x) = \frac{1}{(4\pi\lambda t)^{D/2}} \int d^D x' a(x') \exp\left[-\frac{(x - x')^2}{4\lambda t} \right] . \tag{10.1.19}$$

ii) For the wave equation, we introduce a *wave propagation kernel* which, for simplicity, we also indicate by $D(t, x)$, defined by

$$\left(\frac{1}{c^2} \frac{\partial^2}{\partial t^2} - \Delta \right) D(t, x) = 0 \ ,$$

$$D(0, x) = 0 \ , \quad \dot{D}(0, x) = \delta(x) \ . \tag{10.1.20}$$

Then,

$$\ddot{D}(0, x) = c^2 \Delta D(0, x) = 0 \ ,$$

and the solution $u_{ab}(t, x)$ of the general initial value problem in $\mathbb{R}^D$:

$$\left(\frac{1}{c^2} \frac{\partial^2}{\partial t^2} - \Delta \right) u_{ab}(t, x) = 0 \quad \text{with}$$

$$u_{ab}(0, x) = a(x) \ , \quad \dot{u}_{ab}(0, x) = b(x) \tag{10.1.21}$$

is given by

$$u_{ab}(t, x) = \int d^D x' [a(x')\dot{D}(t, x - x') + b(x')D(t, x - x')] \tag{10.1.22}$$

as can easily be verified. In Sect. 10.2, we will explicitly calculate the form of the wave propagation kernel $D(t, x)$ for $D = 3$.

10.1.3 Inhomogeneous Equations and Green's Functions

The inhomogeneous versions of the equations (10.1.5, 7 and 8)

$$\left(\frac{1}{c^2}\frac{\partial^2}{\partial t^2} - \Delta\right)u(t, x) = h(t, x) \ , \tag{10.1.23}$$

$$\left(\frac{\partial}{\partial t} - \lambda\Delta\right)u(t, x) = h(t, x) \ , \tag{10.1.24}$$

$$\Delta u(x) = h(x) \ , \tag{10.1.25}$$

which describe forced oscillations, heat conduction with heat sources, etc. are of great importance.

Exactly as in the case of oscillations with finitely many degrees of freedom, we can obtain special solutions of the inhomogeneous equation if we construct a Green's function G (see Appendix E) which for the cases (10.1.23–25), respectively, satisfies the following condition:

$$\left(\frac{1}{c^2}\frac{\partial^2}{\partial t^2} - \Delta\right)G(t, x) = \delta(t)\delta(x) \tag{10.1.26}$$

$$\left(\frac{\partial}{\partial t} - \lambda\Delta\right)G(t, x) = \delta(t)\delta(x) \quad \text{or} \tag{10.1.27}$$

$$\Delta G(x) = \delta(x) \ . \tag{10.1.28}$$

The most general solution of the inhomogeneous equation can then be found from

$$u_0(t, x) = \int dt' \int d^D x' h(t', x')\, G(t - t', x - x') \tag{10.1.29}$$

$$u_0(t, x) = \int dt' \int d^D x' h(t', x')\, G(t - t', x - x') \tag{10.1.30}$$

or

$$u_0(x) = \int d^D x' h(x')\, G(x - x') \tag{10.1.31}$$

by adding on the general solution of the corresponding homogeneous equation.

For the heat conduction equation, a special Green's function $G_R(t, x)$ can be found immediately from the heat conduction kernel $D(t, x)$:

$$G_R(t, x) = \Theta(t)D(t, x) \quad \text{with (see Appendix E)} \tag{10.1.32}$$

$$\Theta(t) = \begin{cases} 0 \text{ for } t < 0 \\ 1 \text{ for } t > 0 \end{cases}.$$

We have $G_R(t, x) = 0$ for $t < 0$. G_R is called the *retarded Green's function*.

To show this, we calculate (see Appendix E)

$$\frac{\partial}{\partial t} G_R(t, x) = \dot{\Theta}(t)D(t, x) + \Theta(t)\dot{D}(t, x)$$

$$= \delta(t)D(t, x) + \Theta(t)\dot{D}(t, x)$$

$$= \delta(t)D(0, x) + \Theta(t)\dot{D}(t, x)$$

$$= \delta(t)\delta(x) + \Theta(t)\dot{D}(t, x) \tag{10.1.33}$$

and

$$\left(\frac{\partial}{\partial t} - \lambda\Delta\right) G_R(t, x) = \delta(t)\,\delta(x) + \Theta(t)\left(\frac{\partial}{\partial t} - \lambda\Delta\right)D(t, x)$$

$$= \delta(t)\delta(x) . \tag{10.1.34}$$

We can also find a retarded Green's function for the wave equation from the wave propagation kernel $D(t, x)$:

$$G_R(t, x) = c^2\Theta(t)D(t, x) . \tag{10.1.35}$$

Proof.

$$\dot{G}_R(t, x) = c^2\delta(t)D(t, x) + c^2\Theta(t)\dot{D}(t, x) = c^2\Theta(t)\dot{D}(t, x) , \tag{10.1.36}$$

$$\ddot{G}_R(t, x) = c^2\delta(t)\dot{D}(t, x) + c^2\Theta(t)\ddot{D}(t, x) = c^2\delta(t)\delta(x) + c^2\Theta(t)\ddot{D}(t, x) , \tag{10.1.37}$$

thus

$$\left(\frac{1}{c^2}\frac{\partial}{\partial t^2} - \Delta\right)G_R(t, x) = \delta(t)\delta(x) + \Theta(t)[\ddot{D}(t, x) - c^2\Delta D(t, x)]$$

$$= \delta(t)\delta(x) . \tag{10.1.38}$$

The retarded Green's function for the wave equation will prove very useful in many cases.

10.2   Solutions of the Wave Equation

The wave equation in D dimensions has solutions of the form

$$f(\mathbf{n} \cdot \mathbf{x} - ct) \ , \tag{10.2.1}$$

where $\mathbf{n}$ is an arbitrary D-dimensional unit vector. The function $f(x)$: $\mathbb{R} \to \mathbb{R}$ can be chosen arbitrarily.

Proof. We have

$$\frac{\partial^2}{\partial t^2} f(\mathbf{n} \cdot \mathbf{x} - ct) = c^2 f''(\mathbf{n} \cdot \mathbf{x} - ct) \tag{10.2.2}$$

but also

$$\nabla f(\mathbf{n} \cdot \mathbf{x} - ct) = \mathbf{n} f'(\mathbf{n} \cdot \mathbf{x} - ct) \tag{10.2.3}$$

so that also

$$\Delta f(\mathbf{n} \cdot \mathbf{x} - ct) = \mathbf{n}^2 f''(\mathbf{n} \cdot \mathbf{x} - ct) \ . \tag{10.2.4}$$

Since $\mathbf{n}^2 = 1$, the wave equation is satisfied.

This solution represents travelling waves: If $f(x)$ is maximal at $x = x_0$ then $f(\mathbf{n} \cdot \mathbf{x} - ct)$ is maximal at all (x, t) with

$$\mathbf{n} \cdot \mathbf{x} - ct = x_0 \ .$$

If the z-axis lies in the $\mathbf{n}$-direction, then

$$z = x_0 + ct \ ,$$

that is, the maximum moves with speed c in the direction of greater z values, or, more generally, in the direction of $\mathbf{n}$. $f(\mathbf{n} \cdot \mathbf{x} - ct)$ is then always maximal in the entire plane perpendicular to $\mathbf{n}$. These planes perpendicular to $\mathbf{n}$, propagating with speed c in the $\mathbf{n}$-direction, such that $f(\mathbf{n} \cdot \mathbf{x} - ct)$ is maximal for the points x on each of the planes, are called the *wave fronts*, $\mathbf{n}$ is called the *direction of propagation* of the wave, and c is called the *speed* of the wave.

If f_1 and f_2 are solutions, then so is $f_1 + f_2$.

For the case of one (spatial) dimension $D = 1$, we find immediately that the most general solution has the form:

$$u(t, x) = f(x - ct) + g(x + ct) \tag{10.2.5}$$

for arbitrary functions f and g.

If we are given initial conditions

$$u(0, x) = a(x) \quad \text{and} \quad \dot{u}(0, x) = b(x)$$

then the solution is uniquely determined:

$$u(0, x) = f(x) + g(x) \stackrel{!}{=} a(x) \, , \tag{10.2.6}$$

$$\dot{u}(0, x) \stackrel{\cdot}{=} -cf'(x) + cg'(x) \stackrel{!}{=} b(x) \, , \tag{10.2.7}$$

i.e.

$$g'(x) - f'(x) = \frac{1}{c} b(x) \, ,$$

and thus it follows that

$$g(x) = \frac{1}{2} a(x) + \frac{1}{2c} \int^x b(\zeta)d\zeta \, , \quad f(x) = \frac{1}{2} a(x) - \frac{1}{2c} \int^x b(\zeta)d\zeta \tag{10.2.8}$$

and

$$u(t, x) = f(x - ct) + g(x + ct)$$

$$= \frac{1}{2}[a(x + ct) + a(x - ct)] + \frac{1}{2c} \int\limits_{x-ct}^{x+ct} b(\zeta)d\zeta \, . \tag{10.2.9}$$

For the D-dimensional wave equation, *plane waves*

$$\exp(i\omega t + i\boldsymbol{k} \cdot \boldsymbol{x}) \quad \text{with} \quad |\omega| = |\boldsymbol{k}|c = kc$$

are particularly important solutions. They are solutions of the form given above with the function

$$f(x) = e^{ikx} \, .$$

The direction of $\boldsymbol{k}$ is at the same time the direction of propagation of the wave and the normal to the wave fronts. The time dependence is periodic with period

$$T = \frac{2\pi}{\omega}$$

and frequency

$$v = \frac{\omega}{2\pi} \, .$$

A change in x perpendicular to $\boldsymbol{k}$ leaves the value of

$$\exp(i\omega t + i\boldsymbol{k} \cdot \boldsymbol{x})$$

unchanged. In the $\boldsymbol{k}$-direction, the function is periodic with period

$$\lambda = \frac{2\pi}{k} \, .$$

λ is called the *wavelength* and the vector $\boldsymbol{k}$ which determines the direction of propagation and the wavelength is called the *wave vector*.

It is always true that $|\omega| = kc$ or, written another way, $\lambda = cT$. We note that the propagation speed c of the plane wave is independent of the wavelength. This is a peculiarity of the wave equation

$$\left(\frac{1}{c^2} \frac{\partial^2}{\partial t^2} - \Delta \right) u = 0 \; .$$

In the general case of wave propagation, *dispersion* occurs, i.e. the propagation velocity depends on the wavelength (see also Sect. 14.4). For example, the frequency dependence of the propagation velocity of light causes a white ray of light to divide into many colors in a prism.

Plane waves are very special solutions of the wave equation. They are important because the most general solution of the wave equation can be written as a continuous linear superposition of plane waves. We will now prove this. Along the way, we will find a simple proof of the existence and uniqueness of a solution to the initial value problem.

We therefore look for general solutions $u(t, x)$ of the wave equation. Let us represent, for fixed values of t, the function $u(t, x)$ as a Fourier integral (see Appendix D):

$$u(t, x) = (2\pi)^{-D/2} \int d^D k \, \tilde{u}(t, k) e^{i k \cdot x} \; . \tag{10.2.10}$$

This representation is possible and unique for every "reasonable" function $u(t, x)$. From the inversion formula for the Fourier transform, we have

$$\tilde{u}(t, k) = (2\pi)^{-D/2} \int d^D x \, u(t, x) e^{-i k \cdot x} \; . \tag{10.2.11}$$

Now,

$$\left(\frac{1}{c^2} \frac{\partial^2}{\partial t^2} - \Delta \right) u(t, x) = (2\pi)^{-D/2} \int d^D k \left[\frac{1}{c^2} \ddot{\tilde{u}}(t, k) + k^2 \tilde{u}(t, k) \right] e^{i k \cdot x} = 0 \; ,$$

$$\tag{10.2.12}$$

so that

$$\frac{1}{c^2} \ddot{\tilde{u}}(t, k) + k^2 \tilde{u}(t, k) = 0 \; . \tag{10.2.13}$$

Thus, $\tilde{u}(t, k)$ must be of the form

$$\tilde{u}(t, k) = \alpha(k) e^{i\omega t} + \beta(k) e^{-i\omega t} \tag{10.2.14}$$

with $\omega = \omega(k) = kc$.

Therefore, the most general solution of the wave equation is of the form

$$u(t, x) = (2\pi)^{-D/2} \int d^D k \, [\alpha(k) \exp(i\omega t + i k \cdot x)$$

$$+ \, \beta(k) \exp(- i\omega t + i k \cdot x)] \tag{10.2.15}$$

with arbitrary functions $\alpha(k)$ and $\beta(k)$.

We see that $u(t, x)$ is a continuous superposition (i.e., an integral) of plane waves. We now want to find the solution $u_{ab}(t, x)$ with initial values

$$u_{ab}(0, x) = a(x) \quad \text{and} \quad \dot{u}_{ab}(0, x) = b(x) \; .$$

We find

$$u(0, x) = (2\pi)^{-D/2} \int d^D k \, [\alpha(k) + \beta(k)] \, e^{i k \cdot x} = a(x)$$

$$= (2\pi)^{-D/2} \int d^D k \, \tilde{a}(k) \, e^{i k \cdot x} \tag{10.2.16}$$

where $\tilde{a}$ is the Fourier transform of a and

$$\dot{u}(0, x) = (2\pi)^{-D/2} \int d^D k \, i\omega \, [\alpha(k) - \beta(k)] \, e^{i k \cdot x} = b(x)$$

$$= (2\pi)^{-D/2} \int d^D k \, \tilde{b}(k) \, e^{i k \cdot x} \; . \tag{10.2.17}$$

Then we must have

$$\alpha(k) + \beta(k) = \tilde{a}(k) \; ,$$
$$i\omega \, [\alpha(k) - \beta(k)] = \tilde{b}(k) \quad \text{or} \tag{10.2.18}$$

$$\alpha(k) = \frac{1}{2} \tilde{a}(k) + \frac{1}{2i\omega} \tilde{b}(k) \; ,$$
$$\beta(k) = \frac{1}{2} \tilde{a}(k) - \frac{1}{2i\omega} \tilde{b}(k) \; , \qquad \omega = kc \; . \tag{10.2.19}$$

Thus we have shown that the initial value problem has a unique solution. In particular, we want to determine the wave propagation kernel $D(t, x)$ with the initial conditions $D(0, x) = 0$, $\dot{D}(0, x) = \delta(x)$.

Then, $\tilde{a}(k) = 0$ and $\tilde{b}(k) = (2\pi)^{-D/2}$ so that

$$D(t, x) = (2\pi)^{-D} \int d^D k \, \frac{e^{i\omega t} - e^{- i\omega t}}{2i\omega(k)} \, e^{i k \cdot x} \; . \tag{10.2.20}$$

For the most important case, $D = 3$, the integral can easily be evaluated by introducing polar coordinates in k-space (with the 3-direction parallel to x).

The result is

$$D(t, x) = \frac{1}{4\pi c^2 |x|} \left[\delta\!\left(t - \frac{|x|}{c} \right) - \delta\!\left(t + \frac{|x|}{c} \right) \right] \; . \tag{10.2.21}$$

We also obtain the retarded Green's function for the three dimensional wave equation $G_R(t, x) = c^2 \Theta(t) D(t, x)$:

$$G_R(t, x) = \frac{1}{4\pi|x|} \delta\left(t - \frac{|x|}{c}\right) \Theta(t) \ . \tag{10.2.22}$$

10.3 Boundary Value Problems

10.3.1 Initial Observations

We now turn to the problem in which the wave equation is to be solved in a region V with homogeneous Dirichlet or Neumann boundary conditions:

$$\left(\frac{1}{c^2} \frac{\partial^2}{\partial t^2} - \Delta\right) u(t, x) = 0 \quad \text{in } V \quad \text{and} \tag{10.3.1}$$

$$u(t, x) = 0 \quad \text{for} \quad x \in \partial V \quad \text{or} \tag{10.3.2}$$

$$n \cdot \nabla u(t, x) = 0 \quad \text{for} \quad x \in \partial V \ . \tag{10.3.3}$$

We will now see a close analogy between the wave equation and the harmonic equation

$$\ddot{u}(t) + Ku(t) = 0$$

from Chap. 6, in which $u(t) \in \mathbb{R}^f$ was an f-component vector and K is an $f \times f$ matrix.

In the wave equation, the quantity $u(t, x)$ for fixed value of t is a function of x which satisfies the boundary conditions, thus an element of an infinite dimensional vector space, and in place of the linear mapping K we have the linear differential operator $- c^2 \Delta$.

To solve the harmonic equation, it was most important to find the normal modes and eigenvalues

$$u^{(\alpha)}(t) = e^{i\omega_\alpha t} v^{(\alpha)}$$

where $v^{(\alpha)} \in \mathbb{R}^f$ was a solution of the eigenvalue equation

$$(K - \omega_\alpha^2 \mathbb{1}) v^{(\alpha)} = 0 \ .$$

We obtained the following results:

i) $\omega_\alpha^2 \geq 0$ (and $\omega_\alpha^2 > 0$ if the oscillation occurred about a minimum of the potential).

ii) Orthogonality: $(v^{(\alpha)}, v^{(\beta)}) = 0$ for $\omega_\alpha^2 \neq \omega_\beta^2$.

iii) Completeness: Each solution can be uniquely represented in the form

$$u(t) = \sum_{\alpha=1}^{f} (a_\alpha e^{i\omega_\alpha t} + b_\alpha e^{-i\omega_\alpha t}) v^{(\alpha)}$$

with $a_\alpha,\ b_\alpha \in \mathbb{C}$. (In particular, there were exactly f linearly independent eigenvectors $v^{(\alpha)}$.)

In the same way, we will now look for normal modes of the wave equation with purely harmonic time dependence. We therefore set:

$$u(t, x) = e^{i\omega t} v(x) \ . \tag{10.3.4.}$$

Then, if u is a solution of the wave equation, $v(x)$ must satisfy the equation

$$(\Delta + k^2) v(x) = 0 \ , \quad k^2 = \omega^2/c^2 \ . \tag{10.3.5}$$

This is called the *Helmholtz*[3] *equation.*
In addition, we must have

$$v(x) = 0 \quad \text{for} \quad x \in \partial V \tag{10.3.6}$$

(the Dirichlet boundary conditions)
or

$$n \cdot \nabla v(x) = 0 \quad \text{for} \quad x \in \partial V \tag{10.3.7}$$

(the Neumann boundary conditions).

The Helmholtz equation can be interpreted as an eigenvalue equation for the eigenfunction $v(x)$ with the proper boundary conditions (10.3.6 or 7), and thus we again look for the possible eigenvalues k^2 and eigenfunctions v. We will see that statements (i–iii) hold in an exactly analogous form.

10.3.2 Examples of Boundary Value Problems

a) To orient ourselves, we will begin with the simplest case, the one dimensional wave equation in the interval $[0, L]$ with homogeneous Dirichlet boundary conditions

$$u(t, 0) = u(t, L) = 0 \ .$$

---

[3] *Helmholtz, Hermann von* (*1829 Potsdam, d. 1894 Berlin). Great physiologist and physicist. In 1847, he formulated clearly the law of energy conservation. Works on hydrodynamics (theorems about vortices), thermodynamics, geometry, physiological optics and acoustics, inventor of the ophthalmoscope. "Chancellor of German physics". 1871, professor in Berlin, 1888, president of the "Physikalisch-Technische Riechsaustalt".

This is the problem of determining the vibrations of a string of length L with fixed end points.

In this case, the Helmholtz equation is given by

$$\frac{d^2}{dx^2} v(x) + k^2 v(x) = 0 \tag{10.3.8}$$

with the boundary conditions $v(0) = v(L) = 0$. The most general solution of the Helmholtz equation is

$$v(x) = \alpha \sin(kx) + \beta \cos(kx) \ . \tag{10.3.9}$$

From $v(0) = 0$, it follows that $\beta = 0$, and from $v(L) = 0$, it follows that $kL = n\pi$, with $n = 1, 2, \ldots$. Thus, we obtain normal modes with the proper boundary conditions only for the special real eigenvalues

$$k_n^2 = \left(\frac{n\pi}{L}\right)^2 \ . \tag{10.3.10}$$

The corresponding eigenfrequencies are given by

$$\omega_n^2 = c^2 k_n^2 = \frac{c^2 n^2 \pi^2}{L^2} \ . \tag{10.3.11}$$

We have

$$|\omega_n| = n |\omega_1| \ ,$$

and thus the eigenfrequencies are the multiples of the "basic frequency" ω_1. There is then a countably infinite sequence of eigenvalues.

As we know, for the eigenfunctions v_n

$$\int_0^L dx \, v_n(x) \, v_m(x) = \int_0^L dx \, \sin\left(\frac{n\pi x}{L}\right) \sin\left(\frac{m\pi x}{L}\right) = 0$$

for $m \neq n$ $(m, n \geq 0)$.

If we introduce a scalar product

$$(v, w) := \int_0^L dx \, v(x) \, w(x) \tag{10.3.12}$$

statement (ii) holds in the form

$$(v_m, v_n) = \delta_{mn} \ , \tag{10.3.13}$$

where we have suitably normalized v_n. v_n has exactly $(n - 1)$ zero points (*nodes*) inside the interval $[0, L]$.

Statement (i) clearly holds. The completeness condition (iii) is likewise satisfied, since each function in the interval $[0, L]$ which vanishes at the endpoints can be expanded in a Fourier series of the form

$$f(x) = \sum_{n=1}^{\infty} f_n v_n(x) = \sum_{n=1}^{\infty} f_n \sqrt{\frac{2}{L}} \sin\left(\frac{n\pi x}{L}\right)$$

with $f_n = (v_n, f)$ (see Appendix D).

In order to solve the initial value problem $u(0, x) = a(x)$, $\dot{u}(0, x) = b(x)$ with the boundary conditions $u(t, 0) = u(t, L) = 0$ (of course, we must have $a(0) = a(L) = b(0) = b(L) = 0$), we set

$$u(t, x) = \sum_{n=1}^{\infty} \left[\alpha_n \cos(\omega_n t) + \beta_n \frac{\sin(\omega_n t)}{\omega_n} \right] v_n(x) \tag{10.3.14}$$

with constants α_n and β_n which we then determine from the initial conditions:

$$a(x) = u(0, x) = \sum_{n=1}^{\infty} \alpha_n v_n(x) , \quad \text{thus} \quad \alpha_n = (v_n, a) ,$$

$$\tag{10.3.15}$$

$$b(x) = \dot{u}(0, x) = \sum_{n=1}^{\infty} \beta_n v_n(x) , \quad \text{thus} \quad \beta_n = (v_n, b) .$$

b) Next, we consider the two dimensional wave equation in the rectangle

$$0 \le x \le A , \quad 0 \le y \le B$$

with Dirichlet boundary conditions

$$u(t, 0, y) \equiv u(t, A, y) \equiv u(t, x, 0) \equiv u(t, x, B) \equiv 0 . \tag{10.3.16}$$

The Helmholtz equation reads

$$\left(\frac{\partial^2}{\partial x^2} + \frac{\partial^2}{\partial y^2} + k^2 \right) v(x, y) = 0 . \tag{10.3.17}$$

In order to solve this equation, we use the so-called *separation of variables* method, which yields special solutions to linear differential equations in many important cases:

We assume $v(x, y)$ can be written in the special form

$$v(x, y) = X(x) Y(y) \tag{10.3.18}$$

for some functions X and Y. The Dirichlet boundary conditions then assert

$$X(0) = X(A) = Y(0) = Y(B) = 0 . \tag{10.3.19}$$

Substituting this form for $v(x, y)$ into the Helmholtz equation yields

$$Y(y)\frac{d^2 X}{dx^2} + X(x)\frac{d^2 Y}{dy^2} + k^2 X(x) Y(y) = 0 \; . \tag{10.3.20}$$

Dividing by XY, we obtain

$$\frac{X''}{X} + \frac{Y''}{Y} + k^2 = 0 \; . \tag{10.3.21}$$

This is only possible if both

$$\frac{X''}{X} \quad \text{and} \quad \frac{Y''}{Y}$$

are constant:

$$\frac{X''}{X} = -k_x^2 \; , \quad \text{i.e.} \quad X'' + k_x^2 X = 0 \; , \tag{10.3.22}$$

$$\frac{Y''}{Y} = -k_y^2 \; , \quad \text{i.e.} \quad Y'' + k_y^2 Y = 0 \; . \tag{10.3.23}$$

Then, $k^2 = k_x^2 + k_y^2$ and from the condition $X(0) = Y(0) = 0$, we find immediately

$$X(x) = \alpha_1 \sin(k_x x) \; , \qquad Y(y) = \alpha_2 \sin(k_y y) \; . \tag{10.3.24}$$

The boundary conditions $X(A) = Y(B) = 0$ are only satisfied if

$$k_x = \frac{m\pi}{A} \quad \text{and} \quad k_y = \frac{n\pi}{B} \text{ with } m, n \in \mathbb{N}.$$

Again, we find a countable sequence of eigenvalues

$$k_{mn}^2 = \frac{m^2 \pi^2}{A^2} + \frac{n^2 \pi^2}{B^2} \tag{10.3.25}$$

and eigenfrequencies

$$\omega_{mn}^2 = c^2 k_{mn}^2 > 0 \; . \tag{10.3.26}$$

If we define

$$(v, w) := \int\limits_0^A dx \int\limits_0^B dy \, v(x, y) w(x, y) \tag{10.3.27}$$

then the normalized eigenfunctions

$$v_{mn}(x, y) = \frac{2}{\sqrt{AB}} \sin\left(\frac{m\pi x}{A}\right) \sin\left(\frac{n\pi y}{B}\right) \tag{10.3.28}$$

satisfy the orthogonality condition

$$(v_{mn}, v_{m'n'}) = \delta_{mm'}\,\delta_{nn'} \ . \tag{10.3.29}$$

Completeness is also valid here, since every function defined in a rectangle can be expanded in a double Fourier series (see Appendix D).

The uniquely determined solution to the initial value problem

$$u_{ab}(0, x, y) = a(x, y), \quad \dot{u}_{ab}(0, x, y) = b(x, y)$$

is

$$u_{ab}(t, x, y) = \sum_{m,\,n=1}^{\infty} \left[\alpha_{mn} \cos(\omega_{mn} t) + \beta_{mn} \frac{\sin(\omega_{mn} t)}{\omega_{mn}} \right] v_{mn}(x, y) \tag{10.3.30}$$

with

$$\alpha_{mn} = (v_{mn}, a) \quad \text{and} \quad \beta_{mn} = (v_{mn}, b) \ .$$

c) It should be clear now how we can solve the three dimensional wave equation in a box, $0 \le x \le A, 0 \le y \le B, 0 \le z \le C$. For the problem of sound waves in a box, the Neumann boundary conditions are appropriate.

We start by separating variables, $v(x, y, z) = X(x)\,Y(y)Z(z)$. The Neumann boundary conditions

$$X'(0) = X'(A) = Y'(0) = Y'(B) = Z'(0) = Z'(C) = 0$$

result in eigenfunctions of the form

$$v_{mnr}(x, y, z) = \sqrt{\frac{8}{ABC}} \cos\left(\frac{m\pi x}{A}\right) \cos\left(\frac{n\pi y}{B}\right) \cos\left(\frac{r\pi z}{C}\right) \tag{10.3.31}$$

with $m, n, r \ge 0$, integer numbers.

The corresponding eigenfrequencies are

$$\omega_{mnr}^2 = c^2 \left(\frac{m^2 \pi^2}{A^2} + \frac{n^2 \pi^2}{B^2} + \frac{r^2 \pi^2}{C^2} \right) \ . \tag{10.3.32}$$

With the scalar product

$$(v, w) := \int vw \, d^3x = \int_0^A dx \int_0^B dy \int_0^C dz \, v(x, y, z) w(x, y, z)$$

we have

$$(v_{mnr}, v_{m'n'r'}) = \delta_{mm'}\,\delta_{nn'}\,\delta_{rr'} \ ,$$

and the system of eigenfunctions is again complete, so that the initial value problem is solved uniquely. Thus properties (i–iii) are again satisfied.

10.3.3 The General Treatment of Boundary Value Problems

We will now consider in general the Helmholtz equation with homogeneous Dirichlet or Neumann boundary conditions for an arbitrarily shaped region V in D-dimensional space. We will show that properties (i) and (ii) are valid for the eigensolutions. To prove this, we need to use the identity

$$\mathbf{\nabla}\cdot(v\mathbf{\nabla}w) = \mathbf{\nabla}v\cdot\mathbf{\nabla}w + v\Delta w \ . \tag{10.3.33}$$

If we integrate over the region V and use Gauss' theorem (see Appendix F), we find

$$\int_V d^D x\,\mathbf{\nabla}\cdot(v\mathbf{\nabla}w) = \int_{\partial V} d\mathbf{F}\cdot v\mathbf{\nabla}w = \int_V d^D x\,\mathbf{\nabla}v\cdot\mathbf{\nabla}w + \int_V d^D x\,v\Delta w \ . \tag{10.3.34}$$

If v and w satisfy homogeneous Dirichlet or Neumann boundary conditions, the integral over ∂V always vanishes. If we define a scalar product

$$(v, w) := \int_V d^D x\,vw \tag{10.3.35}$$

then we have shown the identity

$$(v, \Delta w) = -\,(\mathbf{\nabla}v, \mathbf{\nabla}w) \ . \tag{10.3.36}$$

From this, it immediately follows that

a) $(v, \Delta w) = (\Delta v, w) \ ,$ and for $v = w \ ,$ $\hspace{2cm}$ (10.3.37)

b) $(v, \Delta v) = -\,(\mathbf{\nabla}v, \mathbf{\nabla}v) \le 0 \ .$ $\hspace{3.5cm}$ (10.3.38)

Now let v and w be solutions of the Helmholtz equation corresponding to different eigenvalues k^2 and k'^2:

$$(\Delta + k^2)v = 0 \ , \quad (\Delta + k'^2)w = 0 \ . \tag{10.3.39}$$

We now conclude, just as in the case of finite dimensional oscillations

from (b): $(v, \Delta v) = -\,k^2(v, v) = -\,(\mathbf{\nabla}v, \mathbf{\nabla}v) \le 0 \ ,$ $\hspace{1.5cm}$ (10.3.40)

thus $k^2 \ge 0.$

From (a): $(v, \Delta w) = -k'^2(v, w) = (\Delta v, w) = -k^2(v, w)$,   thus   (10.3.41)

$(k^2 - k'^2)(v, w) = 0$ and $(v, w) = 0$ for $k^2 \neq k'^2$.   (10.3.42)

Therefore (i) and (ii) are valid.

Remarks. a) The eigenvalue $-k^2$ of Δ is, in general, degenerate, i.e. there are several eigenfunctions for the same eigenvalue, thus a (finite dimensional) eigenspace corresponds to the eigenvalue $-k^2$. In this eigenspace, we can find a orthonormal system of functions with respect to the scalar product (10.3.35). An important example of this will be considered in Sect. 10.4.

b) In addition, we have answered the question of the uniqueness of the solutions of the inhomogeneous equation

$$\Delta u(x) = h(x)$$

with Dirichlet boundary conditions

$$u = a \quad \text{on } \partial V$$

or Neumann boundary conditions

$$\boldsymbol{n} \cdot \nabla u = b \quad \text{on } \partial V .$$

In particular, if u_1 and u_2 are two solutions of the problem, then $w = u_1 - u_2$ is a solution of

$$\Delta w = 0 \quad \text{and} \quad w = 0 \quad \text{or} \quad \boldsymbol{n} \cdot \Delta w = 0$$

respectively on ∂V.
From (10.3.38), we then have

$$(\nabla w, \nabla w) = 0 , \quad \text{i.e.} \quad \nabla w = 0 \quad \text{in } V .$$

For the Dirichlet boundary conditions this means that $w \equiv 0$ in V, i.e. the solution is unique, whereas for the Neumann boundary conditions this means that $w = $ constant (i.e. uniqueness up to an additive constant).

c) In functional analysis, it is shown that the eigenvalues form a countable series without a finite accumulation point and that "every" function in the region V can be expanded in a series

$$f(x) = \sum_N f_N v_N(x) \quad \text{with} \quad f_N = (v_N, f) .$$ (10.3.43)

Thus, property (iii) is satisfied.

We have thus solved the boundary and initial value problem for the wave equation in V:

$$u_{ab}(t, x) = \sum_N \left[a_N \cos(\omega_N t) + b_N \frac{\sin(\omega_N t)}{\omega_N} \right] v_N(x) \tag{10.3.44}$$

with $a_N = (v_N, a)$ and $b_N = (v_N, b)$. (For $\omega_N = 0$, we set $\sin(\omega_N t)/\omega_N = t$.)

d) At the same time, we have solved the boundary and initial value problem for the heat conduction equation

$$\left(\frac{\partial}{\partial t} - \lambda\Delta \right) u(t, x) = 0$$

for a region V with homogeneous Neumann or Dirichlet boundary conditions. In particular, we expand the position dependence in terms of the eigenfunctions v_N:

$$u(t, x) = \sum_N c_N(t) v_N(x) \; . \tag{10.3.45}$$

The heat conduction equation then reduces to

$$\left(\frac{\partial}{\partial t} - \lambda\Delta \right) u = \sum_N \left[\dot{c}_N(t) + \lambda k_N^2 c_N(t) \right] v_N(t) = 0 \; , \tag{10.3.46}$$

thus

$$\dot{c}_N + \lambda k_N^2 c_N = 0 \; , \quad \text{i.e.} \tag{10.3.47}$$

$$c_N(t) = a_N e^{-\lambda k_N^2 t} \quad \text{and} \quad u(t, x) = \sum a_N e^{-\lambda k_N^2 t} v_N(x) \; . \tag{10.3.48}$$

The initial condition $u(0, x) = a(x)$ leads to

$$a_N = (v_N, a) \; .$$

We see that for $t \to \infty$, since $k_N^2 \geq 0$, only the term with $k_N = 0$ is left over, which corresponds to a constant eigenfunction v_N: as $t \to \infty$, the temperature approaches a uniform distribution.

10.4   The Helmholtz Equation in Spherical Coordinates, Spherical Harmonics, and Bessel Functions

We now want to solve the Helmholtz equation in spherical coordinates. This is a problem with many important applications, since spherical symmetry is present in a great number of physical situations. We will briefly present the special functions which appear in these cases as well as several of their properties. For proofs and further properties see the relevant literature, e.g. [Courant et al., Jackson].

10.4.1 Separation of Variables

The Helmholtz equation

$$(\Delta + k^2)u(r) = 0 \tag{10.4.1}$$

in spherical coordinates

$$r = r(r, \theta, \varphi)$$

has the form (Appendix F)

$$\left(\frac{1}{r}\frac{\partial^2}{\partial r^2} r + \frac{1}{r^2 \sin\theta}\frac{\partial}{\partial\theta}\sin\theta\frac{\partial}{\partial\theta} + \frac{1}{r^2 \sin^2\theta}\frac{\partial^2}{\partial\varphi^2} + k^2\right)u$$

$$= \left[\frac{1}{r}\frac{\partial^2}{\partial r^2} r + k^2 + \frac{1}{r^2}\Lambda\left(\theta, \varphi, \frac{\partial}{\partial\theta}, \frac{\partial}{\partial\varphi}\right)\right]u(r, \theta, \varphi) = 0 \; . \tag{10.4.2}$$

We again try the separation of variables method

$$u(r) = F(r)\, Y(\theta, \varphi) \; . \tag{10.4.3}$$

Then, multiplying by r^2, we find

$$\left(r\frac{\partial^2}{\partial r^2} rF\right) Y + k^2 r^2 F Y + F\Lambda Y = 0$$

or

$$\frac{1}{F(r)}\left[r\frac{\partial^2}{\partial r^2} rF(r) + k^2 r^2 F(r)\right] + \frac{\Lambda Y(\theta, \varphi)}{Y(\theta, \varphi)} = 0 \; . \tag{10.4.4}$$

We have again succeeded in separating the variables, and we obtain the two equations

$$\Lambda Y + \alpha Y \equiv \left(\frac{1}{\sin\theta}\frac{\partial}{\partial\theta}\sin\theta\frac{\partial}{\partial\theta} + \frac{1}{\sin^2\theta}\frac{\partial^2}{\partial\varphi^2}\right) Y(\theta, \varphi) + \alpha Y(\theta, \varphi) = 0 \tag{10.4.5}$$

and

$$\frac{1}{r}\frac{\partial^2}{\partial r^2} rF(r) + \left(k^2 - \frac{\alpha}{r^2}\right) F(r) = 0 \; , \tag{10.4.6}$$

where α is a constant whose value will be determined later. The first of these equations can again be separated if we substitute

$$Y(\theta, \varphi) = P(\theta)\,\phi(\varphi) \tag{10.4.7}$$

into the equation. We find

$$\frac{\partial^2 \phi(\varphi)}{\partial \varphi^2} = -m^2 \phi(\varphi) \tag{10.4.8}$$

where, again, the value of $-m^2$ will be determined later, and

$$\left[\frac{1}{\sin\theta}\frac{\partial}{\partial\theta}\sin\theta\frac{\partial}{\partial\theta} + \left(\alpha - \frac{m^2}{\sin^2\theta}\right)\right]P(\theta) = 0 \ . \tag{10.4.9}$$

We will now consider in sequence the equations for $\phi(\varphi)$, $P(\theta)$, and $F(r)$.

10.4.2 The Angular Equations, Spherical Harmonics

i) The equation for $\phi(\varphi)$

$$\phi'' + m^2\phi = 0$$

has solution

$$\phi(\varphi) = \phi_0 \sin(m\varphi + \delta) \ . \tag{10.4.10}$$

Since $\phi(2\pi) = \phi(0)$, m must be integral. The periodicity of $\phi(\varphi)$ also means that m must be real, and thus $-m^2 \le 0$.

ii) If we substitute the expression $l(l + 1)$ for α in the equation for $P(\theta)$, where l is a constant still to be determined, and if we set $x = \cos\theta$ so that $\sin^2\theta = 1 - x^2$ and

$$\frac{d}{dx} = -\frac{1}{\sin\theta}\frac{d}{d\theta} \ :$$

we find

$$\left[\frac{d}{dx}(1 - x^2)\frac{d}{dx} + l(l + 1) - \frac{m^2}{(1 - x^2)}\right]P(x) = 0 \ . \tag{10.4.11}$$

This equation is called *Legendre's associated differential equation* and its solutions are called the *associate Legendre functions*.

Since m only occurs quadratically, we can assume $m \ge 0$ without loss of generality. First, we consider the behavior of the solutions in the vicinity of the singular points of the differential equation, $x = \pm 1$.

For this purpose, we set

$$P = (x \pm 1)^{\varrho}\left[1 + \sum_{n=1}^{\infty} c_n(x \pm 1)^n\right] \qquad (10.4.12)$$

and try to determine ϱ in such a way that P is a solution of the differential equation.

We find

$$\varrho = \pm\frac{m}{2} .$$

This allows us to write $P(x)$ in the form

$$P(x) = (1 - x^2)^{m/2}h(x) .$$

If we substitute this form for P into the differential equation, we find, after a few calculations, the following differential equation for h

$$(1 - x^2)h'' - 2(m + 1)xh' + [l(l + 1) - m(m + 1)]h = 0 . \qquad (10.4.13)$$

Differentiation with respect to x yields

$$(1 - x^2)h''' - 2(m + 2)xh'' + [l(l + 1) - (m + 1)(m + 2)]h' = 0 . \qquad (10.4.14)$$

This is the same equation with $(m + 1)$ instead of m and h' instead of h.

We then obtain the solution of (10.4.13) for $m \geq 0$ by taking the solution of (10.4.13) with $m = 0$ and differentiating m times. We thus need only consider the equation

$$(1 - x^2)P_l'' - 2xP_l' + l(l + 1)P_l = 0 . \qquad (10.4.15)$$

This is *Legendre's differential equation.*

If P_l is a solution of this equation, then

$$P_l^m(x) = (-1)^m(1 - x^2)^{|m|/2}\frac{d^{|m|}}{dx^{|m|}}P_l(x) \qquad (10.4.16)$$

is a solution of the associated Legendre equation (10.4.11).

The solutions of Legendre's differential equation should be single-valued, finite, and continuous in the interval $-1 \leq x \leq 1$. We therefore assume that we can write P_l in terms of a power series

$$P_l(x) = x^{\beta}\sum_{j=0}^{\infty} a_jx^j = \sum_{j=0}^{\infty} a_jx^{j+\beta} , \qquad (10.4.17)$$

where β is a parameter whose value we must determine. Then

$$P_l'(x) = \sum_{j=0}^{\infty} (\beta + j)a_jx^{\beta+j-1} \qquad (10.4.18)$$

and we then obtain the following relation from the Legendre equation:

$$\sum_{j=0}^{\infty} (\beta + j)(\beta + j - 1)a_j x^{\beta + j - 2}$$

$$- \sum_{j=0}^{\infty} [(\beta + j)(\beta + j + 1) - l(l + 1)]a_j x^{\beta + j} = 0 \ . \tag{10.4.19}$$

The lowest exponent in the second sum is β, in the first sum, $\beta - 2$. The coefficients of $x^{\beta - 2}$ and $x^{\beta - 1}$ in the first sum must therefore vanish. For the $x^{\beta - 2}$ term, this means

$$\beta(\beta - 1)a_0 = 0 \ .$$

Now, from the construction, we can assume $a_0 \neq 0$ (otherwise we can choose a larger value for β and repeat the argument). It then follows that $\beta = 0$ or 1. The coefficient of $x^{\beta - 1}$ reads

$$\beta(\beta + 1)a_1 = 0 \ .$$

For $\beta = 0$, this condition is automatically satisfied. For $\beta = 1$, this condition implies that $a_1 = 0$.

We then obtain the recursion relation

$$(\beta + j + 2)(\beta + j + 1)a_{j+2} = [(\beta + j)(\beta + j + 1) - l(l + 1)]a_j$$

or

$$a_{j+2} = \frac{(\beta + j)(\beta + j + 1) - l(l + 1)}{(\beta + j + 2)(\beta + j + 1)} a_j \ , \quad j = 0, 1, \ldots \ . \tag{10.4.20}$$

From this relation, we can determine $a_2, a_4, \ldots$ from a_0.

If we set $a_0 = 0$, $a_1 \neq 0$, then we must have $\beta = 0$. But this case can be reduced to the case $a_0 \neq 0$, $a_1 = 0$, $\beta = 1$. Without loss of generality, we can then assume that a_1 and thus $a_3, \ldots$ are zero.

It can be shown that:

The series constructed with the a_j determined in this way always converges for $x^2 < 1$. However it diverges for $x = \pm 1$, unless for some $j_0 + 2$, $a_{j_0 + 2} = 0$, so that the series breaks off at that point and becomes a polynomial. The polynomial solutions P_l of the Legendre equation are called the *Legendre polynomials*.

The coefficient $a_{j_0 + 2}$ equal to zero if and only if

$$(\beta + j_0)(\beta + j_0 + 1) = l(l + 1) \ . \tag{10.4.21}$$

Then l must be exactly zero or (without loss of generality) a positive integer number. If l is even, then since j_0 is even, β must be even too, i.e. $\beta = 0$. If l is odd, then $\beta = 1$.

The degree of the polynomial for a given value of l is then $\beta + j_0 = l$.

For $l = 0$, $\beta = 0$ and $j_0 = 0$, so that $P_0 = a_0$. For $l = 1$, $\beta = 1$ and $j_0 = 0$ so that $P_1 = a_0 x$. But for $l = 2$, $\beta = 0$ and $P_2(x) = a_0 - 3a_0 x^2 = a_0(1 - 3x^2)$. If the Legendre polynomials are then normalized so that $P_l(1) = 1$, we obtain

$$P_0(x) = 1 , \quad P_1(x) = x , \quad P_2(x) = \tfrac{1}{2}(3x^2 - 1) \tag{10.4.22}$$

and further

$$P_3(x) = \tfrac{1}{2}(5x^3 - 3x) \tag{10.4.23}$$

$$P_4(x) = \tfrac{1}{8}(35x^4 - 30x^2 + 3) . \tag{10.4.24}$$

To obtain the Legendre polynomials in general, we can use *Rodrigues'*[4] *formula*:

$$P_l(x) = \frac{1}{2^l l!} \frac{d^l}{dx^l}(x^2 - 1)^l . \tag{10.4.25}$$

Thus we have determined the Legendre polynomials as solutions of Legendre's equation. In addition to these polynomials, there are also the so-called Legendre functions of the second type, $Q_l(x)$, which are singular at $x = \pm 1$. Then for arbitrary l, the functions

$$P_l(x) , \quad Q_l(x)$$

are the two independent solutions of the Legendre equation. For integral values of l, $P_l(x)$ becomes a polynomial and is thus regular in the interval $-1 \le x \le 1$.

These two independent solutions form a basis in the solution space, just like $(\exp(im\varphi), \exp(-im\varphi))$ or $(\sin m\varphi, \cos m\varphi)$. This is to say, just as every solution of the differential equation

$$x''(\varphi) + m^2 x(\varphi) = 0$$

can be expressed as a linear combination of $\exp(im\varphi)$ and $\exp(-im\varphi)$, so every solution of the Legendre equation can be expressed as a linear combination of P_l and Q_l. However, in the case of the Legendre equation we forbid the solution Q_l because it does not stay finite in the interval $-1 \le x \le 1$. We will impose this condition later, always from physical considerations.

From (10.4.16),

$$P_l^m(x) = (-1)^m (1 - x^2)^{|m|/2} \frac{d^{|m|}}{dx^{|m|}} P_l(x) , \quad -l \le m \le +l \tag{10.4.26}$$

is a regular solution of Legendre's associated equation.

---

[4] *Rodrigues, Benjamin Olinde* (*1794 Bourdeaux, d. 1851 Paris). French national economist and mathematician. Friend and student of St. Simon. Mathematical works, especially about differential geometry.

We now define

$$Y_{lm}(\theta, \varphi) = \sqrt{\frac{2l+1}{4\pi} \frac{(l-|m|)!}{(l+|m|)!}} \, P_l^m(\cos\theta) e^{im\varphi} ,$$

$$l = 0, 1, \ldots, \quad m = -l, -l+1, \ldots, +l . \tag{10.4.27}$$

These functions are then solutions of the differential equation

$$\left[\frac{1}{\sin\theta} \frac{\partial}{\partial\theta} \sin\theta \frac{\partial}{\partial\theta} + \frac{1}{\sin^2\theta} \frac{\partial^2}{\partial\varphi^2} + l(l+1) \right] Y_{lm}(\theta, \varphi) = 0 . \tag{10.4.28}$$

The Y_{lm} are called *spherical harmonics*, since they are defined parametrically in terms of the coordinates θ and φ on the surface of the unit sphere. They are regular on this surface and the collection of the Y_{lm} forms a complete orthonormal system for the set of functions $f(\theta, \varphi)$ on the surface of the unit sphere, i.e.

$$\langle Y_{lm}, Y_{l'm'} \rangle := \int_{-1}^{+1} d\cos\theta \int_0^{2\pi} d\varphi \, Y_{lm}^*(\theta, \varphi) Y_{l'm'}(\theta, \varphi) = \delta_{ll'} \delta_{mm'} , \tag{10.4.29}$$

and every (reasonable) function $f(\theta, \varphi)$ can be represented as

$$f(\theta, \varphi) = \sum_{l=0}^{\infty} \sum_{m=-l}^{+1} f_{lm} Y_{lm}(\theta, \varphi) \tag{10.4.30}$$

with the coefficients

$$f_{lm} = \langle Y_{lm}, f \rangle \equiv \int_{-1}^{+1} d\cos\theta \int_0^{2\pi} d\varphi \, Y_{lm}^*(\theta, \varphi) f(\theta, \varphi) . \tag{10.4.31}$$

This expansion and the method by which the coefficients can be calculated is again analogous to the Fourier series, which derives from an orthonormal system of solutions of a simpler type of differential equation.

The orthogonality $\langle Y_{lm}, Y_{l'm'} \rangle = 0$ for $(l, m) \neq (l', m')$ can be shown as follows:

First, the orthogonality is immediately clear for $m \neq m'$. For $l \neq l'$, we argue as follows:

$$v(r, \theta, \varphi) = Y_{lm}(\theta, \varphi) \quad \text{and} \quad w(r, \theta, \varphi) = Y_{l'm'}(\theta, \varphi)$$

are functions in the interior of a sphere of radius R which satisfy homogeneous Neumann boundary conditions. Then, as was shown in Sect. 10.3.3,

$$(v, \Delta w) = \int_0^R r^2 dr \int_{-1}^{+1} d\cos\theta \int_0^{2\pi} d\varphi \, v(r, \theta, \varphi) \Delta w(r, \theta, \varphi)$$

$$= (\Delta v, w) . \tag{10.4.32}$$

Now

$$\Delta w = \left(\frac{1}{r}\frac{\partial^2}{\partial r^2}r + \frac{\Lambda}{r^2}\right)w = -\frac{l'(l'+1)}{r^2}w \ , \tag{10.4.33}$$

$$\Delta v = -\frac{l(l+1)}{r^2}v \quad \text{and} \tag{10.4.34}$$

$$(v,\Delta w) = \int_0^R dr\, r^2 \frac{l'(l'+1)}{r^2} \int_{-1}^{+1} d\cos\theta \int_0^{2\pi} d\varphi\, Y_{lm}^* Y_{l'm'}$$

$$= Rl'(l'+1)\langle Y_{lm}, Y_{l'm'}\rangle$$

$$= (\Delta v, w) = Rl(l+1)\langle Y_{lm}, Y_{l'm'}\rangle \ , \tag{10.4.35}$$

so that $\langle Y_{lm'}, Y_{l'm'}\rangle = 0$ for $l \neq l'$.

In particular, since $P_l \sim Y_{l0}$,

$$\int_{-1}^{+1} dx\, P_l(x)\, P_{l'}(x) = 0 \quad \text{for} \quad l \neq l' \ . \tag{10.4.36}$$

10.4.3 The Radial Equation, Bessel Functions

Finally, we need to consider the equation (10.4.6):

$$\frac{1}{r}\frac{d^2}{dr^2} rF(r) + \left(k^2 - \frac{l(l+1)}{r^2}\right)F(r) = 0 \ . \tag{10.4.37}$$

With $F(r) = r^n f(r)$, since

$$\frac{1}{r}(r^{n+1} f)'' = r^n f'' + 2(n+1)r^{n-1} f' + n(n+1)r^{n-2} f$$

we obtain the equation

$$f'' + \frac{2(n+1)}{r}f' + \left[k^2 - \frac{l(l+1) - n(n+1)}{r^2}\right]f = 0 \ . \tag{10.4.38}$$

For $n = -1/2$, if we set $x = kr$ and $f(kr) = u(x)$ so that

$$u' = \frac{du}{dx} = \frac{1}{k}f' \ ,$$

we find:

$$u'' + \frac{1}{x}u' + \left(1 - \frac{v^2}{x^2}\right)u = 0 \tag{10.4.39}$$

with

$$v^2 = l(l + 1) + \tfrac{1}{4} = (l + \tfrac{1}{2})^2 \; . \tag{10.4.40}$$

This is *Bessel's*[5] *differential equation* and its solutions are called *Bessel functions of order v.*

We can find these by again assuming a power series of the form:

$$u(x) = x^\gamma \sum_{j=0}^{\infty} a_j x^j = \sum_{j=0}^{\infty} a_j x^{\gamma+j} \; . \tag{10.4.41}$$

Then

$$u'' = \sum_{j=0}^{\infty} (\gamma + j)(\gamma + j - 1)a_j x^{\gamma+j-2} \; ,$$

and we find the condition

$$\sum_{j=0}^{\infty} [(\gamma + j)(\gamma + j - 1)a_j + (\gamma + j)a_j - v^2 a_j]x^{\gamma+j-2}$$

$$+ \sum_{j=0}^{\infty} a_j x^{\gamma+j} = 0 \; . \tag{10.4.42}$$

The coefficients for negative values of $j - 2$ in the first sum vanish if $\gamma = \pm v$ and if

$$[(\gamma + 1)^2 - v^2]a_1 = 0 \; , \quad \text{i.e.} \quad a_1 = 0 \; .$$

The recursion relation for the $a_j, j \geq 2$ then reads

$$[(\gamma + j)^2 - v^2]a_j + a_{j-2} = 0 \quad \text{or} \tag{10.4.43}$$

$$a_j = \frac{-1}{j^2 + 2\gamma j} a_{j-2}$$

and replacing j by $2j$,

$$a_{2j} = \frac{-1}{4j(j + \gamma)} a_{2(j-1)}$$

---

[5] *Bessel, Friedrich Wilhelm* (*1784 Minden/Westphalia, d. 1846 Königsberg). In 1810 he became professor of astronomy at Königsberg. Performed important astronomical calculations and observations (star catalog, planets, double stars, prediction of a planet outside of Uranus, discovery of the companion of Sirius).

which upon iteration becomes

$$a_{2j} = \frac{(-1)^j \Gamma(\gamma + 1)}{\Gamma(\gamma + j + 1)2^{2j}j!}\, a_0 \, , \tag{10.4.44}$$

where Γ is the gamma function (see Appendix A). The usual choice for a_0 is

$$a_0 = \frac{1}{2^\gamma \Gamma(\gamma + 1)} \, ,$$

so that we find the two solutions (*Bessel functions*)

$$J_\nu(x) = \left(\frac{x}{2}\right)^\nu \sum_{j=0}^\infty \frac{(-1)^j}{j!\Gamma(j + \nu + 1)}\left(\frac{x}{2}\right)^{2j} \tag{10.4.45}$$

and

$$J_{-\nu}(x) = \left(\frac{x}{2}\right)^{-\nu} \sum_{j=0}^\infty \frac{(-1)^j}{j!\Gamma(j - \nu + 1)}\left(\frac{x}{2}\right)^{2j} \, . \tag{10.4.46}$$

If $\nu = m = $ integral, then

$$J_{-m} = (-1)^m J_m \, , \tag{10.4.47}$$

so that J_m and J_{-m} are not independent. Otherwise, the two solutions are clearly linearly independent and again form a basis in solution space. For $F(r)$, we obtain

$$F_+(r) = \frac{J_{l+1/2}(kr)}{\sqrt{kr/2}} \quad \text{or} \tag{10.4.48}$$

$$F_-(r) = \frac{J_{-(l+1/2)}(kr)}{\sqrt{kr/2}} \, . \tag{10.4.49}$$

As $r \to 0$, $F_+(r)$ goes as r^l, whereas $F_-(r)$ goes as r^{-l-1}.
The functions

$$j_l(z) = \sqrt{\frac{\pi}{2z}}\, J_{l+1/2}(z) \, ,$$

$$n_l(z) = \sqrt{\frac{\pi}{2z}}\, N_{l+1/2}(z) \quad \text{with} \tag{10.4.50}$$

$$N_\nu(z) = \frac{J_\nu(z)\cos\nu\pi - J_{-\nu}(z)}{\sin\nu\pi} \tag{10.4.51}$$

are called the *spherical Bessel functions.*

In particular,

$$j_0(z) = \sqrt{\frac{\pi}{2z}} \left(\frac{z}{2}\right)^{1/2} \sum_{j=0}^{\infty} \frac{(-1)^j}{j!\,\Gamma(j+3/2)} \left(\frac{z}{2}\right)^{2j} \tag{10.4.52}$$

It can be shown that

$$j_0(z) = \frac{\sin z}{z} \quad \text{and} \tag{10.4.53}$$

$$j_1(z) = \frac{\sin z}{z^2} - \frac{\cos z}{z} \ , \tag{10.4.54}$$

and, in general,

$$j_l(z) = z^l \left(-\frac{1}{z}\frac{d}{dz}\right)^l \frac{\sin z}{z} \ . \tag{10.4.55}$$

10.4.4 Solutions of the Helmholtz Equation

The solutions of the Helmholtz equation can now be given in the forms

$$v(r, \theta, \varphi) = j_l(kr)\, Y_{lm}(\theta, \varphi) \ ,$$
$$v(r, \theta, \varphi) = n_l(kr)\, Y_{lm}(\theta, \varphi) \ , \tag{10.4.56}$$

respectively, for $l = 0, 1, \ldots, m = -l, \ldots, +l$.

From our considerations, we again find the solutions for $k = 0$ and thus the solution of the Laplace equation in three spatial dimensions

$$\Delta\phi = 0 \ . \tag{10.4.57}$$

While the form of the angular dependence does not change, for the radial component we now obtain the equation

$$\frac{1}{r}\frac{d}{dr^2}\, rF(r) - \frac{l(l+1)}{r^2}\, F(r) = 0 \tag{10.4.58}$$

with solutions

$$F_+(r) = r^l \quad \text{and} \tag{10.4.59}$$

$$F_-(r) = r^{-l-1} \ , \tag{10.4.60}$$

respectively.

The general solution of the Laplace equation

$$\Delta\phi = 0$$

can therefore be written as

$$\phi(r, \theta, \varphi) = \sum_{l=0}^{\infty} \sum_{m=-l}^{+l} (A_l r^l + B_l r^{-l-1}) Y_{lm}(\theta, \varphi) \ . \tag{10.4.61}$$

If we demand regularity at $r = 0$, then the coefficients B_l must vanish. On the other hand, solutions which vanish for $r \to \infty$ are characterized by $A_l = 0$.

10.4.5 Supplementary Considerations

i) Let

$$\hat{\boldsymbol{k}} = (\sin \theta' \cos \varphi', \sin \theta' \sin \varphi', \cos \theta') \ ,$$

$$\hat{\boldsymbol{r}} = (\sin \theta \cos \varphi, \sin \theta \sin \varphi, \cos \theta)$$

be two unit vectors with polar coordinates (θ', φ') and (θ, φ) respectively. Then

$$\hat{\boldsymbol{k}} \cdot \hat{\boldsymbol{r}} = \cos \theta \cos \theta' + \sin \theta \sin \theta' \cos(\varphi - \varphi') \tag{10.4.62}$$

is a number between -1 and $+1$ which we can indicate by $\cos \gamma$.

It can be shown that

$$P_l(\cos \gamma) = \frac{4\pi}{2l + 1} \sum_{m=-l}^{+l} Y_{lm}^*(\theta', \varphi') Y_{lm}(\theta, \varphi) \tag{10.4.63}$$

or, expressed otherwise

$$P_l(\hat{\boldsymbol{k}} \cdot \hat{\boldsymbol{r}}) = \frac{4\pi}{2l + 1} \sum_{m=-l}^{+l} Y_{lm}^*(\hat{\boldsymbol{k}}) Y_{lm}(\hat{\boldsymbol{r}}) \ . \tag{10.4.64}$$

This is the *addition formula* for spherical harmonics.

For $l = 1$, we find

$$P_1(\cos \gamma) = \frac{4\pi}{3} \left[Y_{1-1}^*(\theta', \varphi') Y_{1-1}(\theta, \varphi) \right.$$

$$\left. + Y_{10}^*(\theta', \varphi') Y_{10}(\theta, \varphi) + Y_{11}^*(\theta', \varphi') Y_{11}(\theta, \varphi) \right]$$

$$= \frac{4\pi}{3} \frac{3}{4\pi} \left(\frac{1}{2} \sin \theta' \, e^{i\varphi'} \sin \theta \, e^{-i\varphi} + \cos \theta' \cos \theta \right.$$

$$\left. + \frac{1}{2} \sin \theta' \, e^{-i\varphi'} \sin \theta \, e^{i\varphi} \right) \ . \tag{10.4.65}$$

This is another expression for $\cos \gamma$, compare (10.4.62).

For $\boldsymbol{k} = (0, 0, 1)$,

$$\gamma = \theta \ ,$$

and since

$$Y_{lm}(\theta = 0, \varphi) = \sqrt{\frac{2l + 1}{4\pi}}\, \delta_{m,\,0} \tag{10.4.66}$$

we find with (10.4.63):

$$P_l(\cos\theta) = \frac{4\pi}{2l + 1}\sqrt{\frac{2l + 1}{4\pi}}\, Y_{l0}(\theta, \varphi) \tag{10.4.67}$$

or

$$Y_{l0}(\theta, \varphi) = \sqrt{\frac{2l + 1}{4\pi}}\, P_l(\cos\theta)\ , \tag{10.4.68}$$

which can also be directly verified. A proof of the addition formula can be found in [Jackson].

ii) Clearly, $\exp(i\mathbf{k}\cdot\mathbf{r})$ is also a solution of the Helmholtz equation, so that the exponential function can also be expressed in terms of the solutions

$$j_l(kr)\, Y_{lm}(\theta, \varphi)\ .$$

With $\cos\gamma = \mathbf{k}\cdot\mathbf{r}/kr$ (see [Jackson]), we find

$$e^{i\mathbf{k}\ \mathbf{r}} = \sum_{l=0}^{\infty} i^l(2l + 1)j_l(kr)\, P_l(\cos\gamma)$$

$$= 4\pi \sum_{l=0}^{\infty} \sum_{m=-l}^{+l} i^l j_l(kr)\, Y_{lm}^*(\hat{\mathbf{k}})\, Y_{lm}(\hat{\mathbf{r}})\ , \tag{10.4.69}$$

(the term $l = 0$ reproduces the 1 in the expansion of the exponential function $\exp(i\mathbf{k}\cdot\mathbf{r}) = 1 + \ldots$).

iii) We have seen that the set of functions

$$Y_{lm}\ ,\quad l = 0, \ldots,\quad m = -l, \ldots, l$$

is a complete orthonormal system, that, is every square integrable function $f(\theta, \varphi)$ defined on the unit sphere can be uniquely represented as

$$f(\theta, \varphi) = \sum_{l=0}^{\infty} \sum_{m=-l}^{+l} f_{lm}\, Y_{lm}(\theta, \varphi)\ . \tag{10.4.70}$$

Let us consider in general an interval $[a, b]$ and a complete set of orthonormal functions $v_n(x)$ with

$$(v_n, v_m) = \int_a^b dx\, v_n^*(x)\, v_m(x) = \delta_{nm}\ . \tag{10.4.71}$$

If we want to approximate a function $f(x)$ by a series

$$f(x) = \sum_n a_n v_n(x) \quad \text{then}$$

$$a_n = (v_n, f) = \int_a^b dx' v_n^*(x') f(x') \;, \quad \text{so that}$$

$$f(x) = \sum_n \int_a^b dx' v_n(x) v_n^*(x') f(x') = \int_a^b dx' \delta(x - x') f(x') \;. \tag{10.4.72}$$

Thus we have

$$\sum_n v_n(x) v_n^*(x') = \delta(x - x') \;. \tag{10.4.73}$$

This is called the *completeness relation*. From this consideration, the spherical harmonics satisfy the equation

$$\sum_{l=0}^{\infty} \sum_{m=-l}^{+l} Y_{lm}(\theta, \varphi) Y_{lm}^*(\theta', \varphi') = \delta(\cos\theta - \cos\theta')\delta(\varphi - \varphi') \;. \tag{10.4.74}$$

Problems

10.1 Temperature and Mass Diffusion. Let there be a mixture of two substances between two plates (to be thought of as infinitely extended) at $x = -1$ and $x = +1$ held at constant temperatures T_0 and T_1 respectively. Let us suppose that the temperature distribution $T = T(x)$ and the distribution of mass concentration $\hat{c} = \hat{c}(x)$ satisfy the thermodynamic diffusion equations

$$\frac{\partial T}{\partial t} = \kappa \frac{\partial^2 T}{\partial x^2} + \tilde{\kappa} \frac{\partial^2 \hat{c}}{\partial x^2}$$

$$\frac{\partial \hat{c}}{\partial t} = \tilde{D} \frac{\partial^2 T}{\partial x^2} + D \frac{\partial^2 \hat{c}}{\partial x^2}$$

where the matrix

$$\begin{pmatrix} \kappa & \tilde{\kappa} \\ \tilde{D} & D \end{pmatrix}$$

is assumed to be positive definite; we also assume that the difference in temperature $T_1 - T_0$ is small enough. Formulate boundary conditions for T and $\hat{c}$ at $x = \pm 1$ (disappearance of the thermal diffusion flow at the boundaries) and calculate $T(x)$ and $\hat{c}(x)$ in the stationary state produced as $t \to +\infty$.

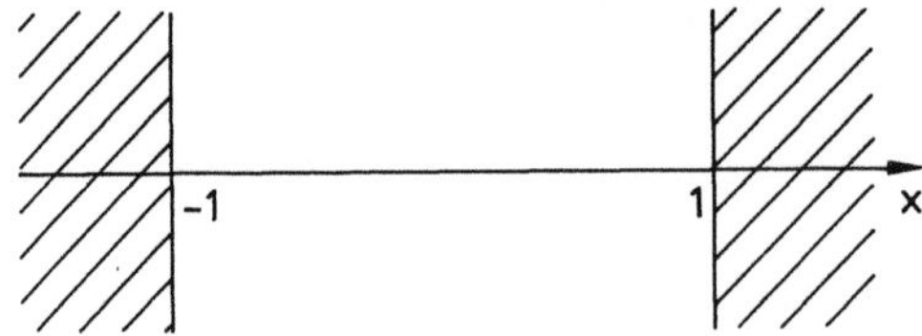

10.2 Solution of the Heat Conduction Equation. Show explicitly that

$$D(t, x) = \left(\frac{1}{4\pi\lambda t}\right)^{3/2} e^{-x^2/4\lambda t}$$

for $t > 0$ is a solution of the three dimensional heat conduction equation

$$\dot{D} = \lambda\Delta D \ .$$

10.3 Permafrost. In order to describe the daily or yearly temperature fluctuations under the surface of the earth, we start with the following model:

The surface of the earth is described as the outer surface $z = 0$ of a half-space $z \leq 0$ which is filled with a material of constant density ϱ, constant thermal conductivity κ, and constant specific heat c.

On the surface – as we very roughly approximate – we are given a temperature distribution $T_0 \cos \omega t$ varying periodically with time, where the period of the temperature fluctuation $\tau = 2\pi/\omega$ is taken to be either a day or a year. The temperature distribution $T = T(t, z)$ is then given by the heat conduction equation

$$\frac{\partial T}{\partial t} - \lambda\frac{\partial^2 T}{\partial z^2} = 0 \quad \text{with} \quad \lambda = \frac{\kappa\varrho}{c}$$

where as a boundary condition as $z = 0$, we set

$$T(t, 0) = T_0 \cos \omega t$$

(in addition, at $z = -\infty$, $T(t, -\infty) = 0$).

Determine, by separating variables, the resulting distribution of temperature.

How great is the depth of penetration z_0 (i.e., the value of z at which the amplitude of $T(t, z)$ is exactly $1/e$ of the magnitude of $T(t, 0)$), and how does z_0 depend on $\tau = 2\pi/\omega$?

Compare z_0 for $\tau = 1$ year with z_0 for $\tau = 1$ day and explain based on this the permafrost found in Siberia or Canada.

10.4 Surface Waves. a) Solve the wave equation

$$\ddot{u} - c^2\Delta u = 0$$

in the half-space $z \leq 0$ with the boundary condition $u(u, x, y, -\infty) = 0$ using separation of variables.

Define (as in the previous exercise) a depth of penetration and determine the possible frequencies and wavelengths for a given z_0.

b) Find all solutions in the region $0 \leq x \leq a,\, 0 \leq y \leq b,\, z \leq 0$ corresponding to a given depth of penetration z_0 which, in addition, satisfy the boundary conditions

$$u(t, 0, y, 0) = 0\ , \qquad u(t, a, y, 0) = 0$$

$$u(t, x, 0, 0) = 0\ , \qquad u(t, x, b, 0) = 0$$

10.5   Green's Functions. a) Show explicitly that

$$G_k(x) = -\frac{1}{4\pi}\frac{e^{ik|x|}}{|x|}$$

is a Green's function of the Helmholtz operator $\Delta + k^2$ (in 3-dimensional space), i.e., that in the distributional sense the equation

$$(\Delta + k^2)G_k(x) = \delta(x)$$

is valid.

Hint: To prove this, both sides must be applied to test functions. Choose polar coordinates and explain why the angular components of the test function yield no contribution to the integral.

b) Show with the help of a) that

$$G_R(t, x) = -\frac{1}{2\pi}\int\limits_{-\infty}^{+\infty} d\omega\, e^{i\omega t}\, G_{\omega/c}(x)$$

is a Green's function of the wave operator

$$\frac{1}{c^2}\frac{\partial^2}{\partial t^2} - \Delta$$

(in 3-dimensional space), i.e., that

$$\left(\frac{1}{c^2}\frac{\partial^2}{\partial t^2} - \Delta\right)G_R(t, x) = \delta(t)\delta(x)\ .$$

Carry out the Fourier transformation explicitly, and convince yourself from this that this is a retarded Green's function, so that $G_R(t, x) = 0$ for $t < 0$.

10.6   Differential Equations with Functional Coefficients. Consider the following differential equation

$$\frac{d^2 f}{dx^2} + p(x)\frac{df}{dx} + q(x)f(x) = 0 \ .$$

a) Show that, by means of the assumption

$$f(x) = v(x)\, Y(x)$$

with a suitable function $v(x)$, we can obtain a differential equation without a first-order term:

$$\frac{d^2 Y}{dx^2} + \hat{q}(x)\, Y(x) = 0 \ .$$

b) Assume that we have found a solution $f_0(x)$ of equation (1).
   Show that by means of the assumption

$$f(x) = f_0(x)w(x)$$

we can construct a second solution.

10.7   Spherical Harmonics. One possible approach to the spherical harmonics is the following:

Consider a homogeneous polynomial of degree l in the variables x_1, x_2, x_3:

$$v(x) = \sum_{i_1, \ldots, i_l = 1}^{3} a_{i_1 i_2 \ldots i_l} x_{i_1} x_{i_2} \ldots x_{i_l} \ .$$

If the polynomial satisfies the Laplace equation

$$\Delta v(x) = 0 \ ,$$

then it can be written as

$$v(x) = r^l \sum_{m=-l}^{l} c_{lm}\, Y_{lm}(\vartheta, \varphi) \ .$$

The functions Y_{lm} are (up to a normalization constant) uniquely determined through their φ-dependence

$$Y_{lm} = P_{lm}(\vartheta)\, e^{im\varphi} \ .$$

a) Show that a polynomial of degree l has in general $\frac{1}{2}(l + 2)(l + 1)$ terms.
b) What conditions must we impose on the coefficients so that the polynomial is a solution of the Laplace equation? How many of these conditions do we obtain?
c) Show that there can only be $2l + 1$ independent solutions of the Laplace equation in the form of homogeneous polynomials of degree l.

11. Electrostatics

Newtonian mechanics is a general theory which allows us to calculate the motion of bodies given a knowledge of force laws. Of the important force laws, which were numerated in Sect. 2.3, so far we have investigated in great depth the Newtonian law of gravitation and linear force laws for small oscillations of a many particle system about its point of equilibrium.

Linear force laws, because of their numerous applications, are important approximations. In actuality, though, they are never exactly valid.

In contrast, Newton's law of gravitation is a fundamental law of nature which describes a well-defined realm of natural phenomena, namely the gravitational forces which bodies exert on each other due to their mass.

In addition to gravitational forces, there are also electromagnetic forces which are not determined by mass, but rather by another type of quality which bodies can possess: electric charge. In contrast to mass, the charge on a body can be either positive or negative and it is quantified, i.e. it is always an integral multiple of an elementary charge e_0, the charge of a proton. It happens that in order to describe electromagnetic forces two fields are necessary, namely an electric field E and a magnetic field B. The electric force qE is exerted on a particle of charge q, even when the particle is at rest. The magnetic Lorentz force $qv \times B$ (2.3.13), on the other hand, is proportional to the speed of the particle. However, the electric and magnetic fields are not merely convenient constructions used to describe the effects of forces between charged particles. As we will see, as physical quantities they also play actively independent roles.

We begin here with electrostatics, the theory of the electric field of charges at rest.

11.1 The Basic Equations of Electrostatics and Their First Consequences

11.1.1 Coulomb's Law and the Electric Field

If we consider two particles at rest with charges q_1 and q_2, then the electric force which particle 2 exerts on particle 1 can be written as

$$F_{12} = k \frac{q_1 q_2}{|r_1 - r_2|^3} (r_1 - r_2) \, , \qquad (11.1.1)$$

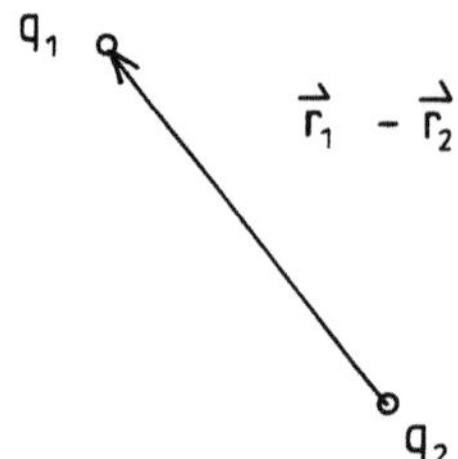

Fig. 11.1.1. The direction of the electrostatic force exerted by particle 2 with charge q_2 on particle 1 with the charge q_1 for the case $q_1 q_2 > 0$. The force is repulsive

where r_1 and r_2 are the position vectors of particles 1 and 2. The quantity $k > 0$ will be discussed later. If $q_1 q_2 > 0$, the force is repulsive (Fig. 11.1.1). The equation (11.1.1) is *Coulomb's law*.

The attractive force arising from the masses of the particles in accordance with the universal law of gravitation is so small in comparison to the electrostatic force that it can be neglected in electrodynamics (see Sect. 2.3).

The forces are immediately measurable quantities. The force effects of the charged particles can also be described in terms of the concept of field, which has been useful in many ways in physics. Conceptually, we can separate the interaction between the two particles into

i) an electric field $E(r)$ which is produced by particle 1 or particle 2 respectively. The particle of charge q_2 located at position r_2 produces at position r a field

$$E_2(r) = kq_2 \, \frac{r - r_2}{|r - r_2|^3} \; ; \tag{11.1.2}$$

ii) The effect of this field on particle 1 which carries the charge q_1, i.e. the force which is exerted on a particle of charge q_1 at point r_1 in a field $E(r)$. This is then given by

$$F_{12} = q_1 E_2(r_1) = kq_1 q_2 \frac{r_1 - r_2}{|r_1 - r_2|^3} \; . \tag{11.1.3}$$

Since we assume that the charge q_2 is always at rest in our frame of reference, $E_2(r)$ is not dependent on time. We then speak of an *electrostatic field*.

This description of electric phenomena with the help of fields was first introduced in the nineteenth century and quantitatively developed by J.C. Maxwell. In ancient times and the middle ages, the only electrical effects known were lightning, Saint Elmo's fire, and the attractive force of a piece of rubbed amber. In the nineteenth century until about 1870–1885, there was a theory of electrodynamics involving action at a distance which was particularly clearly formulated by *W. Weber*[1], but which did not use the field concept.

---

[1] *Weber, Wilhelm Eduard* (*1804 Wittenberg, d. 1891 Göttingen). One of the Göttingen Seven. First works about the theory of oscillations and acoustics. Until his expulsion from the university at Göttingen, he worked with Gauss on the quantitative theory of magnetism. After his return to Göttingen in 1856, he performed a famous experiment with Kohlrausch on the relationship between electric and magnetic force, in which a universal constant with the dimensions of speed appeared and which numerically agreed with the speed of light: a first argument for an electromagnetic theory of light. Weber also proposed a planetary model of the atom to explain spectral lines.

However, further progress was only possible using Maxwell's formulation of electrodynamics as a field theory, which then became a prototype of classical field theory – and is still used as a guide today, for example in the study of another classical field theory, general relativity.

Electrodynamics can also be expanded into a field theory of atomic phenomena, and so soon after the emergence of quantum mechanics, quantum electrodynamics was developed. This is a quantum field theory which describes the interaction of electromagnetic radiation with the matter in atoms. Quantum electrodynamics (QED) is the most developed theory in microscopic physics, and it has made very exact predictions which have been verified by experiment.

Remarks. i) $E(r)$ is a vector field, i.e. at each point in space there is a vector $E(r)$ which points in the direction of the force which a small positive test charge would feel at that point.

ii) The proportionality constant k depends on which units we use to measure the charge q. We can set $k = 1$ and then use Coulomb's law to find the dimensions of q. Then

$$[q] \;= N^{1/2}m = kg^{1/2}\, m^{3/2}\, s^{-1} \sim g^{1/2}\, cm^{3/2}\, s^{-1}$$

$$= \text{e.s.u. (electrostatic unit).}$$

There is a variety of systems of measurement in electrodynamics. In two of these systems, the "Gaussian" system and the "electrostatic" system the value of k is set to 1. In the so-called MKSA system, which today is the standard international SI-system, the *Ampere*[2] (A) is another fundamental quantity. The dimensions of q are then As = C (Coulomb) and in these units, the proportionality constant is given by

$$k = \frac{1}{4\pi\varepsilon_0} \;, \qquad \varepsilon_0 = 8.854 \times 10^{-12}\, C^2\, N^{-1}\, m^{-2} \;. \tag{11.1.4}$$

We will discuss this further in Chap. 13. Here we leave the value of k open.

11.1.2   Electrostatic Potential and the Poisson Equation

If several charges at rest are present, it is found experimentally that the forces which arise from the individual charges add together. The electric field at a point r which arises from the charges q_i at points r_i, $i = 1, \ldots, N$ is then the linear

---

[2] *Ampère, André Marie* (*1775 Poleymieux near Lyon, d. 1836 Marseille). Professor at the Ecole Polytechnique and the Collège de France. His theory of the magnetic effects of flowing charges was inspired by Oersted. In 1822, he developed the theory of magnetism as a result of molecular circular currents. Works on partial differential equations as well as on literary and philosophical themes.

superposition of the individual fields:

$$E(r) = k \sum_i q_i \frac{r - r_i}{|r - r_i|^3} \ . \tag{11.1.5}$$

If the charge carriers are small and dense enough to define a charge density $\varrho(r)$, then

$$E(r) = k \int d^3r' \varrho(r') \frac{r - r'}{|r - r'|^3} \ . \tag{11.1.6}$$

We can now write

$$\frac{r - r'}{|r - r'|^3} = - \nabla \frac{1}{|r - r'|} \quad \text{for} \quad r \neq r' \ , \tag{11.1.7}$$

and therefore

$$E(r) = - \nabla k \int_V d^3r' \frac{\varrho(r')}{|r - r'|} = - \nabla \phi(r) \tag{11.1.8}$$

with

$$\phi(r) = k \int_V d^3r' \frac{\varrho(r')}{|r - r'|} \ . \tag{11.1.9}$$

$E(r)$ can thus be described as the gradient of a scalar field. $\phi(r)$ is called the *electrostatic potential*.

Then, we have immediately

$$\nabla \times E(r) = - \nabla \times \nabla \phi(r) = 0 \ , \tag{11.1.10}$$

$E(r)$ is a conservative field; this is to be expected because of the similarity of Coulomb's law and the law of gravitation.

With the equation

$$\nabla \times E(r) = 0$$

we have already found a *field equation*, that is, an equation for $E(r)$ which states that in the static case E is a gradient field.

We will now derive a second equation for E. This equation clearly must include $\varrho(r)$ since $\varrho(r)$ is the source of the electric field.

If we construct

$$\nabla \cdot E(r) = - \nabla \cdot \nabla \phi = - \Delta \phi \ , \tag{11.1.11}$$

we find

$$\nabla \cdot \boldsymbol{E} = -k \int d^3r' \varrho(\boldsymbol{r}') \Delta \frac{1}{|\boldsymbol{r} - \boldsymbol{r}'|}$$

$$= 4\pi k \int d^3r' \varrho(\boldsymbol{r}') \delta(\boldsymbol{r} - \boldsymbol{r}') \; ,$$

where we have abbreviated

$$\Delta \frac{1}{|\boldsymbol{r} - \boldsymbol{r}'|}$$

by $-4\pi\delta(\boldsymbol{r} - \boldsymbol{r}')$. From the choice of the symbol δ for this function, it will be anticipated that it is identical to the δ-distribution of Chap. 6 and Appendix E. We will now show this:

Without loss of generality we may take $\boldsymbol{r}' = \boldsymbol{0}$. Then for $\boldsymbol{r} \neq \boldsymbol{0}$, we calculate

$$\Delta \frac{1}{r} = \frac{1}{r^2} \frac{d}{dr} r^2 \frac{d}{dr} \frac{1}{r} = \frac{1}{r^2} \frac{d}{dr} r^2 \left(\frac{-1}{r^2} \right) = 0 \; .$$

Thus

$$\Delta \frac{1}{r} = -4\pi\delta(\boldsymbol{r}) = 0 \quad \text{for} \quad \boldsymbol{r} \neq \boldsymbol{0} \; .$$

On the other hand, $\delta(\boldsymbol{r})$ cannot vanish identically, since if we integrate $\Delta(1/r)$ over a sphere with radius R, we find

$$\int\limits_{\text{Sphere } S} d^3r \nabla \cdot \nabla \frac{1}{r} = \int\limits_{\partial K} d\boldsymbol{F} \cdot \nabla \frac{1}{r}$$

$$= R^2 \int d\Omega \frac{\boldsymbol{r}}{r} \cdot \frac{\boldsymbol{r}}{r} \left(\frac{-1}{r^2} \right) \Big|_{r=R} = -4\pi \; ,$$

i.e.,

$$\int\limits_{\text{Sphere } S} d^3r \delta(\boldsymbol{r}) = 1 \; .$$

Thus the function $\delta(\boldsymbol{r})$, defined as $-\Delta(1/4\pi r)$, is identical to the "δ-function" (see Appendix E)

$$\delta(\boldsymbol{r}) = \delta(x)\delta(y)\delta(z) \quad \text{for} \quad \boldsymbol{r} = (x, y, z) \; .$$

Therefore

$$\Delta \frac{1}{|\boldsymbol{r} - \boldsymbol{r}'|} = -4\pi\delta(\boldsymbol{r} - \boldsymbol{r}') \quad \text{and}$$

$$\nabla \cdot \boldsymbol{E}(\boldsymbol{r}) = 4\pi k \varrho(\boldsymbol{r}) \tag{11.1.12}$$

or, with

$$E = -\nabla\phi \ ,$$

we have

$$\Delta\phi(r) = -4\pi k\varrho(r) \ . \tag{11.1.13}$$

This is *Poisson's equation*. It differs from the Laplace equation in that it is inhomogeneous due to the term on the right hand side.

We have therefore formulated the equations

$$\nabla \times E(r) = 0 \ , \tag{11.1.14}$$

$$\nabla \cdot E(r) = 4\pi k\varrho(r) \tag{11.1.15}$$

for the electrostatic field $E(r)$. These equations which arise from special observations are the basic equations of electrostatics.

11.1.3 Examples and Important Properties of Electrostatic Fields

i) If $\varrho(r)$ is the charge distribution corresponding to a point charge q_2 located at position r_2, then

$$\varrho(r) = q_2\delta(r - r_2) \tag{11.1.16}$$

and therefore

$$E_2(r) = kq_2 \int d^3r' \delta(r' - r_2) \frac{r - r'}{|r - r'|^3}$$

$$= kq_2 \frac{r - r_2}{|r - r_2|^3}$$

in agreement with (11.1.2).

ii) The law

$$\nabla \cdot E(r) = 4\pi k\varrho(r)$$

can also be formulated in integral form. Let V be a volume which contains the point r. Then, from Gauss' theorem (see Appendix F)

$$\int_V d^3r \nabla \cdot E = \int_{\partial V} dA \cdot E$$

$$= 4\pi k \int_V d^3r \varrho(r) = 4\pi k Q_V \ ,$$

where Q_V is the charge inside of volume V.

The relation

$$\int_{\partial V} d\mathbf{A} \cdot \mathbf{E} = 4\pi k Q_V \tag{11.1.17}$$

is called *Gauss' Law*. This equation states that flux of $\mathbf{E}$ through the boundary of V is determined by the charge inside of V.

iii) If $\varrho(\mathbf{r})$ is spherically symmetric, i.e. $\varrho(\mathbf{r}) = \varrho(r)$, then

$$\mathbf{E}(\mathbf{r}) = f(r)\frac{\mathbf{r}}{r} \; . \tag{11.1.18}$$

i.e. $E_r = f(r) \neq 0$, but $E_\theta = E_\varphi = 0$.

Let V be a sphere with radius r, then

$$d\mathbf{A} = \frac{\mathbf{r}}{r} r^2 \, d\Omega \; , \quad \mathbf{E} = E_r \frac{\mathbf{r}}{r} \tag{11.1.19}$$

and thus

$$\int_{\partial V} d\mathbf{A} \cdot \mathbf{E} = r^2 \int d\Omega E_r = 4\pi r^2 E_r = 4\pi k Q_V$$

and

$$\mathbf{E} = \frac{\mathbf{r}}{r} \frac{k Q_V}{r^2} \; . \tag{11.1.20}$$

That is, if $\varrho(\mathbf{r})$ is invariant under rotation, then $|\mathbf{E}(\mathbf{r})|$ depends only on the total amount of charge inside the sphere with radius r. If $\varrho(r) = \varrho_0$ in a sphere of radius R, then for $r \geq R$

$$Q_V = Q_R = \frac{4\pi}{3} R^3 \varrho_0 \quad \text{and thus}$$

$$\mathbf{E}(\mathbf{r}) = \frac{k Q_R \mathbf{r}}{r^3} \; . \tag{11.1.21}$$

If $r \leq R$, then

$$Q_V = \frac{4\pi}{3} r^3 \varrho_0 = \frac{r^3}{R^3} Q_R$$

and thus (Fig. 11.1.2)

$$\mathbf{E}(\mathbf{r}) = \frac{k Q_R \mathbf{r}}{R^3} \; . \tag{11.1.22}$$

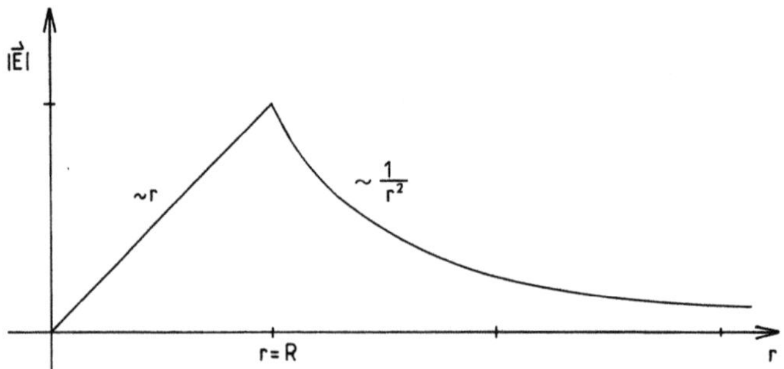

Fig. 11.1.2. $|E|$ as a function of r

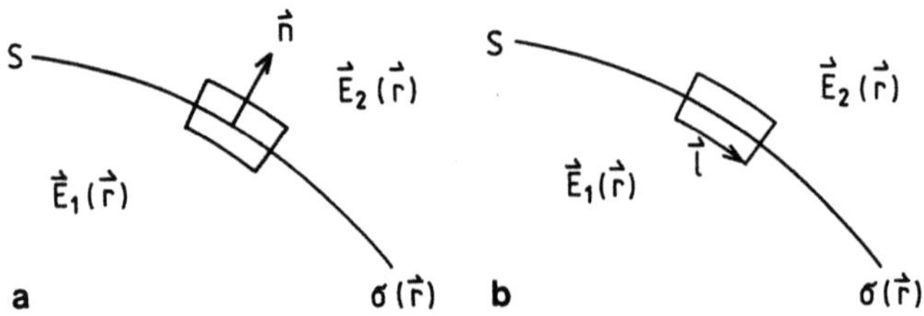

Fig. 11.1.3a, b. The electrostatic fields at a surface S which carries a surface charge density $\sigma(r)$; (a) Cross-section of the box under consideration, (b) The area A considered in v)

iv) Let a surface S be given with surface charge density $\sigma(r)$, and let E_1 be the field on one side of the surface, E_2 the field on the second side. Let n be the normal in the direction of the second side (Fig. 11.1.3a).

Let V be an arbitrary small box with a top and a bottom of area A parallel to the surface S (Fig. 11.1.3a). Then

$$\int_{\partial V} dA \cdot E = \int_A dA\,[n(r) \cdot E_2(r) - n(r) \cdot E_1(r)]$$

$+$ integrals over the sides of the box

and on the other hand

$$4\pi k \int_V d^3r\, \varrho(r) = 4\pi k \int_A dA\, \sigma(r) \ ,$$

where r is a point on the boundary surface of the box.

If we let $A \to 0$, as well as the height of the walls, we are left with

$$n(r) \cdot [E_2(r) - E_1(r)] = 4\pi k \sigma(r) \ , \tag{11.1.23}$$

that is, the change of the normal component of $E(r)$ across a surface with charge density $\sigma(r)$ is given by $4\pi k \sigma(r)$.

v) Let us consider the area A as represented in Fig. 11.1.3b. The boundary ∂A consists of two paths parallel to the surface S and two paths which run perpendicular to S. Then, since $\nabla \times E = 0$,

$$0 = \int_A dA \cdot \nabla \times E = \oint_{\partial A} dr \cdot E = l \cdot (E_1 - E_2) \ ,$$

where l is a vector parallel to the surface S and whose length is unimportant. The integral over the paths perpendicular to the surface can again be made arbitrarily small. It follows that:

The tangential components of E are continuous at a surface which carries charge density $\sigma(r)$.

vi) Let us consider the special case of a conductor. The field strength in the interior of the metal vanishes given a static charge distribution, since metal is characterized by the fact that some of its electrons have great mobility, while the positive ions cannot cause an electric current. If, now, a metal is electrically charged by introducing or removing electrons, the electric charges will move until the field in the interior vanishes. If this field does not vanish, the electrons will still experience forces and continue their motion. Since the field in the interior therefore vanishes, the potential is constant. But since

$$\Delta \phi = - 4\pi k \varrho(r)$$

the fact that $\phi = $ constant inside of the metal means that $\varrho = 0$ inside. Thus, in the case of an isolated metallic body, free charges can only be found on the surface. The free charges on the surface result in a surface charge density $\sigma(r)$.
  Let

$$E_1 = E_{\text{inside}} = 0 \ , \quad E_2 = E_{\text{outside}} \ ,$$

then

$$E_2 = 4\pi k \sigma(r) n \ . \tag{11.1.24}$$

That is, on the surface of the metal there is only a normal component of E and this is determined by the surface charge density $\sigma(r)$.

11.2 Boundary Value Problems in Electrostatics, Green's Functions

11.2.1 Dirichlet and Neumann Green's Functions

In Sect. 11.1.2, we produced the fundamental equations of electrostatics:

$$\nabla \times E(r) = 0 \ , \tag{11.2.1}$$

$$\nabla \cdot E(r) = 4\pi k \varrho(r) \ . \tag{11.2.2}$$

Starting with the first equation, we find that we can write

$$E(r) = - \nabla\phi(r)$$

with a scalar field $\phi(r)$. The second equation implies the Poisson equation for $\phi(r)$,

$$\Delta\phi(r) = - 4\pi k\varrho(r) \ . \tag{11.2.3}$$

In the case in which $\varrho(r)$ is perfectly known for a bounded region, we obtain from Sect. 11.1.2

$$\phi(r) = k \int d^3 r' \, \frac{\varrho(r')}{|r - r'|} \tag{11.2.4}$$

as a solution of the Poisson equation for all $\mathbb{R}^3$. Often though, the area in which the Poisson equation must be valid is bounded, or there is a finite boundary of the region. Think, for example, of some given configuration of metal bodies in free space so that $\Delta\phi = 0$ in between the isolated metal bodies and $\phi = \phi_i$ on the body i with $\phi_i = $ constant.

We then wind up with a typical Poisson equation boundary value problem. Such boundary value problems have already been discussed in a general context in Sect. 10.1. There is a unique solution if $\phi(r)$ is given on a closed surface of the region (the Dirichlet boundary value problem) or when $n \cdot \nabla\phi(r)$ is given on a closed surface (the Neumann boundary value problem). Here n is the normal vector to the surface. (We speak of a mixed boundary value problem if a linear combination of $\phi(r)$ and $n \cdot \nabla\phi(r)$ is given on the closed surface. We will not consider such problems here.)

Examples. a) If $V = \mathbb{R}^3$, then ∂V is the surface of an infinite sphere. The equation

$$\nabla^2\phi(r) = - 4\pi k\varrho(r)$$

has a unique solution if ϕ is given as $r \to \infty$. Let $\phi(r)$ be given as $r \to \infty$ as $\phi(r) = kQ/r$, then

$$\phi(r) = k \int d^3 r' \, \frac{\varrho(r')}{|r - r'|} \quad \text{with} \quad \int_V d^3 r' \varrho(r') = Q$$

is a unique solution in V. Adding an arbitrary solution of the Laplace equation to $\phi(r)$ also satisfies the Poisson equation, but no longer satisfies the condition $\phi(r) \to kQ/r$ as $r \to \infty$.

b) Let V be a compact region without charges enclosed by a conductor. If $\phi = 0$ is given on ∂V, we claim that $\phi = 0$ in all of V. We know that $\phi \equiv 0$ is a solution of the Laplace equation $\Delta\phi = 0$ in V. But given ϕ on the boundary of

V, the solution of Laplace's (or Poisson's) equation is uniquely determined. Therefore for any solution ϕ, we must have $\phi \equiv 0$. This region is called a *Faraday*[3] *cage*.

In order to solve the general boundary value problem for Poisson's equation, we again use the concept of a Green's function. We already introduced this concept in Sect. 10.1 in order to solve inhomogeneous linear differential equations. The Green's function here is a solution of the special Poisson equation

$$\nabla'^2 G(r, r') = -4\pi\delta(r - r') \, , \tag{11.2.5}$$

where for convenience we have introduced the extra factor -4π in contrast to the definition in Sect. 10.1.

In Sect. 11.1.2, we already saw that the function $1/|r - r'|$ satisfies the equations

$$\nabla^2 \frac{1}{|r - r'|} = -4\pi\delta(r - r') \quad \text{or}$$

$$\nabla'^2 \frac{1}{|r - r'|} = -4\pi\delta(r - r') \, .$$

Thus

$$G(r, r') = \frac{1}{|r - r'|} + F(r, r') \tag{11.2.6}$$

with

$$\nabla'^2 F(r, r') = 0 \, . \tag{11.2.7}$$

The form of $F(r, r')$ is determined by the following considerations:
If we consider the first of *Green's identities* (compare (10.3.34))

$$\int_V d^3r' (\nabla'\phi \cdot \nabla'\psi + \phi\nabla'^2\psi) = \int_{\partial V} dA' \cdot \nabla'\psi(r')\phi(r') \equiv \int_{\partial V} dA' \frac{\partial\psi}{\partial n'}\phi$$

or, interchanging ϕ and ψ,

$$\int_V d^3r' (\nabla'\psi \cdot \nabla'\phi + \psi\nabla'^2\phi) = \int_{\partial V} dA'\psi \frac{\partial\phi}{\partial n'} \, ,$$

---

[3] *Faraday, Michael* (*1791 Newington Butts, d. 1867 London). First work in the field of chemistry. After Oersted's discovery, he turned to research in electromagnetism. He was one of the originators of the field concept. Starting with the dynamic representation of the transformation of forces, he discovered the law of induction in 1831. Later work on electrolysis and attempts to find electromagnetic effects of gravitation.

then subtracting the second equation from the first, we find *Green's second identity*:

$$\int\limits_V d^3r'(\phi\nabla'^2\psi - \psi\nabla'^2\phi) = \int\limits_{\partial V} dA'\left(\phi\frac{\phi\psi}{\partial n'} - \psi\frac{\partial\phi}{\partial n'}\right) . \tag{11.2.8}$$

If we choose $\psi(r') = G(r, r')$, $\phi = \phi(r')$ with $\nabla'^2\phi(r') = -4\pi k\varrho(r')$, it follows that

$$\int\limits_V d^3r'\{\phi(r')[-4\pi\delta(r - r')] - G(r, r')[-4\pi k\varrho(r')]\}$$

$$= \int\limits_{\partial V} dA'\left[\phi(r')\frac{\partial G(r, r')}{\partial n'} - G(r, r')\frac{\partial\phi(r')}{\partial n'}\right]$$

or

$$\phi(r) = k\int\limits_V d^3r'G(r, r')\varrho(r') - \frac{1}{4\pi}\int\limits_{\partial V} dA'$$

$$\times\left[\phi(r')\frac{\partial G(r, r')}{\partial n'} - G(r, r')\frac{\partial\phi(r')}{\partial n'}\right] . \tag{11.2.9}$$

In the case of Dirichlet boundary conditions, we choose $F(r, r')$ such that

$$G(r, r') = 0 \quad \text{for} \quad r' \in \partial V .$$

The Green's function defined (uniquely) in this way is called a *Dirichlet Green's function* and is indicated $G_D(r, r')$. Then, from (11.2.9),

$$\phi(r) = k\int\limits_V d^3r'G_D(r, r')\varrho(r') - \frac{1}{4\pi}\int\limits_{\partial V} dA'\phi(r')\frac{\partial G_D(r, r')}{\partial n'} , \tag{11.2.10}$$

i.e. if we know $G_D(r, r')$, then from (11.2.10) the solution $\phi(r)$ is uniquely determined by the value of $\varrho(r')$ in V and by $\phi(r')$ on ∂V.

For the Neumann boundary value problem, we require that the Neumann Green's function $G_N(r, r')$ satisfy

$$\frac{\partial G_N(r, r')}{\partial n'} \equiv n(r')\cdot\nabla'G_N(r, r')$$

$$= -\frac{4\pi}{A} \quad \text{for} \quad r' \in \partial V , \tag{11.2.11}$$

where A is the total area of the boundary ∂V.

We cannot simply require that $\partial G_N(r, r')/\partial n' = 0$, since applying Gauss' theorem to

$$\int\limits_V d^3r'\nabla'^2 G_N(r, r') = \int\limits_V d^3r'[-4\pi\delta(r - r')] = -4\pi$$

yields

$$\int\limits_{\partial V} dA' \cdot \nabla' G_N(r, r') = \int\limits_{\partial V} dA' \frac{\partial G_N(r, r')}{\partial n'} = - 4\pi \tag{11.2.12}$$

in contradiction to $n' \cdot \nabla' G_N(r, r') = 0$, but consistent with (11.2.11).

Thus, for a Neumann boundary value problem, it follows from (11.2.9) that

$$\phi(r) = k \int\limits_{V} d^3r' G_N(r, r') \varrho(r')$$

$$+ \frac{1}{4\pi} \int\limits_{\partial V} dA' G_N(r, r') \frac{\partial \phi(r')}{\partial n'} + \frac{1}{A} \int\limits_{\partial V} dA' \phi(r') \ . \tag{11.2.13}$$

The last term on the right hand side is the average of $\phi(r)$ over the entire surface ∂V. This is merely an irrelevant constant, which in any case is undetermined in the solution of the Neumann boundary value problem.

11.2.2 Supplementary Remarks on Boundary Value Problems in Electrostatics

i) The Green's functions we have defined above are not always easy to determine. It is essential to note the following, though:

$G(r, r')$ does not depend on the sources in V or on the boundary values on ∂V, but only on the geometry of V and the type of boundary value problem.

ii) If in the Dirichlet boundary value problem we can solve the special boundary value problem

$$\varrho(r) = q\delta(r - y) \ , \quad \phi \equiv 0 \quad \text{on} \quad \partial V \tag{11.2.14}$$

another way, then the Green's function for this geometry can be found, since with (11.2.10),

$$\phi(r) = kq \int d^3r' G_D(r, r')\delta(r' - y) = kq G_D(r, y) \ . \tag{11.2.15}$$

This solution $\phi(r)$ is then identical, up to a factor kq with the Green's function $G_D(r, y)$.

Example. Let $V = \mathbb{R}^3$, then ∂V is an infinitely distant spherical surface. For

$$\varrho(r') = q\delta(r' - y) \quad \text{and} \quad \phi \to 0 \quad \text{for} \quad r \to \infty$$

we know the solution

$$\phi(r) = \frac{kq}{|r - y|} \ .$$

Thus, for this geometry, $(V = \mathbb{R}^3)$:

$$G_D(r, r') = \frac{1}{|r - r'|} \;,$$

as we already found.

iii) $G_D(r, r')$ is symmetric, i.e.

$$G_D(r, r') = G_D(r', r) \;, \tag{11.2.16}$$

as can easily be derived from Green's second identity for $\phi(r'') = G_D(r, r'')$ and $\psi(r'') = G_D(r', r'')$.

iv) In the case of a conductor with uniform extent in one direction, if we can consider the problem as being approximately two-dimensional then we have to solve

$$\left(\frac{\partial^2}{\partial x^2} + \frac{\partial^2}{\partial y^2}\right)\phi(x, y) = 0 \quad \text{in} \quad V \tag{11.2.17}$$

with

$$\phi(x, y) = \phi_i \quad \text{on} \quad \partial V_i \;, \qquad \partial V = \bigcup_i \partial V_i$$

Such a two dimensional potential problem can be solved, as in hydrodynamics (Sect. 9.5), with the help of function theory by considering $\phi(x, y)$ as the real part of an analytic function which is chosen to satisfy the boundary conditions (see, for example, [Panofski et al.]).

v) For a region V in which there are no charges, but which is bounded by conductors on which ϕ is constant, we have

$$\phi(r) = -\frac{1}{4\pi} \int_{\partial V} dA' \cdot \nabla' G_D(r, r')\phi(r)$$

$$= \sum_i \phi_j a_j(r) \quad \text{with} \tag{11.2.18}$$

$$a_j(r) = -\frac{1}{4\pi} \int_{\partial L_j} dA' \cdot \nabla' G_D(r, r') \;, \tag{11.2.19}$$

where ϕ_j is the constant potential on the conductor j and the surface integral extends over the boundary of the conductor j. On the other hand, by Gauss' law the charge q_i on the conductor i satisfies

$$q_i = -\frac{1}{4\pi k} \int_{\partial L_i} dA \cdot \nabla \phi(r) = \sum_j C_{ij}\phi_j \quad \text{with} \tag{11.2.20}$$

$$C_{ij} = -\frac{1}{4\pi k} \int\limits_{\partial L_i} dA \cdot \nabla a_j(r) = \frac{1}{4\pi k} \frac{1}{4\pi} \int\limits_{\partial L_i} dA \cdot \nabla \int\limits_{\partial L_j} dA' \cdot \nabla' G_{\mathrm{D}}(r, r') \ . \quad (11.2.21)$$

Clearly $C_{ij} = C_{ji}$. The C_{ij} are called the Maxwell *capacitance coefficients*.

Examples. a) For a sphere with radius a and total charge q,

$$\phi(r) = k\frac{q}{r} \quad \text{thus} \quad \phi(a) = k\frac{q}{a} \ ,$$

i.e.

$$C = \frac{a}{k} \ . \tag{11.2.22}$$

b) For two parallel conducting plates (Fig. 11.2.1) which carry charges q_1 and $-q_1$ (called a parallel plate capacitor) we have, ignoring edge effects:

$$\phi(z) = \alpha z \ , \quad \text{and thus}$$

$$\phi_2 - \phi_1 = \alpha l \ ,$$

where l is the distance between the plates. Further, we have

$$-\frac{d\phi}{dz} = -\alpha = E_z = 4\pi k\sigma = 4\pi k\frac{q_1}{A} \ ,$$

where A is the area of each plate.
Thus we have

$$q_1 = -\frac{A\alpha}{4\pi k} = \frac{A}{4\pi k}\frac{\phi_1 - \phi_2}{l}$$

and therefore

$$C_{11} = -C_{12} = \frac{A}{4\pi k l} \ . \tag{11.2.23}$$

Analogously,

$$C_{22} = -C_{21} = C_{11} \ . \tag{11.2.24}$$

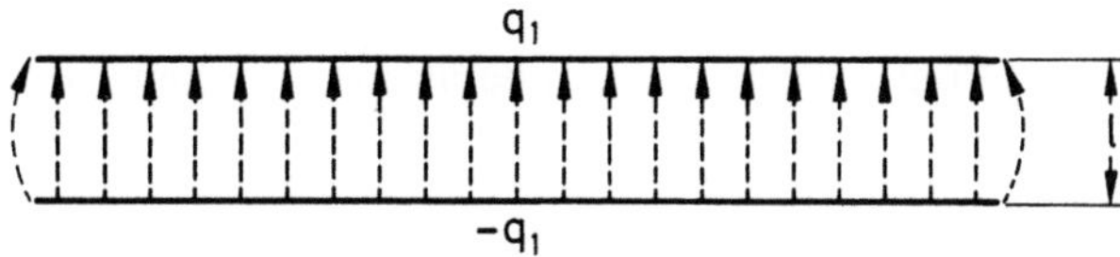

Fig. 11.2.1. Parallel plate capacitor with charges $+q_1$ and $-q_l$ on the plates. Inside the capacitor, to a good approximation the field is homogeneous

We call

$$C = A/4\pi kl \quad (= A\varepsilon_0/l \quad \text{with} \quad k = 1/4\pi\varepsilon_0)$$

the *capacitance* of the parallel plate capacitor.

11.3 The Calculation of Green's Functions, the Method of Images

In some very simple geometric cases, the Dirichlet Green's function can be found using a special method. We will demonstrate this using the example

$$V = \mathbb{R}^3 - \text{a sphere of radius } a \ .$$

We are then looking for a solution to the boundary value problem

$$\Delta\phi(r) = -4\pi kq\delta(r - y) \quad \text{in} \quad \{r \,|\, |r| > a\} \ ,$$

$$\phi \equiv 0 \quad \text{for} \quad r = a \ .$$

Since we are only interested in the solution of the equation for $r \geq a$, we can imagine that for $r < a$ there is such an "image charge" so that for $r = a$ it is exactly true that $\phi(r) = 0$, that is, we set

$$\phi(r) = k\left(\frac{q}{|r - y|} + \frac{q'}{|r - y'|} \right) . \tag{11.3.1}$$

In this equation, we see that the image charge q lies inside the sphere, thus not in V (Fig. 11.3.1). Then, formally, it is true that

$$\Delta\phi(r) = -4\pi kq\delta(r - y) - 4\pi kq'\delta(r - y') \ , \tag{11.3.2}$$

but for $r \geq a$, that is, in V

$$\Delta\phi(r) = -4\pi kq\delta(r - y) \ , \tag{11.3.3}$$

since the second δ-function in (11.3.2) identically vanishes there.

We need to choose a q' and a y' under the condition $|y'| < a$ such that $\phi(r)|_{r=a} = 0$.

Let $y = yn'$, $r = rn$, then from symmetry considerations it follows that $y' = y'n'$. Then,

$$\phi(r)|_{r=a} = k\left(\frac{q}{a\left|n - \dfrac{y}{a}n'\right|} + \frac{q'}{y'\left|n' - \dfrac{a}{y'}n\right|} \right) . \tag{11.3.4}$$

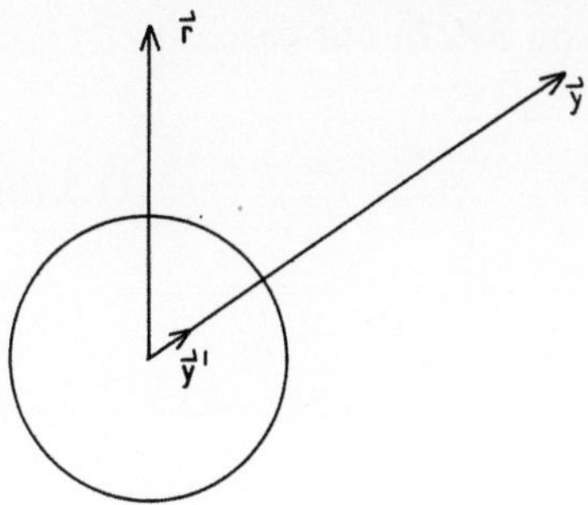

Fig. 11.3.1. A grounded conducting sphere, the point charge at y and the image charge at y'

Note that y/a as well as a/y' is greater than 1. We now choose y' in such a way that

$$\frac{a}{y'} = \frac{y}{a} , \quad \text{therefore} \quad y' = \frac{a^2}{y} \tag{11.3.5}$$

so that

$$\left| n - \frac{y}{a} n' \right|^2 = \left| n' - \frac{a}{y'} n \right|^2 = 1 - 2\frac{a}{y'} n \cdot n' + \left(\frac{a}{y'}\right)^2 . \tag{11.3.6}$$

Further, we choose q' such that

$$\frac{q}{a} = \frac{-q'}{y'} , \quad \text{so that} \quad q' = -\frac{y'}{a} q = -\frac{a}{y} a . \tag{11.3.7}$$

Then we find precisely $\phi(r) = 0$ for $r = a$. In this way, we have determined $\phi(r)$ to be

$$\phi(r) = k\left(\frac{q}{|r - y|} - \frac{qa}{y} \frac{1}{\left| r - \frac{a^2}{y^2} y \right|} \right) \tag{11.3.8}$$

and thus

$$G_{\mathrm{D}}(r, r') = \frac{1}{|r - r'|} - \frac{a}{r'} \frac{1}{\left| r - \frac{a^2}{r'^2} r' \right|} . \tag{11.3.9}$$

In the representation

$$G_{\mathrm{D}}(r, r') = \frac{1}{|r - r'|} + F(r, r')$$

$F(r, r')$ is a function of r which satisfies the homogeneous Laplace equation in V and represents a potential which arises from sources outside of V which ensure that $G_D(r, r')$ satisfies the boundary conditions on ∂V. In our case,

$$F(r, r') = -\frac{a}{r'} \frac{1}{\left| r - \dfrac{a^2}{r'^2} r' \right|} \; . \tag{11.3.10}$$

Writing out, we find

$$G_D(r, r') = \frac{1}{(r^2 + r'^2 - 2rr' \cos \gamma)^{1/2}}$$

$$- \frac{1}{[(r^2 r'^2/a^2) + a^2 - 2rr' \cos \gamma]^{1/2}} \tag{11.3.11}$$

with $\gamma = \sphericalangle\,(r, r')$. $G_D(r, r')$ (seen as a function of r) is the potential which is caused by a unit charge at r' and the spherical conductor. We immediately see that for $r' = a$,

$$G_D(r, r') = 0 \; .$$

It is also true that $G_D(r, r') = G_D(r', r)$ and this symmetry holds for every Green's function G_D (see Sect. 11.2.2). This symmetry can be understood intuitively as expressing the following: The potential at point r caused by a unit charge at r' and by the sphere is equivalent to the potential at point r' caused by a unit charge at r and the sphere. This is plausible.

With the help of $G_D(r, r')$, we can find the potential $\phi(r)$, following (11.2.10), if the surface of the sphere is at some potential $\phi_1(r)$ for $r = a$. For this, we need

$$\frac{\partial G_D}{\partial n'}\bigg|_{r'=a} = n' \cdot \nabla' G_D(r, r')|_{r'=a}$$

$$= -\frac{r'}{r'} \cdot \frac{\partial}{\partial r'} G_D(r, r')|_{r'=a}$$

$$= -\frac{\partial}{\partial r'} G_D(r, r')|_{r'=a}$$

$$= \frac{1}{2}(r^2 + a^2 - 2\,ar \cos \gamma)^{-3/2}\left(2a - 2\frac{r^2}{a} \right)$$

$$= -\frac{r^2 - a^2}{a(r^2 + a^2 - 2\,ar \cos \gamma)^{3/2}} \; . \tag{11.3.12}$$

Thus, in general, for this geometry $V = \{r \mid |r| \geq a\}$:

$$\phi(r) = k \int\limits_V d^3r' \, G_D(r, r') \varrho(r')$$

$$+ \frac{1}{4\pi} \int d\Omega' a^2 \, \frac{r^2 - a^2}{a(r^2 + a^2 - 2\,ar \cos \gamma)^{3/2}} \phi(r' = a, \theta', \varphi') \tag{11.3.13}$$

with

$$r' = a(\sin \theta' \cos \varphi', \sin \theta' \sin \varphi', \cos \theta') \ ,$$

$$d\Omega' = d\varphi' d \cos \theta' \ ,$$

$$r = r(\sin \theta \cos \varphi, \sin \theta \sin \varphi, \cos \theta)$$

and

$$\cos \gamma = \cos \theta \cos \theta' + \sin \theta \sin \theta' \cos(\varphi - \varphi') \ .$$

Remarks. i) We consider again a conducting sphere with $\phi(r) = 0$ for $r = a$. An external point charge at position y produces a potential according to (11.3.8). We want to find the distribution of charges on the surface of the sphere, that is, the surface charge density $\sigma(r)$.

From Sect. 11.1.3, we know that the normal component of $E(r)$ jumps at the surface of the sphere by

$$4\pi k \sigma(r) \ , \quad |r| = a \ .$$

Thus,

$$\sigma(\theta, \varphi) = -\frac{1}{4\pi k} \, n \cdot E = -\frac{1}{4\pi k} \frac{r}{r} \cdot \nabla \phi(r)$$

$$= -\frac{1}{4\pi k} \frac{\partial}{\partial r} \left[\frac{kq}{(r^2 + y^2 - 2ry \cos \gamma)^{1/2}} \right.$$

$$\left. - \frac{kqa}{y \left(r^2 + \dfrac{a^4}{y^2} - 2r \dfrac{a^2}{y} \cos \gamma \right)^{1/2}} \right]\Bigg|_{r=a} \tag{11.3.14}$$

with $\gamma = \angle\,(r, y)$.

We then find, if the z-axis lies in the direction of y,

$$\sigma(\theta, \varphi) = -\frac{q}{4\pi}\left[-\frac{1}{2}(a^2 + y^2 - 2ay\cos\theta)^{-3/2}(2a - 2y\cos\theta)\right.$$

$$\left. -\frac{a}{y}\left(-\frac{1}{2}\right)\left(a^2 + \frac{a^4}{y^2} - 2\frac{a^3}{y}\cos\theta\right)^{-3/2}\left(2a - 2\frac{a^2}{y}\cos\theta\right)\right]$$

$$= -\frac{q}{4\pi}(a^2 + y^2 - 2ay\cos\theta)^{-3/2}$$

$$\times\left[-a + y\cos\theta + \frac{y^2}{a^2}\left(a - \frac{a^2}{y}\cos\theta\right)\right]$$

$$= -\frac{q}{4\pi a}\frac{-a^2 + y^2}{(a^2 + y^2 - 2ay\cos\theta)^{3/2}}$$

$$= -\frac{q}{4\pi a^2}\frac{a}{y}\frac{1 - (a^2/y^2)}{[1 + (a^2/y^2) - (2a/y)\cos\theta]^{3/2}}\,. \tag{11.3.15}$$

If we plot $4\pi\sigma a^2/(-q)$ against θ, we find the curves shown in Fig. 11.3.2, depending on the value of y/a.

The maximal value of the surface charge density occurs directly next to the charge q; of course, it always has opposite sign. The farther away the charge q is, the smaller the effect of the accumulation of charge carriers of opposite charge on the sphere in the direction of the charge.

The total charge on the sphere is

$$\int d\Omega\,\sigma(\theta, \varphi) = 2\pi \int_{-1}^{+1} d\cos\theta\,\sigma(\theta) = -\frac{a}{y}q = q'\,. \tag{11.3.16}$$

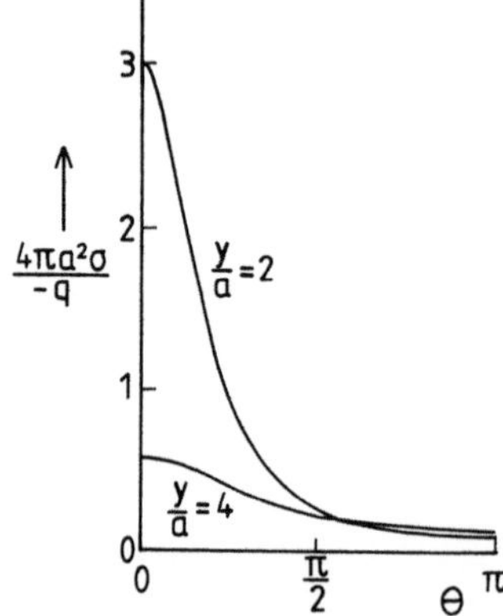

Fig. 11.3.2. Surface charge density $\sigma(\theta)$ as a function of θ for various values of y/a

As $y \to \infty$, q' and $\sigma(\theta, \varphi)$ naturally vanish. The change in q' with increasing y is possible, because the sphere is connected to a "charge reservoir", i.e. it is grounded.

ii) This case in which the potential $\phi(r)$ on the sphere is determined by the connection of the sphere with a "charge reservoir" should be distinguished from the case in which the sphere is isolated and carries a fixed, given charge Q. It is true that the potential on the sphere will remain constant, since it is a conductor, and otherwise charges would move until such a state were produced; however, now the value of the constant potential is not given in advance. Rather, it will depend on the value of Q and on the location and strength of the charge outside of the sphere. To calculate the potential $\phi(r)$ outside of and on the sphere, we use the following considerations:

First, we consider again the grounded sphere. Then, with point charge q at location y:

$$\phi(r) = kq\,G_D(r, y) \ ,$$

where $G_D(r, y)$ is given by (11.3.9). The total charge which is found on the surface is thus

$$q' = -\frac{a}{y}q \ .$$

If we now terminate the connection to the "charge reservoir" and place the additional charge $Q - q'$ on the sphere, the additional charge carriers will distribute themselves uniformly over the surface, since we must always have $\phi = $ constant on the surface. This additional charge will therefore act like a charge $Q - q' = Q + (a/y)q$ at the midpoint of the sphere which causes the additional potential

$$k\frac{Q - q'}{r} \ . \tag{11.3.17}$$

Thus, for this case,

$$\phi(r) = kq\,G_D(r, y) + k\frac{Q + (a/y)q}{r} \ , \tag{11.3.18}$$

and on the surface of the sphere, we have

$$\phi(r)|_{r=a} = k\frac{Q + (a/y)q}{a} \ . \tag{11.3.19}$$

If, as in (11.3.16), we now calculate the total charge on the sphere, we will of course find it to be Q.

iii) The Dirichlet Green's function, which is also the potential of a point charge at r' in the space V between grounded conductors, is of the form

$$G_{\mathrm{D}}(r, r') = \frac{1}{|r - r'|} + F(r, r') \ .$$

Here, $k/|r - r'|$ is the potential of the point charge and $\phi_{r'}(r) = kF(r, r')$ is the potential which arises from the "*induced*" *charges* on the conductors. We have

$$\Delta\phi_{r'}(r) = k\Delta F(r, r') = 0 \quad \text{in} \quad V \ .$$

The total charge Q_i on the i-th conductor and the force F_{B} which the induced charges exert on the point charge at r' can now be easily calculated:

$$Q_i(r') = \frac{1}{4\pi k} \int_{\partial L_i} dF_i \cdot \nabla\phi_{r'}(r) \ , \tag{11.3.20}$$

$$F_{\mathrm{B}}(r') = - k\nabla_r F(r, r')|_{r=r'} \ . \tag{11.3.21}$$

F_{B} is called the *image force*. If the method of imaging is successful, then F is of the form

$$\phi_{r'}(r) = kF(r, r') = \sum_a \frac{ke_a}{|r - r_a(r')|} \ , \tag{11.3.22}$$

where the e_a correspond to the image charges. Then Q_i is the sum of all the image charges in L_i and

$$F_{\mathrm{B}}(r') = \sum_a ke_a \frac{r' - r_a(r')}{|r' - r_a(r')|^3} \ , \tag{11.3.23}$$

is exactly the Coulomb force which is exerted on the unit charge at r' by all of the image charges.

In our example,

$$\phi_{r'}(r) = kF(r, r') = - \frac{qa}{r'} \frac{k}{\left| r - \dfrac{a^2}{r'^2} r' \right|}$$

and thus

$$F_{\mathrm{B}}(r') = - q^2 \frac{ak}{r'^3} \left(1 - \frac{a^2}{r'^2} \right)^{-2} \frac{r'}{r'} \ . \tag{11.3.24}$$

iv) In this chapter, we have considered the case of the conducting sphere. An easier case is that of the conducting plane. Here, we see directly that the Dirichlet boundary conditions can be satisfied if we put an image charge of opposite strength in the mirror-image position.

11.4   The Calculation of Green's Functions, Expansion in Spherical Harmonics

The method of imaging introduced in Sect. 11.3 succeeds only in cases where the geometry is very simple. Even in the case of the Poisson equation in the region between two concentric spheres,

$$V = \{r \,|\, a \le |r| \le b\}$$

this method fails. An infinite number of image charges would be required.

Of course, we could try to solve the differential equation for the Green's function directly. Here, we will sketch an exemplary strategy for such a calculation of the Dirichlet Green's function, in particular for the case of spherically symmetric geometry.

Since for $r \ne r'$, the Green's function $G(r, r')$ must sastisfy the Laplace equation, it is natural to suppose that $G_D(r, r')$ has the form

$$G_D(r, r') = \sum_{l=0}^{\infty} \sum_{m=-l}^{+l} g_l(r, r') Y_{lm}^*(\theta', \varphi') Y_{lm}(\theta, \varphi)$$

$$= \sum_{l=0}^{\infty} g_l(r, r') \frac{2l + 1}{4\pi} P_l(\cos \gamma) \tag{11.4.1}$$

with

$$\gamma = \measuredangle(r, r') , \quad r = r(r, \theta, \varphi) , \quad r' = r'(r', \theta', \varphi') .$$

Since (see Appendix F)

$$\nabla'^2 = \frac{1}{r'} \frac{\partial^2}{\partial r'^2} r' + \frac{1}{r'^2} \left(\frac{1}{\sin \theta'} \frac{\partial}{\partial \theta'} \sin \theta' \frac{\partial}{\partial \theta'} + \frac{1}{\sin^2 \theta'} \frac{\partial^2}{\partial \varphi'^2} \right) \tag{11.4.2}$$

we find for the left side of the differential equation for $G_D(r, r')$

$$\nabla'^2 G_D(r, r') = \sum_{l=0}^{\infty} \sum_{m=-l}^{+l} \left(\frac{1}{r'} \frac{d^2}{dr'^2} r' - \frac{1}{r'^2} l(l + 1) \right)$$

$$\times g_l(r, r') Y_{lm}^*(\theta', \varphi') Y_{lm}(\theta, \varphi) , \tag{11.4.3}$$

while on the right side of the equation, the term $-4\pi\delta(r - r')$ can also be written in the form (compare (10.4.74)):

$$-4\pi\delta(r - r') = -4\pi \frac{1}{r'^2} \delta(r - r')\delta(\cos \theta' - \cos \theta)\delta(\varphi - \varphi')$$

$$-4\pi \frac{1}{r'^2} \delta(r - r') \sum_{l=0}^{\infty} \sum_{m=-l}^{+l} Y_{lm}^*(\theta', \varphi') Y_{lm}(\theta, \varphi) . \tag{11.4.4}$$

Thus, for $g_l(r, r')$, we obtain the differential equation

$$\left[\frac{1}{r'}\frac{d^2}{dr'^2}r' - \frac{l(l+1)}{r'^2}\right]g_l(r, r') = -\frac{4\pi}{r'^2}\,\delta(r - r') \ . \tag{11.4.5}$$

Since this is homogeneous for $r \neq r'$, we can set

$$g_l(r, r') = \begin{cases} Ar'^l + Br'^{-l-1} & \text{for} \quad r' > r \\ A'r'^l + B'r'^{-l-1} & \text{for} \quad r' < r \ , \end{cases} \tag{11.4.6}$$

where A, B, A', B' are to be chosen according to the boundary conditions.

Thus, by knowing the solutions of the Laplace equation in spherical coordinates, we have reduced the problem to the determination of the constants A, A', B, and B'. Of course, these constants can still depend on r.

We want to find how the Green's functions calculated in Sect. 11.3 can be represented in the form (11.4.1) and how the constants A, A', B, B' can be determined.

i) We consider first

$$\phi^{(1)}(r) = \frac{1}{|r - r'|} = \frac{1}{(r^2 + r'^2 - 2rr'\cos\gamma)^{1/2}} \ . \tag{11.4.7}$$

For $\cos\gamma = 1$, on the one hand, from (11.4.1)

$$\phi^{(1)}(r) = \sum_{l=0}^{\infty} g_l(r, r')\frac{2l+1}{4\pi}P_l(1) = \sum_{l=0}^{\infty} g_l(r, r')\frac{2l+1}{4\pi} \ , \tag{11.4.8}$$

on the other hand

$$\phi^{(1)}(r) = \frac{1}{|r - r'|} = \sum_{l=0}^{\infty}\frac{r'^l}{r^{l+1}} \quad \text{for} \quad r' < r$$

$$= \sum_{l=0}^{\infty}\frac{r^l}{r'^{l+1}} \quad \text{for} \quad r' > r \ . \tag{11.4.9}$$

Thus,

$$g_l(r, r') = \frac{4\pi}{2l+1}\frac{r_<^l}{r_>^{l+1}} \tag{11.4.10}$$

with $r_> = \text{Max}\,(r, r')$, $r_< = \text{Min}\,(r, r')$. Of course,

$$g_l(r, r') \to 0 \quad \text{for} \quad r' \to \infty \ , \tag{11.4.11}$$

this is the boundary condition $G_D(r, r') = 0$ for $r' \in \partial V$.

ii) For the geometry

$$V = \mathbb{R}^3 - \text{a sphere of radius } a$$

we have .

$$G_D(r, r') = \frac{1}{|r - r'|} - \frac{a}{r'} \frac{1}{|r - (a^2/r'^2)r'|} \ , \tag{11.4.12}$$

and for $\cos \gamma = 1$, i.e. r parallel to r', we have

$$\frac{a}{r'} \frac{1}{|r - (a^2/r'^2)r'|} = \frac{a}{r'} \frac{1}{r - (a^2/r')}$$

$$= \frac{1}{a} \frac{a^2}{rr'} \frac{1}{1 - (a^2/rr')} = \sum_{l=0}^{\infty} \left(\frac{a^2}{rr'}\right)^{l+1} \frac{1}{a} \ . \tag{11.4.13}$$

Note here that always $(a^2/rr') < 1$.
In this case, we then have

$$g_l(r, r') = \begin{cases} \dfrac{4\pi}{2l + 1} \left[\dfrac{r^l}{r'^{l+1}} - \dfrac{1}{a} \left(\dfrac{a^2}{rr'}\right)^{l+1} \right] & \text{for} \quad r' > r \\[4ex] \dfrac{4\pi}{2l + 1} \left[\dfrac{r'^l}{r^{l+1}} - \dfrac{1}{a} \left(\dfrac{a^2}{rr'}\right)^{l+1} \right] & \text{for} \quad r' < r \ . \end{cases} \tag{11.4.14}$$

For $r' = a$,

$$g_l(r, a) = \frac{4\pi}{2l + 1} \left(\frac{a^l}{r^{l+1}} - \frac{a^l}{r^{l+1}} \right) = 0 \ ,$$

while $g_l(r, r')$ vanishes as $r' \to \infty$. These are again the boundary conditions

$$G_D(r, r') = 0 \quad \text{for} \quad r' \in \partial V \ .$$

iii) We want to figure out in the preceding example how the constants A, A', B and B' can be determined in the still general form for $g_l(r, r')$ in (11.4.6). We require:

$$g_l(r, r') = 0 \quad \text{for} \quad r' = a \quad \text{and} \quad r' \to \infty \ .$$

Then, first,

$$g_l(r, a) = A'a^l + B'a^{-l-1} = 0 \ ,$$

and therefore

$$B' = - A'a^{2l+1} \quad \text{or} \tag{11.4.15}$$

$$g_l(r, r') = A'\left(r'^l - \frac{a^{2l+1}}{r'^{l+1}}\right) \quad \text{for} \quad r' < r \ . \tag{11.4.16}$$

From $g_l(r, r') = 0$ for $r' = \infty$, it follows that

$$A = 0 \ , \quad \text{and thus} \tag{11.4.17}$$

$$g_l(r, r') = Br'^{-l-1} \quad \text{for} \quad r' > r \ . \tag{11.4.18}$$

It remains to determine the constants A' and B which can, of course, still depend on r. The symmetry of $g_l(r, r')$ suggests writing it in the form

$$g_l(r, r') = \begin{cases} C\left(r'^l - \dfrac{a^{2l+1}}{r'^{l+1}}\right) r^{-l-1} & \text{for} \quad r' < r \\[3mm] C\left(r^l - \dfrac{a^{2l+1}}{r^{l+1}}\right) r'^{-l-1} & \text{for} \quad r' > r \ , \end{cases} \tag{11.4.19}$$

so that all that is left is to find the constant C.

If we multiply the equation (11.4.5) by r' and then integrate over r' from $r' = r - \varepsilon$ to $r' = r + \varepsilon$, we then obtain

$$\int_{r-\varepsilon}^{r+\varepsilon} dr'\left[\frac{d^2}{dr'^2}r'g_l(r, r') + \frac{l(l + 1)}{r'}g_l(r, r')\right] = - 4\pi\frac{1}{r} \ . \tag{11.4.20}$$

If $g_l(r, r')$ does not become singular at $r' = r$, then the second integral is of order $O(\varepsilon)$. It follows then that

$$\frac{d}{dr'}(r'g_l(r, r'))$$

jumps by $- 4\pi\dfrac{1}{r}$ at $r' = r$.

Now,

$$\frac{d}{dr'}(r'g_l(r, r')) = \begin{cases} Cr^{-l-1}\left[(l + 1)r'^l + l\dfrac{a^{2l+1}}{r'^{l+1}}\right] & \text{for} \quad r' < r \\[3mm] C\left(r^l - \dfrac{a^{2l+1}}{r^{l+1}}\right)(- l)r'^{-l-1} & \text{for} \quad r' > r \end{cases} \tag{11.4.21}$$

and

$$\frac{d}{dr'}(r'g_l(r,r'))|_{r'=r+\varepsilon} - \frac{d}{dr'}(r'g_l(r,r'))|_{r'=r-\varepsilon}$$

$$= C\left[-(l/r) + l\frac{a^{2l+1}}{r^{2l+2}} - \frac{(l+1)}{r} - l\frac{a^{2l+1}}{r^{2l+2}}\right] = -C\frac{2l+1}{r} . \qquad (11.4.22)$$

Thus, it follows

$$C = \frac{4\pi}{2l+1} . \qquad (11.4.23)$$

We therefore find again

$$g_l(r,r') = \frac{4\pi}{2l+1}\left[\frac{r_<^l}{r_>^{l+1}} - \frac{1}{a}\left(\frac{a^2}{rr'}\right)^{l+1}\right]$$

in agreement with (11.4.14).

Analogously, we can easily find the Green's function, say, for the geometry

$$V = \{r \,|\, a \le |r| \le b\} .$$

We find

$$g_l(r,r') = \frac{4\pi}{2l+1}\frac{1}{1-(a/b)^{2l+1}}\left(r_<^l - \frac{a^{2l+1}}{r_<^{l+1}}\right)\left(\frac{1}{r_>^{l+1}} - \frac{r_>^l}{b^{2l+1}}\right) .$$

For further examples, see [Jackson].

11.5 Localized Charge Distributions, the Multipole Expansion

The expansion of Green's functions in spherical harmonics as demonstrated in Sect. 11.4 leads to a very useful characterization of the solutions of boundary value problems. We will demonstrate using the simplest example ($V = \mathbb{R}^3$).

Here,

$$\phi(r) = k \int d^3r' G_D(r,r')\varrho(r') = k \int d^3r' \frac{\varrho(r')}{|r-r'|} . \qquad (11.5.1)$$

In this expression, let $\varrho(r)$ be a charge distribution which is concentrated in a sphere of radius R. If we now take $r > R$, then from (11.4.1) and (11.4.10) we get

$$\phi(r) = k \sum_{l=0}^{\infty} \sum_{m=-l}^{+l} \frac{4\pi}{2l+1} \int d^3r'\, \varrho(r')r'^l Y_{lm}^*(\theta', \varphi') Y_{lm}(\theta, \varphi) r^{-l-1}$$

$$= k \sum_{l=0}^{\infty} \sum_{m=-l}^{+l} \sqrt{\frac{4\pi}{2l+1}}\, q_{lm} Y_{lm}(\theta, \varphi) r^{-l-1} \tag{11.5.2}$$

with

$$q_{lm} = \sqrt{\frac{4\pi}{2l+1}} \int d^3r'\, \varrho(r')r'^l Y_{lm}^*(\theta', \varphi') \ . \tag{11.5.3}$$

The quantities q_{lm} are called the *multipole moments*[4]. For example,

$$q_{00} = \int d^3r\, \varrho(r) = Q \ . \tag{11.5.4}$$

q_{00} is thus determined by the total charge Q. Further,

$$q_{10} = \sqrt{\frac{4\pi}{3}} \sqrt{\frac{3}{4\pi}} \int d^3r'\, \varrho(r')r' \cos\theta' \ , \tag{11.5.5}$$

$$q_{1\pm1} = \sqrt{\frac{4\pi}{3}} \left(\pm \sqrt{\frac{3}{8\pi}} \right) \int d^3r\, \varrho(r')r' \sin\theta'\, e^{\mp i\varphi'} \ . \tag{11.5.6}$$

Since $r \cos\theta' = z'$, $r' \sin\theta' \exp(\pm i\varphi') = x' + iy'$, the q_{1m} can also be represented as linear combinations of the cartesian components of the *dipole moment*

$$p = \int d^3r\, \varrho(r)r \ . \tag{11.5.7}$$

For two point charges q and $-q$ at positions a and $-a$ respectively, we have, for example,

$$\varrho(r) = -q\delta(r+a) + q\delta(r-a)$$

and thus

$$p = \int d^3r'\, r'[-q\delta(r+a) + q\delta(r-a)]$$

$$= ql \ , \quad l = 2a \quad \text{(the vector from } -q \text{ to } q) \ . \tag{11.5.8}$$

Correspondingly, the five quantities q_{2m}, $m = -2, \ldots, 2$ can be represented as linear combinations of the five linearly independent cartesian components of the

---

[4] Multipole moment (Latin/Greek): Monopole (one pole), dipole (two poles), quadrupole (four poles), octopole (eight poles).

quadrupole moment

$$Q_{ij} = \int d^3r' \, \varrho(r')(3x_i'x_j' - \delta_{ij}r'^2) \ . \tag{11.5.9}$$

(The 3×3 matrix Q_{ij} has only five linearly independent components, because it is symmetric and traceless.)

The expansion of $\phi(r)$ in terms of solutions of the Laplace equation

$$\phi_{lm} = r^{-l-1} Y_{lm} \quad \text{for} \quad r > R$$

is called the *multipole expansion*, since the electrostatic potential of 2^l point charges (also called a 2^l-pole) at great distances from the charges behaves like a linear combination of these ϕ_{lm}:

For $l = 0$,

$$\phi = k\frac{Q}{r} \ ,$$

which is the electrostatic potential of a point charge, a monopole.

For $l = 1$, we have on the one hand from (11.5.2)

$$\phi = \frac{p \cdot r}{r^3} k \ . \tag{11.5.10}$$

If we consider on the other hand a point charge q at the origin and a point charge $-q$ located at a, then

$$\phi(r) = q\left(\frac{1}{r} - \frac{1}{|r - a|}\right)$$

$$= q\left[\frac{1}{r} - \left(\frac{1}{r} - a \cdot \nabla \frac{1}{r} + \ldots\right)\right]$$

$$= qa \cdot \nabla \frac{1}{r} + \ldots = -qa\frac{r}{r^3} = \frac{p \cdot r}{r^3} \quad \text{with} \quad p = -qa \ .$$

A quadrupole field, thus a 2^l field for $l = 2$, can be produced if the field of a dipole at the origin with dipole moment p is superimposed onto the field of a dipole located at a with dipole moment $-p$. Then,

$$\phi(r) = \frac{p \cdot r}{r^3} - \frac{p \cdot (r - a)}{|r - a|^3}$$

$$= \frac{p \cdot r}{r^3} - \left(\frac{p \cdot r}{r^3} - \frac{p \cdot a}{r^3} - p \cdot ra \cdot \nabla \frac{1}{r^3} + \ldots\right)$$

$$= \frac{p \cdot a}{r^3} - \frac{3p \cdot ra \cdot r}{r^5} + O\left(\frac{1}{r^4}\right) \ .$$

Thus, $\phi(r)$ for large r can be written as

$$\phi(r) = \sum_{m=-2}^{+2} Q_m(a) \frac{1}{r^3} Y_{2m}(\theta, \varphi) \ .$$

Correspondingly, a 2^{l+1}-pole field can be produced by superimposing the fields of two 2^l-poles with opposite moments separated by a distance a.

Remark. For a point charge e at r_0,

$$\varrho(r) = e\delta(r - r_0) \ ,$$

thus from (11.5.3)

$$q_{lm} = \sqrt{\frac{4\pi}{2l+1}} \, er_0^l Y_{lm}^*(\theta_0, \varphi_0) \ ,$$

$$r_0 = r_0(r_0, \theta_0, \varphi_0) \ ,$$

that is, a point charge can also have a dipole moment; it all depends on the location of the origin of the coordinate system. Through a proper choice of the origin, all higher moments can be made to vanish. In general, then, we find that the multipole moments depend on the choice of coordinate system. It is easy to show, however, that the order of the smallest multipole moment which does not vanish is independent of such a choice.

Example. Consider two charges q, $-q$ at points r_0 and r_1. Then,

$$q_{00} = \int d^3r' [q\delta(r - r_0) - q\delta(r - r_1)] = 0 \ ,$$

and

$$p = q(r_0 - r_1)$$

is independent on the choice of origin. Of course, the q_{10} and $q_{1\pm 1}$ are dependent on the choice of basis just like the $p_i = q(r_0 - r_1)_i$. The quadrupole moments q_{2m} of this charge configuration also depend on the choice of origin.

11.6 Electrostatic Potential Energy

Let us consider a test charge q in an electrostatic potential field $\phi(r)$. The force exerted on q is

$$F = qE(r) = -q\nabla\phi(r) \ , \tag{11.6.1}$$

and the work which is performed on q if this charge is transported from r_1 to

r_2 is given by

$$W = - \int_{r_1}^{r_2} \boldsymbol{F} \cdot d\boldsymbol{r} = q \int_{r_1}^{r_2} \nabla \phi \cdot d\boldsymbol{r} = q[\phi(r_2) - \phi(r_1)] \ . \tag{11.6.2}$$

The expression $q\phi(r)$ can thus be viewed as the potential energy of a test charge in the field $\phi(r)$.

Correspondingly, for a continuous charge distribution, we find:

The electrostatic potential energy in an external field $\phi(r)$ (that is a field which does not arise from $\varrho(r)$) is given by (cf. Sect. 13.5.5)

$$E_{\text{pot}} = \int d^3 r' \varrho(r')\phi(r') \ . \tag{11.6.3}$$

Now let $\phi(r')$ be only weakly dependent on r' in a region in which $\varrho(r') \neq 0$. Then, choosing a suitable reference point O' inside the charge distribution (Fig. 11.6.1), we can expand ϕ around r: $\phi(r') = \phi(r + b)$. We find:

$$\phi(r') = \phi(r) + \boldsymbol{b} \cdot \nabla \phi(r) + \frac{1}{2} \sum_{i,j} b_i b_j \frac{\partial^2}{\partial x_i \partial x_j} \phi(r) + \ldots \tag{11.6.4}$$

or

$$\phi(r') = \phi(r) - \boldsymbol{b} \cdot \boldsymbol{E}(r) - \frac{1}{2} \sum_{i,j} b_i b_j \frac{\partial}{\partial x_i} E_j(r) + \ldots , \tag{11.6.5}$$

and, since $\nabla \cdot \boldsymbol{E}(r) = 0$ for the external field $\boldsymbol{E}(r)$, it follows also that

$$\phi(r') = \phi(r) - \boldsymbol{b} \cdot \boldsymbol{E}(r) - \frac{1}{6} \sum_{i,j} (3b_i b_j - b^2 \delta_{ij}) \frac{\partial E_j(r)}{\partial x_i} + \ldots , \tag{11.6.6}$$

and thus

$$E_{\text{pot}} = Q\phi(r) - \boldsymbol{p} \cdot \boldsymbol{E}(r) - \frac{1}{6} \sum_{i,j} Q_{ij} \frac{\partial E_j}{\partial x_i} + \ldots \tag{11.6.7}$$

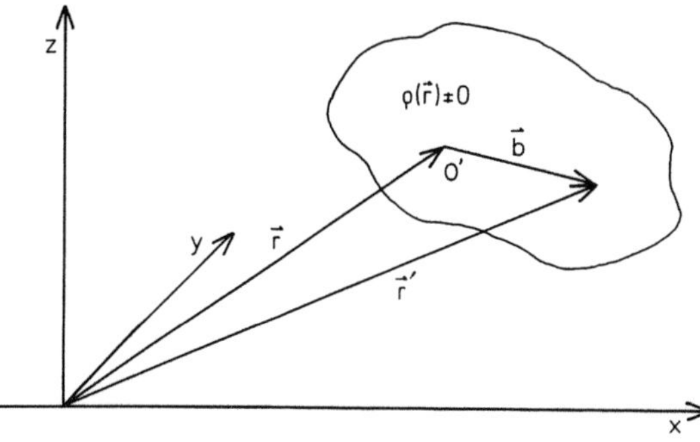

Fig. 11.6.1. A field $\phi(r)$ with a weak r dependence in a region in which the charge distribution does not vanish

with

$$p = \int d^3b\, \varrho(b)b \ , \tag{11.6.8}$$

$$Q_{ij} = \int d^3b(3b_ib_j - \delta_{ij}b^2) \ . \tag{11.6.9}$$

If, then, a charge distribution in the neighborhood of the point r is described by total charge Q, dipole moment p, quadrupole moment Q_{ij}, etc., and in the neighborhood of r there is a field $E(r)$ (or $\phi(r)$), then the potential energy is given by (11.6.7).

Here, p, Q_{ij}, etc. are the multipole moments with respect to a point O' inside of the charge distribution.

Examples. i) Let $\phi(r)$ be produced by a monopole Q_2 at the origin. Then,

$$\phi(r) = k\frac{Q_2}{r} \ .$$

The electrostatic potential energy of another monopole Q_1 in this field at point r is then

$$E_{\text{pot}} = Q_1\phi(r) = k\frac{Q_1Q_2}{r} \ .$$

ii) Let $\phi(r)$ be produced by a dipole with dipole moment p at the origin, i.e.

$$\phi(r) = k\frac{p_2\cdot r}{r^3} \ .$$

Then,

$$E(r) = -\nabla\phi = -k\nabla\frac{p_2\cdot r}{r^3} = k\left(\frac{3p_2\cdot r\, r}{r^4}\frac{1}{r} - \frac{p_2}{r^3}\right) , \tag{11.6.10}$$

since

$$\nabla(p\cdot r) = p \ .$$

The potential energy of a dipole at position r with dipole moment p_1 in this field produced by a dipole with moment p_2 is then given by

$$E_{\text{pot}} = k\left[\frac{p_1\cdot p_2}{r^3} - \frac{3(p_1\cdot r)(p_2\cdot r)}{r^5}\right] \ . \tag{11.6.11}$$

This is called a dipole-dipole interaction.

Problems

11.1   Potential of a Hollow Cylinder. Determine the electrostatic potential Φ of an infinitely long hollow cylinder made up of two halves both inside ($r \leq R$) and outside ($r \geq R$) under the following boundary conditions:

$$\Phi(r = 0) = 0 \, , \quad \Phi(r = -\infty) = 0$$

$$\Phi(R, \phi) = \begin{cases} \quad V & \text{for} \quad 0 < \phi < \pi \\ \quad 0 & \text{for} \quad \phi = 0 \quad \text{and} \quad \phi = \pi \, . \\ -V & \text{for} \quad \pi < \phi < 2\pi \end{cases}$$

11.2   Decomposition of a Vector Field. Every vector field C (which is smooth and which vanishes quickly enough at infinity) in $\mathbb{R}^3$ can be decomposed

$$C = C_{\|} + C_{\perp}$$

into two vector fields (also smooth and vanishing quickly enough at infinity) in $\mathbb{R}^3$ so that

$$\nabla \times C_{\|} = 0 \, , \quad \nabla \cdot C_{\|} = \nabla \cdot C$$

$$\nabla \cdot C_{\perp} = 0 \, , \quad \nabla \times C_{\perp} = \nabla \times C \, .$$

Show that this decomposition is unique and that

$$C_{\|}(x) = -\frac{1}{4\pi} \nabla \left(\int d^3x' \frac{(\nabla \cdot C)(x')}{|x - x'|} \right)$$

$$C_{\perp}(x) = -\frac{1}{4\pi} \nabla \times \left(\int d^3x' \frac{(\nabla \times C)(x')}{|x - x'|} \right) \, .$$

How do the Fourier coefficients $C_{\|}(p), C_{\perp}(p)$ of $C_{\|}(x), C_{\perp}(x)$ relate to the Fourier coefficients $C(p)$ of $C(x)$?

11.3   Self-energy of a Homogeneously Charged Sphere. Calculate the self-energy

$$E = \frac{\varepsilon_0}{2} \int d^3x \, E^2$$

of a homogeneously charged sphere

$$\varrho(r) = \begin{cases} \dfrac{3Q}{4\pi R^3} & \text{for} \quad r \leq R \\[2ex] 0 & \text{for} \quad r > R \, . \end{cases}$$

Hint: Express the self-energy in terms of the potential and the charge distribution.

12. Moving Charges, Magnetostatics

In Chap. 11, we discussed charges at rest. Charges produce fields and charges in a field experience forces. As a consequence of this, charges at rest exert forces on each other which can be measured. These forces, described by Coulomb's law, are the starting point of electrostatics, the theory of charges at rest in a vacuum.

In early times, though, other forces were already known, arising from magnets[1]. These forces were only understood when they were brought into connection with moving charges. The connection between charges in motion and magnetic fields is the subject of this chaper.

12.1 The Biot–Savart Law, the Fundamental Equations of Magnetostatics

12.1.1 Electric Current Density and Magnetic Fields

At first, it was thought that in an electrical current the positive electric charges flowed and thus the direction of the current was defined as the direction of the flow of positive charges. Today, we know that only the negatively charged electrons produce the current in a conductor, thus the electric current is directed oppositely to the actual particle flow.

The current density $J(r, t)$ is thus a vector field whose direction is opposite to the motion of the electrons and whose magnitude is the number of charge carriers flowing through an area element around r in unit time in the direction $-J(r, t)$.

We know that charge is conserved, that is, charge carriers can neither be created nor destroyed. Just like the continuity equation for mass in Sect. 9.2, we can express charge conservation in the equation

$$\frac{\partial}{\partial t} \varrho(r, t) + \nabla \cdot J(r, t) = 0 \tag{12.1.1}$$

so that

$$-\frac{\partial}{\partial t} \int d^3 r \, \varrho(r, t) = \int_V d^3 r \nabla \cdot J(r, t) = \int_{\partial V} dA \cdot J(r, t) \ ,$$

---

[1] Magnetism, magnet: Named after the minerals found in the Thessalian region of Magnesia.

thus,

$$-\frac{\partial}{\partial t} Q_V = \int\limits_{\partial V} dA \cdot J(r,t) \; , \tag{12.1.2}$$

i.e., the amount of total charge lost from V must flow out through the surface. We consider first the case

$$\partial \varrho/\partial t = 0 \; , \quad \text{thus} \quad \nabla \cdot J(r,t) = 0 \; ,$$

i.e.

$$\int\limits_{\partial V} dA \cdot J(r,t) = 0 \; ;$$

Exactly the same amount of charge flows in as flows out. In this chapter, we will still assume that the current density J is independent of time. We speak then of *stationary currents*.

In this case, we often construct the following integral, e.g. in the case of current through a wire,

$$I = \int dA \cdot J(r) \; , \tag{12.1.3}$$

where the surface integral is carried out over the cross section of the conductor. I is then called the *current* flowing through the wire.

A.M. Ampère discovered in 1820 that two parallel conductors which contain currents flowing in the same direction attract each other, i.e. conductors carrying electrical currents exert forces on each other. We can separate the action of this force into:

i) the production of a field by the moving charges, by the currents, and
ii) the effect of the field: if a conductor containing a current is in this field, it experiences a force. We can see that a current produces such a field by noting its effect on a magnet, which will tend to align itself with the field. This was noticed by *Oersted*[2] in 1819 and described in 1820 by *Biot*[3] and *Savart*[4].

---

[2] *Oersted, Hans Christian* (*1777 *Rudkjoping auf Langeland*, d. 1851 *Copenhagen*). Danish physicist, won over to the Romantic Naturphilosophie by J.W. Ritter in Jena, performed experiments with the goal of transforming forces. In 1820, he discovered the effect of an electrical current on a magnetic needle.

[3] *Biot, Jean-Baptiste* (*1774 Paris, d. 1862 Paris). Mathematician, physicist, astronomer, chemist, theorist and historian of science. Famous works about magnetism and the optical characteristics of media.

[4] *Savart, Félix* (*1791 Mézières, d. 1841 Paris). Physicist at the Collège de France. In 1820, the Biot–Savart law was published. He worked also on optics and acoustics and discovered among other things the cog-wheel siren (Savart's wheel) used to determine the frequency of a sound.

These results can be expressed in the following statements:

i) Let a current density $J(r)$ be flowing in a wire. This produces a field, called the *magnetic induction* $B(r)$,

$$B(r) = k' \int d^3r' \frac{J(r') \times (r - r')}{|r - r'|^3} \ . \tag{12.1.4}$$

The factor k' and thus the dimensions of B we leave open for a moment. This is the *Biot–Savart law.*

Take, for example, a current I in an infinitely long wire in the direction of the z-axis. Then,

$$J(r) = I\delta(x)\,\delta(y)\,e_3$$

and thus

$$B(r) = k'I \int dx'dy'\,dz'$$

$$\times \frac{\delta(x')\,\delta(y')\,e_3 \times [(x - x')e_1 + (y - y')e_2 + (z - z')e_3]}{[(x - x')^2 + (y - y')^2 + (z - z')^2]^{3/2}}$$

$$= k'IR \int_{-\infty}^{+\infty} dz' \frac{e(x, y)}{[R^2 + (z - z')^2]^{3/2}} \quad \text{with} \tag{12.1.5}$$

$$R^2 = x^2 + y^2 \ , \quad e(x, y) = (x e_2 - y e_1)/R \ ,$$

$$e \cdot r = 0 \ , \quad e \cdot e_3 = 0 \ , \quad e \cdot e = 1 \ .$$

R is the distance of the point in space r from the wire, e is a unit vector which is perpendicular to both r and e_3 and depends only on (x, y). Then, after evaluating the integral, we find

$$B(r) = k'IRe(x, y) \frac{2}{R^2} = k' \frac{2I}{R} e \ , \tag{12.1.6}$$

that is, we have

$$B(r) \sim e \quad \text{and} \quad |B(r)| \sim I/R \ .$$

The field lines of $B(r)$ are thus circles around the conductor in a plane perpendicular to the conductor (see Fig. 12.1.1). The strength of the field is inversely proportional to the distance from the conductor.

ii) If a current element dl contains a current I and is in a field $B(r)$, then there is a force

$$dF = \gamma I\,dl \times B(r) \tag{12.1.7}$$

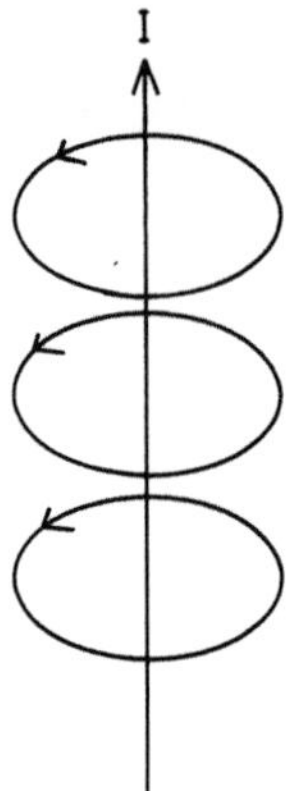

Fig. 12.1.1. Field lines of the magnetic induction caused by the current I

exerted on this current element at position r with a constant γ to be determined later. These two laws (12.1.4) and (12.1.7) are analogous to the corresponding equation of electrostatics:

Magnetostatics
a) A current density causes a field $B(r)$ according to (12.1.4).
b) The field B causes a force on a current I according to (12.1.7).

Electrostatics
a) A charge density causes a field $E(r)$ according to (11.1.6).
b) The field E causes a force on a charge q according to (11.6.1).

In order to derive the equivalent of Coulomb's law, let us consider the field $B(r)$ which is caused by an infinitely long conductor carrying a current I_2. (From (12.1.6),

$$B(r) = k'e\frac{2I_2}{R} \; ,$$

where R is the distance from the conductor. If another conductor is located at this distance carrying a current I_1, then from (12.1.7), the current element $dl = dz\,e_3$ feels a force

$$dF = \gamma I_1 \, dz \, e_3 \times \left(k'e\frac{2I_2}{R} \right) = \gamma k' \, \frac{2I_1 I_2}{R} \, e_3 \times e \, dz \; .$$

Since

$$e_3 \times e = \frac{1}{R} \, e_3 \times (xe_2 - ye_1) = -\frac{1}{R}(xe_1 + ye_2)$$

$e_3 \times e$ must point in the direction of conductor 2. The force is therefore attractive (Fig. 12.1.2) if I_1 and I_2 point in the same direction; otherwise the force is repulsive.

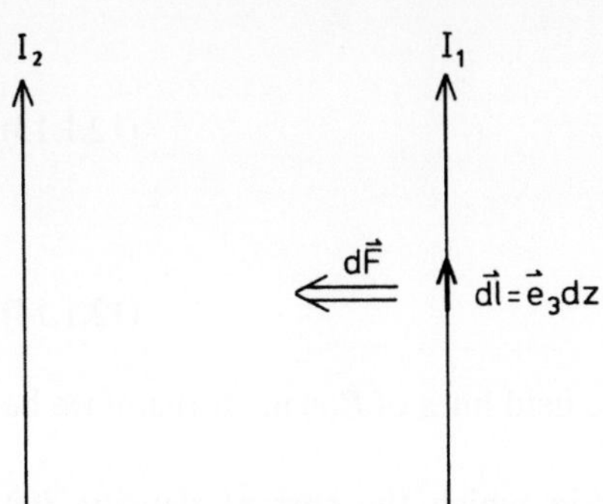

Fig. 12.1.2. If the currents point in the same direction then the force between the current elements is attractive

The force between two parallel currents separated by distance R is thus

$$|dF| = k'\gamma 2 \frac{I_1 I_2}{R} dz \ , \tag{12.1.8}$$

analogous to the force between two charges q_1 and q_2 at a distance R from each other

$$|F| = k \frac{q_1 q_2}{R^2} \ . \tag{12.1.9}$$

12.1.2   The Vector Potential and Ampère's Law

Just as for the electric field $E(r)$ we can also introduce a potential for the magnetic induction $B(r)$. Since

$$\nabla \times \frac{J(r')}{|r - r'|} = \nabla \frac{1}{|r - r'|} \times J(r') = -\frac{(r - r')}{|r - r'|^3} \times J(r')$$

$$= J(r') \times \frac{(r - r')}{|r - r'|^3} \tag{12.1.10}$$

we can write

$$B(r) = \nabla \times A(r) \quad \text{with} \tag{12.1.11}$$

$$A(r) = k' \int d^3 r' \frac{J(r')}{|r - r'|} \ . \tag{12.1.12}$$

$A(r)$ is called the *vector potential* as opposed to the scalar electrostatic potential

$$\phi(r) = k \int d^3 r' \frac{\varrho(r')}{|r - r'|} \ .$$

It then follows that

$$\nabla \cdot \boldsymbol{B} = 0 \qquad (12.1.13)$$

in contrast to the electric field for which

$$\nabla \times \boldsymbol{E} = \boldsymbol{0} \ . \qquad (12.1.14)$$

The magnetic induction has no sources. The field lines of $\boldsymbol{B}(\boldsymbol{r})$ must therefore be closed.

Finally, we want to find an equation in which the current density $\boldsymbol{J}(\boldsymbol{r})$ appears as the cause of the magnetic induction. For this purpose, we use the relation (see Appendix F)

$$\nabla \times \boldsymbol{B} = \nabla \times (\nabla \times \boldsymbol{A}) = -\Delta \boldsymbol{A} + \nabla (\nabla \cdot \boldsymbol{A}) \ . \qquad (12.1.15)$$

First,

$$\nabla \cdot \boldsymbol{A}(\boldsymbol{r}) = k' \int d^3r' \, \boldsymbol{J}(\boldsymbol{r}') \cdot \nabla \frac{1}{|\boldsymbol{r} - \boldsymbol{r}'|}$$

$$= k' \int d^3r' \, \boldsymbol{J}(\boldsymbol{r}') \cdot \left(-\nabla' \frac{1}{|\boldsymbol{r} - \boldsymbol{r}'|} \right) \ ,$$

and after partial integration (Gauss's law), in which the boundary terms vanish, we find

$$\nabla \cdot \boldsymbol{A}(\boldsymbol{r}) = k' \int d^3r' \nabla' \cdot \boldsymbol{J}(\boldsymbol{r}') \frac{1}{|\boldsymbol{r} - \boldsymbol{r}'|} = 0 \ , \qquad (12.1.16)$$

since

$$\nabla' \cdot \boldsymbol{J}(\boldsymbol{r}') = 0$$

On the other hand, since

$$\Delta \frac{1}{|\boldsymbol{r} - \boldsymbol{r}'|} = -4\pi \, \delta(\boldsymbol{r} - \boldsymbol{r}')$$

it is also true that

$$\Delta \boldsymbol{A}(\boldsymbol{r}) = -4\pi k' \boldsymbol{J}(\boldsymbol{r}) \qquad (12.1.17)$$

and thus

$$\nabla \times \boldsymbol{B}(\boldsymbol{r}) = 4\pi k' \boldsymbol{J}(\boldsymbol{r}) \ . \qquad (12.1.18)$$

This is *Ampère's law*. Note again the similarity to the corresponding equation in electrostatics

$$\nabla \cdot E(r) = 4\pi k \varrho(r) \ . \tag{12.1.19}$$

The equations

$$\nabla \cdot B(r) = 0 \ , \quad \nabla \times B(r) = 4\pi k' J(r) \tag{12.1.20}$$

are the fundamental equations of magnetostatics, the study of stationary currents and magnetic fields.

Remark. Ampère's law can also be formulated in integral form:
With the help of Stoke's law (Appendix F)

$$\int\limits_{A} dA \cdot (\nabla \times B(r)) = \int\limits_{\partial A} dr \cdot B(r)$$

and with the help of the equation

$$4\pi k' \int\limits_{A} dA \cdot J = 4\pi k' I_A \ ,$$

where I_A is the current which flows through area A, we find the integral form of Ampère's law:

$$\int\limits_{\partial A} dr \cdot B(r) = 4\pi k' I_A \tag{12.1.21}$$

which is analogous to Gauss's law

$$\int\limits_{\partial V} dA \cdot E(r) = 4\pi k Q_V \ . \tag{12.1.22}$$

Simple Application: Consider a current I through a straight wire. From symmetry considerations, $|B(r)| = B(R)$, $R = $ the distance from the wire. Since the field lines of B are closed (since $\nabla \cdot B(r) = 0$), they must lie in a circle around the wire.

Thus, if we integrate on a circle of radius R around the wire

$$\oint\limits_{\partial K} dr \cdot B(r) = B(R) \oint\limits_{\partial K} dr = 2\pi R B(R) \ .$$

Therefore, we find again

$$B(R) = \frac{1}{2\pi R} 4\pi k' I = 2k' \frac{I}{R}$$

in agreement with (12.1.6).

12.1.3 The SI-System of Units in Electrodynamics

In the SI- or MKSA-system of physical units, equation (12.1.8) is used to define the unit of current, as follows:

A current of 1 *Ampère* is defined as the amount of current flowing in two infinitely long parallel wires of negligible cross section separated by 1 meter which causes a force per unit length of

$$2 \times 10^{-7}\,\text{N/m} \ .$$

Then, from (12.1.8)

$$\frac{dF}{dz} = 2 \times 10^{-7}\,\text{N/m} = \gamma k' \cdot 2\,\frac{1\,\text{A}\,1\,\text{A}}{1\,\text{m}} \ ,$$

so that

$$\gamma k' = 10^{-7}\,\frac{\text{kg m}}{\text{A}^2\text{s}^2} \ . \tag{12.1.23}$$

We also write

$$\gamma k' =: \frac{\mu_0}{4\pi} \ , \quad \text{so that} \tag{12.1.24}$$

$$\mu_0 = 4\pi \times 10^{-7}\,\frac{\text{kg m}}{\text{A}^2\text{s}^2} \ . \tag{12.1.25}$$

Further, we choose $\gamma = 1$, then $k' = \mu_0/4\pi$, and the dimensions of magnetic induction are

$$[B] = \text{kg/A}\,\text{s}^2 =: 1\ \text{Tesla} =: 10^4\ \text{Gauß} \ . \tag{12.1.26}$$

If the Ampère is defined in this way, then the dimensions of charge are

$$[q] = \text{A}\,\text{s} = \text{Coulomb (C)} \ , \tag{12.1.27}$$

and the constant k in Coulomb's law has dimensions

$$[k] = \text{N}\,\text{m}^2/\text{A}^2\text{s}^2 \ . \tag{12.1.28}$$

If we compare this with the dimensions of k', $[k'] = \text{N/A}^2$, we find

$$\frac{[k]}{[k']} = \left(\frac{\text{m}}{\text{s}}\right)^2 \ . \tag{12.1.29}$$

The ratio of the constants k and k' thus has the dimensions of speed squared. This must be a universal speed, and it is determined experimentally that it is c, the speed of light in a vacuum. Thus we have

$$k = k'c^2 = \frac{\mu_0}{4\pi} c^2 \; . \tag{12.1.30}$$

If we set

$$k =: \frac{1}{4\pi\varepsilon_0} \; , \tag{12.1.31}$$

we obtain

$$\varepsilon_0 = \frac{1}{4\pi k} = \frac{1}{\mu_0 c^2} = 8.854 \times 10^{-12} \, \text{C}^2 \text{N}^{-1}/\text{m}^2 \tag{12.1.32}$$

and therefore

$$c = \frac{1}{\sqrt{\varepsilon_0 \mu_0}} \; . \tag{12.1.33}$$

12.2 Localized Current Distributions

In Sect. 11.5, we investigated the electrostatic potential of localized charge distributions and introduced the idea of the (electric) multipole. Here, we will undertake a similar investigation of localized current distributions.

12.2.1 The Magnetic Dipole Moment

We start with the expression for the vector potential

$$A(r) = \frac{\mu_0}{4\pi} \int d^3r' \, \frac{J(r')}{|r - r'|} \; . \tag{12.2.1}$$

Let $J(r') \neq 0$ only in a limited region $|r'| < R$. We consider $A(r)$ for $|r| > R$:

We can again substitute for the Green's function $1/|r - r'|$ the expansion in terms of spherical harmonics from Sect. 11.4. Since we are only interested in the first two terms of this expansion, we can, somewhat more directly, use the Taylor expansion in the form

$$\frac{1}{|r - r'|} = \frac{1}{r} - r' \cdot \nabla \frac{1}{r} + \ldots = \frac{1}{r} + \frac{r' \cdot r}{r^3} + O\left[\frac{1}{r}\left(\frac{r'}{r}\right)^2\right] \tag{12.2.2}$$

to obtain

$$A_i(r) = \frac{\mu_0}{4\pi} \left(\frac{1}{r} \int d^3r'\, J_i(r') + \frac{1}{r^3} \int d^3r'\, J_i(r')\, x_j'\, x_j + \ldots \right) .$$
(12.2.3)

We will show the following in the remarks at the end of Sect. 12.2.2:
Since $\mathbf{\nabla} \cdot \mathbf{J}(r) = 0$, we have

$$\int d^3r\, J_i(r) = 0 \quad \text{and}$$
(12.2.4)

$$\int d^3r\, x_j J_i(r) = - \int d^3r\, x_i J_j(r) .$$
(12.2.5)

From (12.2.4), the first term on the right side of (12.2.5) vanishes, and the second term can be written as

$$\frac{1}{2r^3}\, x_j \int d^3r'\, [x_j' J_i(r') - x_i' J_j(r')]$$

$$= \frac{1}{2r^3} \int d^3r'\, [(r \cdot r')\, J_i(r') - r \cdot J(r')\, x_i']$$

$$= \frac{1}{2r^3} \int d^3r'\, [(r' \times J(r')) \times r]_i \, ,$$
(12.2.6)

since $(r' \times J) \times r = (r \cdot r')\, J - (r \cdot J)\, r'$.

If we define the *magnetic dipole moment*

$$m = \frac{1}{2} \int d^3r\, [r \times J(r)]$$
(12.2.7)

in analogy to the electric dipole moment

$$p = \int d^3r\, r \varrho(r) ,$$
(12.2.8)

then the vector potential of a localized current distribution can be written as

$$A(r) = \frac{\mu_0}{4\pi} \frac{m \times r}{r^3} ,$$
(12.2.9)

which is analogous to the expression for the electrostatic potential of an electric dipole

$$\phi(r) = \frac{1}{4\pi\varepsilon_0} \frac{p \cdot r}{r^3} .$$
(12.2.10)

The magnetic induction $\mathbf{B}(r)$ is then found after some calculation to be

$$B(r) = \mathbf{\nabla} \times A(r) = \frac{\mu_0}{4\pi} \left[\frac{3r(m \cdot r)}{r^5} - \frac{m}{r^3} \right] .$$
(12.2.11)

Compare this to the electric field of an electric dipole (11.6.10).

We now calculate the magnetic dipole moment for a few important charge distributions:

i) Consider a constant current I flowing through a closed path C lying in a plane.

Then, we have

$$m = -\tfrac{1}{2} \int d^3r' [J(r') \times r']$$ (12.2.12)

$$= -\tfrac{1}{2} I \oint_C dr' \times r' \; .$$ (12.2.13)

In going from (12.2.12) to (12.2.13), we have substituted

$$I \, dr$$

for the current density

$$d^3r \, J(r)$$

along a line (the wire is of negligible cross section). This intuitively plausible equation will be justified by a remark at the end of this section.

We then have

$$\tfrac{1}{2} |r' \times dr'| = |dA| \; ,$$ (12.2.14)

where $|dA|$ is the area which is swept out as the vector r becomes $r + dr$. The vector dA is perpendicular to the x_1, x_2-plane in which the current density lies. Then,

$$-\tfrac{1}{2} \oint_C dr' \times r' = A_C e_3 \; ,$$ (12.2.15)

where r' runs around the curve C counterclockwise. Thus we find the magnetic moment of this charge distribution is

$$m = I A_C e_3 \; ,$$ (12.2.16)

where A_C is the area contained inside of C. If we know in addition that C is a circle with radius a, then

$$m = I \pi a^2 e_3 \; .$$ (12.2.17)

ii) If we write the current density as

$$J(r) = \varrho(r) \, v(r) \; ,$$ (12.2.18)

where $v(r)$ is the velocity field of the charge distribution $\varrho(r)$, then the charge current density is analogous to the mass flow density which we saw in Chap. 9

in the introduction to fluid mechanics. Then we find that for the magnetic dipole moment,

$$m = \tfrac{1}{2} \int d^3 r' [r' \times \varrho(r') \, v(r')] \; , \qquad (12.2.19)$$

corresponding to the expression for the mechanical angular momentum

$$L = \int d^3 r' [r' \times \varrho_M(r') \, v(r')] \qquad (12.2.20)$$

of a mass distribution $\varrho_M(r)$ with the same velocity field. If the current is composed of particles of mass M and charge q, then clearly

$$\varrho(r) = \frac{q}{M} \, \varrho_M(r) \quad \text{and thus} \qquad (12.2.21)$$

$$m = \Gamma L \quad \text{with} \qquad (12.2.22)$$

$$\Gamma = \frac{q}{2M} \; . \qquad (12.2.23)$$

Γ is called the *gyromagnetic ratio*[5]. Deviations from this ratio are described through the so-called g-factor in the general form

$$\Gamma = g \frac{q}{2M} \; . \qquad (12.2.24)$$

While in a classical charge distribution we expect $g = 1$, we find strong deviations in quantum physics, particularly in the case of charge distributions "within" individual particles such as electrons, protons, etc. The g-factor for electrons lies near $g = 2$. The deviation from the value 2 is explained by quantum electrodynamics.

12.2.2   Force, Potential, and Torque in a Magnetic Field

Following the consideration of these two important examples of localized current distributions, we want to calculate the forces and torques which are exerted on a charge distribution characterized by a magnetic dipole moment (such a distribution is simply called a magnetic dipole) by an external field.

We thus consider a localized current distribution concentrated around r in an external field $B(r)$. The force on the current distribution is then

$$F = \int d^3 r \, J(r) \times B(r) \; , \qquad (12.2.25)$$

---

[5] Gyromagnetic ratio (from Greek) gyros, curve, circle: ratio between the magnetic moment and the angular momentum.

generalizing (12.1.7),

$$dF = I\,dl \times B(r) \ .\tag{12.2.26}$$

For example, with

$$J(r') = qv\,\delta(r' - r)\tag{12.2.27}$$

it follows immediately that

$$F = qv \times B(r) \ ,\tag{12.2.28}$$

once again the *Lorentz force*.

Let $B(r')$ vary slowly around $r' = r$, and let $r' = r + b$. Then,

$$B(r') = B(r) + (b\cdot\nabla)\,B(r) + \ldots\ .\tag{12.2.29}$$

It then follows that

$$F_i = \int d^3r'\,[\,J(r') \times B(r)]_i + \varepsilon_{ijk}\int d^3b\,J_j(b)\,b_l\nabla_l B_k(r) + \ldots\ .$$

Now, for a vector C independent of b (see (12.2.6)), we have

$$\int d^3b\,J_j(b)\,b_l C_l = (m \times C)_j \ .\tag{12.2.30}$$

Thus we find, using again $\int d^3r'\,J(r') = 0$,

$$F = (m \times \nabla) \times B(r) = \nabla(m\cdot B) - m(\nabla \cdot B)$$

$$\quad = -\nabla V_m(r)\quad\text{with}\tag{12.2.31}$$

$$V_m = -\,m\cdot B(r)\tag{12.2.32}$$

as the potential of a current density with dipole moment m in the external field B at position r, in analogy with the potential energy of an electric dipole p in an external electric field E at position r (Sect. 11.6):

$$E_{\text{pot}} = -\,p\cdot E(r) \ .\tag{12.2.33}$$

Note that we do not call V_m the potential energy of the magnetic dipole, even though this would be closely analogous to the electrostatic situation. In Sect. 13.5.5, we will calculate the energy of a magnetic dipole in an external field and we will find instead that it is given by $-\,V_m$ (cf. [1.2]).

Finally, we will calculate the torque on a magnetic dipole in an external field $B(r)$. The torque on a general current distribution is found using (12.2.25)

$$N = \int d^3r'\,\{r' \times [\,J(r') \times B(r')]\} \ .\tag{12.2.34}$$

If, as an approximation, we substitute $B(r)$ for $B(r')$, where r is a point in the current distribution, in which, say, the field $B(r')$ does not vary very much, we find

$$N = \int d^3r' \, J(r')(r' \cdot B(r)) - \int d^3r'(J(r') \cdot r') \, B(r) \ . \tag{12.2.35}$$

The second term vanishes as will again be shown in the remarks below. Then, with (12.2.30), we obtain

$$N = m \times B(r) \tag{12.2.36}$$

in analogy with the torque that an electric dipole experiences in an external field $E(r)$:

$$N = p \times E(r) \ . \tag{12.2.37}$$

Remarks. i) Proof of the formulas (12.2.4) and (12.2.5):
   We have

$$\nabla \cdot (x_j J) = \nabla_i(x_j J_i) = J_j + x_j \nabla \cdot J = J_j \ . \tag{12.2.38}$$

thus, for an arbitrary $f(r')$, we also find

$$\int d^3r' \left[\nabla' \cdot (x_j^i J(r')) \right] f(r') = \int d^3r' \, J_j(r') f(r') \ . \tag{12.2.39}$$

Partial integration on the left hand side yields

$$- \int d^3r' \, x_j' J(r') \cdot \nabla' f(r') = \int d^3r' \, J_j(r') f(r') \ , \tag{12.2.40}$$

since the boundary term vanishes because $J(r') = 0$ for $|r'| > R$. If we substitute

   a) $f \equiv 1$, we find

$$0 = \int d^3r' \, J_j(r') \ ; \tag{12.2.41}$$

   b) $f \equiv x_k'$, we obtain

$$- \int d^3r' \, x_j' J_k(r') = \int d^3r' \, x_k' J_j(r') \ . \tag{12.2.42}$$

For $j = k$, we find

$$\int d^3r' \, x_j' J_j(r') = 0 \ , \tag{12.2.43}$$

thus it follows that

$$\int d^3r' \, r' \cdot J(r') = 0 \ . \tag{12.2.44}$$

   ii) Reduction of a current density element to a current element:
   Let the current density be limited to a path C (a conducting loop of negligible cross section). Let this path be parametrized as $r = r(s)$. We introduce new coordinates (n_1, n_2, n_3) such that n_1 is the path length s, and n_2 and n_3 remain constant along C. [For example, let C be a circle, then $r = r(r, \theta, \varphi)$ and $r = r(r = a, \theta = \pi/2, \varphi)$ is a circle in the x, y-plane with radius a.]

Then the direction of $J(r)$ coincides with that of $\partial r/\partial n_1$. Since $J(r)$ is non-zero only along the path $n_1 = s$, $n_2 = c_2$, $n_3 = c_3$, we can also write

$$J(r) = A(n_1, n_2, n_3) e_{n_1} \delta(n_2 - c_2)\, \delta(n_3 - c_3) \ . \tag{12.2.45}$$

Now, we need to find the quantity A. We thus calculate

$$I = \int dA \cdot J(r) \ , \quad \text{with}$$

$$dA = e_{n_1} dn_2 dn_3 \sqrt{g_{22} g_{33}} \ ,$$

where g_{22} and g_{33} can still depend on the coordinates (n_1, n_2, n_3) (see Appendix F). Then,

$$I = A \int dn_2 dn_3 \sqrt{g_{22} g_{33}}\, \delta(n_2 - c_2)\delta(n_3 - c_3)$$

thus

$$A = \frac{I}{\sqrt{g_{22} g_{33}}} \ . \tag{12.2.46}$$

For example, for spherical coordinates, with

$$(n_1, n_2, n_3) = (\varphi, r, \theta) \ , \quad g_{22} = g_{rr} = 1 \ , \quad g_{33} = g_{\theta\theta} = r^2 \ ,$$

and for a current in a circular loop with radius a,

$$J(r) = \frac{I}{a}\, e_\varphi \delta(r - a)\delta(\theta - \tfrac{\pi}{2}) \ .$$

In general, then,

$$J(r) = \frac{I}{\sqrt{g_{22} g_{33}}}\, e_{n_1} \delta(n_2 - c_2)\, \delta(n_3 - c_3) \tag{12.2.47}$$

and therefore

$$\int d^3 r' (J(r') \times r') = \int dn_1 dn_2 dn_3 \sqrt{g_{11} g_{22} g_{33}}\, \frac{I}{\sqrt{g_{22} g_{33}}}\, e_{n_1} \times r(n_1, n_2, n_3)$$

$$= I \int \frac{\partial r}{\partial n_1} \times r(n_1, c_2, c_3)\, dn_1 \ , \tag{12.2.48}$$

since $\sqrt{g_{11}} = |\partial r/\partial n_1|$ and thus $\sqrt{g_{11}} e_{n_1} = \partial r/\partial n_1$. Therefore we have

$$\int d^3 r' [J(r') \times r'] = I \int dr' \times r' \ . \tag{12.2.49}$$

13. Time Dependent Electromagnetic Fields

In Chaps. 11 and 12 we worked with time independent fields. The equations

$$\nabla \times \boldsymbol{E}(\boldsymbol{r}) = \boldsymbol{0} \ , \quad \nabla \cdot \boldsymbol{B}(\boldsymbol{r}) = 0 \ ,$$

$$\nabla \cdot \boldsymbol{E}(\boldsymbol{r}) = \frac{1}{\varepsilon_0} \, \varrho(\boldsymbol{r}) \ , \quad \nabla \times \boldsymbol{B}(\boldsymbol{r}) = \mu_0 \, \boldsymbol{J}(\boldsymbol{r})$$

determine independently the electrostatic field $\boldsymbol{E}(\boldsymbol{r})$ and the magnetic induction $\boldsymbol{B}(\boldsymbol{r})$. When considering time dependent phenomena, we will find more generally that these equations become coupled.

13.1 Maxwell's Equations

Faraday made the following observation in 1831:

Consider a conducting loop C in a field $\boldsymbol{B}(\boldsymbol{r}, t)$ and an area A whose boundary ∂A is exactly C. Let us construct the quantity

$$\Phi = \int\limits_{A} d\boldsymbol{A} \cdot \boldsymbol{B}(\boldsymbol{r}, t) \ . \tag{13.1.1}$$

Φ is called the *magnetic flux* through A. The field $\boldsymbol{B}$, the magnetic induction, is for this reason sometimes called the *magnetic flux density*.

If Φ is made to vary with time, for example, if

- $\boldsymbol{B}(\boldsymbol{r}, t)$ varies over time, or
- $d\boldsymbol{A}$ varies, say by a rotation or deformation of the loop C,

then a current flows in C.

Faraday explained this as follows:

As Φ varies with time, a field $\boldsymbol{E}$ is induced in the loop C, which then exerts a force on the charge carriers so that a current flows. In contrast to the stationary case, there is now a field $\boldsymbol{E}(\boldsymbol{r}, t)$ in the conductor which moves the charge carriers and thus produces an *induced current*.

For a current in a conductor, the linear relationship

$$\boldsymbol{J}(\boldsymbol{r}, t) = \sigma \boldsymbol{E}(\boldsymbol{r}, t) \tag{13.1.2}$$

is often observed experimentally, where σ describes the electrical conductivity of the material. The realm of validity of this relation is large for metals and somewhat smaller for semiconductors.

The line integral

$$\mathscr{E} = \oint_{\partial A} dr \cdot E(r, t) \tag{13.1.3}$$

is called the *electromotive force* or *EMF* (although technically it does not have the dimensions of force, the name is used for historical reasons). Then

$$\mathscr{E} = \oint dr \cdot \frac{J(r, t)}{\sigma} = I \oint dr \frac{1}{A\sigma} = IR \tag{13.1.4}$$

with J parallel to dr, $|J| = I/A$ and

$$R = \frac{L}{A\sigma} , \tag{13.1.5}$$

the resistance of a conductor of length L, cross-sectional area A, and conductivity σ.

Faraday's law of induction now reads

$$\mathscr{E} = -k'' \frac{d\Phi}{dt} . \tag{13.1.6}$$

Here, k'' is a constant which must still be determined, and the minus sign, given $k'' > 0$, takes into account the observation that the induction current is oriented in such a way that the magnetic induction produced by the induced current is directed oppositely to the change in magnetic flux density which causes it (*Lenz's*[1] *law*).

In a time independent field $B(r)$, the force on the charges and the current produced by it can be explained by the Lorentz force

$$F = qv \times B$$

which deflects moving charges. If, for example, a conductor which points in the e_1-direction is moved perpendicular to a field $B = Be_3$ in the e_2-direction, then the charges experience a Lorentz force in the e_1-direction, that is, in the direction of the conductor.

We now want to transform Faraday's law from an integral form to a differential form. First of all, we know

$$\oint_{\partial A} dr \cdot E = \int_A dA \cdot (\nabla \times E) = -k'' \frac{d}{dt} \int_A dA \cdot B . \tag{13.1.7}$$

---

[1] *Lenz, Heinrich Friedrich Emil* (*1804 Dorpat, d. 1865 Rome). Physicist from the Baltic area, Professor in St. Petersburg; worked especially on magnetic induction.

We will consider the case in which the area A does not vary with time. Then the magnetic flux changes solely because of the time dependence of B. If (13.1.7) holds for all such areas, even for those which are not bordered by a conductor, it follows that

$$\nabla \times E(r, t) + k'' \frac{\partial B(r, t)}{\partial t} = 0 \; . \tag{13.1.8}$$

This means the following: a B field changing in time produces an E field, even in a vacuum. This is the *law of induction* in the form which Maxwell produced it.

For time dependent areas, dA also depends on t. In this case, the interpretation of equation (13.1.7) leads, among other places, to the behavior of E under a transformation to a moving system of reference, in particular to one in which the area is at rest (see for example [Jackson]). We will not consider this here.

Before we determine k'', we will produce a more general version of another equation, using the same procedure Maxwell did in 1865.

We consider the equation from magnetostatics

$$\nabla \times B(r) = \mu_0 J(r) \; . \tag{13.1.9}$$

This equation can only hold for stationary currents, since the divergence equation yields

$$\nabla \cdot (\nabla \times B) = 0 \quad \text{thus also} \quad \nabla \cdot J = 0 \; .$$

In general, though,

$$\nabla \cdot J(r, t) + \frac{\partial \varrho(r, t)}{\partial t} = 0 \; .$$

But since $\varepsilon_0 \, \nabla \cdot E = \varrho$,

$$\frac{\partial \varrho}{\partial t} = \frac{\partial}{\partial t} (\varepsilon_0 \nabla \cdot E) = \nabla \cdot \left(\varepsilon_0 \frac{\partial E}{\partial t} \right) \; . \tag{13.1.10}$$

Thus, we have also

$$\nabla \cdot \left[J(r, t) + \varepsilon_0 \frac{\partial E(r, t)}{\partial t} \right] = 0 \; . \tag{13.1.11}$$

If we expand equation (13.1.9) to the following form

$$\nabla \times B(r, t) = \mu_0 \left[J(r, t) + \varepsilon_0 \frac{\partial E(r, t)}{\partial t} \right] , \tag{13.1.12}$$

then we find, using $\varepsilon_0 \mu_0 = 1/c^2$,

$$\nabla \times B(r, t) - \frac{1}{c^2} \frac{\partial E(r, t)}{\partial t} = \mu_0 J(r, t) \; . \tag{13.1.13}$$

Thus we have found a generalization of (13.1.9) which ensures that these equations are free of contradictions. Naturally, this generalization must be verified experimentally.

In a region where $J(r, t) = 0$, we then have

$$\nabla \times B(r, t) - \frac{1}{c^2} \frac{\partial E(r, t)}{\partial t} = 0$$

and thus

$$\nabla \times (\nabla \times B) - \frac{1}{c^2} \frac{\partial}{\partial t} (\nabla \times E) = 0 \ , \tag{13.1.14}$$

or, with (13.1.8),

$$\nabla(\nabla \cdot B) - \Delta B - \frac{1}{c^2} \frac{\partial}{\partial t} \left(- k'' \frac{\partial B}{\partial t} \right) = 0$$

or

$$\left(- \Delta + \frac{k''}{c^2} \frac{\partial^2}{\partial t^2} \right) B(r, t) = 0 \ . \tag{13.1.15}$$

Correspondingly, from (13.1.8), we find

$$\nabla \times (\nabla \times E) + k'' \frac{\partial}{\partial t}(\nabla \times B) = 0 \tag{13.1.16}$$

or, with (13.1.13), and if we also have $\varrho \equiv 0$ in the region,

$$\left(- \Delta + \frac{k''}{c^2} \frac{\partial^2}{\partial t^2} \right) E(r, t) = 0 \ . \tag{13.1.17}$$

We find wave equations for $E(r, t)$ and $B(r, t)$. From Sect. 10.2, we know that these equations have solutions which correspond to travelling waves whose wavefronts have speed $c/\sqrt{k''}$. If we consider the speed of light c as the only fundamental constant in electrodynamics, then we must set $k'' = 1$.

13.2   Potentials and Gauge Transformations

Maxwell's equations now read

$$\nabla \times E(r, t) + \frac{\partial B(r, t)}{\partial t} = 0 \ , \tag{13.2.1}$$

$$\nabla \cdot E(r, t) = \frac{1}{\varepsilon_0} \varrho(r, t) \ , \tag{13.2.2}$$

$$\nabla \cdot B(r, t) = 0 \ , \tag{13.2.3}$$

$$\nabla \times B(r, t) - \frac{1}{c^2} \frac{\partial E(r, t)}{\partial t} = \mu_0 J(r, t) \ . \tag{13.2.4}$$

The homogeneous equations (13.2.1) and (13.2.3) can immediately be solved by introducing potentials: From

$$\nabla \cdot B(r, t) = 0$$

it follows that there is a vector potential $A(r, t)$ (see Appendix F) such that:

$$B(r, t) = \nabla \times A(r, t) \ , \tag{13.2.5}$$

and from

$$0 = \nabla \times E(r, t) + \frac{\partial B(r, t)}{\partial t} = \nabla \times \left[E(r, t) + \frac{\partial A(r, t)}{\partial t} \right]$$

it follows that there is a potential ϕ such that

$$E(r, t) + \frac{\partial A(r, t)}{\partial t} = -\nabla \phi(r, t) \quad \text{or}$$

$$E(r, t) = -\nabla \phi(r, t) - \frac{\partial A(r, t)}{\partial t} \ . \tag{13.2.6}$$

Thus the fields $E(r, t)$ and $B(r, t)$ can be expressed in terms of the potentials $\phi(r, t)$ and $A(r, t)$ and the homogeneous Maxwell equations have been taken care of.

Note, however, that the potentials $(\phi(r, t), A(r, t))$ are not uniquely defined. If they are changed through the so-called "*gauge transformations*":

$$\phi \mapsto \phi - \partial \Lambda / \partial t \ , \tag{13.2.7}$$

$$A \mapsto A + \nabla \Lambda \ , \quad \Lambda = \Lambda(r, t) \tag{13.2.8}$$

the fields $E(r, t)$ and $B(r, t)$ stay the same:

$$B \mapsto \nabla \times (A + \nabla \Lambda) = \nabla \times A = B \ , \tag{13.2.9}$$

$$E \mapsto -\nabla \phi + \frac{\partial}{\partial t} \nabla \Lambda - \frac{\partial A}{\partial t} - \frac{\partial}{\partial t} \nabla \Lambda = -\nabla \phi - \frac{\partial A}{\partial t} = E \ . \tag{13.2.10}$$

From the inhomogeneous equations (13.2.2, 4), it then follows

$$-\Delta\phi(r,t) - \frac{\partial}{\partial t}\nabla\cdot A(r,t) = \frac{1}{\varepsilon_0}\varrho(r,t) \tag{13.2.11}$$

and

$$\nabla\times[\nabla\times A(r,t)] + \frac{1}{c^2}\frac{\partial}{\partial t}\left[\nabla\phi(r,t) + \frac{\partial A(r,t)}{\partial t}\right]$$

$$= -\Delta A(r,t) + \frac{1}{c^2}\frac{\partial^2 A(r,t)}{\partial t^2}$$

$$+ \nabla\left[\nabla\cdot A(r,t) + \frac{1}{c^2}\frac{\partial\phi(r,t)}{\partial t}\right] = \mu_0 J(r,t) \ . \tag{13.2.12}$$

We now take advantage of the freedom we have in the choice of (ϕ, A) and impose the additional condition that

$$\nabla\cdot A(r,t) + \frac{1}{c^2}\frac{\partial\phi(r,t)}{\partial t} = 0 \ . \tag{13.2.13}$$

This is called the *Lorentz condition*, and we also speak of the *Lorentz gauge*. This condition can always be satisfied, since if

$$\nabla\cdot A(r,t) + \frac{1}{c^2}\frac{\partial\phi(r,t)}{\partial t} \neq 0 \ ,$$

then we can introduce a gauge transformation

$$\phi\mapsto\phi' = \phi - (\partial\Lambda/\partial t) \ , \quad A\mapsto A' + \nabla\Lambda$$

in which the gauge function $\Lambda(r,t)$ is adjusted in such a way that

$$\nabla\cdot A' + \frac{1}{c^2}\frac{\partial\phi'}{\partial t} = \nabla\cdot A + \frac{1}{c^2}\frac{\partial\phi}{\partial t} + \left(\Delta - \frac{1}{c^2}\frac{\partial^2}{\partial t^2}\right)\Lambda(r,t) = 0 \ .$$

This equation for $\Lambda(r,t)$ is an inhomogeneous wave equation whose solution we will discuss in Sect. 13.4.

The potentials $\phi'(r,t)$ and $A'(r,t)$ then satisfy the Lorentz condition and we then obtain for the inhomogeneous wave equations

$$\left(-\Delta + \frac{1}{c^2}\frac{\partial^2}{\partial t^2}\right)\phi'(r,t) = \frac{1}{\varepsilon_0}\varrho(r,t) \ , \tag{13.2.14}$$

$$\left(-\Delta + \frac{1}{c^2}\frac{\partial^2}{\partial t^2}\right)A'(r,t) = \mu_0 J(r,t) \ . \tag{13.2.15}$$

There is one other well-known gauge, the Coulomb gauge $\nabla\cdot A = 0$, which we will not discuss here.

13.3 Electromagnetic Waves in a Vacuum, the Polarization of Transverse Waves

In Sect. 13.1, we have shown that the solutions of Maxwell's equations in a region in which both $\varrho(r, t)$ and $J(r, t)$ vanish also satisfy the wave equation. This equation, as we have studied in Sects. 10.1 and 10.2, has solutions corresponding to waves which propagate with speed c. The validity of the Maxwell equations then has the immediate consequence that electromagnetic waves should be observed. In fact, such electromagnetic waves were produced and identified for first time in 1886–1888 by *H. Hertz*[2]. Of course, he could not have been aware of the technical importance of these waves, as we know them today in radio and television.

If we consider plane waves, we can attempt to write down the following solution of the wave equation

$$E(r, t) = E_0(k)e^{i(k \cdot r - \omega t)} \ , \qquad B(r, t) = B_0(k)e^{i(k \cdot r - \omega t)} \tag{13.3.1}$$

with $|k| = \omega/c$. This solution indeed satisfies the wave equation. However, from Maxwell equation's, we find the following conditions for the amplitudes $E_0(k)$ and $B_0(k)$

$$\nabla \cdot B(r, t) = 0 \qquad \rightarrow k \cdot B_0(k) = 0 \ , \tag{13.3.2}$$

$$\nabla \cdot E(r, t) = 0 \qquad \rightarrow k \cdot E_0(k) = 0 \ , \tag{13.3.3}$$

$$\nabla \times E(r, t) = -\frac{\partial B(r, t)}{\partial t} \rightarrow k \times E_0(k) = \omega B_0(k) \ , \tag{13.3.4}$$

$$\nabla \times B(r, t) = \frac{1}{c^2}\frac{\partial E(r, t)}{\partial t} \rightarrow k \times B_0(k) = -\frac{\omega}{c^2} E_0(k) \ , \tag{13.3.5}$$

i.e. E_0, B_0, and k must be pairwise orthogonal, and thus they form an orthogonal tripod. Since E and B therefore oscillate perpendicular to k, the direction of propagation, these waves are called *transverse*[3].

Transverse waves exhibit a certain *polarization*:

We set the e_3-direction in the direction of k. If the vector E_0 points in the e_1-direction, then $E(r, t)$ is said to be linearly polarized in the e_1-direction. The electric field $E(r, t)$ then oscillates in the plane perpendicular to k in the direction of e_1.

---

[2] *Hertz, Heinrich* (*1857 Hamburg, d. 1894 Bonn). Fundamental work in electromagnetism. 1886, proof of electromagnetic radiation; 1887, discovery of the photoelectric effect; later, proof that electromagnetic waves are transverse. Hertz excelled both in experiment and theory. His presentation of Maxwell's equations which appeared in 1890 was very influential. He succeeded Clausius in Bonn in 1889.

[3] Transverse (Latin) from *transvertere*: "turn across".

If we now superimpose two solutions, one polarized in the e_1-, one in the e_2-direction, we find a solution of the form

$$E(r, t) = (E_1 e_1 + E_2 e_2)e^{i(k \cdot r - \omega t)} \ . \tag{13.3.6}$$

Here, the amplitudes E_1, E_2 can also be complex so that

$$E_i = |E_i|e^{i\varphi_i} \ , \quad i = 1, 2 \tag{13.3.7}$$

and for the physically relevant real part, we have

$$\operatorname{Re}\{E_i e^{i(k \cdot r - \omega t)}\} = |E_i| \cos(k \cdot r - \omega t + \varphi_i) \ , \quad i = 1, 2 \ . \tag{13.3.8}$$

We consider a few special cases:

a) First, let $\varphi_1 = \varphi_2$, then

$$E(r, t) = (|E_1|e_1 + |E_2|e_2)e^{i(k \cdot r - \omega t + \varphi_1)} \ , \tag{13.3.9}$$

and the solution $E(r, t)$ is now linearly polarized in the direction of

$$|E_1|e_1 + |E_2|e_2$$

(see Fig. 13.3.1a).

b) Now, let $\varphi_2 = \varphi_1 \pm \dfrac{\pi}{2}$, $|E_2| = |E_1|$, then

$$E_2 = \pm iE_1 \ , \tag{13.3.10}$$

and we find the solutions

$$E_{\pm}(r, t) = E_1(e_1 \pm ie_2)e^{i(k \cdot r - \omega t)} \ , \tag{13.3.11}$$

i.e.,

$$\operatorname{Re}\{(E_{\pm})_x\} = |E_1| \cos(k \cdot r - \omega t + \varphi_1) \ , \tag{13.3.12}$$

$$\operatorname{Re}\{(E_{\pm})_y\} = |E_1| \cos(k \cdot r - \omega t \pm \tfrac{\pi}{2} + \varphi_1)$$

$$= \mp |E_1| \sin(k \cdot r - \omega t + \varphi_1) \ , \tag{13.3.13}$$

therefore also

$$[\operatorname{Re}\{E_{\pm}(r, t)\}]^2 = |E_1|^2 \ . \tag{13.3.14}$$

This means the following: Let us set the z-axis in the k-direction, then the physical solution $\operatorname{Re}\{E_{\pm}(r, t)\}$ always has the same amplitude as a function of t, but the direction of $\operatorname{Re}\{E_{\pm}(r, t)\}$ rotates in the xy-plane, and in fact $\operatorname{Re}\{E_-(r, t)\}$ rotates clockwise, while $\operatorname{Re}\{E_+(r, t)\}$ rotates counterclockwise

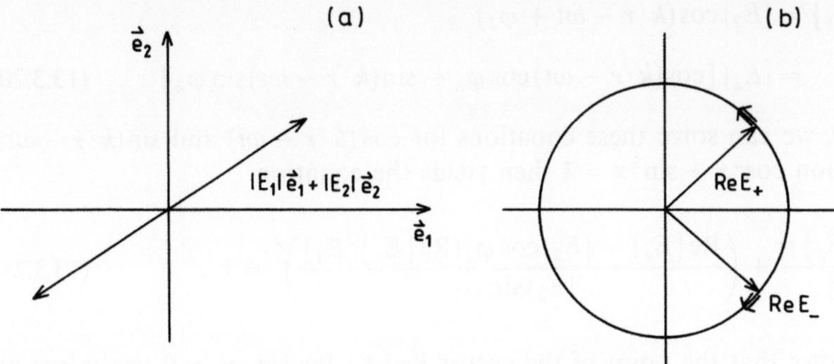

Fig. 13.3.1a, b. The oscillations of the electric field in the plane perpendicular to the direction of propagation, e_3. **(a)** linear polarization, **(b)** circular polarization

(Fig. 13.3.1b). These two waves are said to have *right-handed* and *left-handed* *circular polarization*, respectively. A superposition of linearly polarized waves of equal amplitudes thus yields a circularly polarized wave, if the phase difference of the linearly polarized waves is exactly $\pm\frac{\pi}{2}$.

If we introduce the two vectors

$$e_\pm = \frac{1}{\sqrt{2}}(e_1 \pm ie_2) \tag{13.3.15}$$

as a new basis in the plane perpendicular to k instead of the basis vectors e_1, e_2, then $E(r, t)$ can in general be represented as (this is equivalent to (13.3.6)):

$$E(r, t) = (E_+ e_+ + E_- e_-)e^{i(k \cdot r - \omega t)} \ . \tag{13.3.16}$$

In this representation, circularly polarized waves are obtained when $E_+ = 0$ or $E_- = 0$.

If, though, for example

$$E_- = E_+ \ , \quad \text{then} \tag{13.3.17}$$

$$E(r, t) = E_+(e_+ + e_-)e^{i(k \cdot r - \omega t)}$$

$$= E_+\sqrt{2}e_1 e^{i(k \cdot r - \omega t)} \ , \tag{13.3.18}$$

i.e., the wave is linearly polarized. The same holds true if $E_- = -E_+$. A linearly polarized wave can therefore also be described as a superposition of two circularly polarized waves.

c) For the general case, let us without loss of generality assume that $\varphi_1 = 0$. Then,

$$\mathrm{Re}\{E_x\} = E_1 \cos(k \cdot r - \omega t) \ , \tag{13.3.19}$$

$$\mathrm{Re}\{E_y\} = |E_2|\cos(k\cdot r - \omega t + \varphi_2)$$

$$= |E_2|\left[\cos(k\cdot r - \omega t)\cos\varphi_2 - \sin(k\cdot r - \omega t)\sin\varphi_2\right] . \tag{13.3.20}$$

If $\varphi_2 \neq 0$, we can solve these equations for $\cos(k\cdot r - \omega t)$ and $\sin(k\cdot r - \omega t)$. The relation $\cos^2 x + \sin^2 x = 1$ then yields the equation

$$\frac{(\mathrm{Re}\{E_x\})^2}{E_1^2} + \left(\frac{\mathrm{Re}\{E_y\} - |E_2|\cos\varphi_2(\mathrm{Re}\{E_x\}/E_1)}{|E_2|\sin\varphi_2}\right)^2 = 1 , \tag{13.3.21}$$

which shows that the point of the vector $\mathrm{Re}\{E\}$ for $\sin\varphi_2 \neq 0$ traces out an ellipse. $E(r, t)$ is then said to be *elliptically polarized*. The ellipse degenerates into a circle if E_+ or E_- vanishes and to a line if $E_+ = \pm E_-$.

13.4 Electromagnetic Waves, the Influence of Sources

After our investigation of the general structure of the solution of Maxwell's equations in a region free of sources in Sect. 13.3, we will now turn to the dependence of the solutions on the source terms $\varrho(r, t)$ and $J(r, t)$. We first go back to the inhomogeneous wave equations for the potentials. These have the form

$$\left(-\Delta + \frac{1}{c^2}\frac{\partial^2}{\partial t^2}\right)u(r, t) = h(r, t) . \tag{13.4.1}$$

Here $u(r, t)$ stands for $\phi(r, t)$ and $A(r, t)$, while $h(r, t)$ stands for $\varrho(r, t)/\varepsilon_0$ and $\mu_0 J(r, t)$ respectively.

In Sect. 10.2, we investigated a Green's function for the wave equation. This was

$$G_\mathrm{R}(r, t) = \frac{1}{4\pi r}\delta\left(t - \frac{r}{c}\right)\Theta(t) , \tag{13.4.2}$$

so that the solution of the inhomogeneous wave equation could be written as

$$u(r, t) = \int dt' d^3 r' G_\mathrm{R}(r - r', t - t')h(r', t')$$

$$= \int dt' d^3 r' \frac{1}{4\pi|r - r'|}\delta\left(t - t' - \frac{|r - r'|}{c}\right)h(r', t')$$

$$= \frac{1}{4\pi}\int d^3 r' \frac{1}{|r - r'|}h\left(r', t - \frac{|r - r'|}{c}\right) . \tag{13.4.3}$$

We will now derive this result in a way which appears somewhat different:

We represent the source function as a Fourier integral of the form

$$h(r, t) = \int d\omega\, \tilde{h}(r, \omega) e^{-i\omega t} \tag{13.4.4}$$

(see Appendix D). In the same way, we write the solution as

$$u(r, t) = \int d\omega\, \tilde{u}(r, \omega) e^{-i\omega t} \ . \tag{13.4.5}$$

Then, from the wave equation, it follows that

$$\left(-\Delta + \frac{1}{c^2} \frac{\partial^2}{\partial t^2} \right) u(r, t) = \int d\omega \left(-\Delta + \frac{(-i\omega)^2}{c^2} \right) \tilde{u}(r, \omega) e^{-i\omega t}$$

$$= \int d\omega\, \tilde{h}(r, \omega) e^{-i\omega t} \ . \tag{13.4.6}$$

In this way we are led to the inhomogeneous Helmholtz equation for the Fourier transforms:

$$-(\Delta + k^2)\tilde{u}(r, \omega) = \tilde{h}(r, \omega) \ , \qquad k = \frac{\omega}{c} \ . \tag{13.4.7}$$

In order to solve this equation, we need to use a Green's function for $(\Delta + k^2)$ in all of $\mathbb{R}^3$. An example of such a function is

$$G_k(r, r') = \frac{1}{|r - r'|} e^{\pm ik|r - r'|} \ . \tag{13.4.8}$$

Proof. It is enough to show that

$$-(\Delta + k^2) \frac{e^{\pm ikr}}{r} = 4\pi\delta(r) \ . \tag{13.4.9}$$

Explicitly, we obtain (see Appendix F)

$$\Delta \frac{e^{\pm ikr}}{r} = \left(\frac{\partial^2}{\partial r^2} + \frac{2}{r} \frac{\partial}{\partial r} \right) \frac{e^{\pm ikr}}{r}$$

$$= e^{\pm ikr} \Delta \frac{1}{r} + \frac{1}{r} \left(\frac{\partial^2}{\partial r^2} + \frac{2}{r} \frac{\partial}{\partial r} \right) e^{\pm ikr} + 2 \left(\frac{\partial}{\partial r} \frac{1}{r} \right) \left(\frac{\partial}{\partial r} e^{\pm ikr} \right)$$

$$= -4\pi\delta(r) + (-k^2) \frac{e^{\pm ikr}}{r} \ .$$

$G_k(r, r')$ is a Green's function which vanishes like $1/r$ at infinity. We leave open the sign in the exponent.

The solution of the inhomogeneous Helmholtz equation is then

$$\tilde{u}(r, \omega) = \frac{1}{4\pi} \int d^3 r'\, G_k(r, r') \tilde{h}(r', \omega) \ , \tag{13.4.10}$$

and thus

$$u(r, t) = \int d\omega\, \tilde{u}(r, \omega) e^{-i\omega t}$$

$$= \frac{1}{4\pi} \int d^3 r' \int d\omega\, \frac{e^{\pm ik|r-r'|}}{|r-r'|} e^{-i\omega t} \tilde{h}(r', \omega)$$

$$= \frac{1}{4\pi} \int d^3 r' \int d\omega\, \frac{1}{|r-r'|} e^{-i\omega(t \mp |r-r'|/c)} \tilde{h}(r', \omega)$$

$$= \frac{1}{4\pi} \int d^3 r' \frac{1}{|r-r'|} h\left(r', t \mp \frac{|r-r'|}{c}\right) . \tag{13.4.11}$$

In the expression $(t \mp |r-r'|/c)$, both signs are possible mathematically, however only the upper sign is possible physically, since

$$t' = t - \frac{|r-r'|}{c}$$

is the point in time *before* t, in which the source $h(r', t')$ causes an excitation which then travels the distance $(r - r')$ from r' to r in time interval $|r - r'|/c$, in order to be seen at the point r.

A time variation of the source thus gives rise in turn to a time variation of the fields, however this effect is retarded (i.e. delayed), since the excitation is propagated with a finite speed, namely the speed of light. Such a potential is called a *retarded potential*.

Thus, using Fourier transforms, we have reduced the time dependent problem to the solution of the Helmholtz equation and with the limitation to the upper sign, equation (13.4.11) is identical to the solution (13.4.3).

In general, the integral in these formulas over $h(r', t')$ cannot be calculated analytically. At this point, various approximations and special cases become useful.

We will start with the assumption that $\tilde{h}(r, \omega)$ as a function of ω is nonzero only within a certain fixed region, which, say, is located about ω_0.

The corresponding wavelength $\lambda = 2\pi c/\omega_0$ then represents a measure of the wavelength of the radiation. Now let the size of the source $d \ll \lambda$. Three regions are usually distinguished, depending on the distance r of the receiver from the source:

a) The near zone $d \sim r \ll \lambda$,
b) the intermediate zone $d \ll r \sim \lambda$, and
c) the *far zone* $d \ll \lambda \ll r$, also called the *radiation zone*.

We will study the potentials only in the radiation zone.

With $\mu_0 J(r, \omega)$ for $\tilde{h}(r, \omega)$, and $A(r, \omega)$ for $\tilde{u}(r, \omega)$, we have from (13.4.10)

$$A(r, \omega) = \frac{\mu_0}{4\pi} \int d^3 r' \frac{e^{ik|r-r'|}}{|r-r'|} J(r', \omega) . \tag{13.4.12}$$

For $r' \lesssim d \ll r$, we have

$$|r - r'| = (r^2 + r'^2 - 2r \cdot r')^{1/2} = r\left(1 - 2n \cdot \frac{r'}{r} + \frac{r'^2}{r^2}\right)^{1/2}$$

$$= r\left(1 - n \cdot \frac{r'}{r} + \ldots\right) = r - n \cdot r' + O\left(r\left(\frac{r'}{r}\right)^2\right) . \tag{13.4.13}$$

Then,

$$\frac{1}{|r - r'|} = \frac{1}{r(1 - n \cdot r'/r + \ldots)} = \frac{1}{r} + O\left(\frac{1}{r}\left(\frac{r'}{r}\right)\right) \tag{13.4.14}$$

and thus

$$A(r, \omega) = \frac{\mu_0}{4\pi} \frac{1}{r} \int d^3 r' e^{ikr} e^{-ikn \cdot r'} J(r', \omega)$$

$$= \frac{\mu_0}{4\pi} \frac{e^{ikr}}{r} \sum_{l=0}^{\infty} \frac{1}{l!} \int d^3 r' (-ikn \cdot r')^l J(r', \omega) . \tag{13.4.15}$$

Since for $d \ll \lambda$, we have

$$\frac{d}{\lambda} = \frac{kd}{2\pi} \ll 1 , \quad \text{then} \quad |kn \cdot r'| \lesssim kd \ll 1 ,$$

and thus in the sum of (13.4.15), the term with $l = 0$ is dominant.
Thus, in this approximation (the *dipole approximation*):

$$A(r, \omega) = \frac{\mu_0}{4\pi} \frac{e^{ikr}}{r} \int d^3 r' J(r', \omega) . \tag{13.4.16}$$

For a stationary current, we had $\nabla \cdot J = 0$ and

$$\int d^3 r J(r) = 0 .$$

Now, though, the continuity equation holds in the form

$$\nabla \cdot J(r, t) + \partial \varrho(r, t)/\partial t = 0 ,$$

or, for the Fourier transforms:

$$\nabla \cdot J(r, \omega) - i\omega \varrho(r, \omega) = 0 . \tag{13.4.17}$$

It then follows (cf. (12.2.38))

$$\int d^3 r' J(r', \omega) = -\int d^3 r' r' [\nabla' \cdot J(r', \omega)] = -\int d^3 r' r' i\omega \varrho(r', \omega)$$

$$= -i\omega p(\omega) \tag{13.4.18}$$

with

$$p(\omega) = \int d^3r'\, r'\, \varrho(r', \omega) \ , \tag{13.4.19}$$

the Fourier transform of the time dependent electric dipole moment

$$p(t) = \int d^3r'\, r'\, \varrho(r', t) \ . \tag{13.4.20}$$

Thus, in the dipole approximation,

$$A(r, \omega) = \frac{\mu_0}{4\pi} \frac{e^{ikr}}{r} (-i\omega) p(\omega) \ , \qquad k = \frac{\omega}{c} \ , \tag{13.4.21}$$

and we find for $A(r, t)$

$$A(r, t) = \int d\omega\, A(r, \omega) e^{-i\omega t} = \frac{\mu_0}{4\pi} \frac{1}{r} \frac{\partial}{\partial t} \int d\omega\, e^{ikr} p(\omega) e^{-i\omega t}$$

$$= \frac{\mu_0}{4\pi} \frac{1}{r} \frac{\partial}{\partial t} \int d\omega\, e^{i\omega[t - (r/c)]} p(\omega) = \frac{\mu_0}{4\pi} \frac{1}{r} \dot{p}\left(t - \frac{r}{c}\right) \ . \tag{13.4.22}$$

We can calculate the potential $\phi(r, t)$ from the Lorentz condition. This reads

$$\boldsymbol{\nabla} \cdot A(r, t) + \frac{1}{c^2} \frac{\partial \phi(r, t)}{\partial t} = 0$$

or, for the Fourier transforms

$$\boldsymbol{\nabla} \cdot A(r, \omega) = i(\omega/c^2)\phi(r, \omega) \ . \tag{13.4.23}$$

Thus, for $\phi(r, \omega)$, we find

$$\phi(r, \omega) = -\frac{ic^2}{\omega} \boldsymbol{\nabla} \cdot A(r, \omega) = \frac{\mu_0}{4\pi} \frac{c^2}{\omega} (-\omega) p(\omega) \cdot \boldsymbol{\nabla} \frac{e^{ikr}}{r}$$

$$= -\frac{\mu_0}{4\pi} c^2 p(\omega) \cdot \frac{r}{r} \frac{d}{dr}\left(\frac{e^{ikr}}{r}\right) = -ik \frac{1}{4\pi\varepsilon_0} \frac{p(\omega) \cdot r}{r^2} e^{ikr} + O\left(\frac{1}{r^2}\right)$$

$$= \frac{1}{4\pi\varepsilon_0} \frac{1}{cr^2} r \cdot (-i\omega) p(\omega) e^{ikr} \ , \tag{13.4.24}$$

thus

$$\phi(r, t) = \frac{1}{4\pi\varepsilon_0} \frac{1}{cr^2} r \cdot \dot{p}\left(t - \frac{r}{c}\right) \ . \tag{13.4.25}$$

For the fields E and B, we calculate from the potentials up to terms of order $O(1/r^2)$:

$$B(r, \omega) = \frac{\mu_0}{4\pi}(ik)[i\omega p(\omega) \times n]\frac{e^{ikr}}{r}, \qquad n = \frac{r}{r} \tag{13.4.26}$$

or

$$B(r, t) = \frac{\mu_0}{4\pi cr}\left[\ddot{p}\left(t - \frac{r}{c}\right) \times n\right] \tag{13.4.27}$$

and

$$E(r, t) = c[B(r, t) \times n] . \tag{13.4.28}$$

Remarks. i) We again establish that $B(r, t)$ and $E(r, t)$ are perpendicular to n as well as to each other. $n = r/r$ is however also the direction of propagation of the wave.

ii) The term

$$\frac{1}{r}e^{ikr}$$

is called the spherical wave, since

$$\frac{1}{r}e^{ikr - i\omega t} \tag{13.4.29}$$

is a wave whose phase is the same for all fixed values of r, i.e. points of the same phase lie on a sphere, whereas for the plane wave

$$e^{ik \cdot r - i\omega t} \tag{13.4.30}$$

points with the same phase lie in a plane, namely in the plane perpendicular to k.

The factor $1/r$ in the case of a spherical wave is very important. This factor ensures that the energy density which we can ascribe to such a wave is constant when integrated over the entire sphere. In order to examine this more closely, we will investigate in the next section the energy density of an electromagnetic field.

13.5   The Energy of the Electromagnetic Field

13.5.1   Balance of Energy and the Poynting Vector

In an electromagnetic field, the force

$$F = q[E(r, t) + v(t) \times B(r, t)] \tag{13.5.1}$$

is exerted on a particle of charge q and velocity v at point r. If this then causes a displacement dr, the field will perform work

$$dW = F \cdot dr = F \cdot (dr/dt)dt = F \cdot v(t)\,dt = qE \cdot v(t)\,dt \;, \tag{13.5.2}$$

on the particle. This can also be written as

$$dW = \int d^3 r'\, qv(t)\delta(r' - r)\cdot E(r', t)dt \;. \tag{13.5.3}$$

In this form, we see that more generally

$$dW = \int_V d^3 r'\, J(r', t)\cdot E(r', t)\, dt \tag{13.5.4}$$

is the amount of work that the electromagnetic field performs on charge carriers which produce a current density. Since energy is conserved, this amount of energy must be taken from the electromagnetic field.

We will now rewrite the integrand $J(r, t)\cdot E(r, t)$. We use the Maxwell equation

$$\nabla \times B(r, t) - \frac{1}{c^2}\frac{\partial E(r, t)}{\partial t} = \mu_0 J(r, t) \;,$$

in order to eliminate the current density $J(r, t)$ from the integrand in (13.5.4). This yields:

$$\frac{dW}{dt} = \frac{1}{\mu_0}\int_V d^3 r'\, E(r', t)\left[\nabla' \times B(r', t) - \frac{1}{c^2}\frac{\partial E(r', t)}{\partial t}\right] \;. \tag{13.5.5}$$

On the other hand,

$$\nabla \cdot (E \times B) = (\nabla \times E)\cdot B - (\nabla \times B)\cdot E \;. \tag{13.5.6}$$

Thus, if we use the corresponding Maxwell equation for $\nabla \times E$, we find

$$E \cdot (\nabla \times B) = -B \cdot \frac{\partial B}{\partial t} - \nabla \cdot (E \times B) \;. \tag{13.5.7}$$

Further, $1/c^2 = \varepsilon_0 \mu_0$, so that for dW/dt we finally obtain

$$\frac{dW}{dt} = \frac{1}{\mu_0}\int_V d^3 r\left[-\nabla \cdot (E \times B) - B \cdot \frac{\partial B}{\partial t} - \varepsilon_0 \mu_0 E \cdot \frac{\partial E}{\partial t}\right] \;, \tag{13.5.8}$$

i.e.,

$$\frac{dW}{dt} + \int_V d^3 r\, \nabla \cdot \left(E \times \frac{1}{\mu_0}B\right) + \frac{1}{2}\frac{d}{dt}\int_V d^3 r\left(\frac{1}{\mu_0}B \cdot B + E \cdot \varepsilon_0 E\right) = 0$$

or

$$\frac{dW}{dt} + \int_{\partial V} dA \cdot S(r, t) + \frac{d}{dt} \int_V d^3 r\, e(r, t) = 0 \tag{13.5.9}$$

with

$$S = E \times H , \quad e = \tfrac{1}{2}(B \cdot H + E \cdot D) , \tag{13.5.10}$$

$$D = \varepsilon_0 E , \quad H = \frac{1}{\mu_0} B . \tag{13.5.11}$$

We now consider the dimensions of the newly introduced quantities H, D, S, and e.

From

$$
\begin{aligned}
[E] \ &= \text{N/C, as we see from } F = qE, \\
[B] \ &= \text{Ns/Cm, as we see, e.g., from } F = qv \times B, \text{ and} \\
[\mu_0] \ &= \text{N/A}^2, \\
[\varepsilon_0] \ &= \text{A}^2\text{s}^2/\text{Nm}^2, \text{ we find:} \\
[H] \ &= [B/\mu_0] = \text{A/m}, \\
[D] \ &= [\varepsilon_0 E] = \text{C/m}^2, \\
[S] \ &= [E \times H] = \text{NA/Cm} = \text{Nm/m}^2\text{s} = \text{energy/area} \cdot \text{time}, \\
[e] \ &= [B \cdot H] = [E \cdot D] = \text{Nm/m}^3 = \text{energy/volume}.
\end{aligned}
$$

The quantity $e(r, t)$ represents an energy density, the vector $S(r, t)$ represents an energy flux, i.e. an energy which flows through a surface perpendicular to $S(r, t)$ per unit time and area. Equation (13.5.9) can thus be seen as an energy transport equation and can be read as follows:

The sum of the energy per unit time which is transferred to the particle, plus

– the energy per unit time which flows through ∂V, and
– the change in energy per unit time of the electromagnetic field in V,

is zero.

The quantity $e(r, t)$ is called the *energy density of the electromagnetic field*, $S(r, t)$ is called the *Poynting*[4] *vector* and indicates the energy flux of the electromagnetic field. The total energy of the particles and the electromagnetic field is conserved.

---

[4] *Poynting, John Henry* (*1852 Monton, near Manchester, d. 1914 Birmingham). Professor in Birmingham, worked especially on electrodynamics and radiation pressure, measured the constant of gravitation.

13.5.2 The Energy Flux of the Radiation Field

For simplicity, we consider fields with harmonic time dependence

$$E(r, t) = E(r)e^{-i\omega t} \ , \tag{13.5.12a}$$

$$H(r, t) = H(r)e^{-i\omega t} \ , \tag{13.5.12b}$$

where $E(r)$ and $H(r)$ are still complex and may depend on k or ω respectively. Then, for the physical fields, we have relations such as

$$E_{\text{phys}} = \text{Re}\{E(r, t)\} = \tfrac{1}{2}[E(r, t) + E^*(r, t)] \tag{13.5.13}$$

and thus

$$S = E_{\text{phys}} \times H_{\text{phys}}$$

$$= \tfrac{1}{4}[E(r)e^{-i\omega t} + E^*(r)e^{i\omega t}] \times [H(r)e^{-i\omega t} + H^*(r)e^{i\omega t}] \ .$$

We are interested in the time average of the Poynting vector, i.e.

$$\overline{S(r, t)} = \frac{1}{T} \int_0^T dt\, S(r, t) \ , \tag{13.5.14}$$

where T is given by $\omega T = 2\pi$. This leads to terms of the form

$$\frac{1}{T} \int_0^T dt = 1 \quad \text{and}$$

$$\frac{1}{T} \int_0^T dt\, e^{\pm 2i\omega t} = \frac{1}{\pm 2i\omega T}(e^{\pm 2i\omega T} - 1) = 0 \ .$$

We then obtain

$$\overline{S(r, t)} = \tfrac{1}{4}[E(r) \times H^*(r) + E^*(r) \times H(r)]$$

$$= \tfrac{1}{2}\text{Re}\{E(r) \times H^*(r)\} \ . \tag{13.5.15}$$

In Sect. 13.4, we calculated the electromagnetic field in the radiation zone. We found

$$B(r) = \frac{\mu_0}{4\pi} ck^2[n \times p(\omega)] \frac{e^{ikr}}{r} \ , \tag{13.5.16}$$

$$E(r) = -c[n \times B(r)] \ , \quad n = r/r \ . \tag{13.5.17}$$

In this approximation, we then find the time average of the energy flux to be:

$$\overline{S(r, t)} = -\frac{\mu_0}{8\pi}\left[(n \times (n \times p))c^2 k^2 \frac{e^{ikr}}{r}\right]\left[\frac{1}{4\pi} ck^2 (n \times p)^* \frac{e^{-ikr}}{r}\right]$$

$$= \frac{1}{2}\frac{\mu_0}{(4\pi)^2} k^4 c^3 |n \times p|^2 \frac{1}{r^2} n \ . \tag{13.5.18}$$

If we then want to know the power P, that is, the time average of the energy transmitted in unit time through the surface of a sphere with radius r, we find

$$P = \int dF \cdot \overline{S(r, t)} = \int d\Omega r^2 n \cdot \overline{S(r, t)} \ , \tag{13.5.19}$$

or, since

$$r^2 n \cdot \overline{S(r, t)} = \frac{1}{2}\frac{\mu_0}{(4\pi)^2} k^4 c^3 |n \times p|^2 \tag{13.5.20}$$

it follows that in the dipole approximation

$$\frac{dP}{d\Omega} = r^2 n \cdot \overline{S(r, t)} = \frac{1}{4\pi\varepsilon_0}\frac{k^4 c}{8\pi} |n \times p|^2 \ .$$

In the following, we consider the case that $p(\omega)$ is real, as in the case, for example, of a linearly oscillating dipole (a *Hertzian dipole*). Then, it follows that

$$\frac{dP}{d\Omega} = \frac{1}{4\pi\varepsilon_0}\frac{k^4 c}{8\pi} |p|^2(1 - \cos^2 \theta)$$

$$= \frac{1}{4\pi\varepsilon_0}\frac{k^4 c}{8\pi} |p|^2 \sin^2 \theta \ ; \tag{13.5.21}$$

where θ is the angle between p and the direction of observation n.

The $1/r$-dependence of the fields and the $1/r^2$-dependence of the Poynting vector are responsible for the fact that $dP/d\Omega$ is independent of r. All terms in the fields which fall off faster as $r \to \infty$ lead to terms in $dP/d\Omega$ which vanish as $r \to \infty$, and which can therefore be ignored.

For $dP/d\Omega$, we have the following radiation characteristics:

If we draw a vector of length $dP/d\Omega$ outward from the origin in the direction of n, we obtain Fig. 13.5.1. A charge distribution which oscillates in the z-direction radiates the most energy in a perpendicular direction, and none at all in the z-direction.

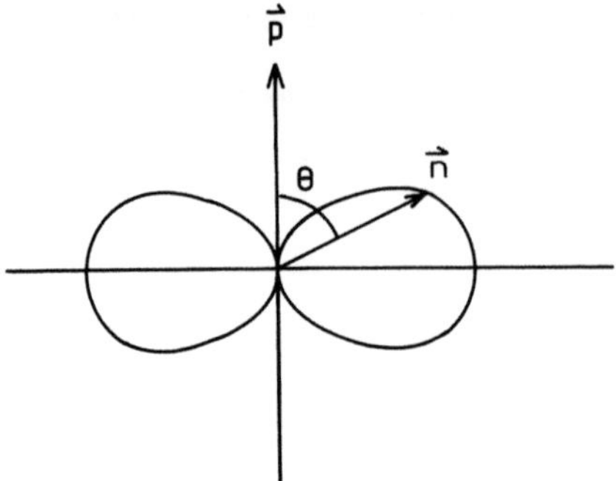

Fig. 13.5.1. Characteristics of electric dipole radiation

For the total radiated power

$$P = \int d\Omega \, \frac{dP}{d\Omega} \tag{13.5.22}$$

we finally obtain

$$P = \frac{1}{4\pi\varepsilon_0} \frac{k^4 c}{8\pi} |p|^2 2\pi \int_{-1}^{+1} d\cos\theta \sin^2\theta = \frac{1}{4\pi\varepsilon_0} \frac{\omega^4 |p|^2}{3c^3} \; . \tag{13.5.23}$$

This ω^4-dependence of the radiated power is responsible, for example, for the frequency dependence of the scattering of light in the atmosphere and thus for the blue color of the sky.

13.5.3 The Energy of the Electric Field

We consider the total energy of an electric field in a volume V. This energy W, using (13.5.10), is given by

$$W_{\mathrm{el}} = \frac{\varepsilon_0}{2} \int d^3 r \, E^2(r, t) \; . \tag{13.5.24}$$

Now, in the static case, $E(r) = -\nabla\phi(r)$ and $\nabla \cdot E = \varrho/\varepsilon_0$, and thus

$$W_{\mathrm{el}} = \frac{\varepsilon_0}{2} \int d^3 r \, E(r) \cdot [-\nabla\phi(r)]$$

$$= \frac{\varepsilon_0}{2} \left[-\int d^3 r \, \nabla \cdot (E\phi) + \int d^3 r \, \phi \nabla \cdot E \right]$$

$$= -\frac{\varepsilon_0}{2} \int_{\partial V} dF \cdot E\phi + \frac{1}{2} \int d^3 r \, \phi(r) \varrho(r) \; . \tag{13.5.25}$$

If $V = \mathbb{R}^3$, then $\phi(r)$ vanishes as $r \to \infty$, and

$$W_{\mathrm{el}} = \tfrac{1}{2} \int d^3 r\, \varrho(r)\phi(r) \tag{13.5.26}$$

is the energy of the static electric field produced by the charge density $\varrho(r)$. Note the additional factor $1/2$ compared to equation (11.6.3) for the potential energy of a charge distribution in an external electric field.

Now let $\varrho(r) \equiv 0$, and let the boundary of V be a set of conductors (and eventually an infinitely distant sphere). Let q_i be the charge on the i-th conductor, then if ∂L_i is the boundary of the i-th conductor and the area element points inwards into V, then

$$\int\limits_{\partial L_i} dA_i \cdot E = \frac{1}{\varepsilon_0}\, q_i \;,$$

and, since we know that $\phi = \phi_i = $ constant on ∂L_i, and considering that the surface elements of ∂V point in the opposite direction to those of ∂L_i, we find

$$W_{\mathrm{el}} = \frac{\varepsilon_0}{2} \sum_i \phi_i \int\limits_{\partial L_i} dA_i \cdot E = \frac{1}{2} \sum_i \phi_i q_i \;. \tag{13.5.27}$$

Thus the energy of the electric field can also be expressed in terms of the charges and potentials on the conductors.

From Sect. 11.2.2, we know that there is a linear relationship between the charges and the potentials on the conductors, namely,

$$q_i = \sum_j C_{ij}\phi_j \tag{13.5.28}$$

with C_{ij}, the capacitance coefficients. The energy of the electrical field is then given by

$$W_{\mathrm{el}} = \tfrac{1}{2} \sum_{i,j} C_{ij}\phi_i\phi_j \;. \tag{13.5.29}$$

In the case of a parallel plate capacitor, we have

$$C_{11} = -C_{12} = -C_{21} = C_{22} = C \tag{13.5.30}$$

and thus

$$W_{\mathrm{el}} = \tfrac{1}{2}C(\phi_1 - \phi_2)^2 = \tfrac{1}{2}CU^2 \;, \tag{13.5.31}$$

where U is the potential difference $\phi_1 - \phi_2$ between the plates, and C the capacitance.

13.5.4 The Energy of the Magnetic Field

We consider a system of conductors in which stationary currents are flowing. From (13.5.10), the energy of the magnetic field is

$$W_{\mathrm{m}} = \frac{1}{2\mu_0} \int d^3 r\, B^2(r) \;, \tag{13.5.32}$$

where V here represents the entire volume in $\mathbb{R}^3$.

Now, we have

$$\boldsymbol{B} = \boldsymbol{\nabla} \times \boldsymbol{A} \quad \text{and}$$

$$\boldsymbol{\nabla} \cdot (\boldsymbol{B} \times \boldsymbol{A}) = (\boldsymbol{\nabla} \times \boldsymbol{B}) \cdot \boldsymbol{A} - (\boldsymbol{\nabla} \times \boldsymbol{A}) \cdot \boldsymbol{B} \;,$$

so it follows that

$$W_{\mathrm{m}} = \frac{1}{2\mu_0} \int d^3 r\, \boldsymbol{B}(r) \cdot [\boldsymbol{\nabla} \times \boldsymbol{A}(r)] = \frac{1}{2\mu_0} \int d^3 r\, [\boldsymbol{\nabla} \times \boldsymbol{B}(r)] \cdot \boldsymbol{A}(r) \;, \tag{13.5.33}$$

since the boundary term, arising from the contribution of $\boldsymbol{\nabla} \cdot (\boldsymbol{B} \times \boldsymbol{A})$ vanishes. With $\boldsymbol{\nabla} \times \boldsymbol{B} = \mu_0 \boldsymbol{J}$, we then find

$$W_{\mathrm{m}} = \tfrac{1}{2} \int d^3 r\, \boldsymbol{J}(r) \cdot \boldsymbol{A}(r) \;. \tag{13.5.34}$$

Let $\boldsymbol{J}_i(r)$ be the current density in the i-th conductor L_i, then, if the current loop is again idealized into a line,

$$W_{\mathrm{m}} = \tfrac{1}{2} \sum_i \int_{L_i} d^3 r'\, \boldsymbol{J}_i(r') \cdot \boldsymbol{A}(r') = \tfrac{1}{2} \sum_i I_i \oint_{L_i} dr' \cdot \boldsymbol{A}(r')$$

$$= \tfrac{1}{2} \sum_i I_i \Phi_i \quad \text{with} \tag{13.5.35}$$

$$\Phi_i = \oint_{L_i} dr \cdot \boldsymbol{A}(r) = \int_{A_i} dA \cdot (\boldsymbol{\nabla} \times \boldsymbol{A}(r)) = \int_{A_i} dA \cdot \boldsymbol{B}(r) \;, \tag{13.5.36}$$

where A_i is the surface bordered by the conducting loop L_i (i.e., ∂A_i is L_i). Φ_i is obviously the magnetic flux through the i-th conducting loop. If we write the vector potential caused by the current in the j-th loop as $\boldsymbol{A}_j(r)$, then from (12.1.12), we find

$$\boldsymbol{A}_j(r) = \frac{\mu_0}{4\pi} I_j \oint_{L_j} \frac{dr'}{|r - r'|} \;, \tag{13.5.37}$$

and we then obtain

$$\Phi_i = \sum_j \oint_{L_i} dr \cdot \boldsymbol{A}_j(r) = \sum_j L_{ij} I_j \quad \text{with} \tag{13.5.38}$$

$$L_{ij} = \frac{\mu_0}{4\pi} \oint_{L_i} \oint_{L_j} \frac{dr \cdot dr'}{|r - r'|} \ .$$

(13.5.39)

L_{ij} is called the *inductance*[5], the quantity L_{ij} for $i \neq j$ is called the *mutual inductance*, and L_{ii} is called the *self-inductance*. Clearly, we have

$$L_{ij} = L_{ji} \ .$$

(13.5.40)

The energy of the magnetic field is then given by

$$W_{\mathrm{m}} = \tfrac{1}{2} \sum_{i,j} L_{ij} I_i I_j \ .$$

(13.5.41)

In the present case of conducting loops with vanishing thickness, equation (13.5.39) for the self-inductance yields logarithmically divergent integrals. For this purpose, then, we must consider the finite thickness of the current loops.

If the field $B(r)$ is known, then (13.5.41) can often provide information about the inductance. We will demonstrate this with an example:

Consider a coil with n windings of radius a, through which a current I flows. Let the path C go through the inside of the coil and close far outside of the coil (Fig. 13.5.2). Then, from Ampère's law,

$$\oint_C dr \cdot B(r) = \mu_0 I_{\mathrm{F}} = \mu_0 n I \ .$$

We need to consider only the field B in the interior of the coil over length l. There, it points in the direction of C. We then find

$$Bl = \mu_0 n I \ , \quad \text{or}$$

$$B = \mu_0 \frac{nI}{l} \ .$$

(13.5.42)

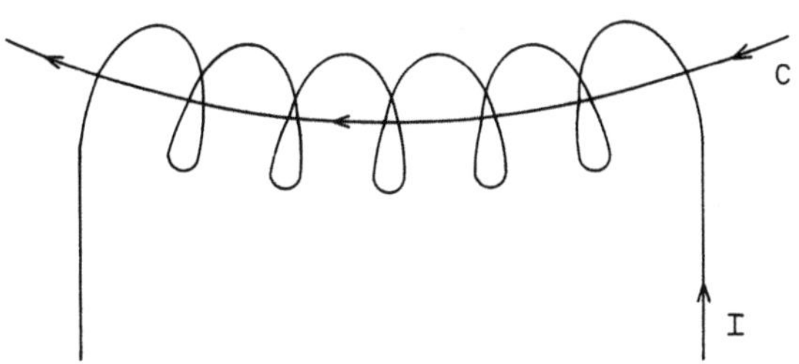

Fig. 13.5.2. The path C through a coil carrying electric current

---

[5] Inductance (Latin/French): ability for induction, which comes from the Latin *inducere*: "introduce, lead in", the creation of a voltage through the change of magnetic flux.

From

$$W_m = \frac{1}{2\mu_0} \int d^3r\, B^2(r)$$

we obtain

$$W_m = \frac{1}{2\mu_0} (\pi a^2 l) \left(\frac{\mu_0 n I}{l} \right)^2 = \frac{1}{2} \mu_0 \pi a^2 \frac{n^2}{l} I^2 \ . \tag{13.5.43}$$

We have therefore, to a good approximation, found the self inductance of this current loop:

$$L = \mu_0 \pi a^2 \frac{n^2}{l} \ . \tag{13.5.44}$$

13.5.5   Self-Energy and Interaction Energy

The total electromagnetic field energy, as we have seen, can be represented as the sum of the electric and magnetic field energies

$$W = W_{el} + W_m = \frac{\varepsilon_0}{2} \int d^3r\, E^2(r) + \frac{1}{2\mu_0} \int d^3r\, B^2(r) \ . \tag{13.5.45}$$

Let us now think of the charge and current distributions as consisting of two spatially separate parts, which are concentrated in the regions V_1 and V_2:

$$\varrho(r) = \varrho_1(r) + \varrho_2(r) \ , \quad J(r) = J_1(r) + J_2(r) \ . \tag{13.5.46}$$

Then, the fields E and B divide into two components which belong to the two separate groupings of charges and currents:

$$E = E_1 + E_2 \ , \quad B = B_1 + B_2 \ . \tag{13.5.47}$$

The electric and magnetic field energy are given by:

$$W_{el} = \frac{\varepsilon_0}{2} \int d^3r\, E_1^2 + \frac{\varepsilon_0}{2} \int d^3r\, E_2^2 + \varepsilon_0 \int d^3r\, E_1 \cdot E_2$$

$$=: W_{el}^{(1)} + W_{el}^{(2)} + W_{el}^{(1,2)} \tag{13.5.48}$$

and

$$W_m = \frac{1}{2\mu_0} \int d^3r\, B_1^2 + \frac{1}{2\mu_0} \int d^3r\, B_2^2 + \frac{1}{\mu_0} \int d^3r\, B_1 \cdot B_2$$

$$=: W_m^{(1)} + W_m^{(2)} + W_m^{(1,2)} \ . \tag{13.5.49}$$

We thus find three and not two terms in the expressions for W_{el} and W_m.

$W_{\text{el}}^{(1)}$, $W_{\text{m}}^{(1)}$, $W_{\text{el}}^{(2)}$, and $W_{\text{m}}^{(2)}$ are the electric and magnetic field energies which would result if there were charge and current distributions only in V_1 or V_2 respectively. These are called the *self-energies* of the distributions.

Additionally, there is also an electric and magnetic *interaction energy* $W_{\text{el}}^{(1,\,2)}$ and $W_{\text{m}}^{(1,\,2)}$, since each of the two distributions finds itself in the electromagnetic field of the other.

In the static case, we obtain, using the relations $E_i = -\nabla\phi_i$, $B_i = \nabla \times A_i$, and $\nabla \times B_i = \mu_0 J_i$, $i = 1, 2$, and Gauss's laws, as described in Sect. 13.5.3, 4:

$$W_{\text{el}}^{(i)} = \tfrac{1}{2}\int d^3 r \varrho_i \phi_i \ , \quad i = 1, 2 \ , \tag{13.5.50}$$

$$W_{\text{m}}^{(i)} = \tfrac{1}{2}\int d^3 r J_i \cdot A_i \ , \quad i = 1, 2 \ , \tag{13.5.51}$$

$$W_{\text{el}}^{(1,\,2)} = \int d^3 r \varrho_1 \phi_2 = \int d^3 r \varrho_2 \phi_1 \ , \tag{13.5.52}$$

$$W_{\text{m}}^{(1,\,2)} = \int d^3 r J_1 \cdot A_2 = \int d^3 r J_2 \cdot A_1 \ . \tag{13.5.53}$$

In each case, we find two equivalent expressions for the interaction energies, since we can either consider the first distribution in the external field of the second, or the second distribution in the external field of the first. (13.5.50–52) agree with the expressions (13.5.26), (13.5.34), and (11.6.3) already calculated.

Applications. We have already found in (11.6.7) the energy of an electric dipole in an external field

$$W_{\text{el}}^{(1,\,2)} = -p \cdot E \ , \tag{13.5.54}$$

which is also given by (13.5.2). We now calculate the energy of a magnetic dipole moment m in an external magnetic field B. From (13.5.53), we find, writing J instead of J_1 and A instead of A_2, assuming that A changes very slowly, and using (12.2.30),

$$W_{\text{m}}^{(1,\,2)} = \int d^3 b J(b) \cdot A(r) + \int d^3 b J(b) \cdot (b \cdot \nabla) A(r) + O(b^2)$$

$$= (m \times \nabla) \cdot A(r) = m \cdot B \ . \tag{13.5.55}$$

This result differs by a sign from the potential V_{m} found in (12.2.32), which can be used to calculate the force on a magnetic dipole held fixed in place, while in the electrostatic case (13.5.54) is identical to (11.6.7).

This difference arises from the fact that the electrostatic system is closed, the magnetic system however is not: In the electrostatic case, $|p| = $ constant can be achieved through constraint forces which do no work, so that the total energy is given by the electrostatic energy, and the force on the dipole can be calculated from the change of energy in a displacement. During the displacement, only the interaction energy is changed. In the magnetostatic case, during a displacement or rotation of the dipole, the current strength in the dipole will be changed by the appearance of voltage due to induction. If, then, $|m|$ is to be held constant, the voltage produced by induction must be compensated for by an external voltage source, and during the motion of the dipole the source of this voltage must give up energy. This energy needs to be taken into account, in addition to the change in the magnetic interaction energy, if we want to calculate the force on the dipole using energy considerations (cf. [Jackson]).

13.6   The Momentum of the Electromagnetic Field

Just as for energy, we can define the momentum of the electromagnetic field.
Let p be the momentum of a particle of charge q, then

$$dp/dt = F = qE + q(v \times B) \ ,$$

and if we add up the momentum of all the particles to find the total momentum P_{mech}

$$dP_{\text{mech}}/dt = \int d^3 r [\varrho(r)E(r, t) + J(r, t) \times B(r, t)] \ .$$

Using the Maxwell equations

$$\nabla \cdot E = \frac{1}{\varepsilon_0} \varrho \ , \qquad \nabla \times B - \frac{1}{c^2} \frac{\partial E}{\partial t} = \mu_0 J \ ,$$

it follows that

$$dP_{\text{mech}}/dt = \int_V d^3 r \left[\varepsilon_0 E (\nabla \cdot E) + \frac{1}{\mu_0} (\nabla \times B) \times B - \frac{1}{\mu_0 c^2} \frac{\partial E}{\partial t} \times B \right]$$

$$= \int d^3 r \left[E (\nabla \cdot D) + \frac{1}{\mu_0} (\nabla \times B) \times B \right.$$

$$\left. - \frac{1}{\mu_0 c^2} \frac{\partial}{\partial t} (E \times B) + \frac{1}{\mu_0 c^2} E \times \frac{\partial B}{\partial t} \right]$$

$$= \int d^3 r \left[E (\nabla \cdot D) + \frac{1}{\mu_0} (\nabla \times B) \times B \right.$$

$$\left. - \frac{1}{\mu_0 c^2} E \times (\nabla \times E) - \frac{1}{\mu_0 c^2} \frac{\partial}{\partial t} (E \times B) \right] \ .$$

If we add

$$(\nabla \cdot B) \frac{1}{\mu_0} B = 0 \ ,$$

to the underlined terms, we can also write

$$\left[E(\nabla \cdot D) + \frac{1}{\mu_0}(\nabla \times B) \times B - \frac{1}{\mu_0 c^2} E \times (\nabla \times E) + \frac{B}{\mu_0}(\nabla \cdot B) \right]_k = \partial_i T_{ik}$$

with

$$T_{ik} = \varepsilon_0 E_i E_k + \frac{1}{\mu_0} B_i B_k - \frac{1}{2}\delta_{ik}\left(\varepsilon_0 E^2 + \frac{1}{\mu_0} B^2 \right) ,$$

since

$$\partial_i T_{ik} = \varepsilon_0 \nabla \cdot E E_k + \frac{1}{\mu_0}\nabla \cdot B B_k$$

$$+ \varepsilon_0 E_i \partial_i E_k + \frac{1}{\mu_0} B_i \partial_i B_k - \frac{1}{2}\partial_k\left(\varepsilon_0 E^2 + \frac{1}{\mu_0} B^2 \right) .$$

On the other hand, though

$$[E \times (\nabla \times E)]_k = \tfrac{1}{2}\partial_k E^2 - E_i \partial_i E_k ,$$

and the same holds for B. Finally, then, we have the following transport equation for momentum

$$(dP_{\text{mech}}/dt)_k + \frac{\partial}{\partial t}\int_V d^3 r(\varepsilon_0 E \times B)_k = \int_V d^3 r \partial_i T_{ik} .$$

T_{ik} is called the *Maxwell stress tensor*, and

$$\varepsilon_0(E \times B) = D \times B$$

is the *momentum density* of the electromagnetic field.

We can then immediately rewrite the transport equation as

$$\frac{d}{dt}(P_{\text{mech}} + P_{\text{Field}})_k = \int_{\partial V} dA_i \cdot T_{ik} = \int_{\partial V} dA n_i T_{ik}$$

where n is the unit vector normal to ∂V, pointing outwards. The quantity $n_i T_{ik}$ is the flux of the k-th component of momentum in direction n, or also the k-th component of the force per unit area which is exerted on the surface dA.

The Maxwell stress tensor is thus the exact analogue of the conductive flow density of the momentum components $-\tau_{ik}$ which we found in fluid mechanics. This momentum flow density arose from the interaction of particles with one another, and at that point we used the following linear approximation,

$$\tau_{ik} = \eta(\nabla_i v_k + \nabla_k v_i - \tfrac{2}{3}\delta_{ik}\nabla \cdot v) + \zeta \nabla \cdot v \delta_{ik} - p\delta_{ik} .$$

Here, $n_i T_{ik}$ represents a momentum flow which occurs due to the electromagnetic field. Particles in the field experience a change in momentum, which is transferred from the field to the particles. The total momentum (of particles and field) thus remains conserved.

Example. We consider an electrostatic case, then

$$\frac{\partial}{\partial t} \int d^3 r \varepsilon_0 (E \times B) = 0$$

and

$$T_{ik} = \varepsilon_0 (E_i E_k - \tfrac{1}{2} \delta_{ik} E^2) \; ,$$

and thus

$$n_i T_{ik} = \varepsilon_0 (n \cdot E) E_k - \tfrac{1}{2} \varepsilon_0 n_k E^2$$

is the force which is exerted on a surface area element $n \cdot dA$.

Let σ be the surface charge density on a conductor, then

$$E = \frac{\sigma}{\varepsilon_0} n$$

is the field on the surface, and thus the force is given by

$$n_i T_{ik} = \frac{1}{\varepsilon_0} \left(\sigma^2 n_k - \frac{1}{2} \sigma^2 n_k \right) = \frac{1}{2\varepsilon_0} \sigma^2 n_k \; ,$$

i.e., the force per unit volume on the surface of the conductor in the direction of n is given by

$$\frac{1}{2\varepsilon_0} \sigma^2 \; .$$

If, for example, the surface charge density is induced by a point charge q, then the image force, which is exerted between the charge and the surface of the conductor, can also be determined, by calculating

$$F = \int_{\partial V} dA \frac{1}{2\varepsilon_0} \sigma^2$$

(see Sect. 11.3, remark (iii)).

14. Elements of the Electrodynamics of Continuous Media

Until now, we have considered the fields E and B which are caused by a charge density ϱ and a current density J in a vacuum. In a fluid or in a rigid body, for example, it is necessary to take into consideration all of the electrons and protons of the material in order to calculate the E- and B-fields.

Such a macroscopic piece of matter consists of approximately 10^{23} electrons and nuclei, which however are in rapid motion, such as thermal motion or motion in bound systems. The fields caused by the individual charge carriers are thus extremely dependent on time and position. These microscopic fields are of no interest to us; rather we will study the fields which arise when we average over regions of space large enough to contain many charge carriers, but small enough so that effects of visible light such as reflection and refraction are not eliminated

The wavelength of visible light is of the order of magnitude of $6000\,\text{Å} = 600$ nm, the order of magnitude of a molecule is $\sim 1\,\text{Å} = 10^{-10}$ m $= 0.1$ nm. We will therefore average the fields over a region with sides of length $L = 10$ nm, that is, over volumes of $L^3 = 10^{-24}\,\text{m}^3$. This will lead us to the macroscopic version of Maxwell's equations.

14.1 The Macroscopic Maxwell Equations

14.1.1 Microscopic and Macroscopic Fields

Let $e(r, t)$ be the *microscopic electric field*, which we have until now indicated by $E(r, t)$. We then introduce

$$E(r, t) = \langle e(r, t) \rangle = \int d^3 c f(-c) e(r + c, t) \tag{14.1.1}$$

as the spatially averaged electric field, which we will call the *macroscopic electric field*. In this definition, c should range over the neighborhood of r, and accordingly $f(c)$ should be a function with angular symmetry which satisfies the normalization condition

$$\int d^3 c f(c) = 1 \tag{14.1.2}$$

and looks something like Fig. 14.1.1. Analogously, we introduce $b(r, t)$ and $B(r, t)$ as the *microscopic* and *macroscopic magnetic induction fields*.

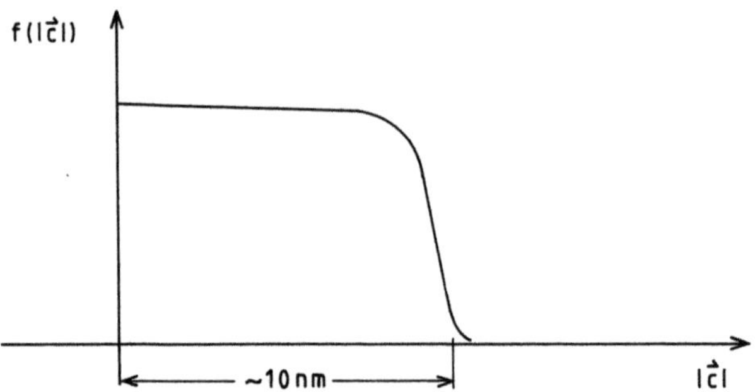

Fig. 14.1.1. Typical behavior of the averaging function $f(|c|)$

In a volume of $10^{-24}\,\mathrm{m}^3$ there are, as a rule, something like 10^6 nuclei and electrons. We will average, then, over the fields produced by these charge carriers. Microscopic time dependencies will then be simultaneously averaged out. Electromagnetic radiation with wavelengths of $1\,\text{Å}$ (X-rays) cannot be treated with such a theory. For such radiation, a microscopic theory is necessary.

As in Chap. 9, we will treat matter here as a continuum. We are thus developing the theory of electrodynamics of continuous media.

It is clear that

$$\frac{\partial}{\partial x_i}\langle e(r,t)\rangle = \left\langle \frac{\partial}{\partial x_i} e(r,t) \right\rangle \tag{14.1.3}$$

and

$$\frac{\partial}{\partial t}\langle e(r,t)\rangle = \left\langle \frac{\partial}{\partial t} e(r,t) \right\rangle . \tag{14.1.4}$$

Maxwell's equations for the microscopic fields, which we will now call the microscopic Maxwell equations, now read

$$\nabla \cdot b(r,t) = 0 \;, \quad \nabla \times e(r,t) + \frac{\partial b(r,t)}{\partial t} = 0 \;,$$

$$\nabla \cdot e(r,t) = \frac{1}{\varepsilon_0}\eta \;, \quad \nabla \times b(r,t) - \frac{1}{c^2}\frac{\partial e(r,t)}{\partial t} = \mu_0 j \;, \tag{14.1.5}$$

where we have written η and j instead of ϱ and J for the microscopic charge and current density.

After averaging, we obtain

$$\nabla \cdot B(r, t) = 0 \ , \quad \nabla \times E(r, t) + \frac{\partial B(r, t)}{\partial t} = 0 \ ,$$

$$\nabla \cdot E(r, t) = \frac{1}{\varepsilon_0} \langle \eta \rangle \ ,$$

$$\nabla \times B(r, t) - \frac{1}{c^2} \frac{\partial E(r, t)}{\partial t} = \mu_0 \langle j \rangle \ . \tag{14.1.6}$$

We now need to study the averages of the charge and current densities.

14.1.2 The Average Charge Density and Electric Displacement

Matter consists of electrons, which are free, or more generally, of free charges, and of bound charges, which build up larger units such as molecules. The charge density can thus be written as

$$\eta(r, t) = \eta_{\text{free}}(r, t) + \eta_{\text{bound}}(r, t) \tag{14.1.7}$$

with

$$\eta_{\text{free}}(r, t) = \sum_i q_i \delta(r - r_i(t)) \ , \tag{14.1.8}$$

$$\eta_{\text{bound}}(r, t) = \sum_n \eta_n(r, t) \quad \text{and} \tag{14.1.9}$$

$$\eta_n(r, t) = \sum_j q_{nj} \delta(r - r_{nj}(t)) \ . \tag{14.1.10}$$

$\eta_{\text{free}}(r, t)$ contains the charges q_i of the free charges. The charge density of the n-th molecule $\eta_n(r, t)$ contains the charges q_{nj} of the constituents at points r_{nj}.
Then we have

$$\langle \eta_n \rangle = \int d^3 c f(-c) \eta_n(r + c, t)$$

$$= \sum_j q_{nj} \int d^3 c f(-c) \delta(r + c - r_{nj}(t))$$

$$= \sum_j q_{nj} f(r - r_{nj}(t)) \ . \tag{14.1.11}$$

Now let

$$r_{nj} = r_n + d_{nj} \ , \tag{14.1.12}$$

i.e. let the position vector of the j-th constituent of the molecule n, r_{nj}, be written as the sum of the position vectors of the center of the molecule, r_n, and the vector from this center to the constituent, d_{nj}. This vector d_{nj} is the length ~ 0.1 nm, and thus a Taylor expansion of $f(r - r_n - d_{nj})$ around $r - r_n$ will converge quickly. It follows then that

$$\langle \eta_n \rangle = \sum_j q_{nj} f(r - r_n - d_{nj})$$

$$= \sum_j q_{nj} f(r - r_n) - \sum_j q_{nj} d_{nj} \cdot \nabla f(r - r_n) - \ldots$$

$$= q_n f(r - r_n) - p_n \cdot \nabla f(r - r_n) - \ldots . \tag{14.1.13}$$

Here

$$q_n = \sum_j q_{nj} \tag{14.1.14}$$

is the charge, and

$$p_n(t) = \sum_j q_{nj} d_{nj}(t) \tag{14.1.15}$$

is the dipole moment of the molecule n. The further terms in the Taylor expansion contain the higher multipole moments of the molecule.

With

$$q_n f(r - r_n) = \int d^3 c f(- c) q_n \delta(r - r_n + c)$$

$$= \langle q_n \delta(r - r_n) \rangle , \tag{14.1.16}$$

$$p_n f(r - r_n) = \int d^3 c f(- c) p_n \delta(r - r_n + c)$$

$$= \langle p_n \delta(r - r_n) \rangle \tag{14.1.17}$$

we can also write $\langle \eta_n \rangle$ as

$$\langle \eta_n \rangle = \langle q_n \delta(r - r_n) \rangle - \nabla \cdot \langle p_n \delta(r - r_n) \rangle . \tag{14.1.18}$$

Finally, we obtain

$$\langle \eta \rangle = \left\langle \sum_i q_i \delta(r - r_i(t)) \right\rangle + \left\langle \sum_n q_n \delta(r - r_n(t)) \right\rangle$$

$$- \nabla \cdot \sum_n \langle p_n(t) \delta(r - r_n(t)) \rangle + \ldots$$

$$= \varrho(r, t) - \nabla \cdot P(r, t) + \ldots \tag{14.1.19}$$

with

$$\varrho(r, t) = \left\langle \sum_i q_i \delta(r - r_i(t)) \right\rangle + \left\langle \sum_n q_n \delta(r - r_n(t)) \right\rangle \qquad (14.1.20)$$

as the macroscopic charge density, in which the molecule is considered as a unit and

$$P(r, t) = \sum_n \langle p_n(t) \delta(r - r_n(t)) \rangle \qquad (14.1.21)$$

as the average dipole moment per unit volume. $P(r, t)$ is thus the *dipole moment density* at point r. This is also called the *macroscopic polarization*.

It then follows that

$$\nabla \cdot E = \frac{1}{\varepsilon_0} (\varrho - \nabla \cdot P + \dots)$$

or

$$\nabla \cdot (\varepsilon_0 E + P + \dots) = \varrho , \qquad (14.1.22)$$

and with

$$D = \varepsilon_0 E + P + \dots \qquad (14.1.23)$$

we find the macroscopic Maxwell equation

$$\nabla \cdot D(r, t) = \varrho(r, t) . \qquad (14.1.24)$$

Here, then, $\varrho(r, t)$ is the charge density which is produced by the free charge carriers and the charges in the bound systems. The polarization P is the sum of the individual dipole moments of the individual molecules per unit volume. The vector D is called the *electric displacement* or the electric displacement density, even though this term better describes the vector P.

The average dipole moment depends, of course, on the electric field E, and it is often true in the stationary case if the medium is isotropic that

$$P(r) = \varepsilon_0 \chi E(r) . \qquad (14.1.25)$$

We will find a more general relationship, which is also valid for time dependent fields, in Sect. 14.4.1.

χ is called the electric *susceptibility* of the medium. We can then write

$$D(r) = \varepsilon_0 (1 + \chi) E(r) = \varepsilon_0 \varepsilon E(r) \qquad (14.1.26)$$

with

$$\varepsilon = 1 + \chi . \qquad (14.1.27)$$

ε is called the *dielectric*[1] *constant*. The relation

$$P(r) = \varepsilon_0 \chi E(r) \;, \quad \text{in general} \quad P = P(E)$$

thus characterizes the polarizability of the material. The Maxwell equations in the static case

$$\nabla \cdot D(r) = \varrho(r) \;, \quad \nabla \times E(r) = 0 \tag{14.1.28}$$

are now no longer complete, but need to be supplemented by a relation between D and E such as (14.1.26).

14.1.3 The Average Current Density and the Magnetic Field Strength

We have first

$$j(r, t) = j_{\text{free}}(r, t) + \sum_n j_n(r, t) \tag{14.1.29}$$

with the free charge current density

$$j_{\text{free}}(r, t) = \sum_i q_i v_i \delta(r - r_i(t)) \tag{14.1.30}$$

and the current density of the bound charges in the n-th molecule

$$j_n(r, t) = \sum q_{nj} v_{nj} \delta(r - r_{nj}) \;. \tag{14.1.31}$$

For the current density of the n-th molecule, we obtain, with $v_{nj} = v_n + \dot{d}_{nj}$:

$$\langle j_n(r, t) \rangle = \left\langle \sum_j q_{nj}(v_n + \dot{d}_{nj}) \delta(r - r_n - d_{nj}) \right\rangle$$

$$= \sum_j q_{nj}(v_n + \dot{d}_{nj}) f(r - r_n - d_{nj})$$

$$= \sum_j q_{nj}(v_n + \dot{d}_{nj})[f(r - r_n) - d_{nj} \cdot \nabla f(r - r_n(t)) - \dots] \;. \tag{14.1.32}$$

We can rewrite the individual terms as follows:

$$\sum_j q_{nj} v_n f(r - r_n) = \langle q_n v_n \delta(r - r_n(t)) \rangle \;, \tag{14.1.33}$$

$$\sum_j q_{nj} \dot{d}_{nj} f(r - r_n) = \langle \dot{p}_n(t) \delta(r - r_n(t)) \rangle \;, \tag{14.1.34}$$

$$- \sum_j q_{nj} v_n (d_{nj} \cdot \nabla) f(r - r_n) = - \langle v_n (p_n \cdot \nabla) \delta(r - r_n(t)) \rangle \;. \tag{14.1.35}$$

---

[1] Dielectric (Greek) *dia*: through, against, something like electric penetrability, or counter-electricity.

The sum of the right hand sides of (14.1.34 and 35) differs from the expression

$$\frac{\partial}{\partial t}\langle p_n(t)\delta(r - r_n(t))\rangle = \langle \dot{p}_n(t)\delta(r - r_n(t))\rangle$$

$$- \langle p_n(t)(v_n \cdot \nabla)\delta(r - r_n)\rangle \qquad (14.1.36)$$

by the term

$$\langle [v_n(p_n \cdot \nabla) - p_n(v_n \cdot \nabla)]\delta(r - r_n)\rangle = \nabla \times \langle (v_n \times p_n)\delta(r - r_n)\rangle \ . \qquad (14.1.37)$$

If we compare this term after summation over n with the time derivative of $P(r, t)$

$$\frac{\partial}{\partial t} P(r, t) = \int d^3 k \, d\omega(-i\omega)P(k, \omega)e^{ik \cdot r - i\omega t} \ , \qquad (14.1.38)$$

we find that it is smaller by the factor

$$\sim \frac{k}{\omega}\bar{v}_n = \frac{\bar{v}_n}{\bar{c}}$$

where $\bar{v}_n$ is an average speed of the bound system, and $\bar{c}$ is given by

$$\omega = k\bar{c}$$

in analogy with the relation $\omega = kc$ for electromagnetic waves in a vacuum. In Sect. 14.4.2, we will see that $\bar{c}$ is the speed of light in matter and since $\bar{v}_n/\bar{c} \ll 1$, the term (14.1.37) can be neglected.

Finally, we still have the term

$$-\sum_j q_{nj}\dot{d}_{nj}(d_{nj} \cdot \nabla)f(r - r_n) = \langle \nabla \times m_n \delta(r - r_n(t))\rangle \ , \qquad (14.1.39)$$

where

$$m_n = \frac{1}{2}\sum_j q_{nj}(d_{nj} \times \dot{d}_{nj}) \qquad (14.1.40)$$

represents the magnetic dipole moment which is caused by the charges in the n-th molecule, since

$$\nabla \times \frac{1}{2}(d_{nj} \times \dot{d}_{nj}) = \frac{1}{2}d_{nj} \cdot (\dot{d}_{nj} \cdot \nabla) - \frac{1}{2}\dot{d}_{nj}(d_{nj} \cdot \nabla)$$

$$= -\dot{d}_{nj}(d_{nj} \cdot \nabla) + \frac{\partial}{\partial t}\left[\frac{1}{2}d_{nj}(d_{nj} \cdot \nabla)\right] \ . \qquad (14.1.41)$$

The first term in (14.1.41) is the desired one, the second belongs to the time derivative of the electric quadrupole moment. Since we need only consider the dipole moments, we can neglect this second term.

We then find

$$\langle j(r, t)\rangle = J(r, t) + \frac{\partial}{\partial t}(P(r, t) + \ldots) + \nabla \times M(r, t) + \ldots, \tag{14.1.42}$$

where $J(r, t)$ is now the current produced by the free charge carriers and the bound molecules, $P(r, t)$ is the polarization of the bound systems, and $M(r, t)$ is the *magnetization*, i.e. the sum of the magnetic moments of the individual bound systems per unit volume.

It follows then that

$$\nabla \times B - \frac{1}{c^2}\frac{\partial E}{\partial t} = \mu_0\left(J + \frac{\partial P}{\partial t} + \nabla \times M + \ldots\right) \tag{14.1.43}$$

or

$$\nabla \times (B - \mu_0 M + \ldots) - \frac{1}{c^2}\frac{\partial}{\partial t}(E + c^2\mu_0 P + \ldots) = \mu_0 J . \tag{14.1.44}$$

If we now introduce in addition to the electric displacement

$$D(r, t) = \varepsilon_0 E(r, t) + P(r, t) + \ldots$$

the *magnetic excitation* or the *magnetic field strength*

$$H(r, t) = \frac{1}{\mu_0} B(r, t) - M(r, t) + \ldots \tag{14.1.45}$$

we obtain the next macroscopic Maxwell equation

$$\nabla \times H(r, t) - \frac{\partial D(r, t)}{\partial t} = J(r, t) . \tag{14.1.46}$$

Together with (14.1.24),

$$\nabla \cdot D(r, t) = \varrho(r, t)$$

these equations represent the macroscopic inhomogeneous Maxwell equations.

Once again, we need to find a material equation for the magnetization and the magnetic field $H(r, t)$, that is, we need a relation between H and B (and eventually E), in order to find a closed system of equations. With the assumption

$$M = \chi_m H \tag{14.1.47}$$

in the static case, analogous to the assumption

$$P = \varepsilon_0 \chi E$$

we find

$$B = \mu_0(H + M) = \mu_0(H + \chi_m H) = \mu_0 \mu H \tag{14.1.48}$$

with the *magnetic permeability*

$$\mu = 1 + \chi_m \tag{14.1.49}$$

analogous to

$$\varepsilon = 1 + \chi \ .$$

It is ascertained experimentally that:

Typical values of ε are as follows: $\varepsilon = 3 - 8$ for amber or mica, 81.6 for water, and up to 2500 for ceramic materials.

Two types of molecules are distinguished:

a) Molecules whose dipole moments are first produced by an external field, and
b) Molecules with their own dipole moment.

Molecules of type (a) are called Class A dielectrics (H_2, N_2) or diamagnetic[2] substances (zinc, gold, mercury). For diamagnetic substances, $\mu \lesssim 1$.

Molecules of the second type are called polar dielectrics (e.g. H_2O, NH_3) or paramagnetic substances (Na, K, O_2, NO, Platinum). For paramagnetic materials, $\mu \gtrsim 1$.

Crystalline materials with particularly high polarizability are called ferroelectrics (Rochelle salt, BaTiO), and materials which have particularly high magnetization are called ferromagnetic[3] materials (iron, cobalt, nickel). Even without an external field, they possess macroscopic magnetic moments inside so-called magnetic domains.

14.2 Electrostatic Fields in Continuous Media

Let us again consider time independent fields, then Maxwell's equations for the electric fields read

$$\mathbf{V} \cdot D(r) = \varrho(r) \quad \text{and} \quad \mathbf{V} \times E(r) = 0 \ . \tag{14.2.1}$$

Let us assume $D(r) = \varepsilon \varepsilon_0 E(r)$ as the material equation.

---

[2] Paramegnetism, diamagnetism (Greek), something like pro- and anti-magnetic: the field produced by the polarization is in the same direction as or oppositely directed to the external field.
[3] Ferromagnetism (Latin/Greek) *ferrum*: iron. Magnetic properties like iron.

From the two Maxwell equations, it follows, in analogy with electrostatics in a vacuum, that the normal component of $D(r)$ at the boundary surface between two media which carries a surface charge density $\sigma(r)$ has a discontinuity of amount $\sigma(r)$ and that the tangential component of E is continuous, i.e.

$$[D_2(r) - D_1(r)] \cdot n(r) = \sigma(r) \quad and$$

$$[E_2(r) - E_1(r)] \times n(r) = 0 \ . \tag{14.2.2}$$

Here, $n(r)$ is again the normal vector to the surface at point r. These equations represent boundary conditions for electrostatic problems in polarizable media. We will demonstrate this in a typical application:

Consider a dielectric sphere with radius a and dielectric constant ε in a homogeneous field E_0. Let there be no free charges present, so that $\varrho(r) = 0$. Since $\nabla \times E = 0$, we again have

$$E = -\nabla\phi(r) \ , \tag{14.2.3}$$

and since $\nabla \cdot D(r) = 0$, $D = \varepsilon\varepsilon_0 E$, it follows again for the potential $\phi(r)$ that

$$\Delta\phi(r) = 0 \tag{14.2.4}$$

inside and outside the sphere.

We assume a solution of the following form:

for $r \le a$:

$$\phi = \phi_i(r, \theta) = \sum_{l=0}^{\infty} A_l r^l \mathrm{P}_l(\cos\theta) \ , \tag{14.2.5}$$

for $r \ge a$:

$$\phi = \phi_a(r, \theta) = \sum_{l=0}^{\infty} (B_l r^l + C_l r^{-l-1})\mathrm{P}_l(\cos\theta) \ . \tag{14.2.6}$$

The constants A_l, B_l, and C_l will now be determined using the boundary conditions.

Let us write the homogeneous field E_0 as $E_0 = E_0 e_3$, then we must have

a) as $r \to \infty$: $\phi_a(r, \theta) \to -E_0 z$. $\tag{14.2.7}$

It then follows

$$B_l = 0 \quad for \quad l \ne 1, \quad B_1 = -E_0 \ . \tag{14.2.8}$$

b) At $r = a$, the normal component of D increases by 0, i.e., since

$$n(r) \cdot D(r) = -\frac{r}{r} \cdot \varepsilon\varepsilon_0 \nabla\phi(r) = -\varepsilon\varepsilon_0 \frac{\partial\phi}{\partial r} \tag{14.2.9}$$

it follows that

$$-\varepsilon_0 \frac{\partial \phi_a}{\partial r}\bigg|_{r=a} = -\varepsilon\varepsilon_0 \frac{\partial \phi_i}{\partial r}\bigg|_{r=a} \tag{14.2.10}$$

c) The tangential component of $E(r)$

$$E(r)\cdot e_\theta(r) = -\frac{1}{a}\frac{\partial \phi}{\partial \theta} \tag{14.2.11}$$

is continuous, i.e.,

$$-\frac{1}{a}\frac{\partial \phi_i}{\partial \theta}\bigg|_{r=a} = -\frac{1}{a}\frac{\partial \phi_a}{\partial \theta}\bigg|_{r=a} \tag{14.2.12}$$

From (14.2.10), we get

$$-\frac{\partial}{\partial r}\left[-E_0 r P_1(\cos\theta) + \sum_l C_l r^{-l-1} P_l(\cos\theta)\right]_{r=a}$$

$$= E_0 P_1(\cos\theta) + \sum_l (l+1)C_l a^{-l-2} P_l(\cos\theta)$$

$$= -\varepsilon\sum_l l A_l a^{l-1} P_l(\cos\theta) , \tag{14.2.13}$$

so that for

$$l = 1: E_0 + 2C_1 a^{-3} = -\varepsilon A_1, \quad \text{for} \tag{14.2.14}$$

$$l \neq 1: (l+1)C_l a^{-l-2} = -\varepsilon l A_l a^{l-1} , \tag{14.2.15}$$

while from (14.2.12), we find:

$$\sum_l A_l a^l P_l'(\cos\theta) = \sum_l C_l a^{-l-1} P_l'(\cos\theta) - E_0 a P_1'(\cos\theta) . \tag{14.2.16}$$

This is an equation of the form

$$\sum_{l=1}^{\infty} c_l P_l'(\cos\theta) = 0 ,$$

from which it follows, though, that

$$\sum c_l \frac{1}{\sin\theta}\frac{\partial}{\partial\theta}\left[\sin\theta\frac{\partial}{\partial\theta}P_l(\cos\theta)\right] = -\sum_l c_l l(l+1)P_l(\cos\theta) = 0$$

and thus by the linear independence of the Legendre polynomials, $c_l = 0$. We can compare then the coefficients in (14.2.16) and we find for $l \neq 1$:

$$A_l a^l = C_l a^{-l-1} \, , \tag{14.2.17}$$

and with (14.2.15)

$$(l + 1)C_l a^{-l-2} = - \varepsilon l A_l a^{l-1} = - \varepsilon l C_l a^{-l-2} \tag{14.2.18}$$

or

$$(l + 1)C_l = - \varepsilon l C_l \, , \tag{14.2.19}$$

so that $C_l = 0$ for all $l \neq 1$. For $l = 1$, we have

$$A_1 = C_1 a^{-3} - E_0 \, , \tag{14.2.20}$$

and with (14.2.14)

$$E_0 + 2C_1 a^{-3} = - \varepsilon A_1 = - \varepsilon (C_1 a^{-3} - E_0) \tag{14.2.21}$$

or

$$(1 - \varepsilon)E_0 a^3 = - (2 + \varepsilon)C_1 \, , \tag{14.2.22}$$

i.e.,

$$C_1 = - \frac{1 - \varepsilon}{2 + \varepsilon} a^3 E_0 \tag{14.2.23}$$

and thus

$$A_1 = C_1 a^{-3} - E_0 = E_0\left(\frac{\varepsilon - 1}{\varepsilon + 2} - 1\right) \, , \tag{14.2.24}$$

so that

$$A_1 = - \frac{3}{\varepsilon + 2} E_0 \, . \tag{14.2.25}$$

Thus, the complete solution for $r \leq a$ reads:

$$\phi(r, \theta) = - \frac{3}{2 + \varepsilon} E_0 r \cos \theta = - \frac{3}{2 + \varepsilon} E_0 z \, , \tag{14.2.26}$$

and for $r \geq a$,

$$\phi(r, \theta) = -E_0 z + \frac{\varepsilon - 1}{\varepsilon + 2} E_0 \frac{a^3}{r^2} \cos\theta \ . \tag{14.2.27}$$

This means the following:

In the interior,

$$E_i = \frac{3}{\varepsilon + 2} E_0 e_3 \ , \tag{14.2.28}$$

and thus points in the direction of the external field. But since $\varepsilon > 1$, the E-field is weaker inside. As $\varepsilon \to \infty$, E_i goes to $\mathbf{0}$.

Outside the sphere,

$$\phi_a = -E_0 z + \frac{1}{4\pi\varepsilon_0} \frac{\mathbf{p} \cdot \mathbf{r}}{r^3} \tag{14.2.29}$$

with

$$\mathbf{p} = 4\pi\varepsilon_0 \frac{\varepsilon - 1}{\varepsilon + 2} E_0 a^3 e_3 \ . \tag{14.2.30}$$

The field is a superposition of the original field and the field of a dipole in the z-direction. This dipole moment $\mathbf{p}$ is produced by the polarization of the bound systems in the inside of the sphere.

Inside, we have

$$\mathbf{D} = \varepsilon\varepsilon_0 \mathbf{E} = \varepsilon_0 \mathbf{E} + \mathbf{P} \qquad \text{thus}$$

$$\mathbf{P} = \varepsilon_0(\varepsilon - 1)\mathbf{E} = \frac{3}{4\pi} 4\pi\varepsilon_0 \frac{\varepsilon - 1}{\varepsilon + 2} E_0 e_3 \ . \tag{14.2.31}$$

The dipole moment density $\mathbf{P}$ is constant in this case, and the total dipole moment of the sphere then agrees with (14.2.30):

$$\mathbf{p} = \frac{4\pi}{3} a^3 \mathbf{P} = 4\pi\varepsilon_0 \frac{\varepsilon - 1}{\varepsilon + 2} E_0 a^3 e_3 \ . \tag{14.2.32}$$

The external field causes the polarization of the bound systems within the dielectric. The P-field which is produced in this way is parallel to the E-field, so that the polarized bound complexes produce an oppositely directed field which weakens the original field (Fig. 14.2.1).

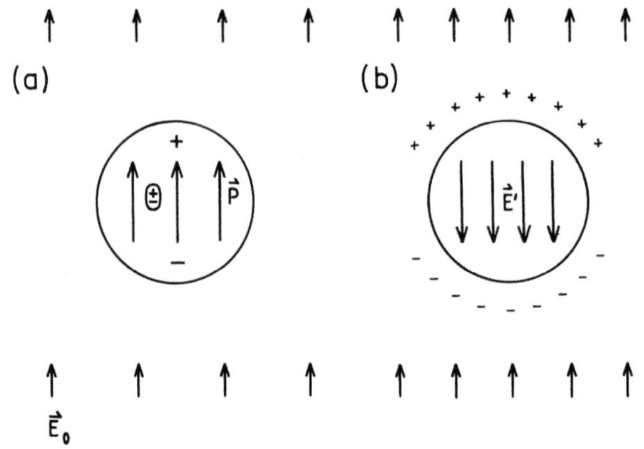

Fig. 14.2.1a, b. In an electric field, the bound systems of charges in a dielectric are polarized. A polarization density **P** arises, and the separated charges produce an oppositely directed field **E′** which weakens the original field

14.3 Magnetostatic Fields in Continuous Media

Maxwell's equations for magnetostatics read

$$\nabla \cdot \boldsymbol{B}(\boldsymbol{r}) = 0 \ , \quad \nabla \times \boldsymbol{H}(\boldsymbol{r}) = \boldsymbol{J}(\boldsymbol{r}) \ . \tag{14.3.1}$$

We assume that the material equation is

$$\boldsymbol{B}(\boldsymbol{r}) = \mu\mu_0 \boldsymbol{H}(\boldsymbol{r}) \ .$$

Most of the time, magnetostatic problems can be solved in a manner completely analogous to electrostatic problems. We will illustrate this in the following examples or statements.

i) It follows from $\nabla \cdot \boldsymbol{B}(\boldsymbol{r}) = 0$ that:

*On the boundary surface between two media with different permeabilities μ, the normal component of **B** is continuous. This is shown in the same way as the corresponding statement for **D**(r).*

*In the absence of surface currents, the tangential component of **H** is constant. This follows from $\nabla \times \boldsymbol{H} = \boldsymbol{0}$.*

ii) Consider a sphere with permeability μ in a homogeneous magnetic induction field $\boldsymbol{B}_0$, located in non-permeable space. Then the system of equations to be solved is

$$\nabla \cdot \boldsymbol{B}(\boldsymbol{r}) = 0 \ , \quad \nabla \times \boldsymbol{H}(\boldsymbol{r}) = \boldsymbol{0} \ ,$$

$$\boldsymbol{B} = \mu\mu_0 \boldsymbol{H} = \mu_0 \boldsymbol{H} + \mu_0 \boldsymbol{M} \ . \tag{14.3.2}$$

We now compare this to the problem considered in Sect. 14.2.

$$\nabla \cdot D(r) = 0 \ , \qquad \nabla \times E(r) = 0 \ ,$$
$$D = \varepsilon\varepsilon_0 E = \varepsilon_0 E + P \ . \tag{14.3.3}$$

We can immediately solve the magnetostatic problem by substituting the corresponding quantities into the solution of the electrostatic problem

$$D \to B \ , \quad E \to H \ , \quad P \to \mu_0 M \ , \quad \varepsilon, \varepsilon_0 \to \mu, \mu_0 \ . \tag{14.3.4}$$

We then find

$$\mu_0 M = \frac{3}{4\pi} 4\pi\mu_0 \frac{\mu - 1}{\mu + 2} H_0 \ , \quad \text{thus} \tag{14.3.5}$$

$$M = 3\frac{\mu - 1}{\mu + 2} H_0 \ ; \tag{14.3.6}$$

$$H_i = \frac{3}{\mu + 2} H_0 \ , \quad \text{thus} \tag{14.3.7}$$

$$B_i = \frac{3\mu}{\mu + 2} B_0 \ . \tag{14.3.8}$$

iii) A further, very illustrative example is the following:

In space, at first empty, let there be a field B_0. We then introduce a hollow spherical shell into the space whose permeability is μ. We expect a screening of the magnetic field in the interior of the shell.

Let the inner radius be a, the outer radius b. We must solve the system of equations (14.3.2) outside the shell, within the shell, and inside the shell.

From $\nabla \times H(r) = 0$ we conclude: there exists a ϕ_{M} with

$$H = -\nabla\phi_{\mathrm{M}} \ , \tag{14.3.9}$$

and from $\nabla \cdot B(r) = 0$, it immediately follows that

$$\Delta\phi_{\mathrm{M}} = 0 \ . \tag{14.3.10}$$

We assume solutions for ϕ_{M} of the form:

$$r \leq a: \quad \phi_{\mathrm{M}} = \sum_l \delta_l r^l \mathrm{P}_l(\cos\theta) \ , \tag{14.3.11}$$

$$a \leq r \leq b: \quad \phi_{\mathrm{M}} = \sum_l (\beta_l r^l + \gamma_l r^{-l-1})\mathrm{P}_l(\cos\theta) \ , \tag{14.3.12}$$

$$b \leq r: \quad \phi_{\mathrm{M}} = \sum_l \alpha_l r^{-l-1}\mathrm{P}_l(\cos\theta) - H_0 r\cos\theta \ , \tag{14.3.13}$$

where we have chosen the z-direction in the direction of $\boldsymbol{H}_0$. From (i), we know that the tangential components of $\boldsymbol{H}$ and the normal components of $\boldsymbol{B}$ must be continuous. It follows that

$$\frac{\partial \phi_{\mathrm{M}}}{\partial \theta}\bigg|_{b+0} = \frac{\partial \phi_{\mathrm{M}}}{\partial \theta}\bigg|_{b-0} \,, \qquad \frac{\partial \phi_{\mathrm{M}}}{\partial \theta}\bigg|_{a+0} = \frac{\partial \phi_{\mathrm{M}}}{\partial \theta}\bigg|_{a-0} \,, \tag{14.3.14}$$

for the tangential components of $\boldsymbol{H}(\boldsymbol{r})$ and

$$\frac{\partial \phi_{\mathrm{M}}}{\partial r}\bigg|_{b+0} = \mu \frac{\partial \phi_{\mathrm{M}}}{\partial r}\bigg|_{b-0} \,, \qquad \mu \frac{\partial \phi_{\mathrm{M}}}{\partial r}\bigg|_{a+0} = \frac{\partial \phi_{\mathrm{M}}}{\partial r}\bigg|_{a-0} \,, \tag{14.3.15}$$

for the normal components of $\boldsymbol{B}(\boldsymbol{r})$.

We have four equations here from which we can determine the $\alpha_l, \beta_l, \ldots$. It turns out that the coefficients vanish for all $l \neq 1$, and for $l = 1$, we find for example, that

$$\alpha_1 = \frac{(2\mu + 1)(\mu - 1)}{(2\mu + 1)(\mu + 2) - 2(a^3/b^3)(\mu - 1)^2}(b^3 - a^3)H_0 \,, \tag{14.3.16}$$

$$\delta_1 = \frac{-9\mu}{(2\mu + 1)(\mu + 2) - 2(a^3/b^3)(\mu - 1)^2}H_0 \,, \tag{14.3.17}$$

so that, for $r \geq b$

$$\phi_{\mathrm{M}} = -H_0 z + \alpha_1 \frac{\mathrm{P}_1(\cos\theta)}{r^2} = -H_0 z + \frac{\boldsymbol{m} \cdot \boldsymbol{r}}{r^3} \quad \text{with} \tag{14.3.18}$$

$$\boldsymbol{m} = \alpha_1 \boldsymbol{e}_3 \,. \tag{14.3.19}$$

The external field again is a superposition of the original field and a new dipole field. As $a \to 0$, we obtain a result which we already know from the electrostatic case.

Within the shell, we find

$$\phi_{\mathrm{M}} = \delta_1 r \cos\theta = \delta_1 z$$

$$= -\frac{9\mu}{(2\mu + 1)(\mu + 2) - 2(a^3/b^3)(\mu - 1)^2}H_0 z \,, \tag{14.3.20}$$

that is, the internal field is parallel to the external field and is of the order μ^{-1}. A material with $\mu \approx 10^3$ to 10^6 thus produces quite a strong magnetic screening. Of course, for $\mu = 1$, $\phi_{\mathrm{M}} = -H_0 z$.

For β_1 and γ_1, we find

$$\beta_1 = \frac{-3(1 + 2\mu)}{(\mu + 2)(1 + 2\mu) - 2(a^3/b^3)(\mu - 1)^2} H_0 \;, \tag{14.3.21}$$

$$\gamma_1 = \frac{-3(1 - \mu)}{(\mu + 2)(1 + 2\mu) - 2(a^3/b^3)(\mu - 1)^2} a^3 H_0 \;. \tag{14.3.22}$$

We can easily verify that as $a \to 0$, the field in the sphere is again given by:

$$\phi_{\mathrm{M}} = - \frac{3}{\mu + 2} H_0 z \quad \text{or}$$

$$\tag{14.3.23}$$

$$H_i = \frac{3}{\mu + 2} H_0$$

in agreement with (ii).

14.4 Plane Waves in Matter, Wave Packets

In Sect. 14.1.2, we noted that for isotropic materials in the stationary case, the linear equation

$$P(r) = \varepsilon_0 \chi E(r) \tag{14.4.1}$$

is often valid. For time dependent fields, we expect a polarization which oscillates with the frequency of $E(r, t)$ but with a phase displacement, which as we have seen in Sect. 6.5.3 is the general response of a linear system to an external periodic force. To best formulate the more general connection between polarization and electrical field, we use the corresponding Fourier transforms. If we also consider anisotropic media, say polymers, in which some direction can be privileged due to the long chains in the molecular structure, the general relation will read

$$P_i(k, \omega) = \varepsilon_0 \chi_{ij}(\omega) E_j(k, \omega) + \varepsilon_0 \int d^3 k' d\omega'$$

$$\times \chi_{ijk}(k, k', \omega, \omega') E_j(k', \omega')$$

$$\times E_k(k - k', \omega - \omega') + \dots \,, \tag{14.4.2}$$

where we have also considered nonlinear terms. These terms play a role, for instance, in *nonlinear optics*, in which electric fields are considered which are no longer small in comparison to the forces inside of the molecules. We will neglect these nonlinear terms in the following sections.

The susceptibility becomes a tensor when considering anisotropic materials, and taking into account time dependence it becomes a complex function of the frequency ω.

14.4.1   The Frequency Dependence of Susceptibility

Here, we will consider a very simple model for the frequency dependence of susceptibility and the dielectric constant ε.

Let $r(r)$ be the displacement from equilibrium of a charge carrier with charge q and mass m due to the field $E(r, t) = E_0\exp(-i\omega t)$. Then, from Chap. 6, a linear approximation for $r(r)$ is given by

$$m[\ddot{r}(t) + \gamma\dot{r}(t) + \omega_0^2 r(t)] = qE_0 e^{-i\omega t} \; . \tag{14.4.3}$$

Here, $\gamma\dot{r}$ is a damping term which describes the damping of the motion by collisions, radiation, etc. ω_0 here is the characteristic frequency of the oscillation. With $r(t) = r_0\exp(-i\omega t)$, the amplitude r_0 must satisfy

$$m(-\omega^2 - i\gamma\omega + \omega_0^2)r_0 = qE_0 \; , \tag{14.4.4}$$

and we have a dipole moment $p(t) = qr(t) = p_0\exp(-i\omega t)$ with

$$p_0 = qr_0 = \frac{q^2 E_0}{m}\frac{1}{\omega_0^2 - \omega^2 - i\gamma\omega} \; . \tag{14.4.5}$$

Let f_j charge carriers be present in the bound system with characteristic frequency ω_j and damping constants γ_j. Then, it follows, if there are n bound systems per unit volume, that

$$P = \frac{nq^2}{m}\sum_j f_j \frac{1}{\omega_j^2 - \omega^2 - i\gamma_j\omega} E(t) = \varepsilon_0\chi E(t) \; . \tag{14.4.6}$$

Thus we find for a wave of fixed frequency

$$\varepsilon(\omega) = 1 + \chi(\omega) = 1 + \frac{nq^2}{m\varepsilon_0}\sum_j f_j \frac{1}{\omega_j^2 - \omega^2 - i\gamma_j\omega} \; . \tag{14.4.7}$$

The dielectric constant ε is thus in general complex. Quite often the frequency dependence and the imaginary part of the magnetic permeability can be ignored.

The quantity

$$\alpha = 2\,\mathrm{Im}\{\sqrt{\mu\varepsilon(\omega)}\}\cdot\omega/c \tag{14.4.8}$$

is called the *absorption coefficient* and

$$n = \mathrm{Re}\{\sqrt{\mu\varepsilon(\omega)}\} \tag{14.4.9}$$

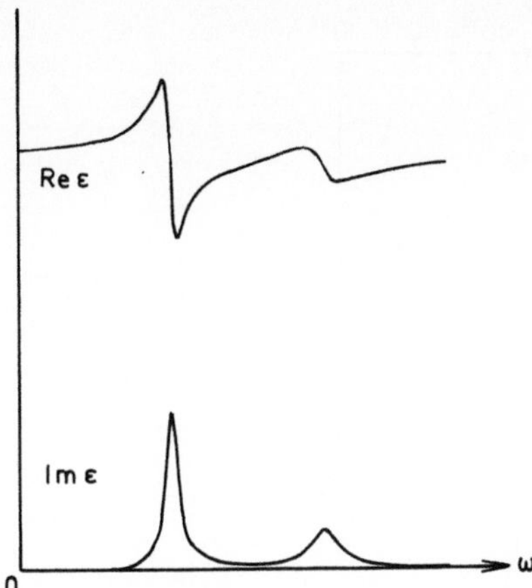

Fig. 14.4.1. Typical behavior of the real and imaginary parts of the dielectric constant $\varepsilon(\omega)$

is called the *index of refraction* (the reason for this name will become clear below). The imaginary part is, of course, particularly large when $\omega \approx \omega_j$, that is, at a frequency which corresponds to the characteristic frequency of the system. The system, which can oscillate, will then absorb much energy. For example, water has a minimal absorption coefficient for visible light, but a maximal coefficient for microwaves.

In Fig. 14.4.1, typical curves for $\mathrm{Re}\{\varepsilon\}$ and $\mathrm{Im}\{\varepsilon\}$ are shown. In the frequency regions where absorption cannot be ignored, we speak of an *anomalous dispersion*, otherwise of a *normal dispersion*. For anomalous dispersion,

$$dn/d\omega < 0 \ .$$

The corresponding curves for water are displayed in Fig. 14.4.2. The static value, that is, the value of ε for static fields, can be found at $\omega = 0$. We have

$$\lim_{\omega \to \infty} \mathrm{Im}\{\varepsilon(\omega)\} = 0 \ , \tag{14.4.10}$$

and when we speak simply of the dielectric constant, most of the time we mean $\mathrm{Re}\{\varepsilon(0)\}$.

So far, we have assumed, also in the simple model, that the system can only absorb energy. As in the model, this implies that $\mathrm{Im}\{\varepsilon(\omega)\} \geq 0$. In the following, we will always maintain this assumption, although it should be noted that lasers and masers radiate energy at resonance frequencies.

We now consider the time dependent Maxwell equations in terms of the Fourier-transformed functions. In regions in which ϱ and $\boldsymbol{J}$ vanish, we have

$$\boldsymbol{k} \cdot \boldsymbol{B}(\boldsymbol{k}, \omega) = 0 \ , \quad \boldsymbol{k} \times \boldsymbol{E}(\boldsymbol{k}, \omega) - \omega \boldsymbol{B}(\boldsymbol{k}, \omega) = \boldsymbol{0} \ ,$$

$$\boldsymbol{k} \cdot \boldsymbol{D}(\boldsymbol{k}, \omega) = 0 \ , \quad \boldsymbol{k} \times \boldsymbol{H}(\boldsymbol{k}, \omega) + \omega \boldsymbol{D}(\boldsymbol{k}, \omega) = \boldsymbol{0} \ . \tag{14.4.11}$$

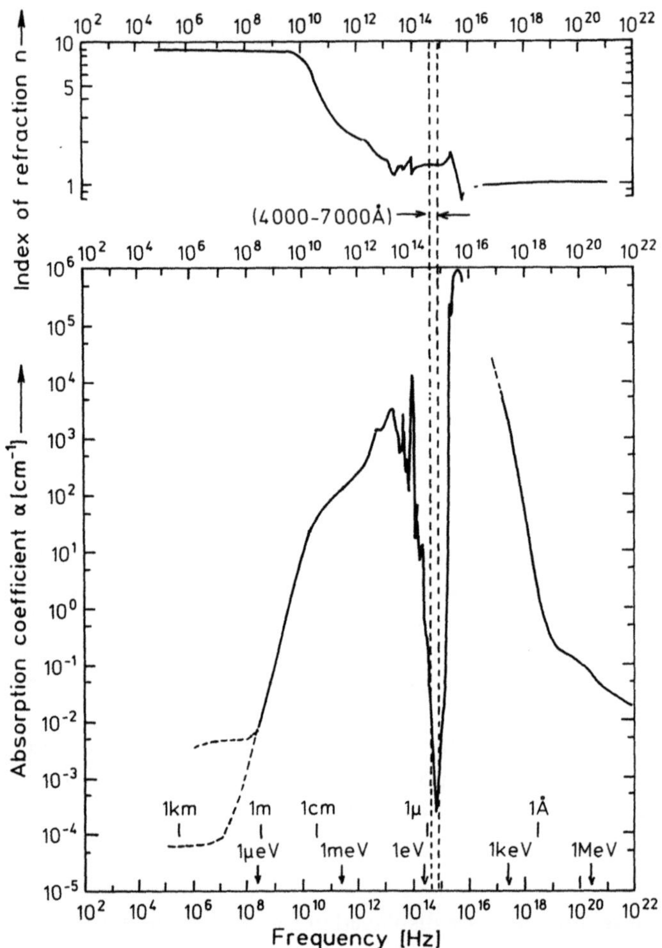

Fig. 14.4.2. Index of refraction and absorption coefficient for water

If we use the material equations

$$D(k, \omega) = \varepsilon_0 \varepsilon(\omega) E(k, \omega) \quad \text{and}$$

$$H(k, \omega) = B(k, \omega)/\mu\mu_0$$

we find now from Maxwell's equations:

$$k \times [k \times E(k, \omega)] = -k^2 E(k, \omega)$$

$$= -\frac{\omega^2 \mu\varepsilon(\omega)}{c^2} E(k, \omega) \ , \tag{14.4.12}$$

that is, we obtain a relation between ω and k similar to the equation $k = \omega/c$

$$k = \frac{\omega}{c} \sqrt{\mu \varepsilon(\omega)} \; . \tag{14.4.13}$$

The right hand side is a complex function of ω. In the equation

$$E(r, t) \sim e^{ik \cdot r - i\omega t} \tag{14.4.14}$$

with $k = k n$, the imaginary part of k signifies an exponentially decaying or growing factor. An exponential decay describes the absorption of radiation by the medium.

14.4.2   Wave Packets, Phase and Group Velocity

In the following, we will limit ourselves to the frequency domain of normal dispersion, so that we ignore the imaginary part α. Then, k is also real.

If we think of the equation (14.4.13) as solved for ω, we find

$$\omega = \omega(k) \; , \tag{14.4.15}$$

and for the time dependent field $E(r, t)$, for example, we find a plane wave of the form

$$E(r, t) = E(k) e^{i(k \cdot r - i\omega(k)t)} \; .$$

The *phase velocity* of this wave, that is, the speed with which a point of constant phase moves along is then

$$v_{\mathrm{P}} = \frac{\omega(k)}{k} \; .$$

Plane waves with a single fixed wave vector k are called *monochromatic*[5]. Monochromatic waves are unbounded in time and space. In nature, waves are always bounded in time and space and can be represented as superpositions of monochromatic waves. We then speak of a *wave packet*:

$$E(r, t) = \int d^3 k \, E(k) e^{i(k \cdot r - \omega(k)t)} \; .$$

The phase velocity, given by $v_{\mathrm{P}} = \omega(k)/k$ is different for different values of k. In a wave packet, the relative phases of the individual components vary. This has consequences for the wave train. We will study this for the one-dimensional

---

[5] Monochromatic (Greek) *monos*: single; *chroma*: color, thus "monochromatic" light is a single color since it contains waves of a single frequency.

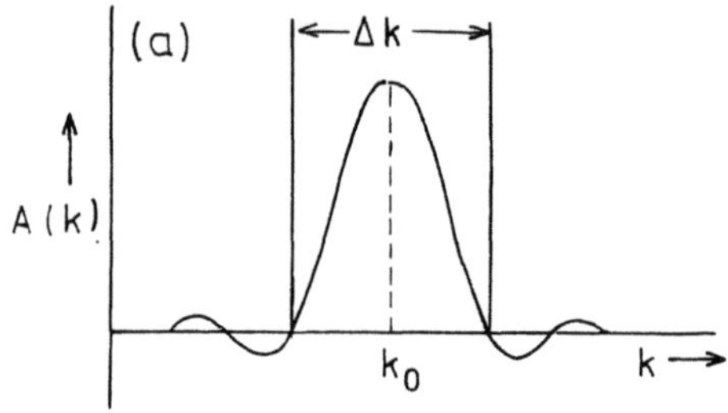

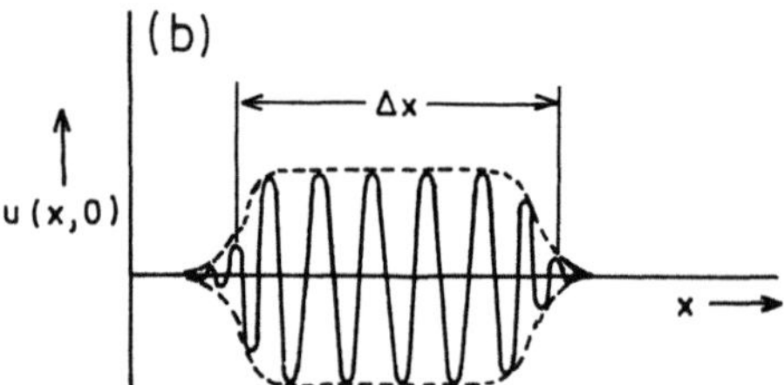

Fig. 14.4.3. A Fourier transform $A(k)$ with limited width and the corresponding wave packet

example

$$u(x, t) = \frac{1}{\sqrt{2\pi}} \int dk\, A(k) e^{i(kx - \omega(k)t)} \tag{14.4.16}$$

which will introduce us to three important characteristics of wave packets.

a) The group velocity:

We assume that $A(k)$ be a function which has a maximum at k_0, but which falls off quickly with increasing distance from k_0 (Fig. 14.4.3a).

We then expand $\omega(k)$ in the exponential term in (14.4.16) around k_0:

$$\omega(k) = \omega(k_0) + (k - k_0)v_G + O((k - k_0)^2) \tag{14.4.17}$$

with

$$v_G = \frac{d\omega(k)}{dk}\bigg|_{k_0}, \tag{14.4.18}$$

and thus we find for $u(x, t)$:

$$u(x, t) = \frac{1}{\sqrt{2\pi}} \exp[ik_0 v_G t - i\omega(k_0)t]$$

$$\times \int dk\, A(k) \exp[ik(x - v_G t)] . \tag{14.4.19}$$

If we ignore the terms of order $(k - k_0)^2$, we have

$$u(x, t) = u(x - v_G t, 0) \exp[ik_0 v_G t - i\omega(k_0)t] . \tag{14.4.20}$$

$u(x, t)$ thus represents an excitation which propagates with speed v_G and maintains the form of $u(x, 0)$. v_G is called the group velocity. We see immediately that for $\varepsilon\mu = 1$, which means $\omega = ck$, that

$$\frac{d\omega(k)}{dk} = c \ ,$$

and the group velocity is identical to the phase velocity $\omega(k)/k$. In any medium with $\varepsilon\mu \neq 1$, though, the two velocities are different.

For the phase velocity, we find then

$$v_P = \frac{\omega(k)}{k} = \frac{c}{n} \quad \text{with} \quad n = \sqrt{\varepsilon\mu} \ , \tag{14.4.21}$$

and v_P is smaller than or greater than c according to whether n is smaller or greater than 1.

For the group velocity v_G, we find from

$$\omega = \frac{ck}{n} \quad \text{or} \quad n(\omega)\omega = ck, \quad \text{respectively}$$

by differentiating with respect to k

$$\left[n(\omega) + \omega\frac{dn}{d\omega} \right]\frac{d\omega}{dk} = c \ , \quad \text{that}$$

$$v_G = \frac{d\omega}{dk} = \frac{c}{n(\omega) + \omega\, dn/d\omega} \ . \tag{14.4.22}$$

For normal dispersion, $dn/d\omega > 0$ and $n > 1$, so that

$$v_G < v_P = c/n < c \ . \tag{14.4.23}$$

For anomalous dispersion, that is, if $dn/d\omega < 0$, the imaginary part of $\sqrt{\mu\varepsilon}(\omega)$ can no longer be ignored, and the group velocity is no longer a physically useful quantity.

b) Let $A(k)$ and $u(x, 0)$ be bounded wave packets with widths Δx and Δk as shown, say, in Fig. 14.4.3. Then, we find the following result:
There is a lower bound for the product

$$\Delta k\, \Delta x \ ,$$

so that, e.g., Δx must increase for smaller values of Δk. The following are the limiting cases:

i) Let $A(k) = \delta(k - k_0) + \delta(k + k_0)$, then $u(x, 0) = e^{ik_0 x} + e^{-ik_0 x} \sim \cos k_0 x$, and thus $u(x, 0)$ is unbounded, i.e. $\Delta x = \infty$, while $\Delta k = 0$.

ii) Let $A(k) = 1$, i.e., $A(k)$ is unbounded, $\Delta k = \infty$, so that

$$u(x, 0) \sim \delta(x) , \quad \text{i.e.} \quad \Delta x = 0 .$$

In the theory of Fourier transforms, the following result can be shown: Let

$$\langle x^n \rangle = \int dx \, |u(x, 0)|^2 x^n , \tag{14.4.24}$$

$$\langle k^n \rangle = \int dk \, |A(k)|^2 k^n \tag{14.4.25}$$

be given with

$$u(x, 0) = \frac{1}{\sqrt{2\pi}} \int dk \, A(k) e^{ikx} ,$$

and let the width of the wave packets Δx and Δk be defined as

$$(\Delta x)^2 = \langle x^2 \rangle - \langle x \rangle^2 = \langle (x - \langle x \rangle)^2 \rangle , \tag{14.4.26}$$

$$(\Delta k)^2 = \langle k^2 \rangle - \langle k \rangle^2 = \langle (k - \langle k \rangle)^2 \rangle , \tag{14.4.27}$$

then

$$\Delta x \, \Delta k \geq \tfrac{1}{2} . \tag{14.4.28}$$

The proof for this inequality is usually given in quantum mechanics. There, the wave vector k is connected to the momentum operator of a particle by the de Broglie equation

$$p = \hbar k . \tag{14.4.29}$$

The result

$$\Delta p \, \Delta x \geq \hbar/2 \tag{14.4.30}$$

is then the famous uncertainty principle. This implies that if we want to construct a theory in which this kind of relationship holds between the variables x and p, the basic equation must presumably be a wave equation. The connection between the momentum vector and the wave vector then implies that the momentum can also be represented as

$$p = \frac{\hbar}{i} \frac{\partial}{\partial r} .$$

c) A wave packet with $v_G \neq v_p$ disperses, i.e. the width Δx increases with time, and in fact, the smaller Δx is at $t = 0$, the faster it disperses. This process is called *dispersion*. We will study one example of this (see, for instance, [10.5]).

Let

$$A(k) = \frac{L}{2} \exp[-L^2(k - k_0)^2/2] \tag{14.4.31}$$

and

$$\omega(k) = v(1 + \tfrac{1}{2}a^2k^2) . \tag{14.4.32}$$

Then $u(x, t)$ can be explicitly calculated

$$u(x, t) = \frac{1}{\sqrt{2\pi}} \int dk\, A(k) \exp[ikx - iv(1 + a^2k^2/2)t]$$

$$= \frac{1}{2} \frac{\exp[-(x - v_G t)^2/2L^2 R^2(t)]}{R(t)} \exp[ik_0 x - i\omega(k_0)t] \tag{14.4.33}$$

with

$$R(t) = \left(1 + i\frac{a^2 vt}{L^2}\right)^{1/2} , \qquad v_G = va^2 k_0 . \tag{14.4.34}$$

We have

$$u(x, 0) = \tfrac{1}{2}e^{-x^2/2L^2}e^{ik_0 x} , \qquad \text{thus}$$

$$\mathrm{Re}\{u(x, 0)\} \sim e^{-x^2/2L^2} \cos k_0 x . \tag{14.4.35}$$

We find:

i) $\Delta x = L/\sqrt{2}$, $\Delta k = 1/L\sqrt{2}$, and $\Delta x \Delta k = \tfrac{1}{2}$ for $t = 0$.

ii) For $t > 0$, the maximum moves with group velocity v_G.

iii) The width Δx changes with time. It is given by

$$\Delta x \sim L(t) \equiv \frac{L}{\sqrt{2}}[R(t)R^*(t)]^{1/2} = \frac{L}{\sqrt{2}}\left[1 + \left(\frac{a^2 vt}{L^2}\right)^2\right] , \tag{14.4.36}$$

that is, $L(t)$ becomes greater, and the smaller $L \sim L(0)$ is in comparison to a, the faster the pace at which this occurs (Fig. 14.4.4). For $a \ll L$, this dispersion is only noticeable over a very large period of time. Then, for smaller time intervals, everything that was said in (a) is valid. For smaller values of L, $u(x, 0)$ becomes narrower, the $A(k)$ curve wider, and the terms neglected in remark (a) in (14.4.17) become relevant.

The more rapid dispersion of the wave packet for smaller values of L is understandable, since the smaller L is, the larger the region in k-space must be which contributes to the wave packet. Thus, waves with very different phase velocities contribute to the wave packet.

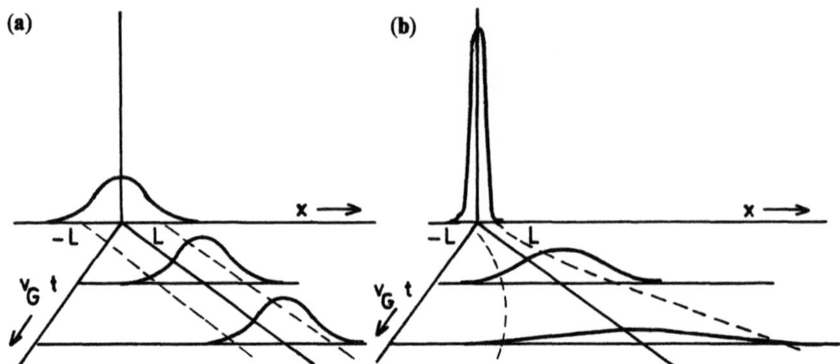

Fig. 14.4.4a, b. Dispersion of a wave packet. The smaller the width of the wave packet is at the beginning, the stronger the dispersion

14.5 Reflection and Refraction at Plane Boundary Surfaces

14.5.1 Boundary Conditions, the Laws of Reflection and Refraction

In this section, we will study how electromagnetic waves behave at plane boundary interfaces between two different dielectrics. We assume that the boundary surface is the $z = 0$ plane and we consider the incidence of the plane waves

$$E(r, t) = E_0 e^{i(k \cdot r - \omega t)} \tag{14.5.1}$$

$$B(r, t) = B_0 e^{i(k \cdot r - \omega t)} \tag{14.5.2}$$

on this boundary surface (Fig. 14.5.1). Since

$$\partial B/\partial t = - \nabla \times E = - ik \times E = - i\omega B \tag{14.5.3}$$

it follows for $\mathrm{Im}\{\mu\varepsilon(\omega)\} = 0$ that

$$B_0 = \frac{n}{c} \frac{1}{k} (k \times E_0) , \tag{14.5.4}$$

since

$$\omega = \frac{kc}{n} .$$

The incident wave continues on partly as a refracted wave, partly as a reflected wave. We write for the refracted wave

$$E'(r, t) = E_0' e^{i(k' \cdot r - \omega' t)} , \tag{14.5.5}$$

$$B'(r, t) = \frac{n'}{c} \frac{1}{k'} (k' \times E') \tag{14.5.6}$$

and for the reflected wave

$$E''(r, t) = E_0'' e^{i(k'' \cdot r - \omega'' t)} \; , \tag{14.5.7}$$

$$B''(r, t) = \frac{n}{c} \frac{1}{k''} (k'' \times E'') \tag{14.5.8}$$

with $\omega' = \omega(k')$, $\omega'' = \omega(k'')$. In Sect. 11.1.3, in the case of electrostatics, we concluded from the equation $\nabla \times E(r) = 0$ that the tangential component of $E(r)$ at the boundary surface between two media must be continuous. Here, we have

$$\nabla \times E(r, t) = -\frac{\partial B(r, t)}{\partial t} \; .$$

Integrating over a surface like the one in Fig. 11.1.3b yields

$$\int_A dA \cdot (\nabla \times E) = \oint_{\partial A} dr \cdot E = -\frac{d}{dt} \int_A dA \cdot B = -\dot{\Phi} \; , \tag{14.5.9}$$

but the additional contribution $-\dot{\Phi}$ is of the order of magnitude of the area and thus vanishes if we make it arbitrarily small, as in the argumentation of Sect. 11.1.3. Thus, if follows here too, that

At all times and all points on the plane, the tangential component of E at $z = 0$ must be continuous.

This means that for $z = 0$,

$$e_3 \times [E_0 e^{i(k \cdot r - \omega t)} + E_0'' e^{i(k'' \cdot r - \omega'' t)}] = (e_3 \times E_0') e^{i(k' \cdot r - \omega' t)} \; . \tag{14.5.10}$$

It follows that

a) All frequencies must be equal: $\omega = \omega' = \omega''$, i.e.

$$\frac{k}{n} = \frac{k'}{n'} = \frac{k''}{n} \quad \text{or} \tag{14.5.11}$$

$$k = k'' \; , \quad \frac{k'}{k} = \frac{n'}{n} \; . \tag{14.5.12}$$

b) Let the x-direction be such that k lies in the xz-plane. Then, writing $k = (k_1, k_2, k_3)$, etc., we have, since the position dependence of the exponential must be the same on both sides,

$$k_1 x_1 = k_1'' x_1 + k_2'' x_2 = k_1' x_1 + k_2' x_2 \; , \tag{14.5.13}$$

so that

$$k_2' = k_2'' = 0 \; , \tag{14.5.14}$$

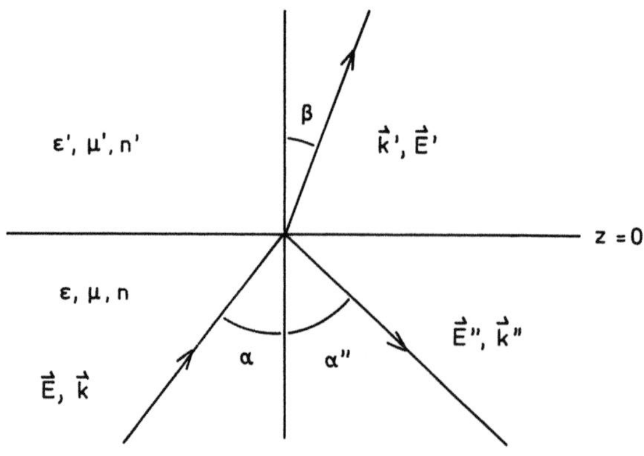

Fig. 14.5.1. A plane wave E with wave vector k is incident on a plane boundary surface between two media. The result is a reflected wave E'' and a refracted wave E'

i.e., the vectors k, k', and k'' lie in a single plane and

$$k_1 = k_1' = k_1'' \ . \tag{14.5.15}$$

c) Furthermore, let (see Fig. 14.5.1).

$$k_1 = k \sin \alpha \ , \tag{14.5.16}$$

$$k_1' = k' \sin \beta \ , \tag{14.5.17}$$

$$k_1'' = k'' \sin \alpha'' \ . \tag{14.5.18}$$

It then follows from $k_1 = k_1''$ that

$$\alpha = \alpha'' \ , \tag{14.5.19}$$

and hence we find the law

angle of incidence = angle of reflection.

From $k_1 = k_1'$, it follows further that

$$\frac{\sin \alpha}{\sin \beta} = \frac{k'}{k} = \frac{n'}{n} \ . \tag{14.5.20}$$

This is *Snell's*[6] *law.*

14.5.2   Fresnel's Equations

Since we have taken care of the exponential terms, we now examine the amplitudes. They must satisfy:

$$e_3 \times [E_0 + E_0'' - E_0'] = 0 \ . \tag{14.5.21}$$

---

[6] *Snellius (Snell van Roigen, Willibrord)* (*1561 Leiden, d. 1626 Leiden). Dutch astronomer and mathematician, professor in Leiden. He discovered his law of refraction in the year 1621.

Now, the tangential component of H must be continuous, since

$$H = \frac{1}{\mu\mu_0} B = \frac{1}{\mu\mu_0} \frac{n}{ck} (k \times E) \ . \tag{14.5.22}$$

It then follows that

$$e_3 \times \left[\frac{1}{\mu} k \times E_0 + \frac{1}{\mu} k'' \times E_0'' - \frac{1}{\mu'} k' \times E_0' \right] = 0 \ . \tag{14.5.23}$$

Note here that

$$\frac{n}{k} = \frac{n'}{k'} = \frac{n}{k''} \ .$$

Finally, the normal component of B is continuous:

$$e_3 \cdot [k \times E_0 + k'' \times E_0'' - k' \times E_0'] = 0 \ , \tag{14.5.24}$$

and the normal component of D must also be continuous:

$$e_3 \cdot [\varepsilon(E_0 + E_0'') - \varepsilon' E_0'] = 0 \ . \tag{14.5.25}$$

In (14.5.21, 23–25), we have eight equations for the unknowns E_0' and E_0''. To solve these equations, we consider the following cases:

a) Let E be linearly polarized, orthogonal to the plane of incidence, which is spanned by k and e_3 (the xz-plane), i.e. let

$$E_0 = E_2 e_2 \ . \tag{14.5.26}$$

With

$$E_0' = \sum_{i=1}^{3} E_i' e_i \ , \tag{14.5.27}$$

$$E_0'' = \sum_{i=1}^{3} E_i'' e_i \ , \tag{14.5.28}$$

it follows from (14.5.21) that

$$E_2 + E_2'' - E_2' = 0 \quad \text{and} \tag{14.5.29}$$

$$E_1'' - E_1' = 0 \ . \tag{14.5.30}$$

From (14.5.23) it then follows that

$$\frac{1}{\mu} (k_2 E_3 - k_3 E_2) + \frac{1}{\mu} (k_2'' E_3'' - k_3'' E_2'') - \frac{1}{\mu'} (k_2' E_3' - k_3' E_2') = 0 \tag{14.5.31}$$

and

$$\frac{1}{\mu}(k_3 E_1 - k_1 E_3) + \frac{1}{\mu}(k_3'' E_1'' - k_1'' E_3'') - \frac{1}{\mu'}(k_3' E_1' - k_1' E_3') = 0 \ . \qquad (14.5.32)$$

Now, we have assumed that $E_1 = E_3 = 0$, $E_1' = E_1''$, $k_2 = k_2' = k_2'' = 0$, and since $\mathbf{V} \cdot \mathbf{E} = 0$, we must also have

$$k_1' E_1' = -k_3' E_3' \ , \qquad k_1'' E_1'' = -k_3'' E_3'' \ .$$

Then, in (14.5.32), all the E_i' and E_i'' can be expressed in terms of E_1'' and it follows that

$$E_1' = E_1'' = E_3' = E_3'' = 0 \ . \qquad (14.5.33)$$

From (14.5.31) we then find

$$-\frac{1}{\mu} k_3 E_2 - \frac{1}{\mu} k_3'' E_2'' + \frac{1}{\mu'} k_3' E_2' = 0 \ , \qquad (14.5.34)$$

or, since

$$k_3 = k \cos\alpha = \frac{\omega}{c} n \cos\alpha \ , \qquad (14.5.35)$$

$$k_3'' = -k \cos\alpha = -\frac{\omega}{c} n \cos\alpha \ , \qquad \text{and} \qquad (14.5.36)$$

$$k_3' = k' \cos\beta = \frac{\omega}{c} n' \cos\beta \qquad (14.5.37)$$

it follows, with (14.5.29), that:

$$-\frac{1}{\mu} k \cos\alpha E_2 + \frac{1}{\mu} k \cos\alpha E_2'' + \frac{1}{\mu'} k' \cos\beta (E_2 + E_2'') = 0 \qquad (14.5.38)$$

or

$$\frac{E_2''}{E_2} = \frac{|E_0''|}{|E_0|} = \frac{n \cos\alpha - (\mu/\mu') n' \cos\beta}{n \cos\alpha + (\mu/\mu') n' \cos\beta} \qquad (14.5.39)$$

and then

$$\frac{E_2'}{E_2} = \frac{|E_0'|}{|E_0|} = 1 + \frac{E_2''}{E_2} = \frac{2n \cos\alpha}{n \cos\alpha + (\mu/\mu') n' \cos\beta} \ . \qquad (14.5.40)$$

The equations (14.5.24) and (14.5.25) yield no additional information. With $\mu \approx \mu' \approx 1$ for optical frequencies, it then follows that for E_0 orthogonal to the plane of incidence:

$$\frac{|E_0''|}{|E_0|} = \frac{\cos\alpha - (n'/n)\cos\beta}{\cos\alpha + (n'/n)\cos\beta} \; , \tag{14.5.41}$$

$$\frac{|E_0'|}{|E_0|} = \frac{2\cos\alpha}{\cos\alpha + (n'/n)\cos\beta} \; . \tag{14.5.42}$$

b) If E_0 lies in the plane of incidence, then analogously we can derive:

$$\frac{|E_0''|}{|E_0|} = \frac{(n'/n)\cos\alpha - \cos\beta}{(n'/n)\cos\alpha + \cos\beta} \; , \tag{14.5.43}$$

$$\frac{|E_0'|}{|E_0|} = \frac{2\cos\alpha}{(n'/n)\cos\alpha + \cos\beta} \; . \tag{14.5.44}$$

The equations (14.5.41–44) are called *Fresnel's*[7] *equations.*

14.5.3 Special Effects of Reflection and Refraction

a) The Brewster[8] Angle

We pose the following question: Is there an incident angle α_B for which there is no reflected wave?

a1) Let E_0 be parallel to the plane of incidence, then from (14.5.43) it follows for α_B that

$$\frac{n'}{n}\cos\alpha_B = \cos\beta \quad\text{or}$$

$$\frac{n'^2}{n^2}\cos^2\alpha_B = \cos^2\beta = 1 - \sin^2\beta = 1 - \frac{n^2}{n'^2}\sin^2\alpha_B \; ,$$

since from Snell's law: $\sin\beta = (n/n')\sin\alpha$. It follows that

$$\frac{n'^2}{n^2} = \frac{1}{\cos^2\alpha_B} - \frac{n^2}{n'^2}\tan^2\alpha_B = 1 + \tan^2\alpha_B - \frac{n^2}{n'^2}\tan^2\alpha_B$$

---

[7] *Fresnel, Augustin Jean* (*1788 in Broglie, Normandy, d. 1827 Ville-d'Avray near Paris). Engineer and physicist. He made fundamental contributions to optics, in particular to the proof of the transverse wave nature of light: experiments on interference, diffraction, and polarization.

[8] *Brewster, David* (*1781 Jedburg/Scotland, d. 1868 Allerby/Scotland). Scottish physicist, originally theologian, later physics professor at St. Andrews. He discovered his law of reflection in 1818. Brewster was known as the writer of biographical and popular scientific works.

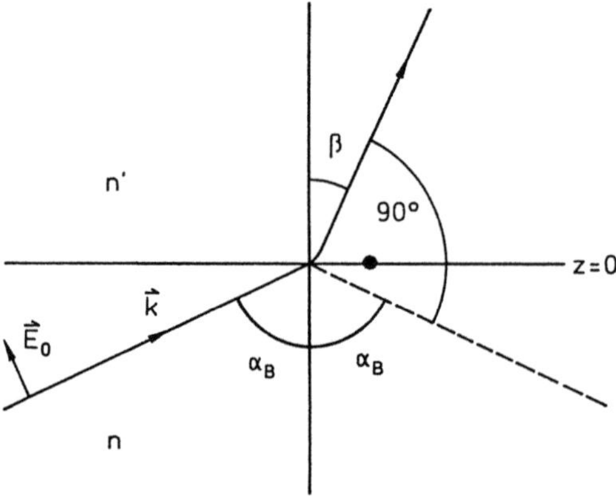

Fig. 14.5.2. If E_0 lies in the incident plane, there is no reflected wave if its wave vector would form an angle of $\pi/2$ with the refracted ray

and thus the *Brewster angle* α_B satisfies:

$$\tan^2 \alpha_B = \frac{(n'^2/n^2) - 1}{1 - (n^2/n'^2)} = \frac{n'^2}{n^2} \, ,$$

$$\tan \alpha_B = \frac{n'}{n} \quad \text{or} \quad \alpha_B = \arctan\left(\frac{n'}{n}\right) .$$

(14.5.45)

From Snell's law

$$\sin \alpha_B / \sin \beta = n'/n(= \tan \alpha_B)$$

it follows also that

$$\alpha_B + \beta = \frac{\pi}{2} \, .$$

(14.5.46)

Thus: *If E_0 lies in the plane of incidence, then the reflected ray vanishes, if its direction would be complementary to the refracted ray (that is, if the sum of the angle of refraction and the angle of reflection would be $\pi/2$). If waves with mixed polarization are incident at this angle, then the reflected ray is totally linearly polarized perpendicular to the incident plane (Fig. 14.5.2).*

a2) Let E_0 be orthogonal to the plane of incidence:
     Then if $E_0'' = 0$, we must have $n \cos \alpha_B = n' \cos \beta$, so that

$$\frac{n^2}{n'^2} \cos^2 \alpha_B = 1 - \sin^2 \beta = 1 - \frac{n^2}{n'^2} \sin^2 \alpha_B$$

or

$$\frac{n^2}{n'^2} = 1 \, .$$

It follows that $n = n'$, and this is a trivial case.

b) Total Reflection

Let $n' < n$, then from Snell's law

$$\sin\alpha = \frac{n'}{n}\sin\beta \ . \tag{14.5.47}$$

Here, we must always have $\beta > \alpha$ and in particular

$$\beta = \frac{\pi}{2} \quad \text{i.e.} \quad \sin\alpha = \frac{n'}{n} < 1 \ , \quad \text{i.e. for} \tag{14.5.48}$$

$$\alpha = \alpha_0 = \arcsin\left(\frac{n'}{n}\right) \ . \tag{14.5.49}$$

In the case $n' < n$ there is thus an angle of incidence α_0 for which no refracted wave appears in the medium with the index of refraction n'.

If we make α greater than α_0, then with $\sin\alpha_0 = n'/n$

$$\sin\beta = \frac{\sin\alpha}{(n'/n)} = \frac{\sin\alpha}{\sin\alpha_0} > 1 \ ,$$

so that β must be imaginary, and thus also

$$\boldsymbol{k}' = k(\sin\beta, 0, \cos\beta) \ ,$$

i.e. the plane wave $\exp(\mathrm{i}\boldsymbol{k}' \cdot \boldsymbol{r})$ must contain an exponentially decaying factor.

c) The Bending of the Path of Light in an Inhomogeneous Medium

Since in Snell's law

$$\frac{\sin\alpha}{\sin\beta} = \frac{n'}{n} \ ,$$

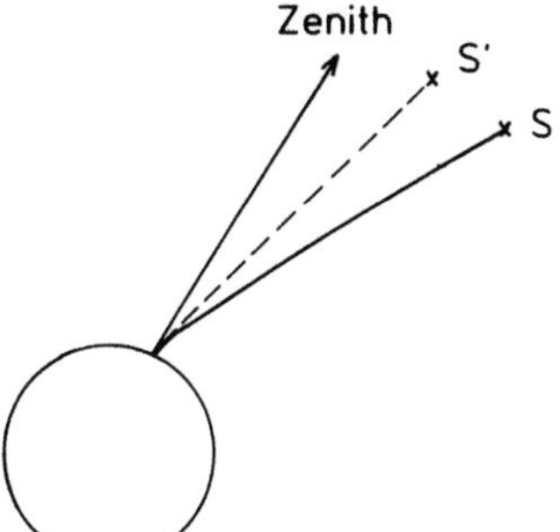

Fig. 14.5.3. Astronomical refraction: The light enters into denser and denser layers of air and is thus always deflected towards the zenith. A star S is observed at the point S'

it follows for $n' > n$ that

$$\alpha > \beta \; ,$$

i.e., upon entering an optically denser medium ($n' > n$) the ray is deflected towards a line perpendicular to the medium. The amplitude depends on the polarization.

In the case of refraction of light in an inhomogeneous medium in which the index of refraction n varies from point to point, the path of the light therefore curves. In particular, for astronomical refraction, the situation is as sketched in Fig. 14.5.3. The angle between the location of a star and the zenith point appears less than it really is.

Appendices

A. The Γ-Function

The Γ-function is a generalization of $n! = \prod_{k=1}^{n} k$ to real or even complex values of n. Clearly,

$$(n + 1)! = (n + 1)\cdot n! \ .$$

For integers $x \geq 0$

$$\Gamma(x + 1) = x! \ . \tag{A.1}$$

It is a convention that $x + 1$ rather than x is the argument of Γ in this equation. In order to assure $\Gamma(x)$ is an extension of $(x - 1)!$ to complex arguments, we must require

$$\Gamma(x + 1) = x\Gamma(x) \ . \tag{A.2}$$

This equation, together with a more technical regularity requirement, is enough to uniquely determine Γ. It is given by

$$\Gamma(x) = \int_0^\infty dt\, t^{x-1} e^{-t} \ . \tag{A.3}$$

We see first that $\Gamma(x)$ is defined for all real $x > 0$, and is even an analytic function for all complex x with $\mathrm{Re}\,\{x\} > 0$.

For $\mathrm{Re}\{x\} \leq 0$, the values of the function are determined through analytic continuation. By partial integration, we can show that for $\mathrm{Re}\{x\} > 0$ we have

$$\Gamma(x + 1) = \int_0^\infty dt\, t^x e^{-t}$$

$$= \int_0^\infty dt \left[-\frac{d}{dt}(t^x e^{-t}) + xt^{x-1} e^{-t} \right] = x\Gamma(x) \ . \tag{A.4}$$

It can be shown that:

$\Gamma(x)$ can be given an analytic continuation for all values of x with $x \neq -n, n = 0, 1, \ldots$.

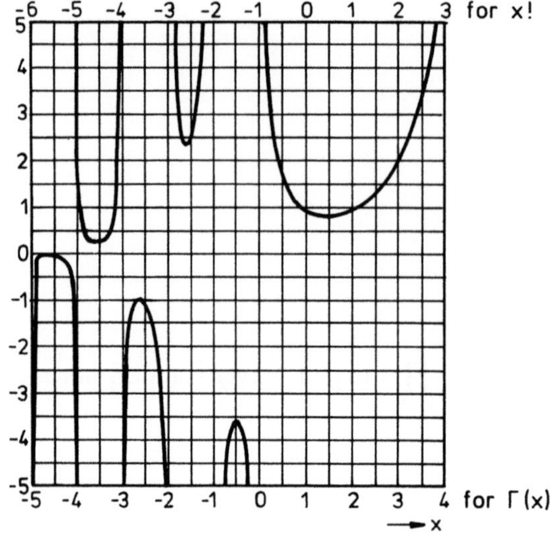

Fig. A.1. The Gamma function $\Gamma(x)$

As $x \to -n$, $\Gamma(x)$ has a pole with residue $(-1)^n/n!$:

$$\Gamma(x) \underset{x \to -n}{\sim} \frac{(-1)^n}{n!} \cdot \frac{1}{x+n}\,, \quad \text{and} \tag{A.5}$$

$\Gamma(x+1) = x\Gamma(x)$ holds for all x for which $\Gamma(x)$ is defined.

Figure A.1 shows the graph of the Γ-function. Of particular importance for us is an approximation for $\Gamma(x)$ for large positive values of x. We have

$$\ln \Gamma(n) = \ln(n-1)! = \sum_{\nu=1}^{n-1} \ln \nu \ . \tag{A.6}$$

This can be seen as a discrete approximation with step $\Delta\nu = 1$ for the Riemann integral

$$\int_1^n d\nu \ln \nu \ .$$

For large values of n, since the increase of $\ln x$ approaches zero as x grows, the error caused by the size of the step becomes smaller and smaller. We thus expect

$$\ln \Gamma(x) \underset{x \to \infty}{\sim} \int_1^x d\nu \ln \nu = x(\ln x - 1) \tag{A.7}$$

or

$$\Gamma(x) \underset{x \to \infty}{\sim} \left(\frac{x}{e}\right)^x \ .$$

A more exact consideration yields *Stirling's formula*

$$\Gamma(x) = \sqrt{\frac{2\pi}{x}}\, e^{-x} x^x \left[1 + O\left(\frac{1}{x}\right) \right],$$ (A.8)

i.e.

$$\ln \Gamma(x) = x(\ln x - 1) - \frac{1}{2}\ln x + \frac{1}{2}\ln 2\pi + O\left(\frac{1}{x}\right).$$ (A.9)

This formula is quite important in statistical mechanics.

Here, $x \sim 10^{23}$, so that $\ln x \sim 50$ and the relative error in the approximation

$$\ln \Gamma(x) \approx x(\ln x - 1)$$ (A.10)

is of the order of magnitude $< 10^{-21}$.

We also note a special value of Γ:

$$\Gamma(\tfrac{1}{2}) = \sqrt{\pi} \; .$$ (A.11)

B. Conic Sections

a) An ellipse can be characterized in the following manner: The sum of the distances between any point P on the ellipse and the two foci F_1 and F_2 is always $2a$ (Fig. B.1).

Thus, we must have

$$\overline{F_1 P} + \overline{F_2 P} = 2a \; .$$ (B.1)

Let

$$e = \overrightarrow{MF_1} \;, \quad \overrightarrow{F_1 P} = r \;, \quad \text{then}$$
$$\overrightarrow{F_2 P} = r + 2e \;,$$ (B.2)

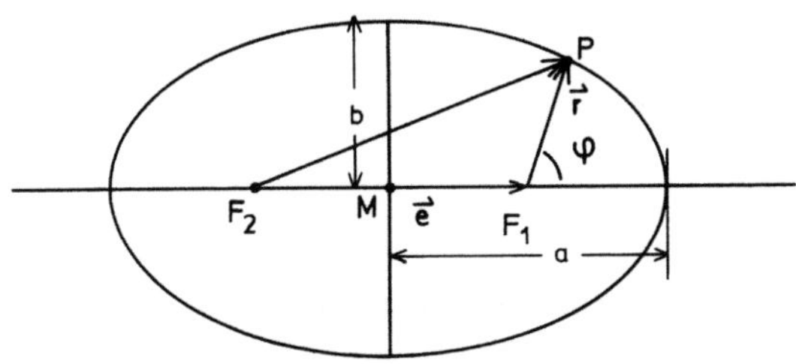

Fig. B.1. An ellipse is characterized by the fact that the sum of the lengths $\overline{F_2 P}$ and $\overline{F_1 P}$ is always exactly $2a$

and thus we find

$$|r| + |r + 2e| = 2a \quad \text{or}$$

$$(r + 2e)^2 = (2a - r)^2 , \quad \text{i.e.} \tag{B.3}$$

$$r^2 + 4e \cdot r + 4e^2 = 4a^2 - 4ar + r^2 .$$

Let

$$|e| = \varepsilon a , \quad a^2 - e^2 = b^2 , \quad \varphi = \angle (e, r) . \tag{B.4}$$

Then we find from (B.3)

$$4a\varepsilon r \cos \varphi + 4e^2 = 4a^2 - 4ar \quad \text{or}$$

$$ra(1 + \varepsilon \cos \varphi) = b^2 , \quad \text{i.e.}$$

$$r = \frac{(b^2/a)}{1 + \varepsilon \cos \varphi} = \frac{p}{1 + \varepsilon \cos \varphi} \quad \text{with} \quad p = \frac{b^2}{a} . \tag{B.5}$$

b) The relationship defining a hyperbola with foci F_1 and F_2 (Fig. B.2) is:

$$\overline{F_2 P} - \overline{F_1 P} = \pm 2a \tag{B.6}$$

for the right and left branches respectively. From this, as in the case of the ellipse, we can derive

$$r = \frac{\pm p}{1 \mp \varepsilon \cos \varphi} , \quad \varepsilon = \frac{e}{a} > 1 , \tag{B.7}$$

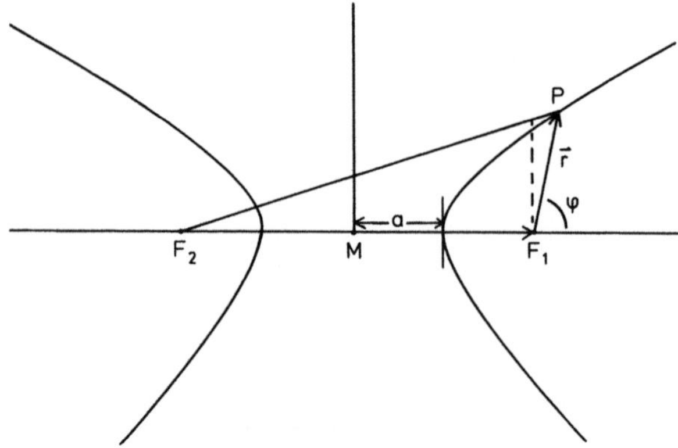

Fig. B.2. A hyperbola is characterized by the fact that the difference of the distances $\overline{F_1 P}$ and $\overline{F_2 P}$ is always equal to $\pm 2a$ (for the right and left branches respectively)

where again $e = |\overline{MF_1}|$, and we have set the origin at F_1 and

$$b^2 = e^2 - a^2 \ , \quad p = \frac{b^2}{a} \ . \tag{B.8}$$

The equations for the ellipse and hyperbola in Cartesian coordinates centered at the origin are given by

$$\frac{x^2}{a^2} + \frac{y^2}{b^2} = 1 \quad \text{and} \quad \frac{x^2}{a^2} - \frac{y^2}{b^2} = 1 \ . \tag{B.9}$$

c) Finally, a parabola is defined as the set of all points which are equidistant from a line, called the fixed line, and a point F, the focus (Fig. B.3).

If p is the distance from F to the fixed line, we have

$$r = - r \cos \varphi + p \ , \quad \text{i.e.} \tag{B.10}$$

$$r = \frac{p}{1 + \cos \varphi} \ . \tag{B.11}$$

With the tip of the parabola as origin, the equation in Cartesian coordinates is given by

$$y^2 = - 2px \ . \tag{B.12}$$

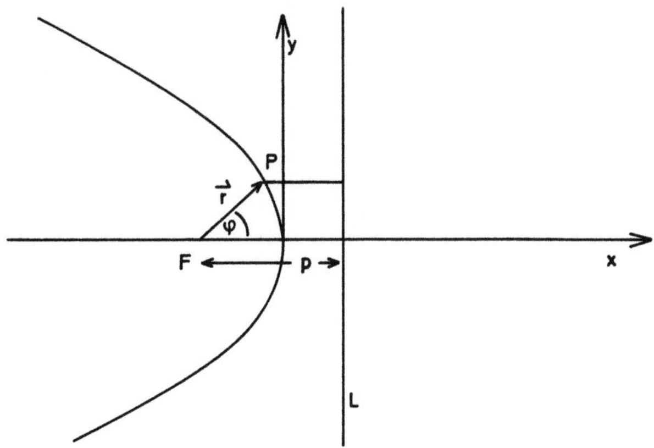

Fig. B.3. A parabola is the set of points P which are of equal distance from a point F and a fixed line L

C. Tensors

We limit ourselves first to the case which is most important for us, namely tensors on a finite dimensional Euclidean vector space. Let V then be a Euclidean vector space, and let $e_1, \ldots, e_n$ be a positively oriented orthonormal basis in V. The set of bilinear forms on V forms a vector space $T^2 V$, the *second rank tensor space* over V. Addition and multiplication by scalars are easily defined in this space by

$$(\alpha A + \beta B)(x, y) = \alpha A(x, y) + \beta B(x, y) \tag{C.1}$$

with

$$\alpha, \beta \in \mathbb{R} \; ; \quad A, B \in T^2 V ; \quad x, y \in V \; .$$

We define a *tensor product* $a \otimes b$ of vectors from V, which associates with the two vectors a and b a bilinear form $a \otimes b \in T^2 V$ as follows:

$$(a \otimes b)(x, y) := (a \cdot x)(b \cdot y) \; , \quad (a, b, x, y \in V) \; , \tag{C.2}$$

where $a \cdot x$ and $b \cdot y$ are scalar products in V. With this definition, the following properties clearly hold:

$$a \otimes (b + c) = a \otimes b + a \otimes c \; , \tag{C.3}$$

$$(a + b) \otimes c = a \otimes c + b \otimes c \; , \tag{C.4}$$

$$a \otimes (\alpha b) = (\alpha a) \otimes b = \alpha(a \otimes b) \; , \tag{C.5}$$

$$a \otimes b = 0 \Rightarrow a = 0 \text{ or } b = 0 \; . \tag{C.6}$$

In general, $a \otimes b \neq b \otimes a$.

The tensors $e_{ij} := e_i \otimes e_j$ satisfy

$$e_{ij}(e_r, e_s) = \delta_{ir} \delta_{js} \; . \tag{C.7}$$

They form a basis for $T^2 V$, since for each bilinear form $A \in T^2 V$, we have

$$A(x, y) = A\left(\sum_i x_i e_i, \sum_j y_j e_j \right) = \sum_{i,j} x_i y_j A(e_i, e_j)$$

$$= \sum_{i,j} (e_i \cdot x)(e_j \cdot y) A_{ij} = \left(\sum_{i,j} A_{ij} e_i \otimes e_j \right)(x, y) \tag{C.8}$$

for all $x, y \in V$ with

$$A_{ij} = A(e_i, e_j) \; , \tag{C.9}$$

and this representation is unique.

The dimension of T^2V is thus n^2. The n^2 quantities A_{ij} are called components of the tensor A with respect to the basis $e_1, \ldots, e_n$ of V. Expressed in terms of the components, addition and multiplication are simply given by

$$(A + B)_{ij} = A_{ij} + B_{ij} \; ; \quad (\alpha A)_{ij} = \alpha A_{ij} \; . \tag{C.10}$$

The inertia tensor I can be written in the body-fixed basis $e_1(t), e_2(t), e_3(t)$ as

$$I = \sum_{i,j=1}^{3} I_{ij} e_i(t) \otimes e_j(t) \; .$$

If we go from one orthonormal basis $e_1, \ldots, e_n$ to another $e'_1, \ldots, e'_n$, with $e'_i = \sum e_j D_{ji}$, then the basis vectors e_{ij} and the components A_{ij} transform as follows

$$e'_{ij} = e'_i \otimes e'_j = \sum_{r,k} e_{rk} D_{ri} D_{kj} \tag{C.11}$$

and

$$A'_{ij} = \sum_{r,k} D_{ri} D_{kj} A_{rk} \; .$$

The *identity tensor*

$$1 = \sum_{i,j} \delta_{ij} e_i \otimes e_j = \sum_i e_i \otimes e_i \tag{C.12}$$

has in *every* orthonormal system the components

$$(1)_{ij} = \delta_{ij} \; .$$

Every tensor $A \in T^2V$ can be seen as a linear mapping $V \to V$, and, in fact, the two can be identified. To see this, we first define for each $x \in V$

$$(a \otimes b)(x) = a(b \cdot x) \in V \; , \tag{C.13}$$

and in general for $x = \sum x_k e_k$,

$$A(x) = \sum_{i,j} (A_{ij} e_i \otimes e_j)(x)$$

$$= \sum_{i,j} A_{ij} e_i (e_j \cdot x) = \sum_{i,j} e_i A_{ij} x_j \; . \tag{C.14}$$

Instead of $A(x)$, we generally write Ax or $A \cdot x$.

(For example, for the inertial tensor, $I \cdot \Omega = \sum_{i,j} e_i I_{ij} \Omega_j$.)

The components of the tensor A with respect to the basis $e_1, \ldots, e_n$ correspond also to the matrix of the linear mapping which is associated with A (with respect to the same basis). If we compose the two mappings A and B, the corresponding matrix for $A \cdot B$ is given by

$$(A \cdot B)_{ij} = \sum_k A_{ik} B_{kj} \ . \tag{C.15}$$

Thus, it is the matrix product of the matrices for A and B.

Tensors of rank k can be defined analogously as k-linear forms over V. They form vector spaces $T^k V$ of dimension n^k. In particular, $T^1 V = V$. Multiple tensor products are defined analogously, for example

$$(a \otimes b \otimes c)(x, y, z) = (a \cdot x)(b \cdot y)(c \cdot z) \ . \tag{C.16}$$

The quantities

$$e_{i_1, \ldots, i_k} := e_{i_1} \otimes \ldots \otimes e_{i_k}$$

form a basis of $T^k V$. For example, each tensor $C \in T^3 V$ can be written uniquely in the form

$$C = \sum_{i, j, k} C_{ijk} e_i \otimes e_j \otimes e_k \ .$$

If V is a three dimensional oriented Euclidean space, there is a special tensor

$$E = \sum_{i, j, k} \varepsilon_{ijk} e_i \otimes e_j \otimes e_k \in T^3 V \ , \tag{C.17}$$

whose components in every positively oriented orthonormal basis are given by the ε-symbol which we defined in Sect. 2.1.

If V is not a Euclidean vector space, we must distinguish between V and the dual space V^* (the space of all linear forms $V \to \mathbb{R}$).

We then define

a) covariant tensors of rank k as k-linear forms on V^* and
b) contravariant tensors of rank k as k-linear forms on V.

Mixed tensors can also be defined.

An r-fold covariant and s-fold contravariant tensor is an $(r + s)$-fold linear mapping A

$$A \colon \ \underbrace{V^* \times \ldots \times V^*}_{r \text{ times}} \times \underbrace{V \times \ldots \times V}_{s \text{ times}} \to \mathbb{R}$$

$$(v^{*1}, \ldots, v^{*r}, w_1, \ldots, w_s) \mapsto A(v^{*1}, \ldots, v^{*r}, w_1, \ldots, w_s) \ . \tag{C.18}$$

The vector space of all r-fold covariant, s-fold contravariant tensors over V will be designated by $T^r_s V$.

The tensor product of two tensors $A \in T^r_s V$ and $B \in T^{r'}_{s'} V$ is a tensor $A \otimes B \in T^{r+r'}_{s+s'} V$, which we again simply define by multiplying the values of A and B:

$$(A \otimes B)(v^{*1}, \ldots, v^{*r+r'}, w_1, \ldots, w_{s+s'}) := A(v^{*1}, \ldots, v^{*r}, w_1, \ldots, w_s)$$

$$\times B(v^{*r+1}, \ldots, v^{*r+r'}, w_{s+1}, \ldots, w_{s+s'}) . \tag{C.19}$$

The tensor product $A \otimes B$ is bilinear in A and B and associative, but not commutative.

The double dual space V^{**} (the space of all linear forms $V^* \to \mathbb{R}$) and higher dual spaces do not need to be considered, since there is a canonical isomorphism, that is, an isomorphism defined independently of any basis, $\tau: V \to V^{**}$.

In particular, let $w^* \in V^*$ be a linear form on V and $v \in V$, then we can associate v with a linear form $\tau(v)$ on V^* as follows:

$$\tau(v)(w^*) := w^*(v) \tag{C.20}$$

τ is clearly a linear isomorphism.

We now will introduce bases in $T^r_s V$.

Let $(e_1, \ldots, e_n)$ be a basis for V. We define a dual basis $(e^{*1}, \ldots, e^{*n})$ of V^* by

$$e^{*i}(e_j) = \delta^i_j .$$

We then see easily that a basis for $T^r_s V$ is given by

$$e^{j_1 \ldots j_s}_{i_1 \ldots i_r} = e_{i_1} \otimes \ldots \otimes e_{i_r} \otimes e^{*j_1} \otimes \ldots \otimes e^{*j_s} , \tag{C.21}$$

so that every tensor $A \in T^r_s V$ can be written uniquely in the form

$$A = \sum_{\substack{i_1, \ldots, i_r \\ j_1, \ldots, j_s}} A^{i_1 \ldots i_r}_{j_1 \ldots j_s} e_{i_1} \otimes \ldots \otimes e_{i_r} \otimes e^{*j_1} \otimes \ldots \otimes e^{*j_s} . \tag{C.22}$$

We often leave off the summation sign, according to the Einstein summation convention, in which it is understood that every repeated index should be summed over.

In calculations performed with indices, we indicate a tensor by its components with respect to a pregiven basis. It is easy to transform to other bases. We can then write out tensor summation and the tensor product as follows

$$(A + B)^{i_1 \ldots i_r}_{j_1 \ldots j_s} = A^{i_1 \ldots i_r}_{j_1 \ldots j_s} + B^{i_1 \ldots i_r}_{j_1 \ldots j_s} ,$$

$$(A \otimes B)^{i_1 \ldots i_{r+r'}}_{j_1 \ldots j_{s+s'}} = A^{i_1 \ldots i_r}_{j_1 \ldots j_s} \cdot B^{i_{r+1} \ldots i_{r+r'}}_{j_{s+1} \ldots j_{s+s'}} . \tag{C.23}$$

Remark. Finally, we will show why there is no need to distinguish between V and V^* for Euclidean vector spaces. In this case, with the aid of the scalar product we can define a (linear) isomorphism $\iota: V \to V^*$.

Each $v \in V$ is simply associated with a linear form $\iota(v)$ according to

$$\iota(v)(x) = v \cdot x \ . \tag{C.24}$$

If $(e_1, \ldots, e_k)$ is an orthonormal basis of V then the basis dual to it in V^* is given by $(\iota(e_1), \ldots, \iota(e_k))$:

$$\iota(e_i)(e_j) = e_i \cdot e_j = \delta_{ij} \ . \tag{C.25}$$

If we identify the spaces V and V^* using the isomorphism ι, then orthonormal bases are dual to themselves.

D. Fourier Series and Fourier Integrals

D.1 Fourier Series

The theory of Fourier series works with the problem of approximating functions defined in an interval $[0, a]$, or equivalently, periodic functions f of period a, with functions of the form

$$g_N(t) = \sum_{n=-N}^{+N} a_n e^{2\pi i n t/a} \ . \tag{D.1}$$

We thus want to represent arbitrary functions as superpositions of particular harmonic basis functions. First, we have the basic

Theorem: Every function f which is continuous in the interval $[0, a]$ can be approximated by functions

$$g_N(t) = \sum_{n=-N}^{+N} a_n e^{2\pi i n t/a} \ .$$

In other words, for any $\varepsilon > 0$, there is an $N(\varepsilon)$ and coefficients a_n with $-N(\varepsilon) < n < N(\varepsilon)$ such that

$$\left| f(t) - \sum_{n=-N}^{N} a_n e^{2\pi i n t/a} \right| < \varepsilon \tag{D.2}$$

holds for all $t \in [0, a]$.

This theorem is a corollary to the well-known Weierstrass approximation theorem.

If we are given N, we then might ask: How must the coefficients a_n in approximation (D.1) look, so that $f(t)$ is particularly well approximated? One measure of the accuracy of the approximation is, for example, the size of the mean square deviation

$$\int_0^a dt \left| f(t) - \sum_{n=-N}^{+N} a_n e^{2\pi int/a} \right|^2 \tag{D.3}$$

We then look for coefficients a_n $(-N \leq n \leq N)$ such that the mean square deviation becomes minimal.

To solve this problem, we introduce the following notation:

$$\langle f, g \rangle := \int_0^a dt\, f^*(t) g(t) \ . \tag{D.4}$$

We have

$$\langle f, g \rangle = \langle g, f \rangle^* \ ,$$

$$\langle f, \alpha_1 g_1 + \alpha_2 g_2 \rangle = \alpha_1 \langle f, g_1 \rangle + \alpha_2 \langle f, g_2 \rangle \ , \qquad \alpha_1, \alpha_2 \in \mathbb{C}$$

and $\langle f, f \rangle > 0$ for $f \neq 0$, continuous.

The expression $\langle f, g \rangle$ is a (Hermitian) scalar product in the vector space of functions continuous on $[0, a]$.

For the functions

$$\varphi_n(t) = \frac{1}{\sqrt{a}} e^{2\pi int/a} \tag{D.5}$$

we then have

$$\langle \varphi_n, \varphi_m \rangle = \delta_{nm} \ , \tag{D.6}$$

The functions thus form an orthonormal system.

In this new formulation, our problem now reads: we look for coefficients b_n $(-N \leq n \leq N)$ such that

$$\left\langle f - \sum_{n=-N}^{+N} b_n \varphi_n, \ f - \sum_{n=-N}^{+N} b_n \varphi_n \right\rangle \tag{D.7}$$

is minimal. Now, with $c_n = \langle \varphi_n, f \rangle$, we have

$$\left\langle f - \sum_{n=-N}^{+N} b_n \varphi_n, \ f - \sum_{n=-N}^{+N} b_n \varphi_n \right\rangle$$

$$= \langle f, f \rangle - \sum_{n=-N}^{+N} b_n^* \langle \varphi_n, f \rangle - \sum_{n=-N}^{+N} b_n \langle f, \varphi_n \rangle + \sum_{m,n=-N}^{+N} b_m^* b_n \langle \varphi_m, \varphi_n \rangle$$

$$= \langle f, f \rangle - \sum_{n=-N}^{+N} (b_n^* c_n - b_n c_n^* - b_n^* b_n - c_n^* c_n + c_n^* c_n)$$

$$= \langle f, f \rangle - \sum_{n=-N}^{+N} |c_n|^2 + \sum_{n=-N}^{+N} |b_n - c_n|^2 \ . \tag{D.8}$$

We see that the minimum is achieved exactly for the choice of coefficients $b_n = c_n \equiv \langle \varphi_n, f \rangle$. In addition, the coefficients c_n ($-N \leq n \leq N$) are independent of N. Thus, if we want to improve the approximation by increasing N, we do not need to replace the coefficients determined before, just as in the case of Taylor series.

The series

$$\sum_{n=-\infty}^{+\infty} c_n \varphi_n(t) = \sum_{n=-\infty}^{+\infty} \langle \varphi_n, f \rangle \varphi_n(t) = \frac{1}{\sqrt{a}} \sum_{n=-\infty}^{+\infty} c_n e^{2\pi i n t/a} \tag{D.9}$$

is called the *Fourier series* of the function f.

It can be shown that with $b_n = c_n$, the mean square deviation converges to zero as $N \to \infty$:

$$\lim_{N \to \infty} \left\langle f - \sum_{n=-N}^{+N} c_n \varphi_n, f - \sum_{n=-N}^{+N} c_n \varphi_n \right\rangle = 0 \ . \tag{D.10}$$

We say then that the Fourier series converges in *quadratic mean* to the function f. This will still occur even if $f(t)$ is not continuous. We do have the

Theorem: For every periodic function f with period a which is continuous and has continuous derivatives up to a finite number of discontinuities, the Fourier series converges at all points of continuity and even uniformly outside of arbitrarily small open intervals around the discontinuities. At the discontinuities, the Fourier series converges to

$$\lim_{\varepsilon \to +0} \tfrac{1}{2}(f(t_i + \varepsilon) + f(t_i - \varepsilon)), \qquad t_i = point\ of\ discontinuity.$$

The coefficients c_n are called the *Fourier coefficients*, and the representation of a function $f(t)$ by a Fourier series is called *Fourier analysis*.

The set $\{\varphi_n | n = 0, \pm 1, \pm 2, \dots\}$ thus represents a complete orthonormal system of functions. The Fourier series

$$f = \sum_{n=-\infty}^{+\infty} \varphi_n \langle \varphi_n, f \rangle \tag{D.11}$$

is completely analogous to the expansion of a vector with respect to an orthonormal basis. In particular, we have

$$\langle f, f \rangle = \int_0^a dt\, f^*(t) f(t) = \sum_{n=-\infty}^{+\infty} c_n^* c_n \ . \tag{D.12}$$

Other complete orthonormal systems in which any piecewise smooth function in

$[0, a]$ can be expanded are:

a) $\dfrac{1}{\sqrt{a}}, \sqrt{\dfrac{2}{a}}\cos\left(\dfrac{2\pi nt}{a}\right), \sqrt{\dfrac{2}{a}}\sin\left(\dfrac{2\pi nt}{a}\right);$

b) $\sqrt{\dfrac{2}{a}}\sin\left(\dfrac{n\pi t}{a}\right);$

c) $\sqrt{\dfrac{1}{a}}, \sqrt{\dfrac{2}{a}}\cos\left(\dfrac{n\pi t}{a}\right).$

The functions in (a) are simply the normalized real and imaginary parts of the function $\varphi_n(t)$. We find systems (b) and (c) when we assign to each function on $[0, a]$ an antisymmetric or symmetric function on $[-a, a]$, respectively, and note that in the Fourier expansion of such functions only antisymmetric or symmetric functions arise.

Expressed in terms of the orthonormal system (a), the Fourier series

$$f(t) = \frac{1}{\sqrt{a}}\sum_{-\infty}^{+\infty} c_n e^{2\pi int/a} \tag{D.13}$$

takes on the following form

$$f(t) = \frac{1}{\sqrt{a}}\sum_{n=0}^{\infty}\left[a_n\cos\left(\frac{2\pi nt}{a}\right) + b_n\sin\left(\frac{2\pi nt}{a}\right)\right]$$

with

$$a_0 = c_0 = \frac{1}{\sqrt{a}}\int_0^a f(t)\,dt\ ,$$

$$a_n = (c_n + c_{-n}) = \frac{2}{\sqrt{a}}\int_0^a dt\, f(t)\cos\left(\frac{2\pi nt}{a}\right)\ ,\qquad n > 0\ ,$$

$$b_n = i(c_n - c_{-n}) = \frac{2}{\sqrt{a}}\int_0^a dt\, f(t)\sin\left(\frac{2\pi nt}{a}\right)\ ,\qquad n > 0\ .$$

We will give two examples:

a) Let

$$f(t) = \begin{cases} 0 & \text{for} \quad -\pi \le t < 0\ , \\ \pi & \text{for} \quad 0 \le t < \pi\ . \end{cases}$$

With the orthonormal system $\varphi_n(t) = \dfrac{1}{\sqrt{2\pi}}\, e^{int}$, we have

$$c_n = \langle \varphi_n, f \rangle = \int\limits_{-\pi}^{+\pi} dt\, \varphi_n^*(t) f(t) = \frac{\pi}{\sqrt{2\pi}} \int\limits_0^\pi dt\, e^{-int}$$

$$= \sqrt{\frac{\pi}{2}}\,\frac{i}{n}\, e^{-int}\Big|_0^\pi$$

$$= \sqrt{\frac{\pi}{2}}\,\frac{i}{n}\,[(-1)^n - 1] = \begin{cases} \dfrac{-\sqrt{2\pi}\,i}{n} & \text{for odd } n \\[2mm] 0 & \text{for } n > 0 \text{ and even} \end{cases}$$

Furthermore,

$$c_0 = \frac{\pi^2}{\sqrt{2\pi}}\ .$$

Thus,

$$f(t) = \frac{1}{\sqrt{2\pi}} \sum_{n=-\infty}^{+\infty} c_n e^{int} = \frac{\pi}{2} + 2 \sum_{r=0}^{\infty} \frac{\sin(2r+1)t}{(2r+1)}$$

is the representation of the function $f(t)$ in a Fourier series. We note that for $t = 0$, the Fourier series converges to the value

$$\frac{\pi}{2} = \frac{1}{2}\,[f(0_+) + f(0_-)]$$

b) Let $f(t) = t$ for $-\pi < t \le \pi$. Then

$$c_n = \frac{1}{\sqrt{2\pi}} \int\limits_{-\pi}^{+\pi} dt\, t\, e^{-int} = \sqrt{2\pi}\,i\,\frac{(-1)^n}{n}$$

for $|n| > 0$; $c_0 = 0$. It follows that

$$f(t) = i \sum_{n=1}^{\infty} \frac{(-1)^n}{n}\,(e^{int} - e^{-int})\ ,$$

i.e.

$$f(t) = 2 \sum_{n=1}^{\infty} (-1)^{n+1}\,\frac{\sin nt}{n} \quad \text{for} \quad -\pi < t < \pi\ ,$$

For $t = \pi$, the series converges to 0.

The convergence of the Fourier series is shown in Figs. D.1 and D.2.

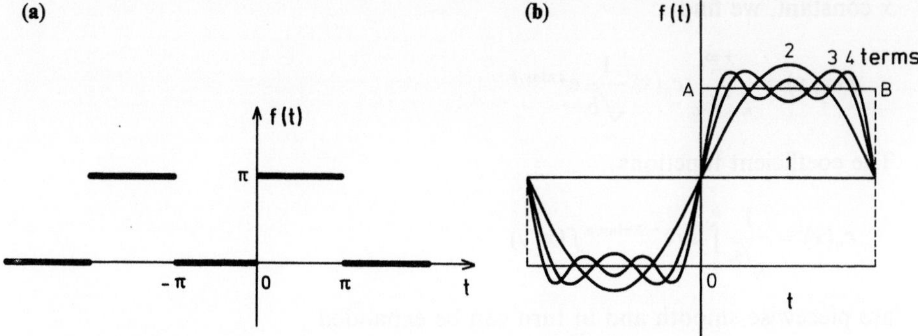

Fig. D.1. The approximation of the function $f(t)$ see (**a**) by the first two, three, and four terms of the Fourier series (**b**)

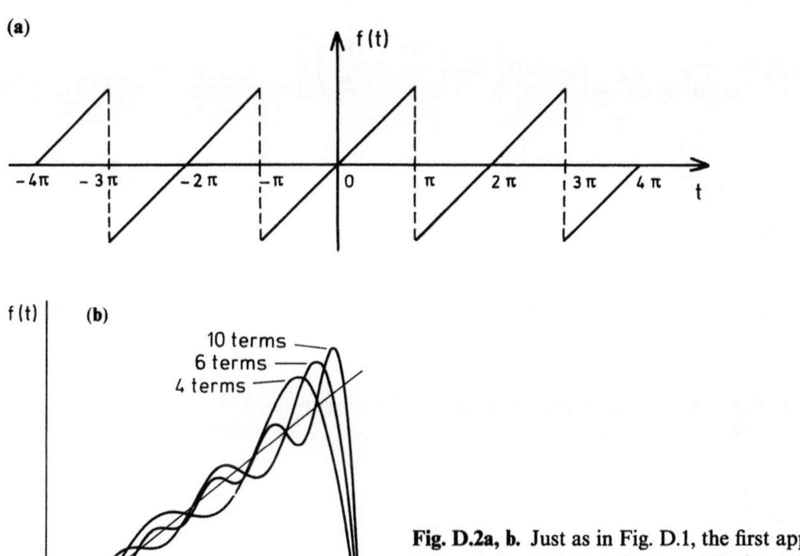

Fig. D.2a, b. Just as in Fig. D.1, the first approximations of the function $f(t)$ (see (**a**)) through partial sums of the Fourier series, the first four, six, and ten terms (**b**)

It can be shown that the Fourier series of sums, products, derivatives, and integrals of functions are given by the addition, multiplication, termwise differentiation, and integration of the corresponding Fourier series (for differentiation, of course, we need suitable smoothness conditions).

Functions of many variables can also be expanded in Fourier series. Let $f(x, y)$ be piecewise smooth in the rectangle $0 \leq x \leq a,\ 0 \leq y \leq b$. Holding

x constant, we find

$$f(x, y) = \sum_{n=-\infty}^{+\infty} c_n(x) \frac{1}{\sqrt{b}} e^{2\pi iny/b} \ .$$

The coefficient functions

$$c_n(x) = \frac{1}{\sqrt{b}} \int_0^b dy \, e^{-2\pi iny/b} f(x, y)$$

are piecewise smooth and in turn can be expanded

$$c_n(x) = \frac{1}{\sqrt{a}} \sum_{m=-\infty}^{+\infty} c_{mn} e^{2\pi imx/a} \ .$$

Substituting, we find

$$f(x, y) = \frac{1}{\sqrt{ab}} \sum_{m,n=-\infty}^{+\infty} c_{mn} \exp\left[2\pi i \left(\frac{mx}{a} + \frac{ny}{b} \right) \right]$$

$$= \sum_{m,n=-\infty}^{+\infty} c_{mn} \varphi_{m,n}(x, y) \ . \tag{D.14}$$

The functions

$$\varphi_{m,n}(x, y) = \frac{1}{\sqrt{ab}} \exp\left[2\pi i \left(\frac{mx}{a} + \frac{ny}{b} \right) \right] \tag{D.15}$$

form a complete orthonormal system on the rectangle:

$$\int_0^a dx \int_0^b dy \, \varphi_{m,n}^*(x, y) \varphi_{m',n'}(x, y) \equiv \langle \varphi_{m,n}, \varphi_{m',n'} \rangle = \delta_{mm'} \delta_{nn'} \ . \tag{D.16}$$

For the coefficients c_{mn}, we find

$$c_{mn} = \langle \varphi_{mn}, f \rangle \ . \tag{D.17}$$

D.2 Fourier Integrals and Fourier Transforms

If in the expression

$$f(t) = \sum_{n=-\infty}^{+\infty} c_n e^{inwt}$$

we plot $|c_n|^2$ against nw, we find something like the picture shown in Fig. D.3.

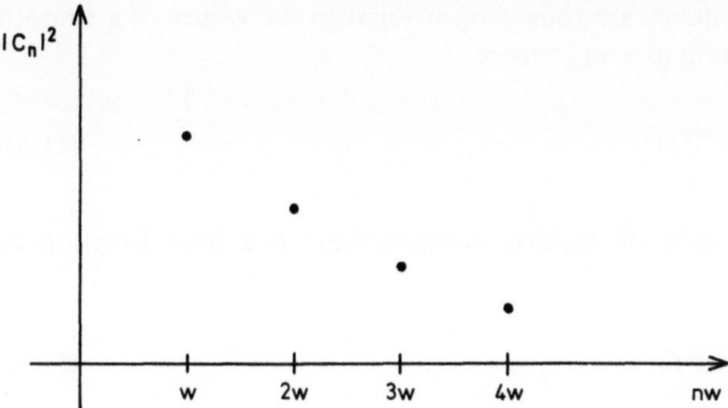

Fig. D.3. Typical frequency spectrum of a function $f(t)$

The diagram represented in Fig. D.3 is called the *frequency spectrum*. A *power spectrum* is such a diagram in which $\ln|c_n|^2$ is plotted against nw.

These spectra are discrete, since each term of the Fourier series has a frequency which is a multiple of the base frequency w, and w is related to the period T by $w = 2\pi/T$.

If we now let T get larger and larger, in order to postpone for as long as possible the periodic repetition of the function, w will get smaller and smaller and the spectrum will contain points which lie closer and closer together. As $T \to \infty$, w goes to 0, and the power spectrum, like the frequency spectrum, will become a continuous function.

If we work out this line of thought exactly we arrive at the theory of *Fourier transforms* and the *Fourier integral.*

We will discuss the Fourier expansion for the limiting case of infinite interval length. As our base interval, we will take $[-a, a]$ (i.e. $T = 2a$, $w = \pi/a$), and consider the limiting case $a \to \infty$.

A complete orthonormal system on $[-a, a]$ is given by

$$\varphi_n(t) = \frac{1}{\sqrt{2a}} e^{in\pi t/a} , \qquad n = 0, \pm 1, \pm 2, \ldots .$$

A Fourier expansion yields

$$f(t) = \sum_{n=-\infty}^{+\infty} c_n \frac{1}{\sqrt{2a}} e^{in\pi t/a}$$

$$= \sum_{n=-\infty}^{+\infty} \frac{c_n}{\sqrt{2a}} e^{i\omega_n t} , \qquad \omega_n = \frac{n\pi}{a} \tag{D.18}$$

with

$$c_n = \frac{1}{\sqrt{2a}} \int_{-a}^{+a} dt \, e^{-i\omega_n t} f(t) \equiv \sqrt{\frac{\pi}{a}} \tilde{f}_a(\omega_n) . \tag{D.19}$$

The Fourier coefficients c_n are thus proportional to the values of a smooth function $\tilde{f}_a(\omega)$ at the points $\omega = \omega_n$, where

$$\tilde{f}_a(\omega) = \frac{1}{\sqrt{2\pi}} \int\limits_{-a}^{+a} dt \, e^{-i\omega t} f(t) \ . \tag{D.20}$$

For a function f which falls off quickly enough, there is a limit function as $a \to \infty$:

$$\tilde{f}(\omega) = \frac{1}{\sqrt{2\pi}} \int\limits_{-\infty}^{+\infty} dt \, e^{-i\omega t} f(t) \ . \tag{D.21}$$

The function $\tilde{f}(\omega)$ defined in this way is called the *Fourier transform* of the function f. The correspondence $f \mapsto \tilde{f}$ is clearly a linear map.

If we substitute c_n from (D.19) into (D.18), we find

$$f(t) = \sum_{n=-\infty}^{+\infty} \sqrt{\frac{\pi}{2}} \frac{1}{a} \tilde{f}_a(\omega_n) e^{i\omega_n t} = \frac{1}{\sqrt{2\pi}} \sum_{n=-\infty}^{+\infty} \frac{\pi}{a} \tilde{f}_a(\omega_n) e^{i\omega_n t}$$

$$= \frac{1}{\sqrt{2\pi}} \sum_{n=-\infty}^{+\infty} (\omega_{n+1} - \omega_n) \tilde{f}_a(\omega_n) e^{i\omega_n t} \ . \tag{D.22}$$

In the limit $a \to \infty$, the sum becomes an integral, and we find

$$f(t) = \frac{1}{\sqrt{2\pi}} \int\limits_{-\infty}^{+\infty} d\omega \, e^{i\omega t} \tilde{f}(\omega) \ . \tag{D.23}$$

Together with

$$\tilde{f}(\omega) = \frac{1}{\sqrt{2\pi}} \int\limits_{-\infty}^{+\infty} dt \, e^{-i\omega t} f(t) \tag{D.24}$$

this is the basic equation of the Fourier transform. While (D.24) defines the Fourier transform $\tilde{f}$ of f, (D.23) yields the inversion formula for the Fourier transform, in which the function f which corresponds to $\tilde{f}$ is given. The Fourier transform $\tilde{f}$ of a function f will always exist if f vanishes quickly enough at infinity. The correspondence $f \mapsto \tilde{f}$ is also written $\tilde{f} = F[f]$, where F is a linear functional. Often $\tilde{f}(\pm \omega)$ with some other factors before it is defined as the Fourier transform of $f(t)$.

Fourier integrals may be differentiated, so that

$$f'(t) = \frac{1}{\sqrt{2\pi}} \int\limits_{-\infty}^{+\infty} d\omega \, e^{i\omega t} i\omega \tilde{f}(\omega) \ .$$

The Fourier transform of the function $f'(t)$ is thus given by $i\omega \tilde{f}(\omega)$: differentiation corresponds to multiplication in Fourier space.

We now present a few rules about the correspondence between functions and their Fourier transforms:

If $\tilde{f}(\omega)$ is the Fourier transform of $f(t)$ [written $f(t) \leftrightarrow \tilde{f}(\omega)$], we then have:

i) $\;f'(t) \leftrightarrow i\omega\tilde{f}(\omega)\;$,

ii) $\;-itf(t) \leftrightarrow \tilde{f}'(\omega)\;$,

iii) $\;f(t + a) \leftrightarrow e^{i\omega a}\tilde{f}(\omega)\;$,

iv) $\;e^{iat}f(t) \leftrightarrow \tilde{f}(\omega - a)\;$,

v) $\;f(at) \leftrightarrow \dfrac{1}{|a|}\tilde{f}\!\left(\dfrac{\omega}{a}\right)\;$,

vi) $\;f^*(t) \leftrightarrow \tilde{f}^*(-\omega)\;$.

From (v) we see that the more sharply $f(t)$ is localized around $t = 0$, the less sharply $\tilde{f}(\omega)$ is around $\omega = 0$, and vice versa.

In order to formulate another important rule for Fourier transforms, we define the *convolution* $f * g$ of two functions as

$$(f*g)(t) := \frac{1}{\sqrt{2\pi}} \int_{-\infty}^{+\infty} ds\, f(t - s)g(s)\;. \tag{D.25}$$

Then, $\widetilde{f*g} = \tilde{f}\tilde{g}\;,\quad$ i.e.,

vii) $\;(f*g)(t) \leftrightarrow \tilde{f}(\omega)\tilde{g}(\omega)$, ("*The convolution theorem*").

Proof:

$$\widetilde{f*g}(\omega) = \frac{1}{2\pi} \int_{-\infty}^{+\infty} dt\, e^{-i\omega t} \int_{-\infty}^{+\infty} ds\, f(t - s)g(s)$$

$$= \frac{1}{2\pi} \int dt \int ds\, e^{-i\omega(t-s)} f(t - s)\, e^{-i\omega s} g(s)$$

$$= \frac{1}{\sqrt{2\pi}} \int d(t - s)\, e^{-i\omega(t-s)} f(t - s)\, \frac{1}{\sqrt{2\pi}} \int ds\, e^{-i\omega s} g(s)$$

$$= \tilde{f}(\omega) \cdot \tilde{g}(\omega)$$

E. Distributions and Green's Functions

In order to solve the inhomogeneous differential equation $Lx = f$, it would be nice to find an inverse L^{-1} for L. Then we could simply write $x = L^{-1}f$. Since, however, the homogeneous equation $Lx = 0$ has nontrivial solutions, L is not injective, and thus does not have an inverse. However, L is surjective, since the differential equation $Lx = f$ has solutions for each function f. Thus there is a linear mapping G which is a right inverse to L, so that $LG = 1$. (To construct G we need a basis in the image space, and then for each basis vector e we choose an element of the original space which is mapped to e by L as the value of Ge). Since L is not injective, there are indeed many such linear mappings G with this property. A solution $x^{(0)}$ of the inhomogeneous equation $Lx^{(0)} = f$ is then given by $x^{(0)} = Gf$. Indeed, $Lx^{(0)} = LGf = 1\,f = f$.

G is called a *Green's function* for L. In order to actually find G, we must work somewhat more closely with linear mappings on function spaces. In particular, we consider the linear functionals on function spaces $\mathscr{F}$, more precisely the continuous mappings

$$l : \mathscr{F} \to \mathbb{C} \quad \text{with} \quad l(c_1 \varphi_1 + c_2 \varphi_2) = c_1 l(\varphi_1) + c_2 l(\varphi_2)$$

for all $\varphi_1, \varphi_2 \in \mathscr{F}, c_1, c_2 \in \mathbb{C}$.

Good examples of function spaces $\mathscr{F}$ are the spaces

$\mathscr{D}$: The space of infinitely differentiable complex-valued functions with compact supports;

and, above all,

$\mathscr{S}$: The space of infinitely differentiable complex-valued functions which, along with all of their derivatives, fall to zero more strongly than any power law.

In order to give sense to the term "continuity," these spaces must be given suitable metrics. We will suppress this question along with other mathematical fine points.

E.1 Distributions

Examples of continuous linear functionals are

a) $l(\varphi) = \int_{-\infty}^{\infty} dt\, f(t)\, \varphi(t)$, where f is a fixed continuous (polynomially bounded) function. Such functionals are called *regular*.

b) $\delta(\varphi) = \varphi(0)$, this functional is called the *δ-functional*.

Not every continuous linear function is regular in the form (a). There is however a theorem which states:

For every continuous linear functional l (on $\mathscr{D}$ or $\mathscr{S}$), there is a sequence (f_n), $n = 1, 2, \ldots$ of continuous functions such that

$$l(\varphi) = \lim_{n \to \infty} \int_{-\infty}^{+\infty} dt\, f_n(t)\varphi(t) \quad \text{for all} \quad \varphi \in \mathscr{D} \quad \text{or} \quad \mathscr{S} \;. \tag{E.1}$$

If l is not regular, the limit $f_n(t)$ as $n \to \infty$ will not exist for every t – the theorem asserts only the existence of a sequence (f_n), for which the sequence of integrals above converge. In any case, we can always identify linear functionals with a certain sequence of continuous functions. We write *formally*

$$l(\varphi) = \int_{-\infty}^{+\infty} dt\, f(t)\varphi(t) \tag{E.2}$$

even for non-regular functionals. In this case, the expression $f(t)$ does not signify a function, but a *distribution*, and the left hand side of the equation is understood merely as a different way of writing the right hand side. Furthermore, in this case, we say that $f_n(t) \to f(t)$ as $n \to \infty$ in the *distribution sense*, where we simply mean the relation (E.1). (In addition, the linear functional is itself often called a distribution.)

The δ-functional is obviously not regular. A corresponding function δ would have to vanish everywhere but $t = 0$, but on the other hand would have to satisfy

$$\int_{-\infty}^{+\infty} dt\, \delta(t) = 1$$

which is impossible. There are, however, various ways of presenting sequences of continuous functions which in the distribution sense converge to the δ-distribution.

The δ-distribution is thus defined by

$$\int_{-\infty}^{+\infty} dt\, \delta(t)\varphi(t) = \varphi(0) \quad \text{for all} \quad \varphi \in \mathscr{D} \quad \text{or} \quad \mathscr{S} \;. \tag{E.3}$$

One example of a sequence $f_n(t)$ with $f_n(t) \to \delta(t)$ in the distribution sense is

$$f_n(t) = \frac{1}{\pi} \frac{n}{1 + n^2 t^2} \;. \tag{E.4}$$

Other sequences which converge to $\delta(t)$ in the distribution sense are

$$f_n(t) = n e^{-n^2 \pi t^2} \;, \tag{E.5}$$

$$f_n(t) = \frac{n}{\pi}\left(\frac{\sin nt}{nt}\right)^2 \, , \tag{E.6}$$

$$f_n(t) = \frac{1}{\pi}\frac{\sin nt}{t} \, . \tag{E.7}$$

We see that in all cases the $f_n(t)$, as $n \to \infty$, become more and more concentrated on the point $t = 0$ and that at the point $t = 0$, a maximum is formed which becomes unbounded as n grows. The limit value of $f_n(0)$ does not exist, however,

$$\int\limits_{-\infty}^{+\infty} dt\, f_n(t) = 1 \quad \text{for all} \quad n > 0 \, . \tag{E.8}$$

Let $l(\varphi) = \int dt\, f(t)\, \varphi(t)$ be a regular functional with f a continuously differentiable function. Then the functional

$$l'(\varphi) := \int dt\, f'(t)\varphi(t) \, . \tag{E.9}$$

is also regular, and partial integration yields

$$l'(\varphi) = -\int dt\, f(t)\varphi'(t) = -l(\varphi') \, . \tag{E.10}$$

In general, we define for (regular and irregular) linear continuous functionals a *distributional derivative* by

$$l'(\varphi) = -l(\varphi') \quad \text{for all} \quad \varphi \in \mathcal{D} \quad \text{or} \quad \mathcal{S} \, . \tag{E.11}$$

Since the function φ in $\mathcal{D}$ or $\mathcal{S}$ is differentiable arbitrarily often, each (continuous) linear functional obviously has *distributional derivatives* of arbitrarily high orders. If $g(t)$ is a distribution, we then have

$$\int g'(t)\varphi(t)dt := -\int g(t)\varphi'(t)dt \, . \tag{E.12}$$

For a continuously differentiable $g(t)$, $g'(t)$ agrees with the conventional derivative.

Example. We have

$$\delta'(\varphi) = -\varphi'(0) \quad \text{or} \quad \int dt\, \delta'(t)\varphi(t) = -\int dt\, \delta(t)\varphi'(t) = -\varphi'(0) \, . \tag{E.13}$$

The following functions appear often in physical applications:

a) The Θ-function (the step function)

$$\Theta(t) = \begin{cases} 0 & \text{for} \quad t < 0 \, , \\ 1 & \text{for} \quad t > 0 \, . \end{cases} \tag{E.14}$$

(The Θ-function has a discontinuity at $t = 0$);

b) the function $\alpha(t) = t\Theta(t)$

$$\alpha(t) = \begin{cases} 0 & \text{for} \quad t < 0 \ , \\ t & \text{for} \quad t \geq 0 \ . \end{cases} \tag{E.15}$$

In the distribution sense, we have

   i) $\alpha'(t) = \Theta(t)$ and

   ii) $\Theta'(t) = \delta(t)$, so that $\alpha''(t) = \delta(t)$.

Proof:

For (i):

$$\int_{-\infty}^{+\infty} dt\,\alpha'(t)\varphi(t) := - \int_{-\infty}^{+\infty} dt\,\alpha(t)\varphi'(t) = - \int_{0}^{\infty} dt\,t\,\varphi'(t)$$

$$= \int_{0}^{\infty} dt\,\varphi(t) = \int_{-\infty}^{+\infty} dt\,\Theta(t)\varphi(t) \ .$$

For (ii):

$$\int_{-\infty}^{+\infty} dt\,\Theta'(t)\varphi(t) := - \int_{-\infty}^{+\infty} dt\,\Theta(t)\varphi'(t) = - \int_{0}^{\infty} dt\,\varphi'(t) = \varphi(0)$$

$$= \int_{-\infty}^{+\infty} dt\,\delta(t)\varphi(t) \ .$$

Thus, $\delta(t)$ is the second distributional derivative of the continuous function $\alpha(t)$.
In general, we have the

Theorem: Each distribution on $\mathscr{S}$ can be represented as a distributional derivative of finite order of a continuous function.

Now, a few rules of calculation for the δ-distribution:
We define $\delta(t - t_0)$ by

$$\int_{-\infty}^{+\infty} dt\,\delta(t - t_0)\varphi(t) = \varphi(t_0) \ . \tag{E.16}$$

It follows that

   i) $\delta(t - t_0)f(t) = \delta(t - t_0)f(t_0)$, $\qquad$ (E.17)

in particular,

   ii) $t\delta(t) = 0$, $\qquad$ (E.18)

   iii) $\delta(t) = \delta(-t)$, $\qquad$ (E.19)

iv) $\Theta'(t - t_0) = \delta(t - t_0)$,   (E.20)

v) $\delta(at) = \dfrac{1}{|a|}\,\delta(t)$,   (E.21)

or, more generally,

$$\delta(g(t)) = \sum_i \frac{1}{|g'(t_i)|}\,\delta(t - t_i) ,$$ (E.22)

(here, t_i are the zeroes of g).

The inversion formula of the Fourier transform

$$f(t) = \frac{1}{\sqrt{2\pi}} \int_{-\infty}^{+\infty} d\omega\; e^{i\omega t}\,\tilde{f}(\omega) ,$$ (E.23)

$$\tilde{f}(\omega) = \frac{1}{\sqrt{2\pi}} \int_{-\infty}^{+\infty} dt'\, e^{-i\omega t'} f(t')$$ (E.24)

yields a very useful representation of the δ-functional: Substitution of (E.24) in (E.23) gives

$$f(t) = \frac{1}{2\pi} \int_{-\infty}^{+\infty} dt' \int_{-\infty}^{+\infty} d\omega\; e^{i\omega(t - t')} f(t')$$

$$= f(t) = \int_{-\infty}^{+\infty} dt'\, \delta(t - t') f(t') ,$$

thus

$$\delta(t - t') = \frac{1}{2\pi} \int_{-\infty}^{+\infty} d\omega\; e^{i\omega(t - t')} .$$ (E.25)

Indeed,

$$\frac{1}{2\pi} \int_{-\infty}^{+\infty} d\omega\, e^{i\omega(t - t')} = \lim_{n \to \infty} \frac{1}{2\pi} \int_{-n}^{+n} d\omega\, e^{i\omega(t - t')}$$

$$= \lim_{n \to \infty} \frac{1}{\pi} \frac{\sin n(t - t')}{t - t'}$$

is precisely one of the sequences given above, which converge in the distribution sense to the δ-distribution.

We can also express the identity

$$\frac{1}{2\pi} \int d\omega\; e^{i\omega t} = \delta(t)$$

in the form

$$\delta(t) \leftrightarrow \frac{1}{\sqrt{2\pi}} \; .$$

The Fourier transform of the δ-distribution is a constant.

E.2   Green's Functions

We are now ready to formulate the definition of a Green's function of a normal linear differential operator L:

A distribution $G(t, t')$ which depends on a parameter t' is called a Green's function for a linear differential operator L if, in the distribution sense, we have

$$LG(t, t') = \delta(t - t') \; . \tag{E.26}$$

Obviously, if $G(t, t')$ is a Green's function, then so is $G(t, t') + u_{t'}(t)$ if $Lu_{t'} = 0$.
A particular solution $x^{(0)}$ of the differential equation $Lx = h$, is then

$$x^{(0)}(t) = \int\limits_{-\infty}^{+\infty} dt' \, G(t, t') h(t') \; , \tag{E.27}$$

since

$$Lx^{(0)}(t) = \int\limits_{-\infty}^{+\infty} dt' \, LG(t, t') h(t')$$

$$= \int\limits_{-\infty}^{+\infty} dt' \, \delta(t - t') h(t') = h(t) \; .$$

We will now find a Green's function for the operator

$$L = \frac{d^2}{dt^2} + a(t) \frac{d}{dt} + b(t) \; . \tag{E.28}$$

Let $Z_{t'}(t)$ be a solution of the homogeneous equation

$$\ddot{Z}_{t'}(t) + a(t) \dot{Z}_{t'}(t) + b(t) Z_{t'}(t) = 0 \tag{E.29}$$

with the initial values $Z_{t'}(t') = 0$, $\dot{Z}_{t'}(t') = 1$.
Then,

$$G(t, t') = \Theta(t - t') Z_{t'}(t) \tag{E.30}$$

is a Green's function for L.

To prove this, we calculate

$$\frac{d}{dt} G(t, t') = \Theta(t - t')\dot{Z}_{t'}(t) + \delta(t - t')Z_{t'}(t)$$

$$= \Theta(t - t')\dot{Z}_{t'}(t) \quad \text{and}$$

$$\frac{d^2}{dt^2} G(t, t') = \Theta(t - t')\ddot{Z}_{t'}(t) + \delta(t - t')\dot{Z}_{t'}(t)$$

$$= \Theta(t - t')\ddot{Z}_{t'}(t) + \delta(t - t') \ ,$$

and thus

$$LG(t, t') = \delta(t - t') \ ,$$

where we have used the fact that $LZ_{t'}(t) = 0$

Example. We consider the differential equation with constant coefficients:

$$Lx(t) = \ddot{x}(t) + 2\varrho\dot{x}(t) + \omega_0^2 x(t) = f(t) \ , \qquad \varrho^2 < \omega_0^2 \ .$$

The solution to $LZ_{t'}(t) = 0$ with the initial values

$$Z_{t'}(t') = 0 \ , \qquad \dot{Z}_{t'}(t') = 1$$

reads

$$Z_{t'}(t) = \frac{1}{\Omega} e^{-\varrho(t - t')} \sin[\Omega(t - t')] \quad \text{with}$$

$\Omega^2 = \omega_0^2 - \varrho^2$. Thus the Green's function is

$$G(t, t') = \Theta(t - t')\frac{1}{\Omega} e^{-\varrho(t - t')} \sin[\Omega(t - t')] \ .$$

A particular solution of the differential equation is then

$$x^{(0)}(t) = \int\limits_{-\infty}^{+\infty} dt' G(t, t')f(t')$$

$$= \frac{1}{\Omega} e^{-\varrho t} \int\limits_{-\infty}^{t} dt' e^{\varrho t'} \sin[\Omega(t - t')]f(t') \ . \tag{E.31}$$

For the operator

$$L\left(\frac{d}{dt}\right) = \sum_{r=0}^{N} L_r\left(\frac{d}{dt}\right)^r$$

with constant coefficients, the Green's function is calculated as follows: Let

$$G(t - t') = \frac{1}{2\pi} \int_{-\infty}^{+\infty} d\omega \, e^{i\omega(t-t')} \tilde{g}(\omega) \, ,$$

then the defining equation reads

$$L\left(\frac{d}{dt}\right) G(t - t') \equiv \frac{1}{2\pi} \int_{-\infty}^{+\infty} d\omega \, e^{i\omega(t-t')} L(i\omega) \tilde{g}(\omega)$$

$$\stackrel{!}{=} \delta(t - t') \equiv \frac{1}{2\pi} \int_{-\infty}^{+\infty} d\omega \, e^{i\omega(t-t')} \, .$$

It then follows that

$$L(i\omega) \tilde{g}(\omega) = 1 \quad \text{and thus}$$

$$\tilde{g}(\omega) = 1/L(i\omega) \equiv Y(i\omega) \, , \quad \text{hence} \tag{E.32}$$

$$G(t - t') = \frac{1}{2\pi} \int_{-\infty}^{+\infty} d\omega \, e^{i\omega(t-t')} Y(i\omega)$$

in agreement with Sect. 6.5.4.

Linear inhomogeneous *systems* can be treated analogously or with the introduction of normal coordinates.

Remark. The defining equation

$$LG(t, t') = \delta(t - t')$$

of the Green's function can be interpreted in a simple, intuitive manner:

The Green's function $G(t, t')$ describes the response of the linear system to a short "jolt of force" $\delta(t - t')$ at time t', thus to a force which vanishes except for an arbitrarily small neighborhood of the point of time t'. Since any external force $f(t)$ can be written

$$f(t) = \int dt' \, \delta(t - t') f(t')$$

i.e., as a linear superposition of such jolts of force, the Green's function can furnish the solution to the inhomogeneous equation for any arbitrary inhomogeneous f.

F. Vector Analysis and Curvilinear Coordinates

In this appendix, we will briefly present the most important rules for our purposes in the calculus of vector fields. In particular, we will define line, surface, and volume integrals over fields, explain the integral theorems of Stokes and

Gauss for these integrals, and find expressions for the gradient, divergence, and curl of fields in curvilinear coordinates.

F.1 Vector Fields and Scalar Fields

Scalar fields are mappings $\varphi\colon \mathbb{R}^3 \to \mathbb{R}$ which assign to each point in space a numerical value $\varphi(r) \in \mathbb{R}$. Examples include: a temperature field $T(r)$, a density field $\varrho(r)$, etc. Under a rotation, $R\colon r \mapsto Rr$, a scalar field transforms as follows

$$\varphi \mapsto \varphi^{\mathrm{R}} \quad \text{with} \quad \varphi^{\mathrm{R}}(r) := \varphi(R^{-1}r) \ . \tag{F.1}$$

Vector fields are mappings $A\colon \mathbb{R}^3 \to \mathbb{R}^3$, which assign to each point in space $r \in \mathbb{R}^3$ a vector from the (Euclidean) vector space $\mathbb{R}^3$. (Here, as later, we often simply identify a vector with the collection of its components with respect to a particular fixed basis.) Under a (proper) rotation R, a vector field transforms as $A \mapsto A^{\mathrm{R}}$ with

$$A^{\mathrm{R}}(r) = RA(R^{-1}r) \ . \tag{F.2}$$

Examples of vector fields include a force field $F(r)$, an electric field $E(r)$, a magnetic field $B(r)$, a velocity field $v(r)$ of a flowing medium, and a flow density field $j(r)$ of a quantity such as mass or charge. Here, $j(r)$ is the amount of flow per unit time at a point r through a unit area with normal vector in the $j(r)$ direction.

All of the above named fields can, of course, have an additional dependence on time.

F.2 Line, Surface, and Volume Integrals

a) The line integral of a vector field along an oriented curve γ is defined in terms of the limit of a sequence of finer and finer approximations of γ by polygons as follows:

$$\int_{\gamma} A(r) \cdot dr = \lim \sum_{i} A(r_i) \cdot \Delta r_i \ . \tag{F.3}$$

The calculation is carried out by introducing an oriented parametrization of $\gamma\colon [a, b] \subset \mathbb{R} \to \mathbb{R}^3$, $t \mapsto r(t)$ with $r(a)$ the starting point, and $r(b)$ the endpoint of γ. Then,

$$\int_{\gamma} A(r) \cdot dr = \int_{a}^{b} dt\, A(r(t)) \cdot \dot{r}(t) \ , \tag{F.4}$$

independent of the parametrization.

If $A(r)$ is a force field, then $\int_{\gamma} A(r) \cdot dr$ is the work done by the field along γ.

b) The surface integral of a vector field on an oriented surface S:

Let $S \subset \mathbb{R}^3$ be a surface which is oriented through a continuous set of normal unit vectors given at each point on the surface. We define a surface integral using a sequence of finer and finer approximations of S by polyhedra:

$$\int_S A(r) \cdot dS = \lim \sum_i A(r_i) \cdot \Delta S_i \ . \tag{F.5}$$

Here, ΔS_i is the area vector of the i-th planar face S_i of the polyhedral surface defined as follows:

$|\Delta S_i|$ is the area of S_i and ΔS_i points in the direction of the normal vector.

We now calculate the surface integral for an oriented parametrized surface: Let the surface be described by a mapping of a rectangle into $\mathbb{R}^3$:

$$[a_1, a_2] \times [b_1, b_2] \to \mathbb{R}^3 \ , \quad (u_1, u_2) \mapsto r(u_1, u_2)$$

(see Fig. F.1).

Holding either u_2 or u_1 fixed and using the other variable as a parameter, we obtain curves on the surface using $r(u_1, u_2)$ which are called u_1- or u_2-lines.

$$\frac{\partial r(u_1, u_2)}{\partial u_1} \quad \text{and} \quad \frac{\partial r(u_1, u_2)}{\partial u_2}$$

are tangent vectors to the u_1 and u_2-lines respectively.

Let the parametrization be so chosen that $(\partial r/\partial u_1) \times (\partial r/\partial u_2)$ always points in the direction of the normal to S. The area of the image of the piece $[u_1, u_1 + \Delta u_1] \times [u_2, u_2 + \Delta u_2]$ of the rectangle in parameter space is

$$\left| \frac{\partial r}{\partial u_1} \times \frac{\partial r}{\partial u_2} \, \Delta u_1 \, \Delta u_2 \right| + \mathrm{o}(\Delta u_1 \, \Delta u_2) \ .$$

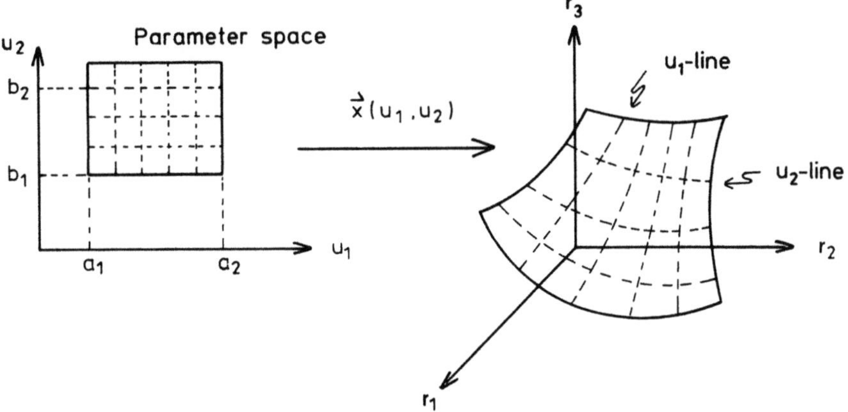

Fig. F.1. Parametrization of a surface element

We then have

$$\int_S A(r)\cdot dS = \int_{a_1}^{a_2} du_1 \int_{b_1}^{b_2} du_2\, A(r(u_1,u_2))\cdot\left(\frac{\partial r}{\partial u_1}\times\frac{\partial r}{\partial u_2}\right) \tag{F.6}$$

independent of the parametrization.

Example. We now calculate $\int_S (r/|r|^3)\cdot dS$ where S is the surface of a sphere with radius R and center at 0. We parameterize using spherical coordinates

$$r(\theta,\varphi) = R\begin{pmatrix} \sin\theta\cos\varphi \\ \sin\theta\sin\varphi \\ \cos\theta \end{pmatrix} \; ;$$

$$dS = \frac{\partial r}{\partial\theta}\times\frac{\partial r}{\partial\varphi}\,d\theta\,d\varphi = R^2\sin\theta\begin{pmatrix} \sin\theta\cos\varphi \\ \sin\theta\sin\varphi \\ \cos\theta \end{pmatrix} d\theta\,d\varphi \; .$$

Thus

$$\int_S \frac{r}{|r|^3}\cdot dS = \int_0^\pi d\theta \int_0^{2\pi} d\varphi\,\sin\theta = 2\pi\int_0^\pi d\theta\,\sin\theta = 4\pi \; .$$

c) The volume integral of a scalar or vector field:
For a volume V, we define the volume integral

$$\int_V \varphi(r)\,dV = \lim\sum\varphi(r_i)\,\Delta V_i$$

and calculate it using the parametrization

$$(u_1, u_2, u_3) \mapsto r(u_1, u_2, u_3) \quad \text{with}$$

$$a_1 \le u_1 \le a_2 \; , \qquad b_1 \le u_2 \le b_2 \; , \qquad c_1 \le u_3 \le c_2 \; .$$

Let the parametrization be chosen so that

$$\left(\frac{\partial r}{\partial u_1}\times\frac{\partial r}{\partial u_2}\right)\cdot\frac{\partial r}{\partial u_3} = \det\left(\frac{\partial r}{\partial u_1},\frac{\partial r}{\partial u_2},\frac{\partial r}{\partial u_3}\right) \ge 0 \; .$$

Then,

$$\int_V \varphi(r)\,dV = \int_{a_1}^{a_2} du_1\int_{b_1}^{b_2} du_2\int_{c_1}^{c_2} du_3\,\varphi(r(u_1,u_2,u_3))\det\left(\frac{\partial r}{\partial u_1},\frac{\partial r}{\partial u_2},\frac{\partial r}{\partial u_3}\right) \tag{F.7}$$

independent of the parametrization.

F.3 Stokes's Theorem

Let R be a rectangle in the 1–2 plane with normal vector pointing in the 3-direction, and let ∂R be its boundary oriented as shown in Fig. F.2. After introducing the parametrization $u_1 = r_1$, $u_2 = r_2$, we calculate $\int_{\partial R} A \cdot dr$ as follows:

$$\int_{\partial R} A \cdot dr = \int_{a_1}^{a_2} dr_1 \left[A_1(r_1, b_1) - A_1(r_1, b_2) \right]$$

$$+ \int_{b_1}^{b_2} dr_2 \left[A_2(a_2, r_2) - A_2(a_1, r_2) \right]$$

$$= \int_{a_1}^{a_2} dr_1 \int_{b_1}^{b_2} dr_2 \left[\frac{\partial A_2}{\partial r_1}(r_1, r_2) - \frac{\partial A_1}{\partial r_2}(r_1, r_2) \right]$$

$$=: \int_R (\nabla \times A) \cdot dS \ . \tag{F.8}$$

Here, the vector field $\nabla \times A$ is defined by

$$\nabla \times A = \begin{pmatrix} \partial_2 A_3 - \partial_3 A_2 \\ \partial_3 A_1 - \partial_1 A_3 \\ \partial_1 A_2 - \partial_2 A_1 \end{pmatrix}, \quad \text{i.e.}$$

$$(\nabla \times A)_i = \varepsilon_{ijk} \partial_j A_k \quad \text{with} \quad \partial_j := \frac{\partial}{\partial r_j}.$$

We also write $\nabla \times A = \operatorname{curl} A$. Under a rotation R, $(\nabla \times A)^R = \nabla \times A^R$. In general, for an oriented surface S with oriented boundary ∂S, we have *Stokes's theorem*,

$$\int_{\partial S} A \cdot dr = \int_S (\nabla \times A) \cdot dS \ . \tag{F.9}$$

For a parametrized surface with parameters varying in a rectangle, a proof of the

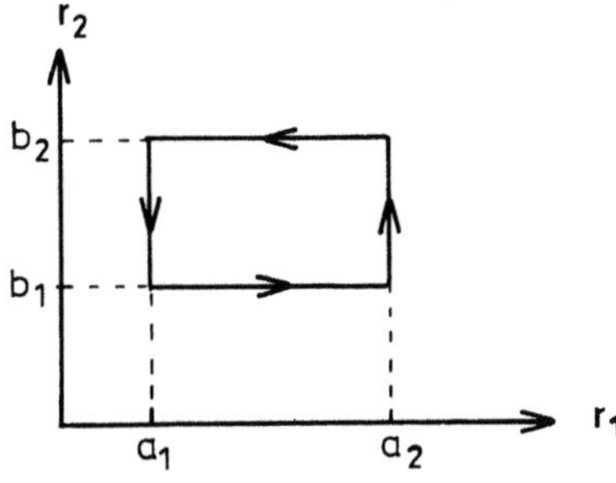

Fig. F.2. Integration over the boundary of a rectangle in the x_1–x_2 plane

theorem follows from the formula

$$\mathbf{V} \times A(r(u_1, u_2)) \cdot \left(\frac{\partial r}{\partial u_1} \times \frac{\partial r}{\partial u_2} \right)$$

$$= \frac{\partial}{\partial u_2} \left[\frac{\partial r}{\partial u_2} \cdot A(r(u_1, u_2)) \right] - \frac{\partial}{\partial u_2} \left[\frac{\partial r}{\partial u_1} \cdot A(r(u_1, u_2)) \right] \tag{F.10}$$

by reducing it to the case just considered to a rectangle in the $u_1 - u_2$ plane.

More general surfaces can be constructed out of such parametrized surfaces. The next observation gives us an intuitive interpretation of $\mathbf{V} \times A$:

The integral $\int_\gamma A \cdot dr$ along a closed curve γ is a measure of the strength of the circulation or vorticity of A around an axis going through γ. Now, for surfaces S with normal vector n,

$$n \cdot (\mathbf{V} \times A) = \lim_{|S| \to 0} \frac{1}{|S|} \int_S (\mathbf{V} \times A) \cdot dS = \lim_{|S| \to 0} \frac{1}{|S|} \int_{\partial S} A \cdot dr \ . \tag{F.11}$$

The quantity $n \cdot (\mathbf{V} \times A)$ thus measures the surface density of the vorticity of A with respect to an axis in the n direction.

F.4   Gauss's Theorem

For a parallelepiped Q whose boundary ∂Q is oriented so that the normal vectors point outward, we have (Fig. F.3)

$$\int_{\partial Q} A \cdot dS = \int_{a_1}^{a_2} dr_1 \int_{b_1}^{b_2} dr_2 [A_3(r_1, r_2, c_2) - A_3(r_1, r_2, c_1)]$$

$$+ \int_{b_1}^{b_2} dr_2 \int_{c_1}^{c_2} dr_3 [A_1(a_2, r_2, r_3) - A_1(a_1, r_2, r_3)]$$

$$+ \int_{c_1}^{c_2} dr_3 \int_{a_1}^{a_2} dr_1 [A_2(r_1, b_2, r_3) - A_2(r_1, b_1, r_3)]$$

$$= \int_{a_1}^{a_2} dr_1 \int_{b_1}^{b_2} dr_2 \int_{c_1}^{c_2} dr_3 \left[\frac{\partial A_1}{\partial r_1} + \frac{\partial A_2}{\partial r_2} + \frac{\partial A_3}{\partial r_3} \right] = \int_Q \mathbf{V} \cdot A \, dv \ . \tag{F.12}$$

The scalar field $\mathbf{V} \cdot A = \partial_i A_i$ is called the *divergence* of A (written also div A).

In general, for a compact volume $V \subset \mathbb{R}^3$, whose (piecewise smooth) boundary ∂V is oriented in such a way that the normal vectors point outward, *Gauss's theorem* holds:

$$\int_{\partial V} A \cdot dS = \int_V \mathbf{V} \cdot A \, dV \ . \tag{F.13}$$

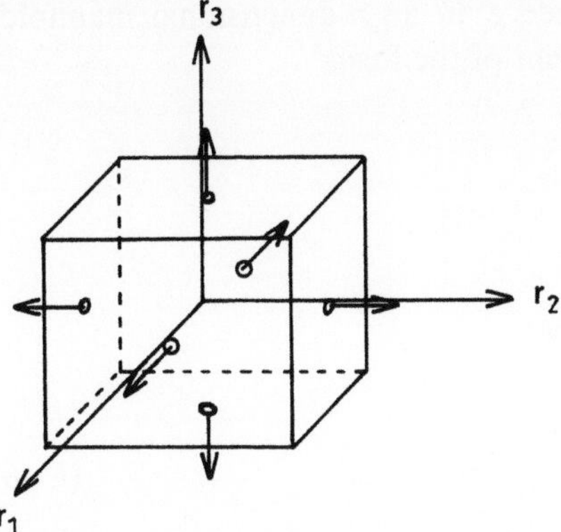

Fig. F.3. Integration over the surface of a parallelepiped. The normal vectors are included in the picture

For parametrized volume elements, as in Sect. F.2c, Gauss's theorem is a consequence of the identity

$$\nabla \cdot A(r(u_1, u_2, u_3)) \det\left(\frac{\partial r}{\partial u_1}, \frac{\partial r}{\partial u_2}, \frac{\partial r}{\partial u_3}\right)$$

$$= \frac{\partial}{\partial u_1}\left[A \cdot \left(\frac{\partial r}{\partial u_2} \times \frac{\partial r}{\partial u_3}\right)\right] + \frac{\partial}{\partial u_2}\left[A \cdot \left(\frac{\partial r}{\partial u_3} \times \frac{\partial r}{\partial u_1}\right)\right]$$

$$+ \frac{\partial}{\partial u_3}\left[A \cdot \left(\frac{\partial r}{\partial u_1} \times \frac{\partial r}{\partial u_2}\right)\right] . \tag{F.14}$$

The interpretation of $\nabla \cdot A$: If A is a flow density field, then the flux $\Phi = \int_{\partial V} A \cdot dS$ is the amount flowing out of V per unit time, thus a measure of the strength of the sources inside of V. Now,

$$\nabla \cdot A = \lim_{|V| \to 0} \frac{1}{|V|} \int_V \nabla \cdot A \, dV = \lim_{|V| \to 0} \frac{1}{|V|} \int_{\partial V} A \cdot dS , \tag{F.15}$$

so that $\nabla \cdot A$ can be interpreted as the volume density of sources of A.

Gauss's theorem changes a volume integral into a surface integral, and Stokes's theorem turns a surface integral into a line integral.

The trivial identity

$$\int_\gamma \nabla \varphi \cdot dr = \varphi(r_2) - \varphi(r_1) =: \int_{\partial \gamma} \varphi ,$$

where r_2 and r_1 are the end and beginning points of γ, represents a transformation of a line integral into a zero-dimensional "integral."

(In general, for a k-dimensional surface Σ in an n-dimensional manifold, there is a generalization of Stokes's theorem of the form

$$\int_{\partial\Sigma} \omega = \int_{\Sigma} d\omega \ .)$$

F.5   Applications of the Integral Theorems

a) We can immediately show that

$$\nabla \times (\nabla\varphi) = 0 \quad \text{and} \quad \nabla\cdot(\nabla\times A) = 0 \ . \tag{F.16}$$

These intuitive results also follow directly from the integral theorems

$$\int_{S} (\nabla\times\nabla\varphi)\cdot dS = \int_{\partial S} \nabla\varphi\cdot dr = 0 \ ,$$

since ∂S is closed; thus $\nabla\times\nabla\varphi = 0$, since S is arbitrary,

$$\int_{V} \nabla\cdot(\nabla\times A)\, dV = \int_{\partial V} (\nabla\times A)\cdot dS = \int_{\partial\partial V} A\cdot dr = 0 \ , \quad \text{since}$$

$$\partial\partial V = \emptyset \ ; \quad \text{thus} \quad \nabla\cdot(\nabla\times A) = 0 \ ,$$

since V is arbitrary.

b) Since

$$\int_{S} (\nabla\times A)\cdot dS = \int_{\partial S} A\cdot dr$$

we see that $\int_{S}(\nabla\times A)\cdot dS$ depends only on the boundary of S. The flow of the field $\nabla\times A$ is thus the same for all surfaces with the same boundary.

c) Sufficient condition for the existence of a potential for a force field F:

We saw earlier that a potential ϕ with $F = -\nabla\phi$ can exist for F only if $\nabla\times F = 0$. This condition turns out to be also sufficient for the existence of a potential for a (smooth) field A defined on all of $\mathbb{R}^3$.

Every closed curve γ in $\mathbb{R}^3$ is, in particular, the boundary of a surface $S \subset \mathbb{R}^3$, $\gamma = \partial S$.

But then, for closed curves

$$\int_{\gamma} A\cdot dr = \int_{S} (\nabla\times A)\cdot dS = 0 \ .$$

Thus the function

$$\phi(r) = -\int_{r_0}^{r} A(r')\cdot dr'$$

is well-defined, since it depends only an a (fixed) initial point r_0 and an endpoint r of the curve, and we can easily show that $A = -\nabla\phi$.

We can also show that for every vector field B defined on all of $\mathbb{R}^3$ such that $\nabla \cdot B = 0$, there is a vector field A (a vector potential for B) such that $B = \nabla \times A$ (*Poincaré's theorem*).

d) Identities for the gradient, divergence, and curl: The following can be verified by immediate calculation

i) $\nabla \times \nabla\varphi = 0$, $\qquad\qquad\qquad\qquad\qquad\qquad\qquad\qquad$ (F.17)

$\qquad \nabla \cdot (\nabla \times A) = 0$, $\qquad\qquad\qquad\qquad\qquad\qquad\qquad$ (F.18)

$\qquad \nabla \cdot (\nabla\varphi) = \Delta\varphi$, $\qquad\qquad\qquad\qquad\qquad\qquad\qquad$ (F.19)

$\qquad \nabla \times (\nabla \times A) = \nabla(\nabla \cdot A) - \Delta A$, $\qquad\qquad\qquad\qquad$ (F.20)

where $\Delta = \partial_i\partial_i = \sum_i (\partial^2/\partial r_i)$ is called the *Laplacian*.

ii) $\nabla \cdot (\varphi A) = \varphi\nabla \cdot A + A \cdot \nabla\varphi$, $\qquad\qquad\qquad\qquad$ (F.21)

$\qquad \nabla \cdot (A \times B) = B \cdot (\nabla \times A) - A \cdot (\nabla \times B)$. $\qquad\qquad$ (F.22)

iii) $\nabla \times (\varphi A) = \varphi\nabla \times A + (\nabla\varphi) \times A$, $\qquad\qquad\qquad$ (F.23)

$\qquad \nabla \times (A \times B) = A(\nabla \cdot B) - B(\nabla \cdot A) + (B \cdot \nabla)A - (A \cdot \nabla)B$. $\qquad$ (F.24)

F.6 Curvilinear Coordinates

In many physical situations, it is useful to introduce curvilinear coordinates which fit the symmetry of the problem.

We then need to transform our formulas for $\nabla\varphi$, $\nabla \times A$, $\nabla \cdot A$, and $\Delta\varphi$ from Cartesian to curvilinear coordinates. Let us then be given the curvilinear coordinates u_1, u_2, u_3 in Euclidean space $\mathbb{R}^3$:

$$r = r(u_1, u_2, u_3) \ .$$

Examples include:

Cylindrical coordinates (Fig. F.4)

$$u_1 = r, \qquad u_2 = \varphi \ , \qquad u_3 = z$$

$$r = \begin{pmatrix} r\cos\varphi \\ r\sin\varphi \\ z \end{pmatrix} ,$$

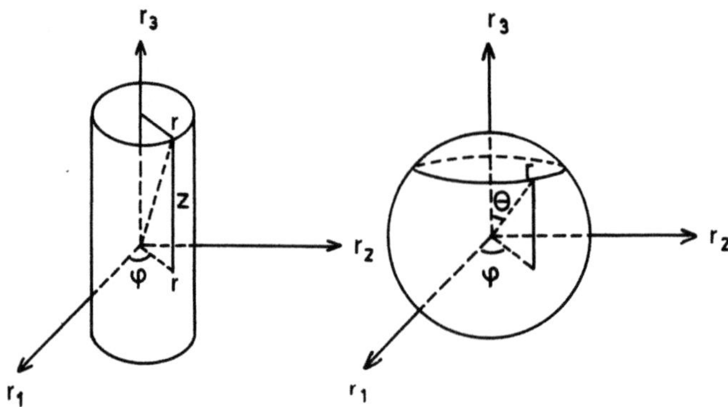

Fig. F.4. Cylindrical and spherical coordinates

Spherical coordinates (Fig. F.4)

$$u_1 = r, \quad u_2 = \theta, \quad u_3 = \varphi$$

$$r = \begin{pmatrix} r \sin \theta \cos \varphi \\ r \sin \theta \sin \varphi \\ r \cos \theta \end{pmatrix} .$$

The u_1-lines are the curves $r(u_1, u_2, u_3)$ with u_2 and u_3 held fixed (analogously, we define the u_2- and u_3-lines), and the u_1- surfaces are the surfaces $r(u_1, u_2, u_3)$ with u_1 held fixed (analogously for the u_2- and u_3-surfaces).

Thus, for example, for spherical coordinates:

r-lines:	Rays starting at the origin
θ-lines:	Longitude semi circles.
φ-lines:	Latitude circles with axis in the 3-direction.
r-surfaces:	Spherical surfaces
θ-surfaces:	Cones with tip in the origin and axes in the 3-direction,
φ-surfaces:	Half planes which contain the 3-axis.

The vectors $(\partial r/\partial u_i)(u_1, u_2, u_3)$ are tangential to the u_i-lines at the point $r(u_1, u_2, u_3)$. We define at each point $r(u_1, u_2, u_3)$ the unit vectors

$$e_i = \frac{1}{h_i}\frac{\partial r}{\partial u_i} \quad \text{with} \quad h_i(u_1, u_2, u_3) = \left| \frac{\partial r}{\partial u_i}(u_1, u_2, u_3) \right| . \tag{F.25}$$

These unit vectors are defined and are linearly independent if the parametrization is smooth and invertible.

Orthogonal curvilinear coordinates are defined by the condition $e_i \cdot e_j = \delta_{ij}$, i.e. the u_i-lines are pairwise perpendicular to each other. Spherical and cylindrical coordinates, as we see immediately, are orthogonal curvilinear coordinates. The component $A^{(i)}$ of a vector field in the e_i direction is

$$A^{(i)} = e_i \cdot A \quad \text{thus} \quad A = \sum_{i=1}^{3} A^{(i)} e_i \ . \tag{F.26}$$

For surface and volume elements in orthogonal curvilinear coordinates, we find, applying the expressions in Sect. F.2b, c to this particular case:

$$df^{(1)} = h_2 h_3 du_2 du_3 \ , \quad df^{(2)} = h_3 h_1 du_3 du_1 \ ,$$
$$df^{(3)} = h_1 h_2 du_1 du_2 \ , \quad dV = h_1 h_2 h_3 du_1 du_2 du_3 \ . \tag{F.27}$$

The e_i component of $\nabla \phi$ at the point $r(u_1, u_2, u_3)$ is

$$(\nabla \phi)^{(i)} = e_i \cdot \nabla \phi = \frac{1}{h_i} \frac{\partial r}{\partial u_i} \cdot \nabla \phi = \frac{1}{h_i} \frac{\partial \phi}{\partial u_i}, \quad [\phi = \phi(r(u))] \ . \tag{F.28}$$

Applying

$$n \cdot (\nabla \times A) = \lim_{|S| \to 0} \frac{1}{|S|} \int A \cdot dr$$

and

$$\nabla \cdot A = \lim_{|V| \to 0} \frac{1}{|V|} \int_{\partial V} A \cdot dS$$

to small rectangles with normal vectors in the e_i-direction and small cubes with sides in the e_1, e_2, and e_3-directions, we obtain the following expressions for $\nabla \times A$ and $\nabla \cdot A$ in orthogonal curvilinear coordinates:

$$(\nabla \times A)^{(1)} = \frac{1}{h_2 h_3} \left[\frac{\partial}{\partial u_2} (h_3 A^{(3)}) - \frac{\partial}{\partial u_3} (h_2 A^{(2)}) \right] ,$$

$$(\nabla \times A)^{(2)} = \frac{1}{h_3 h_1} \left[\frac{\partial}{\partial u_3} (h_1 A^{(1)}) - \frac{\partial}{\partial u_1} (h_3 A^{(3)}) \right] , \tag{F.29}$$

$$(\nabla \times A)^{(3)} = \frac{1}{h_1 h_2} \left[\frac{\partial}{\partial u_1} (h_2 A^{(2)}) - \frac{\partial}{\partial u_2} (h_1 A^{(1)}) \right] ,$$

$$\nabla \cdot A = \frac{1}{h_1 h_2 h_3} \left[\frac{\partial}{\partial u_1} (h_2 h_3 A^{(1)}) + \frac{\partial}{\partial u_2} (h_3 h_1 A^{(2)}) + \frac{\partial}{\partial u_3} (h_1 h_2 A^{(3)}) \right] \ . \tag{F.30}$$

The same formulas can also be obtained by applying the identities for parametrized surface and volume elements which we used in Sects. F.3 and F.4 to prove Stokes's and Gauss's theorems.

Finally, by substituting $(\nabla\phi)^{(i)} = (1/h_i)(\partial\phi/\partial u_i)$, we find for the Laplacian

$$\nabla\cdot\nabla\phi = \Delta\phi = \frac{1}{h_1 h_2 h_3}\left[\frac{\partial}{\partial u_1}\frac{h_2 h_3}{h_1}\frac{\partial}{\partial u_1} + \frac{\partial}{\partial u_2}\frac{h_3 h_1}{h_2}\frac{\partial}{\partial u_2} + \frac{\partial}{\partial u_3}\frac{h_1 h_2}{h_3}\frac{\partial}{\partial u_3}\right]\phi \ .$$

$$\text{(F.31)}$$

We can rewrite these expressions in a mnemonic form by introducing the following notation:

$$A_i = A\cdot\frac{\partial r}{\partial u_i} \ , \tag{F.32}$$

$$g_{ij} = \frac{\partial r}{\partial u_i}\cdot\frac{\partial r}{\partial u_j} \ , \qquad g = \det(g_{ij}) \ , \tag{F.33}$$

g^{ij}: the inverse matrix to g_{ij}, thus

$$g^{ik}g_{kj} = \delta^i_j \ . \tag{F.34}$$

In orthogonal curvilinear coordinates, we then have

$$A_i = h_i A^{(i)} \ , \tag{F.35}$$

$$g_{ij} = h_i^2\delta_{ij} \ , \qquad g = h_1^2 h_2^2 h_3^2 \ , \tag{F.36}$$

$$g^{ij} = \frac{1}{h_i^2}\delta_{ij} \tag{F.37}$$

and thus

$$(\nabla\times A)_i = \frac{1}{\sqrt{g}}g_{ij}\varepsilon^{jkl}\frac{\partial}{\partial u^k}A_l \ , \tag{F.38}$$

$$\nabla\cdot A = \frac{1}{\sqrt{g}}\frac{\partial}{\partial u^i}\sqrt{g}\,g^{ij}A_j \ , \tag{F.39}$$

$$\Delta\phi = \frac{1}{\sqrt{g}}\frac{\partial}{\partial u^i}\sqrt{g}g^{ij}\frac{\partial}{\partial u_j}\phi \ . \tag{F.40}$$

These expressions are also valid even for arbitrary, and not necessarily orthogonal, curvilinear coordinates.

In particular, for *cylindrical coordinates*, we have

$$g^{11} = 1, \quad g^{22} = \frac{1}{r^2} \ , \quad g^{33} = 1 \ ; \quad \sqrt{g} = r \ ; \tag{F.41}$$

$$\Delta\phi = \frac{1}{r}\left[\frac{\partial}{\partial r} r \frac{\partial}{\partial r} + \frac{1}{r}\frac{\partial^2}{\partial\varphi^2} + \frac{\partial}{\partial z} r \frac{\partial}{\partial z}\right]\phi$$

$$= \left[\frac{\partial^2}{\partial r^2} + \frac{1}{r}\frac{\partial}{\partial r} + \frac{1}{r^2}\frac{\partial^2}{\partial\phi^2} + \frac{\partial^2}{\partial z^2}\right]\phi \tag{F.42}$$

and for *spherical coordinates*

$$g^{11} = 1, \quad g^{22} = \frac{1}{r^2}, \quad g^{33} = \frac{1}{r^2 \sin^2\theta} ;$$

$$\sqrt{g} = r^2 \sin\theta ; \tag{F.43}$$

$$\Delta\phi = \frac{1}{r^2\sin\theta}\left[\frac{\partial}{\partial r} r^2 \sin\theta \frac{\partial}{\partial r} + \frac{\partial}{\partial\theta}\sin\theta\frac{\partial}{\partial\theta} + \frac{\partial}{\partial\varphi}\frac{1}{\sin\theta}\frac{\partial}{\partial\varphi}\right]\phi$$

$$= \left[\frac{\partial^2}{\partial r^2} + \frac{2}{r}\frac{\partial}{\partial r} + \frac{1}{r^2\sin\theta}\frac{\partial}{\partial\theta}\sin\theta\frac{\partial}{\partial\theta} + \frac{1}{r^2\sin^2\theta}\frac{\partial^2}{\partial\varphi^2}\right]\phi$$

$$=: \left[\frac{\partial^2}{\partial r^2} + \frac{2}{r}\frac{\partial}{\partial r} + \frac{1}{r^2}\varLambda\right]\phi . \tag{F.44}$$

Problems

A.1 Divergence, Gradient, and Curl. Show through explicit calculation using the δ_{ij} and ε_{ijk} symbols:

$$\boldsymbol{\nabla}\times(\boldsymbol{\nabla}\times\boldsymbol{A}) = \boldsymbol{\nabla}(\boldsymbol{\nabla}\cdot\boldsymbol{A}) - \Delta\boldsymbol{A}$$

$$\boldsymbol{\nabla}\cdot(\phi\boldsymbol{A}) = (\boldsymbol{\nabla}\phi)\cdot\boldsymbol{A} + \phi(\boldsymbol{\nabla}\cdot\boldsymbol{A})$$

$$\boldsymbol{\nabla}\times(\phi\boldsymbol{A}) = (\boldsymbol{\nabla}\phi)\times\boldsymbol{A} + \phi(\boldsymbol{\nabla}\times\boldsymbol{A})$$

$$\boldsymbol{\nabla}\cdot(\boldsymbol{A}\times\boldsymbol{B}) = (\boldsymbol{\nabla}\times\boldsymbol{A})\cdot\boldsymbol{B} - \boldsymbol{A}\cdot(\boldsymbol{\nabla}\times\boldsymbol{B})$$

$$\boldsymbol{\nabla}\times(\boldsymbol{A}\times\boldsymbol{B}) = (\boldsymbol{B}\cdot\boldsymbol{\nabla})\boldsymbol{A} - \boldsymbol{B}(\boldsymbol{\nabla}\cdot\boldsymbol{A}) - (\boldsymbol{A}\cdot\boldsymbol{\nabla})\boldsymbol{B} + \boldsymbol{A}(\boldsymbol{\nabla}\cdot\boldsymbol{B}) .$$

Prove the following identity:

$$[\boldsymbol{\nabla}\cdot\boldsymbol{A}(r(u_1, u_2, u_3))]\det\left(\frac{\partial\boldsymbol{r}}{\partial u_1}, \frac{\partial\boldsymbol{r}}{\partial u_2}, \frac{\partial\boldsymbol{r}}{\partial u_3}\right)$$

$$= \frac{\partial}{\partial u_1}\left[\boldsymbol{A}\cdot\left(\frac{\partial\boldsymbol{r}}{\partial u_2}\times\frac{\partial\boldsymbol{r}}{\partial u_3}\right)\right] + \frac{\partial}{\partial u_2}\left[\boldsymbol{A}\cdot\left(\frac{\partial\boldsymbol{r}}{\partial u_3}\times\frac{\partial\boldsymbol{r}}{\partial u_1}\right)\right]$$

$$+ \frac{\partial}{\partial u_3}\left[\boldsymbol{A}\cdot\left(\frac{\partial\boldsymbol{r}}{\partial u_1}\times\frac{\partial\boldsymbol{r}}{\partial u_2}\right)\right] .$$

A.2 Electrical Flux. Calculate using explicit integration the flux

$$\int_S \boldsymbol{E} \cdot d\boldsymbol{f} \quad \text{of the Coulomb field} \quad \boldsymbol{E}(\boldsymbol{r}) = q\,\frac{\boldsymbol{r}}{r^3}$$

of a point-charge q at the origin through the surface S of a cylinder with radius R and height $2h$.

References

Chapter 2

Duffey, G.H.: *Theoretical Physics. Classical and Modern Views* (Krieger, Melbourne 1980)

Greub, W.H.: *Linear Algebra*, Graduate Texts in Mathematics Vol. 23, 4th ed. (Springer, New York 1991)

Harris, E.G.: *Introduction to Modern Theoretical Physics, Vol. 1. Classical Physics & Relativity* (Books on Demand, Ann Arbor, MI 1975)

Moore & Yaqub: *First Course Linear Algebra*, 2nd ed. (HarperCollins, New York 1992)

Roman, S.: *Advanced Linear Algebra*, Graduate Texts in Mathematics, Vol. 135 (Springer, New York 1992)

Robinson, D.J. (Ed.): *A Course in Linear Algebra with Applications* (World Scientific, Singapore 1991)

Robinson, D.J.: *Course in Linear Algebra with Applications: Solutions to the Exercises* (World Scientific, Singapore 1992)

Stauffer, D. Stanley, H.E.: *From Newton to Mandelbrot. A Primer in Theoretical Physics* (Springer, Berlin, Heidelberg 1991)

Trainor, L.E. Wise, M.B.: *From Physical Concept to Mathematical Structure. An Introduction to Theoretical Physics*. Mathematical Expositions Ser. Vol. 22 (Books on Demand, Ann Arbor, MI 1979)

Chapter 3

Haken, H.: *Synergetics. An Introduction*, Springer Ser. Syn., Vol. 1, 3rd ed. (Springer, Berlin, Heidelberg 1983)

Landau, L.D., Lifshitz, E.M.: *A Shorter Course of Theoretical Physics*. Vol. 1. Mechanics & Electrodynamics. Vol. 2. Quantum Mechanics (Franklin Elkins Park, PA 1992)

Lichtenberg, A.J., Liebermann, M.A.: *Regular and Stochastic Motion*, Applied Mathematical Sciences, Vol. 38 (Springer, Berlin, Heidelberg, New York 1983)

Schmid, E.W., Spitz, G. and Lösch, W.: *Theoretical Physics on the Personal Computer*, 2nd ed. (Springer, Berlin, Heidelberg 1990)

Schuster, H.G.: *Deterministic Chaos: An Introduction*, 2nd ed. (VCH Publishers, New York 1987)

Sommerfeld, A.: *Lectures on Theoretical Physics, Vol. 1. Mechanics* (Academic Press, SanDiego 1964)

Chapter 4

Constant, F.W.: *Theoretical Physics. Mechanics of Particles, Rigid & Elastic Bodies & Heat Flow* (Krieger, Melbourne, FL 1978)
Landau, L.D., Lifshitz, E.M.: *Course of Theoretical Physics, Vol. 1. Mechanics,* 3rd ed. (Pergamon, Oxford 1982)

Chapter 5

Abraham, R Marsden, J.E.: *Foundations of Mechanics. A Mathematical Exposition of Classical Mechanics With an introduction to the Qualitative Theory of Dynamical Systems & Applications to the Three-Body Problem,* 2nd ed. (Addison-Wesley, Reading 1978)
Arnol'd, V.I.: *Mathematical Methods of Classical Mechanics,* translated by K. Vogtmann, A. Weinstein, Graduate Texts in Mathematics, Vol. 60 (Springer, Berlin, Heidelberg 1991)
Goldstein, H.: *Classical Mechanics,* 2nd ed. (Addison-Wesley, Reading 1980)
Thirring, W.: *A Course in Mathematical Physics I: Classical Dynamical systems,* 2nd ed. (Springer, Wien, New York 1991)
Whittaker, E.T.: *A Treatise on the Analytical Dynamics of Particles and Rigid Bodies* (Cambridge University Press, Cambridge 1989)

Chapter 6

Butenin, N.V.: *Elements of the Theory of Nonlinear Oscillations* (Blanschell, New York 1965)
Champeney, D.C.: *Fourier Transforms and Their Physical Applications* (Academic, London 1973)
Cullen, C.G.: *Linear Algebra & Differential Equations,* 2nd ed. (PWS-KENT, Boston, MA 1991)
Drazin, P.G. Reid, W.H.: *Hydrodynamic Stability* (Cambridge University Press, Cambridge 1984)
Goode, S.W.: An Introduction to Differential Equations & *Linear Algebra* (Prentice Hall, Englewood Cliffs, NJ 1991)
Haken, H.: *Synergetics. An Introduction,* Springer Ser. Syn., Vol. 1, 3rd ed. (Springer, Berlin, Heidelberg 1983)
Jeffrey: *Linear Algebra & Ordinary Differential Equations* (CRC Press, Boca Raton, FL 1991)
Nayfeh, A.H. Mook, D.T.: *Nonlinear Oscillations* (Wiley, New York 1979)
Schwartz, L.: *Théorie des distributions I, II* (Hermann, Paris 1957–1959)

Chapter 7

Landau, L.D., Lifshitz, E.M.: *Course of Theoretical Physics, Vol. 5. Statistical Physics* (Pergamon, Oxford 1980)
McQuarrie, D.A.: *Statistical Mechanics* (Harpe Collins, New York 1976)

Chapter 8

Barrow, G.M.: *Physical Chemistry*, 5th ed. (McGraw-Hill, New York 1988)
Landau, L.D., Lifshitz, E.M.: *Course of Theoretical Physics, Vol. 5. Statistical Physics* (Pergamon, Oxford 1969)
Reichl, L.E.: *A Modern Course in Statistical Physics* (University of Texas Press, Austin, TX 1980)
Thirring, W.: *A Course in Mathematical Physics IV: Quantum Mechanics of Large Systems* (Springer, Wien, New York 1982)

Chapter 9

Bird, R.B., Armstrong, R.C., Hassanger, O.: *Dynamics of Polymeric Liquids Vol. I, Fluid Mechanics* (Wiley, New York 1977)
Houghton, J.T.: *The Physics of Atmospheres* 2nd ed. (Cambridge University Press, Cambridge 1986)
Landau, L.D., Lifshitz, E.M.: *Course of Theoretical Physics, Vol. 6. Fluid Mechanics* (Pergamon, Oxford 1989)
Landau, L.D., Lifshitz, E.M., and Kosevich, A.M.: *Course of Theoretical Physics, Vol. 7. Theory of Elasticity*, 3rd ed. (Pergamon, Oxford 1986)
Pedlosky, J.: *Geophysical Fluid Dynamics*. Springer Study Edition, 2nd ed. (Springer, New York 1992)

Chapter 10

Courant, R., Hilbert, D.: *Methods of Mathematical Physics, Vols. 1 & 2* (Wiley, New York 1989)
Jackson, J.D.: *Classical Electrodynamics*, 2nd ed. (Wiley, New York 1975)

Chapter 11

Jackson, J.D.: *Classical Electrodynamics*, 2nd ed. (Wiley, New York 1975)
Panofsky, W., Phillips, M.: *Classical Electricity and Magnetism*, 2nd ed. (Addison-Wesley, Reading, MA 1962)

Chapter 13

Landau, L.D. Lifshitz, E.M.: *Course of Theoretical Physics, Vol. 2. The Classical Theory of Fields*, 4th ed. (Pergamon, Oxford 1980)

Chapter 14

Landau, L.D., Lifshitz, E.M.: *Course of Theoretical Physics, Vol. 2. The Classical Theory of Fields*, 4th ed.; *Vol. 8. Electrodynamics of Continuous Media*. (Pergamon, Oxford 1980, 1984)
Thirring, W.: *Lehrbuch der Mathematischen Physik, Bd. 2, Klassische Feldtheorie* (Springer, Wien, New York 1978)

Name and Subject Index

MIX
Papier aus verantwortungsvollen Quellen
Paper from responsible sources
FSC® C105338

If you have any concerns about our products,
you can contact us on
ProductSafety@springernature.com

In case Publisher is established outside the EU,
the EU authorized representative is:
Springer Nature Customer Service Center GmbH
Europaplatz 3, 69115 Heidelberg, Germany

Printed by Libri Plureos GmbH
in Hamburg, Germany